Maximum Entropy and Bayesian Methods

Fundamental Theories of Physics

An International Book Series on The Fundamental Theories of Physics: Their Clarification, Development and Application

Volume 53

Maximum Entropy and Bayesian Methods

Paris, France, 1992

Proceedings of the Twelfth International Workshop on Maximum Entropy and Bayesian Methods

edited by

A. Mohammad-Djafari

and

G. Demoment

Laboratoire des Signaux et Systèmes,
CNRS-ESE-UPS,
Gif-sur-Yvette, France

KLUWER ACADEMIC PUBLISHERS

DORDRECHT / BOSTON / LONDON

Library of Congress Cataloging-in-Publication Data

Maximum entropy and Bayesian methods : Paris, France, 1992 / edited by
Ali Mohammad-Djafari.
p. cm. -- (Fundamental theories of physics ; v. 53)
Includes index.

1. Maximum entropy method--Congresses. 2. Bayesian statistical
decision theory--Congresses. 3. Signal processing--Congresses.
4. Image processing--Congresses. I. Mohammad-Djafari, Ali.
II. Series.
QC370.M39 1993
501'.154--dc20 93-2024

ISBN 978-90-481-4272-9

Published by Kluwer Academic Publishers,
P.O. Box 17, 3300 AA Dordrecht, The Netherlands.

Kluwer Academic Publishers incorporates
the publishing programmes of
D. Reidel, Martinus Nijhoff, Dr W. Junk and MTP Press.

Sold and distributed in the U.S.A. and Canada
by Kluwer Academic Publishers,
101 Philip Drive, Norwell, MA 02061, U.S.A.

In all other countries, sold and distributed
by Kluwer Academic Publishers Group,
P.O. Box 322, 3300 AH Dordrecht, The Netherlands.

Printed on acid-free paper

PREFACE

The Twelfth International Workshop on Maximum Entropy and Bayesian Methods in Sciences and Engineering (MaxEnt 92) was held in Paris, France, at the Centre National de la Recherche Scientifique (CNRS), July 19–24, 1992. It is important to note that, since its creation in 1980 by some of the researchers of the physics department at the Wyoming University in Laramie, this was the second time that it took place in Europe, the first time was in 1988 in Cambridge.

The two specificities of MaxEnt workshops are their spontaneous and informal characters which give the participants the possibility to discuss easily and to make very fruitful scientific and friendship relations among each others.

This year's organizers had fixed two main objectives:

i) to have more participants from the European countries, and

ii) to give special interest to maximum entropy and Bayesian methods in signal and image processing.

We are happy to see that we achieved these objectives:

i) we had about 100 participants with more than 50 per cent from the European countries,

ii) we received many papers in the signal and image processing subjects and we could dedicate a full day of the workshop to the image modelling, restoration and reconstruction problems.

As in the two last workshops, we also started this workshop by a tutorial day to give the possibility to all the participants, and more specifically to the newcomers, to have some synthetic and state of the art talks on the foundations of maximum entropy and Bayesian methods in sciences and engineering. We want to thank all the persons who gave these tutorial talks: Ray Smith, Tom Loredo, Larry Bretthorst, Steve Gull, Antony Garrett and Tom Grandy.

During the workshop we had many speakers who gave excellent lectures. We regret, however, that some of these lectures were not written down for publication. We particularly think of the talks given by A. Tarantola, J. Skilling and N. Rivier.

We particularly wish to thank the universities and corporations which helped us in organizing this workshop, namely,

- Centre National de la Recherche Scientifique (CNRS), France,
- Defense Ministry, Direction de la Recherche et Développement (DRED), France,
- Shounan Institute of Technology, Japan,
- Centre National d'Études des Télécommunications (CNET) (France),
- The Group Research GDR 134 of the CNRS, France,
- Laboratoire des Signaux et Systèmes (LSS), France, and finally,
- the Kluwer Editions, the publisher of this book.

Finally, we would like to thank our colleagues [1] who helped us with the organization, translation of the abstracts in French for the booklet of abstracts, and the administration [2] and secretarial staff [3] who helped us with the material organization and typesetting of the booklet of abstracts and some of the final papers in this book.

Ali Mohammad–Djafari and Guy Demoment

Groupe Problèmes Inverses
Laboratoire des Signaux et Systèmes
(CNRS–ESE–UPS)
91192 Gif-sur-Yvette Cedex, France,

April 1993.

[1] Ch. Bendjaballah, J.F. Bercher, F. Champagnat, J.F. Giovannelli, J. Idier, C. Kulcsar, G. Le Besnerais, M. Nguyen, M. Nikolova and D. Prémel.
[2] Mrs. Rimoldi
[3] Mrs. Groen

CONTENTS

Chapter 6. Image Restoration and Reconstruction 373

Chapter 1

Bayesian Inference and Maximum Entropy

INTERSUBJECTIVITY, RELATIVISTIC COVARIANCE, AND CONDITIONALS (A NEW AXIOMATIZATION)

O. Costa De Beauregard
Institut Henri Poincaré
11 rue Pierre et Marie Curie,75005 Paris

ABSTRACT. Conceived in a pre-field-theoretic era, the Bayesian axiomatization of conditional probabilities disregards the spacetime propagation of causality as expressed by the transition probabilities of physics. We show that the conditionality concept is indeed implied in the latter, and make this explicit by redefining conditionals. A Landé style "correspondence" exists between this scheme and the Born-Jordan one of quantum mechanics. Relativistic and spacetime reversal invariance of the formalism is "manifest".

The claim, by a physicist, that a re-conceptualization and re-formalization of the doctrine of "conditional probabilities" and of "the probability of causes" is needed in physics may seem presumptuous.

The point is that in an *essentially probabilistic physics* (as ours is assumed to be) *mutual stochastic dependence* of two occurrences is synonymous to *existence of an interaction* between them. So, in a broad sense, *conditional and transition probability are synonymous* [1], and *should be amenable to a common formalization* [2,3].

An essentially probabilistic physics has two basically connected features : *intersubjectivity* (the substitute of the former objectivity) and *relativistic covariance* (underlying "subjective" or "relative" probabilities). *All observers agree* as to which physical occurrence is observed. And *a conditional probability has intersubjectivity*, that is, physicality.

Therefore, the *conceptualization-and-formalization of the physical conditional probabilities definitely needs a relativistically and intersubjectively covariant expression.* As it happily turns out, it is thus brought into coincidence with the existing one of *physical transition probabilities.*

Let us first make clear that, *issued in a pre-field-theoretic era, the Bayes-Laplace concept of conditional probability, and of the probability of causes, fails to render the propagation of causality troughout spacetime.* Thus *it misrepresents the interaction between physical occurrences.*

Familiar P symbols being dropped for brevity, the *two inverse* Bayes-Laplace expressions of the *joint probability* of two occurrences A and C are

$$| A) \cap (C | \equiv | C) \cap (A | = | A | C)(C | = | C | A)(A | \equiv | A)(A | C | \qquad (1)$$

implying the *conditional probabilities* "of A if C" and "of C if A", and the *prior probabilities* of C and A.

A. Mohammad-Djafari and G. Demoments (eds.), Maximum Entropy and Bayesian Methods, 9–14.

Case of two spacelike separated occurrences. Recently [4] some fifteen neutrinos of the swarm emitted by a galactic supernova explosion were detected in laboratories, some of which distant by thousands of miles. *What sense does it make* to express the *joint probability* of two such spacelike separated detections by either one of the "inverse" Bayesian expressions (1)? Would not *one synthetic expression, symmetric* in A and C, be preferable? Also this formula completely disregards how *causality propagates* from the neutrino source to the detectors at A and C, and how the detection efficiencies are attenuated by the distance. *So, this formula is indeed nearer to nominalism than to physics.*

Case of two timelike separated occurrences. What is the *joint probability* that, having seen a lightning, I will hear the *associated* thunder roar? The Bayes' formula (1) gives for it *two inverse answers*, a predictive and a retrodictive one, in which the estimation of both "conditionals" is blurred by not severing *propagation* from *detection* of the *signals.* If I happen to catch a "summer lightning", I will never hear the thunder roar — if only because when the phonons arrive I will have lost interest in them. And, if being indoors at night, I hear a thunder roar, I will not have seen the lightning. Ligthtning is not the cause of thunder; both proceed from a common cause, an electric discharge emitting "jointly" photons and phonons.

Enunciated in a pre-field-theoretic and pre-relativistic era of physics, the classical axiomatization of "conditionals" completely disregards the physical propagation of causality, that is, also of information. It thus overlooks completely an essential feature of the physics of probability.

What then should be done? First of all, is not the very epithet *joint* an invitation to treat *symmetrically* the two occurrences at stake, and to have *both* of them explicitly present in the formula? Thus we write, instead of (1),

$$| A).(C | \equiv | C).(A | = | A)(A | C)(C | \equiv | C)(C | A)(A | \tag{2}$$

and call it the (un-normalized) *joint number of chances* of the two occurrences of A and C, displaying *action-reaction reciprocity* of causality.

Assuming that, in both cases (1) and (2), the "priors" $| A) \equiv (A |$ and $(C | \equiv | C)$ are normalized to unity, and that, in Bayes' formula (1), the "conditionals" are normalized via

$$\mathrm{S_A} | A | C) = \mathrm{S_C}(A | C | = 1, \tag{3}$$

the *joint probability* comes out as normalized via

$$\mathrm{SS} | A) \cap (C | = 1. \tag{4}$$

Identification of the concepts (1) and (2) is *forbidden*, as entailing $(A | C) \equiv \delta_{AC}$, that is, independence [5] of the occurrences A and C. This I did not make clear in my early presentations of the matter [2, 3].

Thus in the *explicitly covariant* scheme I am proposing, it is the *intrinsic reversible conditional probability "of A if C" or "of C if A"*

$$(A | C) \equiv (C | A) \tag{5}$$

that is normalized according to (compare with Landé [6])

$$\mathrm{S_A}(A \mid C) = \mathrm{S_C}(A \mid C) = 1. \tag{6}$$

An important advantage of the new scheme is that *it has the group generating formula* (summation being over mutually exclusive states)

$$(A \mid C) = \mathrm{S}(A \mid B)(B \mid C). \tag{7}$$

Formulas (2) and (7) find a straightforward application in the two previous exemples; the priors $\mid A)$ and $(C \mid$ measure the detection probabilities; the reversible conditionals $(A \mid B)$ and $(B \mid C)$ are "propagators"; $(A \mid C)$ expresses the *reversible intrinsic conditional probability* of the A and C detections. Finally, $\mid A).(C \mid$ denotes the "un-normalized extrinsic, or dressed, transition probability".

So, *this re-conceptualization and re-formalization of conditionals makes it coincide with that of the physical transition probabilities.* As expressed in terms of spacetime geometry, it "manifestly" has *relativistic covariance* and *time reversal symmetry.*

An other significant advantage of this new scheme is that it is in *manifest correspondence* with the 1926 Born [7] and Jordan [8] quantal probability calculus. Here again my path meets Landé's footsteps.

While the *classical probabilities calculus* obeys the two Laplacian rules of adding partial and multiplying independent probabilities, the Born–Jordan *wavelike amplitudes calculus* adds partial and multiplies independent (complex Hermitian) transition amplitudes

$$\langle A \mid C\rangle \equiv \langle C \mid A\rangle^*; \tag{8}$$

these "propagators" are the "square roots" of the *quantal transition probabilities*

$$(A \mid C) = \langle A \mid C\rangle\langle C \mid A\rangle \equiv \mid \langle A \mid C\rangle \mid^2 . \tag{9}$$

Corresponding to formulas (2) and (7) we respectively have

$$\mid A\rangle.\langle C \mid \equiv \mid A\rangle\langle A \mid C\rangle\langle C \mid \tag{10}$$

and

$$\langle A \mid C\rangle \equiv \mathrm{S}\langle A \mid B\rangle\langle B \mid C\rangle; \tag{11}$$

(10) expresses the *dressed transition amplitude* in terms of the *intrinsic* one and the *occupation amplitudes*; (11), the "generating formula of Landé chains", is the *propagators composition law.*

This whole scheme has manifest *micro-relativistic invariance*, that is, *Lorentz plus Lüders (CPT) invariance*. Geometrical spacetime reversal PT, and particle–antiparticle exchange C, are respectively expressible as

$$\langle A \mid C\rangle \rightleftharpoons \langle C \mid A\rangle \tag{12}$$

$$\langle A \mid C\rangle \rightleftharpoons \langle A \mid C\rangle^* \tag{13}$$

so, the Hermitian symmetry (8) expresses the product operation CPT=1.

Landé significantly emphasizes that the quantal probability calculus has more self-consistency than the classical one; this we now verify.

While a back-and-forth Markov chain, expressed as

$$1 = (A \mid A) = \mathbf{S}(A \mid B)(B \mid A) \equiv \mathbf{S_B}(A \mid B)^2 \tag{14}$$

contradicts the classical normalization rule, a back-and-forth Landé chain expressed as

$$1 = \langle A \mid A \rangle = \mathbf{S}\langle A \mid B \rangle \langle B \mid A \rangle \equiv \mathbf{S_B}(A \mid B) \tag{15}$$

yields the correct normalization of probabilities.

More general than Markov or Landé chains, they are "concatenations" with more than two propagators attached to a vertex A; these are guidelines for computing transition probabilities of a process. Known as Feynman graphs in quantum mechanics, they could be used just as well in classical statistical mechanics. Had Boltzmann likened his colliding molecules to rotating ogival shells rather than to spherical bullets, Loschmidt would have stated a CPT style rather than a PT invariance.

Cause–effect reversibility, known to Laplace, but more explicitly evidenced in this presentation, has truly mind stretchting aspects in the *wavelike and CPT invariant* quantal probability calculus. The *conditionality* concept is there instrumental for unravelling these "paradoxes".

The cross terms generated by substituing (11) in (9) express the quantal "interference of probability amplitudes", entailing (algebraic) "non separability" and (geometrical) "non-locality". While the classical $\mid B)(B \mid$ sums were thought of as over "real hidden states", the quantal $\mid B\rangle\langle B \mid$'s *cannot.* Whence, among others, the EPR [10], and the Wheeler [11] smoky dragon, "paradoxes".

Conditionality implies that *a transition probability holds if and only if the incoming and outgoing particles are actually prepared and measured as the formula expresses.* Disregard of this warning has caused [12] many loose, or even erroneous statements concerning EPR correlations. Even Einstein's [10] prejudice in favour of hidden realism is "falsifiable" on these grounds.

Formula (11) yields the amplitude of two YES answers to *independent* questions put at distant places A and C upon particles issuing from a common source B; it *certainly* does *not* mean that an answer $\mid A\rangle$ at A enforces at C the strictly associated answer! The formula is *symmetric* in A and C; so *which of the two questions should enforce the other answer?*

Formula (11) says that the *correlated answers* $\mid A\rangle$ and $\mid C\rangle$ *cannot* preexist as such in the source B; as the *independent questions* put at A and C are *arbitrary*, this is *unequivocal proof that causality is arrowless* at the micro-level. "Retrocausation" has indeed been displayed in "delayed choice experiments".

Something similar shows up with Wheeler's [11] "smoky dragon". Formula (11) expresses a quantal system evolving from its preparation $\mid A\rangle$ to its measurement $\mid C\rangle$ as a "superposition of virtual states" $\mid B\rangle\langle B \mid$.

Wheeler likens it to a "smoky dragon" living above our empirical spacetime, letting hang down here only its tail held at A and its mouth biting at C. As the question put at C is arbitrary, there is "retrocausation", and delayed choice experiments have proved it.

Causality identified with conditional probability: Laplace's 1818 Memoir on the probability of causes hints to this, and, as it turns out, field–theorical physics affords an affirmative answer.

To this identification Jaynes [13] objects that , in formal logic, if A implies B, non–B implies non–A; then says he, if A $\Rightarrow B$ were interpreted as "is the cause of B", there would be difficulty in accepting that "non–B is the cause of non–A".

Non–A means of course *observation of non–A*, *not* the weaker "A non observed"! So, that non–A can cause non–B is trivial if the AB separation as "future–timelike". If the AB separation is spacelike, or past–time like, the EPR and the smoky dragon delayed choice experiments, both vindicating the wavelike Born–Jordan formalism, do display *causality as essentially arrowless.* This disposes of Jaynes' objection, which is analogous to an earlier one by Renninger [14].

To conclude, this paper aims *not* at disputing the usefulness of Bayesian methods for partial information handling and decision making.

Its emphasis is upon 1) the *intersubjectivistic* character of *physical probabilities*; 2) the *necessary requirement* that their conceptualization and formalization be relativistic–invariant and 3) in "correspondence" with quantum mechanics; finally 4) that the *physical propagation of causality* be taken care of.

References

[1] L. Accardi, *"The probabilistic roots of the quantum mechanical paradoxes*, in *The Wave Particle Dualism*, S.Diner et al. eds, Reidel, Dordrecht, pp. 297–330 (1984).

[2] O. Costa De Beauregard, *" Causality as identified with conditional probability and the quantal non separability*, in *Microphysical reality and the Quantum Formalism*, A.van der Merwe et al. eds, Kluwer, Dordrecht, pp. 219–232 (1988).

[3] O. Costa De Beauregard, *"Relativity and probability, classical and quantal*, in *Bell's Theorem and Conceptions of the Universe*, M.Kafatos ed.,Kluwer, Dordrecht, pp. 117–126 (1989).

[4] D. Kleppner , *A lesson in humility* in *Physics Today*, December 1991, pp. 9–11.

[5] O. Costa De Beauregard , Found. Phys. Lett. *5* , pp. 291 (1992).

[6] A. Landé, in *New foundations of quantum Mechanics*, Cambridge University Press, Chap VI (1965).

[7] M. Born, in *Zeits. Phys. 38*, 803 (1926).

[8] P. Jordan, in *Zeits. Phys. 40*, 809 (1926).

[9] G. Lüders, in *Zeits. Phys. 133*, 525 (1952).

[10] A. Einstein, B. Podolsky, N. Rosen, *Phys Rev. 47* , 77 (1935).

[11] W. A. Miller, J. A. Wheeler, *Delayed choice experiments and Bohr's elementary quantum phenomenon* in *Foundations of Quantum Mechanics in the Light of New Technology*, S.Kamefuchi et al. eds, Phys. Soc. Japan, pp. 140–152 (1988).

[12] O. Costa De Beauregard, *Found. Phys. 15*, 873, (1985).

[13] E. T. Jaynes , *Clearing up Mysteries* in *Maximum Entropy and Bayesian Methods*, J. Skilling ed., Kluwer Academic, (1989).

[14] W. Renninger, in *Zeits. Phys. 158*, 417 (1960).

GAME THEORETICAL EQUILIBRIUM,
MAXIMUM ENTROPY
AND MINIMUM INFORMATION DISCRIMINATION

Flemming Topsøe
Institute of Mathematics, University of Copenhagen
Universitetsparken 5
DK-2100 Copenhagen
topsoe@euromath.dk

ABSTRACT. Games are considered which rely on the concept of a code and which focus on the interplay between the observer and the system being observed. The games lead to specific principles of *Game Theoretical Equilibrium*. From these, one readily deduces the well known principles of *Maximum Entropy* and *Minimum Information Discrimination*. The game theoretical principles apply in certain situations where the classical principles do not.

1. Overview of main points

For the sake of abbreviation, let GTE stand for the principles of *Game Theoretical Equilibrium* which we shall introduce. Likewise, let ME and MID stand for "Maximum Entropy" and "Minimum Information Discrimination", respectively. The main points made are the following:

- the principles of GTE imply those of ME and MID;
- GTE gives more information than ME or MID, and in certain cases GTE is applicable whereas ME or MID are not;
- the role of the observer is stressed (via the concept of coding);
- the mathematics becomes simpler than is usually the case, e.g., concrete standard problems can be handled directly without recourse to Lagrange multipliers;
- the partition functions of the physisists occur naturally (in the search for certain special codes);
- the approach of GTE offers a new conceptual base with new objects to worry about and new problems to look into.

From discussions at MaxEnt-92 it appeared that, for the solution of the type of inference problems at hand, the idea has been around to incorporate more actively concepts from Information Theory other than just the entropy function itself, in particular, dr. C.R. Smith pointed to the paper (Rissanen, 1988). We shall take a step in this direction by introducing codes in the discussion. However, this is done in a primitive fashion, forgetting the actual structure of codes, only retaining information about the length of individual code words.

A. Mohammad-Djafari and G. Demoments (eds.), Maximum Entropy and Bayesian Methods, 15–23.

The point made above about Lagrange multipliers was met at MaxEnt-92 with some scepticism. Recall, that the Lagrange multiplier technique is a general mathematical tool applicable to almost any constrained optimization problem. There is, however, no need for any "multipliers" when dealing with the *special* optimization problems we have in mind. These problems are solved in a better way without, using an intrinsic method specifically adapted to the situation at hand. The fact that the multipliers often have a special interpretation (free energy, ...) of course point to your wanting to introduce them anyhow - but there is no need to do so for the actual solution of the optimization problems in question.

For a more technical discussion the reader is referred to (Topsøe, 1979). That paper contains the main ideas, however, the useful Theorems 2 and 4 were overlooked (Theorem 2 was observed later and published in the elementary paper (Topsøe, 1978) in danish). The paper (Csiszár, 1975) was an important source of inspiration in formulating the general game theoretical principles. The reader may also wish to consult (Jupp and Mardia, 1983).

2. The absolute game

We consider a primitive model (I, C) of a (physical) *system*. Here, I is a discrete, countable set, the *state space* (or the set of *pure states*, if you wish). And C, the *preparation*, is a subset of the set M of all probability distributions over I. The distributions in C are said to be *consistent* (with the preparation of the system). On a few instances we refer to the topology on M. It is clear what is meant by this as there is only one sensible topology on M (note that the pointwise and uniform topologies coincide on M).

The key problem is to make intelligent inferences about an unknown distribution based only (or primarily) on the incomplete information that this distribution is consistent. We attack the problem by introducing, artificially, two "players", "nature" and "the observer", and let these players interact via a certain two person zero sum game, denoted by $\gamma = \gamma(C)$ and called the *absolute game* associated with C. Nature participates in γ by choosing a consistent distribution, and the observer chooses a way of observing the system, formalized via the concept of a *code*, cf. below. It is the objective of the observer to minimize the expected observation time - in our setting equal to the *expected code length* -, whereas we assume that nature has the opposite objective, to maximize this quantity.

To us, a *code* is a map $\kappa : I \to [0, \infty]$ such that

$$\sum_{i \in I} e^{-\kappa_i} = 1 . \tag{1}$$

Those familiar with elementary Information Theory will realize that (1) is an idealized form of Kraft's inequality (with the mathematically convenient base e rather than base 2, with the possibility of non-integer κ_i - a technically necessary idealization, which can be defended by reference to the noicelcss coding theorem -, and with equality rather than inequality to reflect coding without superfluous symbols). Thus, though this is only of conceptual importance, we may imagine that, corresponding to κ, there exists a *codebook* listing all the *codewords* corresponding to the possible states. The length of the codeword corresponding to $i \in I$ is κ_i, and this may be interpreted as the time needed by the observer in order to locate the actual state of the system in case this state is i, or, as another possible interpretation, as the time it takes to communicate to a "receiver" the actual state of the system.

There exists a natural bijection between M and K, expressed notationally by $\mu \leftrightarrow \kappa$ or $\kappa \leftrightarrow \mu$, and given by

$$\kappa_i = -\log \mu_i \; ; \; \mu_i = e^{-\kappa_i} \quad (i \in I) . \tag{2}$$

When these relations hold, we also say that κ is *adapted* to μ, or that μ is *associated* with κ.

If $\mu \in M$ and f is a function on I, $\langle f, \mu \rangle$ denotes expectation of f w.r.t. μ (when well defined, allowing for $+$ or $-\infty$)

The *cost function* for the game $\gamma = \gamma(C)$ is the function on $K \times C$ with values in $[0, \infty]$ given by

$$(\kappa, \mu) \mapsto \langle \kappa, \mu \rangle \; ; \; (\kappa, \mu) \in K \times C . \tag{3}$$

Note the interpretation of the cost function as *expected code length* (or observation time). Clearly,

$$\sup_{\mu \in C} \inf_{\kappa \in K} \langle \kappa, \mu \rangle \leq \inf_{\kappa \in K} \sup_{\mu \in C} \langle \kappa, \mu \rangle . \tag{4}$$

Following normal terminology, the common number determined by (4) in case of equality is the *value* of the game.

For $\kappa \in K$, the *risk* $R(\kappa) = R(\kappa \mid C)$ is defined as

$$R(\kappa) = \sup_{\mu \in C} \langle \kappa, \mu \rangle . \tag{5}$$

The *minimum risk* $R_{\min} = R_{\min}(C)$ is

$$R_{\min} = \inf_{\kappa \in K} R(\kappa) . \tag{6}$$

If $R(\kappa_0) = R_{\min}$, κ_0 is a *minimum risk code.* This is nothing but an optimal strategy for the observer for the game γ.

The code κ_0 is *cost stable* if $\langle \kappa_0, \mu \rangle$ is finite and independent of $\mu \in C$, i.e. if $R(\kappa_0) < \infty$ and $\langle \kappa_0, \mu \rangle = R(\kappa_0)$ for all $\mu \in C$.

The quantities related to nature which are analogous to those given by (5) and (6) are the following

$$H(\mu) = \inf_{\kappa \in K} \langle \kappa, \mu \rangle \tag{7}$$

$$H_{\max} = H_{\max}(C) = \sup_{\mu \in C} H(\mu) . \tag{8}$$

It is well known (and also follows from (12) below) that $H(\mu)$ is the *entropy* of μ:

$$H(\mu) = -\sum_{i \in I} \mu_i \log \mu_i . \tag{9}$$

Thus, an optimal strategy for nature in the game γ is the same as a *maximum entropy distribution,* i.e. a consistent distribution μ_0 with $H(\mu_0) = H_{\max}(C)$.

Not surprisingly, we shall also need to consider the *divergence* (*information distance, discrepancy* or *relative entropy*) between two distributions μ and η in M. This quantity is defined by

$$D(\mu\|\eta) = \sum_{i\in I} \mu_i \log \frac{\mu_i}{\eta_i} \tag{10}$$

which is always a well defined number in the interval $[0, \infty]$. As is well known, $D(\mu\|\eta) = 0$ if and only if $\mu = \eta$.

The divergence has an illuminating interpretation. Indeed, let κ and ρ be the codes adapted to μ, respectively η. Then

$$D(\mu\|\eta) = \sum_{i\in I} \mu_i(\rho_i - \kappa_i) , \tag{11}$$

hence $D(\mu\|\eta)$, seen from the point of view of the observer, is the average gain or improved efficiency resulting from the information that a distribution believed by the observer to be η has, in fact, changed to μ.

A key formula connecting expected code length, entropy and divergence is the following trivial identity, just a way of rewriting (11):

$$\langle \rho, \mu \rangle = D(\mu\|\eta) + H(\mu) \; ; \; \mu, \eta \in M \; , \; \rho \leftrightarrow \eta \, . \tag{12}$$

The principle of *Game Theoretical Equilibrium* (GTE) which we are led to, amounts to consider the game γ and search for optimal strategies, both for nature and for the observer in attempting to solve the basic problem of inference. To state something precise, we might say that GTE *applies ideally* to the game $\gamma = \gamma(C)$ if the following conditions hold:

(i) the value of γ exists and is finite;

(ii) there exists a unique minimum risk code;

(iii) there exists a unique maximum entropy distribution.

Thus, if GTE applies ideally, we have

$$\textit{minimum risk} = \textit{maximal entropy} \, . \tag{13}$$

We also note that GTE leads to the Maximum Entropy Principle, ME, which corresponds to looking only for optimal strategies for one of the players in the game, nature.

If there exists a unique minimum risk code, we denote this code by κ_C. For reasons which will become clear shortly, we call the distribution associated with κ_C, the *centre of attraction* of C. We shall denote this special distribution by μ_C.

In our first theorem we exhibit the close to complete information of a general and basic nature which is available on the games $\gamma(C)$. Together with the preparation C, its convex hull, $\mathrm{co}\, C$, will also be considered.

Theorem 1. *(GTE, general results for the absolute game).*

(i) $R_{\min}(C) = H_{\max}(\mathrm{co} C)$.

(ii) The value of the game $\gamma(C)$ *exists and is finite if and only if* $R_{\min}(C) < \infty$ *and* $H_{\max}(C) = H_{\max}(\mathrm{co} C)$ *hold.*

(iii) There exists a unique minimum risk code for the game $\gamma(C)$ *if and only if* $R_{\min}(C) < \infty$.

(iv) *Assume that $R_{\min}(C) < \infty$ so that κ_C and μ_C are well defined. Then every sequence (μ_n) of distributions in co$C$ with $H(\mu_n) \to H_{\max}(\text{co}C)$, converges to the centre of attraction: $\mu_n \to \mu_C$.*

(v). *Assume that $R_{\min}(C) < \infty$ and $H_{\max}(C) = H_{\max}(\text{co}C)$ hold. If there exists a maximum entropy distribution for C, it must be the centre of attraction, and this is the case if and only if $\mu_C \in C$ and $H(\mu_C) = H_{\max}(C)$ hold.*

Proof: (i): The result in the one direction follows from

$$H_{\max}(C) \le H_{\max}(\text{co}\, C) \le R_{\min}(\text{co}\, C) = R_{\min}(C)\,,$$

and the result in the other direction follows from Theorem 3 of (Topsøe, 1979). Property (ii) follows from (i). The non-trivial part of (iii) follows by combining (i) and Theorem 3, loc.cit. Property (iv) follows from (i) and from Theorems 1 and 3, loc.cit. The last property is an immediate consequence of (iv).

Remarks. The necessity of considering convex hulls in Theorem 1 may be illustrated by taking as C the set of deterministic distributions (both cases I finite and I infinite are illuminating).

Theorem 1 may be taken to support the point of view that the principle of GTE is superior to that of ME. Indeed, under very mild conditions, the value of the game considered exists and so does the unique minimum risk code. In contrast to this, the maximum entropy distribution need not exist. However, there is almost always, in particular if C is convex, a unique candidate to consider, to which any attempt to find a maximum entropy distribution must converge.

In case C is not convex (and $H_{\max}(C) < H_{\max}(\text{co}\, C)$), the theorem does not give complete information. Note that it is easy to construct examples such that a unique maximum entropy distribution with finite entropy exists but does not coincide with the centre of attraction, which need not even exist (take C the set of deterministic distributions with one of the distributions slightly distorted). In spite of this, the result appears satisfactory as one can claim that, for naturally occurring applications, if certain distributions are consistent, then so is any mixture. Therefore, the naturally occurring preparations appear to be convex.

You may view the essential property (iv) as providing a theoretical method for finding μ_C, hence for finding the maximum entropy distribution in case it exists. However, in certain situations, the result below provides a much simpler method.

It must be stressed that both conditions in (v) may fail (even for convex sets). The failure of $\mu_C \in C$ is perhaps not so serious (though it should not be avoided by simply assuming that C is closed since there are many applications with non-closed preparations). It is worse that $H(\mu_C) < H_{\max}(C)$ may happen. This case can be described as *collapse of entropy* (recall that the entropy function is only lower semi-continuous, not continuous). For concrete examples, we refer to (Ingarden and Urbanik, 1962) or to (Topsøe, 1979). The author is unaware if this phenomenon can occur in realistic physical systems.

Finally, we remark that the uniqueness assertions of Theorem 1 can be derived from an interesting strengthening of the trivial inequality $H_{\max}(C) \le R_{\min}(C)$. Indeed, if $R_{\min}(C) < \infty$, if $\mu \in \text{co}\, C$, $\kappa \in K$ and $\kappa \leftrightarrow \eta$, then results from (Topsøe, 1979) imply that

$$H(\mu) + D(\mu\|\mu_C) + D(\mu_C\|\eta) \le R(\kappa \mid C)\,. \tag{14}$$

Theorem 2. *(GTE, special result for the absolute game). Assume that $\kappa^* \in K$ and $\mu^* \in M$ satisfy the following conditions:*
(i) $\kappa^* \leftrightarrow \mu^*$,
(ii) κ^* is cost stable (i.e. $\exists\, h < \infty \forall\, \mu \in C : \langle \kappa^*, \mu \rangle = h$),
(iii) $\mu^* \in C$.

Then κ^* is the unique minimum risk code for the game $\gamma(C)$ and μ^* is the unique maximum entropy distribution for C.

Proof: (from Topsøe, 1978). Clearly, $H(\mu^*) = \langle \kappa^*, \mu^* \rangle = h$. For $\mu \in C$ with $\mu \neq \mu^*$, we find by (12) that

$$H(\mu) < D(\mu\|\mu^*) + H(\mu) = \langle \kappa^*, \mu \rangle = h\,.$$

It follows that μ^* is the unique maximum entropy distribution.

Clearly, $R(\kappa^* \mid C) = h$, and for any code $\kappa \neq \kappa^*$ with associated distribution η we have, again by (12),

$$R(\kappa \,|\, C) \geq \langle \kappa, \mu^* \rangle = D(\mu^*\|\eta) + H(\mu^*) > H(\mu^*) = h\,.$$

We conclude that κ^* is the unique minimum risk code.

Though it is easy to find examples where this result does not apply, as κ_C need not be cost stable, the result does apply in a number of important cases.

Example 1. If I is finite and $C = M$, the uniform distribution is the maximum entropy distribution. This follows from Theorem 2 as the code adapted to the uniform distribution is cost stable.

Example 2. If μ_j is a distribution on the countable set I_j ; $j = 1, 2$, then the code adapted to the product measure $\mu_1 \otimes \mu_2$ is cost stable for the preparation of measures on $I_1 \times I_2$ with μ_1 and μ_2 as marginals. It follows that $\mu_1 \otimes \mu_2$ is the maximum entropy distribution in this case.

Example 3. Let $E : I \to [0, \infty]$ be a function (the *energy function*) on the countable set I, and let $C(\overline{E})$ be the preparation consisting of all distributions μ with $\langle E, \mu \rangle = \overline{E}$. Clearly, any code κ of the form

$$\kappa = \alpha + \beta E \tag{15}$$

with α and β two constants, is cost stable. By (1), we may write (15) in the form $\kappa = \kappa_\beta$ with

$$\kappa_\beta = \log Z(\beta) + \beta E\,, \tag{16}$$

where

$$Z(\beta) = \sum_{i \in I} e^{-\beta E_i}\,. \tag{17}$$

This formula is the well known expression for the *partition function*. In case we can determine β such that the associated distribution μ_β is consistent, then it follows by Theorem 2 that this distribution is the maximum entropy distribution (the *canonical distribution* of the physicists).

We note that

$$R(\kappa_\beta | C(\overline{E})) = \log Z(\beta) + \beta \overline{E}\,,$$

and in case $\mu_\beta \in C(\overline{E})$, this value is the same as $H(\mu_\beta)$ which equals $H_{\max}(C(\overline{E}))$. The further investigations follow a well known pattern and are not given here. However, we do wish to point out that entropy collapse may occur for preparations of the simple type considered. This happens if there exists a number t (necessarily positive) such that $Z(t)$ is finite, $Z(s)$ is infinite for $s < t$ and such that $\langle E, \mu_t \rangle = E_{\text{crit}}$, the *critical energy*, is finite. Then, if $\overline{E} > E_{\text{crit}}$, the centre of attraction is μ_t. This distribution is not consistent and the phenomenon of collapse occurs as

$$H_{\max} - H(\mu_t) = t(\overline{E} - E_{\text{crit}}) > 0 \,. \tag{18}$$

Generalization to more constraints of the form $\langle f, \mu \rangle = \overline{f}$, is straight forward (and leads to the *grand canonical distribution*).

In all the above examples, the maximum entropy distribution (and the minimum risk code) were found directly without recourse to Lagrange multipliers. All we did was to search for cost stable codes.

3. The relative game

In order to simplify the discussion, we consider only the discrete case, thus the situation with I a countable set and $C \subseteq M$ is unchanged. For a more technical and comprehensive discussion, see (Topsøe, 1979). In certain cases, e.g. if $H_{\max}(C) = \infty$, it is sensible to choose a reference distribution η (the *a priori* distribution). Denote by ρ the code adapted to η. This is the code used a priori by the observer. For the sake of simplicity, the reader may assume that $\eta_i > 0$ for all $i \in I$. If $\kappa \in K$, the *code improvement* associated with κ is the function $\rho - \kappa$.

We may then consider the *relative game* $\gamma = \gamma(C \mid \eta)$ which has two players, "nature" and the "observer" who may choose a consistent distribution and a code, respectively. If $\mu \in C$ and $\kappa \in K$ are chosen, the associated *pay-off* to the observer is the *expected code improvement*, i.e. the quantity $\langle \rho - \kappa, \mu \rangle$. Assuming that the observer attempts to maximize the pay-off and that nature has the opposite objective, we again see that we are dealing with a two person zero sum game. The principle to investigate this game in much the same way as we investigated the absolute game, constitutes the GTE principle in the setting of relative games.

We stress that the pay-off $\langle \rho - \kappa, \mu \rangle$ is only considered when this quantity is a well defined extended real number. It is noted that this is the case if and only if the expression $D(\mu\|\eta) - D(\mu\|\mu_\kappa)$ is not of the indeterminate form $\infty - \infty$. Here, μ_κ denotes the distribution associated with κ. If the expression considered is not indeterminate, the following identity, analogous to, in fact derivable from (12), holds:

$$\langle \rho - \kappa, \mu \rangle = D(\mu\|\eta) - D(\mu\|\mu_\kappa) \,. \tag{19}$$

It follows that

$$D(\mu\|\eta) = \sup_{\kappa \in K} \langle \rho - \kappa, \mu \rangle \tag{20}$$

(with the understanding that really, κ ranges only over those codes for which the pay-off $\langle \rho - \kappa, \mu \rangle$ is well defined; similar interpretations are understood below).

We put

$$D_{\min}(C\|\eta) = \inf_{\mu \in C} D(\mu\|\eta) \tag{21}$$

and realize that an optimal strategy for nature in the relative game is the same as a *minimum information discrimination* distribution. We realize, therefore, that, in this case, the GTE principle leads to that of MID. We refer to (Barndorff-Nielsen, 1978), (Csiszár, 1975) and (Kullback, 1959) regarding MID.

Corresponding to $\kappa \in K$, we consider the *pay-off* of κ, defined by

$$P(\kappa|C,\eta) = \inf_{\mu \in C} \langle \rho - \kappa, \mu \rangle \ , \tag{22}$$

and the *maximal pay-off* defined by

$$P_{\max}(C,\eta) = \sup_{\kappa \in K} P(\kappa|C,\eta) \ . \tag{23}$$

For the game $\gamma(C|\eta)$, κ is an optimal strategy for the observer, or a *maximum pay-off strategy* if $P(\kappa|C,\eta) = P_{\max}(C,\eta)$.

We always have

$$P_{\max}(C\|\eta) \le D_{\min}(C\|\eta) \tag{24}$$

and when equality holds, the number determined is the *value* of the game $\gamma(C|\eta)$. When this is the case, we have the principle

$$\textit{maximum pay-off} = \textit{minimum information discrimination.} \tag{25}$$

A code κ is *pay-off stable* for the relative game $\gamma(C,\eta)$ if $P(\kappa|C,\eta)$ is finite and equal to $\langle \rho - \kappa, \mu \rangle$ for all $\mu \in C$.

From (Topsøe, 1979) we obtain the following result.

Theorem 3. *(GTE, general results for the relative game). Assume that C is convex and $D_{\min}(C\|\eta) < \infty$. Then the value of the game $\gamma(C|\eta)$ exists and there exists $\mu^* \in M$ such that $\mu_n \to \mu^*$ for any sequence (μ_n) of consistent distributions with $D(\mu_n\|\eta) \to D_{\min}(C\|\eta)$. Furthermore, the code κ^* adapted to μ^* is the unique maximum pay-off strategy for the observer.*

The distribution μ^* is the *relative centre of attraction* for $\gamma(C|\eta)$. Trivial examples show that GTE for the relative game may apply in cases where GTE for the absolute game does not (as $H_{\max} = \infty$). Also, in cases where GTE for both types of games apply, the relative centre of attraction may of course be quite distinct from the (absolute) centre of attraction. However, if the a priori distribution η has an extra property, viz. that the code adapted to it is cost stable, and if C is convex, say, then the two centres of attraction coincide, as is easily seen.

Theorem 4. *(GTE, special result for the relative game). Assume that $\kappa^* \leftrightarrow \mu^*$, that $\mu^* \in C$ and that κ^* is a pay-off stable strategy for the observer in the game $\gamma(C|\eta)$. Then the value of $\gamma(C|\eta)$ exists, μ^* is the unique minimum information discrimination distribution and κ^* is the unique maximum pay-off strategy.*

We leave it to the reader to derive this result from (19). This can be done in much the same way as Theorem 2 was derived from (12).

REFERENCES

Barndorff-Nielsen, O.: 1978, *Information and Exponential Families in Statistical Theory,* John Wiley, New York.

Csiszár, I.: 1975, 'I-divergence geometry of probability distributions and minimization problems', *Annals of Probability* **3**, 146–158.

Ingarden, R.S. and Urbanik, K.: 1962, 'Quantum Informational Thermodynamics', *Acta Physica Polonica* **21**, 281–304.

Jupp, P.E. and Mardia, K.U.: 1983, 'A note on the Maximum Entropy Principle', *Scand. J. Stat., Theory Appl.* **10**, 45–47.

Kullback, S.: 1959 *Information Theory and Statistics* Wiley, New York.

Rissanen, J.: 1988, 'Stochastic Complexity and the Maximum Entropy Principle', in G.J. Erickson and C.R. Smith (eds.) *Maximum Entropy and Bayesian Methods in Science and Engineering* **1**, Kluwer, Dordrecht.

Topsøe, F.: 1979, 'Information theoretical optimazation techniques', *Kybernetika* **15**, 8–27.

Topsøe, F.: 1978, 'Et informationsteoretisk spil i tilknytning til maksimum-entropiprincippet', *Nordisk Matematisk Tidskrift* **25-26**, 156–172.

ENTROPY, MOMENTS AND PROBABILITY THEORY

Sławomir Kopeć
Jagellonian University
Institute of Physics
Reymonta 4
30-059 Kraków, Poland.

G. Larry Bretthorst
Washington University
Department of Chemistry
1 Brookings Drive
Saint Louis, Missouri 63130 USA.

ABSTRACT. Bayesian probability theory, using an entropy prior, is applied to the moment problem when the data are noisy. When the noise level tends to zero, the Bayesian solution tends to the classic maximum entropy (MaxEnt) solution. The uncertainty in the estimated function is derived and a numerical example is presented to illustrate the calculation.

1. Introduction

The MaxEnt approach to solving the problem of moments was studied in detail by Mead and Papanicolau (Mead, 1984). Their analysis assumed the availability of exact moments. This is a severe limitation of the technique. An attempt to apply MaxEnt to the noisy moment problem was made by Ciulli *et al.* (Ciulli, 1991). Their solution, although essentially correct, left some important questions unanswered. In particular, an estimate of the uncertainty in the estimated function was not presented. In this paper, the noisy moment problem is addressed using Bayesian probability theory. Many of the details omitted here are contained in (Bretthorst, 1992), where the deconvolution problem is studied.

In the moment problem, there are some known moments or data, d_i, which are related to an unknown function $x(t)$:

$$d_i = \int_0^1 dt\, \omega_i(t)\, x(t) + n_i, \quad i = 1, 2, \ldots N, \tag{1}$$

where $\omega_i(t)$ are linearly independent known functions, and n_i represents the measurement error in the ith moment.

2. Method of Solution

To solve this problem using Bayesian probability theory the continuous function $x(t)$ is first replaced with a set of discrete values $\{x_k\}, k = 1, 2, \ldots, M$, with $x_k := x(t_k)$, and

A. Mohammad-Djafari and G. Demoments (eds.), Maximum Entropy and Bayesian Methods, 25–30.

$dt \to \Delta t = 1/M$. Using an entropy prior and assigning a gaussian prior probability for the noise, the posterior probability for one of the x_k is then given by

$$P(x_k|D,\beta,\sigma,I) \propto \int dx_1 \cdots dx_{k-1} dx_{k+1} \cdots dx_M \exp\{S_\beta(x)\} \tag{2}$$

with

$$S_\beta(x) := \frac{2\beta}{M}\sum_{j=1}^{M}[x_j(\log x_j - 1) + 1] + \sum_{i=1}^{N}\left[d_i - \sum_{j=1}^{M}\frac{\omega_i(t_j)x_j}{M}\right]^2, \tag{3}$$

where D denotes the collection of moments, or data, $D \equiv \{d_1, \ldots, d_N\}$, σ is the standard deviation of the noise, and β is a measure of the relative importance of the prior information.

The integrals are evaluated in the Gaussian approximation. To make this approximation, the values of the x_k that maximize $S_\beta(x)$ are denoted as $\hat{x}_k$. Equation (2) has a maximum when the derivatives of $S_\beta(x)$ vanish. So $\hat{x}_k$ must satisfy

$$\left.\frac{\partial S_\beta(x)}{\partial x_k}\right|_{\hat{x}_k} = \beta \log \hat{x}_k - \sum_{i=1}^{N}\left[d_i - \sum_{j=1}^{M}\frac{\omega_i(t_j)}{M}\hat{x}_j\right]\omega_i(t_k) = 0, \quad k = 1, \ldots, M. \tag{4}$$

Because the $\{\omega_i\}$ are linearly independent, $\log \hat{x}$ is a linear combination of functions ω_i. Thus $\hat{x}$ is given by the solution to

$$\beta\alpha_i = d_i - \sum_{j=1}^{M}\frac{\omega_i(t_j)\hat{x}_j}{M}, \quad i = 1, \ldots, N, \tag{5}$$

$$\hat{x}_j = \exp\left\{\sum_{k=1}^{N}\alpha_k\omega_k(t_j)\right\}, \tag{6}$$

where the α_k play the role of generalized Lagrange multipliers.

Note that vanishing of σ implies vanishing of β. Thus when the noise level tends to zero, the left-hand side of (5) vanishes, and Eqs. (5) and (6) reduce to the maximum entropy solution to the problem. However, for noisy moments the maximum entropy solution is only an approximation to $\hat{x}$.

To complete the gaussian approximation of Eq. (2), a second order Taylor expansion is made about $\hat{x}_j$ to obtain

$$P(x_k|D,\beta,\sigma,I) \propto \int dx_1 \cdots dx_{k-1} dx_{k+1} \cdots dx_M \exp\left\{-\sum_{jl=1}^{M}\frac{R_{jl}(x_j - \hat{x}_j)(x_l - \hat{x}_l)}{2\sigma^2}\right\} \tag{7}$$

where

$$R_{jl} := \frac{\beta\delta_{jl}}{M\hat{x}_k} + \sum_{i=1}^{N}\frac{\omega_i(t_j)\omega_i(t_l)}{M^2}. \tag{8}$$

Evaluating the integrals one obtains:

$$P(x_k|D,\beta,\sigma,I) \propto \exp\left\{-\frac{(x_k-\hat{x}_k)^2}{2\sigma^2}\left[R_{kk}-\sum_{lmn}\frac{e'_{lm}e'_{ln}R_{km}R_{kn}}{\lambda'_l}\right]\right\}, \tag{9}$$

where λ'_l and e'_{lm} are the eigenvalues and eigenvectors of the mth component of the lth eigenvector of the kth cofactor of R_{lm}. The kth cofactor of R_{lm} is formed by deleting the kth row and column from R_{lm}. If one adopts the convention that the cofactor's rows and columns are indexed just like R_{lm}, then the sums are over all values of the index except $l = k$ or $m = k$ or $n = k$.

The above result is valid, provided σ is known. In the event σ is unknown, we can eliminate it from the problem by assigning a Jeffrey's prior and integrating. However, note that several terms were dropped from Eq. (9). This could be done because when σ is known, these terms are constants. Recovering these terms and eliminating σ as a nuisance parameter one obtains

$$P(x_k|D,\beta,I) \propto \left[S_\beta(\hat{x})+\left(R_{kk}-\sum_{lmn}\frac{e'_{lm}e'_{ln}R_{km}R_{kn}}{\lambda'_l}\right)(x_k-\hat{x}_k)^2\right]^{-\frac{N+1}{2}}. \tag{10}$$

3. Estimating the Uncertainty

An estimate of the uncertainty in the x_k may be obtained in the (mean $\pm$ standard deviation) approximation. To compute this, one must compute both $\langle x_k\rangle$ and $\langle x_jx_k\rangle$, the expectation values of, respectively, x_k and x_jx_k. The expectation value of x_k is equal to $\hat{x}_k$, while

$$\langle x_jx_k\rangle = \hat{x}_j\hat{x}_k + \sigma^2\sum_{l=1}^{M}\frac{1}{\lambda_l}e_{lj}e_{lk}. \tag{11}$$

The (mean $\pm$ standard deviation) approximation is then given by

$$(x_k)_{\rm est} = \hat{x}_k \pm \sqrt{\langle\sigma^2\rangle}\left[\sum_{l=1}^{M}\frac{(e_{lk})^2}{\lambda_l}\right]^{1/2}, \quad k=1,\ldots,M. \tag{12}$$

For finite N, this expression tends to infinity when the number of intervals tends to infinity. When M grows, our N moment equations become more and more insufficient to determine the approximate solution. In other words macroscopic data (on moments) cannot affect our knowledge of microscopic structure. This result occurs because neither the prior probability nor the likelihood introduce any point to point correlations in x_k, so on any infinitely small intervals, the function $x(t)$ could be doing wild things, long as the weighted average (represented by the moments) are satisfied.

4. Estimating the Parameters β and σ

In the above, β is assumed known. However, in general, β will not be known. The rules of probability theory tell one how to proceed. One should simply multiply Eq. (9) by an appropriate prior probability and integrate over β. This procedure will yield the posterior probability for x_k, independent of β. Unfortunately, β appears in these equations in a very nonlinear fashion and the integrals are not available in a closed form. However, a good approximation is available, provided the moments are not very noisy. When the moments are not very noisy, the integral over β is well approximated by a delta function, and δ functions just fix the value of the parameter. So if one knew where the integrand peaked as a function of β, one could simply constraint β to its maximum value.

Fortunately, the posterior probability for β can be computed. The value of β for which this probability is maximum is essentially identical to the value of β for which joint probability of x_j and β is maximum. The posterior probability for β, $P(\beta|D,\sigma,I)$ is computed from the joint probability for β and the $\{x\}$:

$$P(\beta|D,\sigma,I)\int dx_1\cdots dx_M P(\beta,\{x\}|\sigma,DI)$$
$$=\int dx_1\cdots dx_M P(\beta|I)P(\{x\}|\beta,\sigma,D,I).$$

Assuming Jeffrey's prior, $P(\beta|I)\propto 1/\beta$, the posterior probability for β is given by

$$P(\beta|D,\sigma,I)\propto \beta^{\frac{M}{2}}\exp\left\{-\frac{1}{2\sigma^2}S_\beta(\{\hat{x}\})\right\}\prod_{l=1}^{M}(\lambda_l\hat{x}_l)^{-\frac{1}{2}}. \tag{13}$$

The expected value of the noise variance, $\langle\sigma^2\rangle$, should be replaced by its true value in Eq. (12), if σ is known. However, in typical applications the noise level is unknown. We then treat σ as a nuisance parameter and eliminate it from the formalism. In Eq. (12) this resulted in the expected value of σ^2 showing up in the calculation, so to use Eq. (12) the expected value of σ^2 must be computed. This is given by

$$\langle\sigma^2\rangle=\frac{1}{N-2}S_\beta\left(\{\hat{x}\}\right). \tag{14}$$

5. Numerical Example

To demonstrate that for a fixed M the errors remain finite as the noise goes to zero, a simple example is given. In this example the function $\exp\{-5t^2\}$ was normalized over the interval zero to one and its first five moments, $\omega_i(t)=t^{i-1}$, were computed. These exact moments were then used in the numerical example. The posterior probability for β was first computed and is shown in the logarithmic plot in Fig. 1A. Note that as β goes to zero this posterior probability should go to infinity. However, we were unable to compute the posterior probability for values of β smaller than 10^{-16}. In spite of this, the fully normalized posterior probability, Fig. 1B, is a fair representation to a delta function. Because the probability for β is so sharply peaked, constraining β to its peak value is an extremely good approximation in Eq. 5. Using $\beta=10^{-16}$ in Eq. 5 one obtains the (mean $\pm$ standard deviation) estimates of the function. These are shown in Fig. 2. On this scale there is no observable difference between the estimated function and the (mean $\pm$ standard deviation) estimates. Because of this, we have expanded the region around the origin and

Figure 1: Probability for β

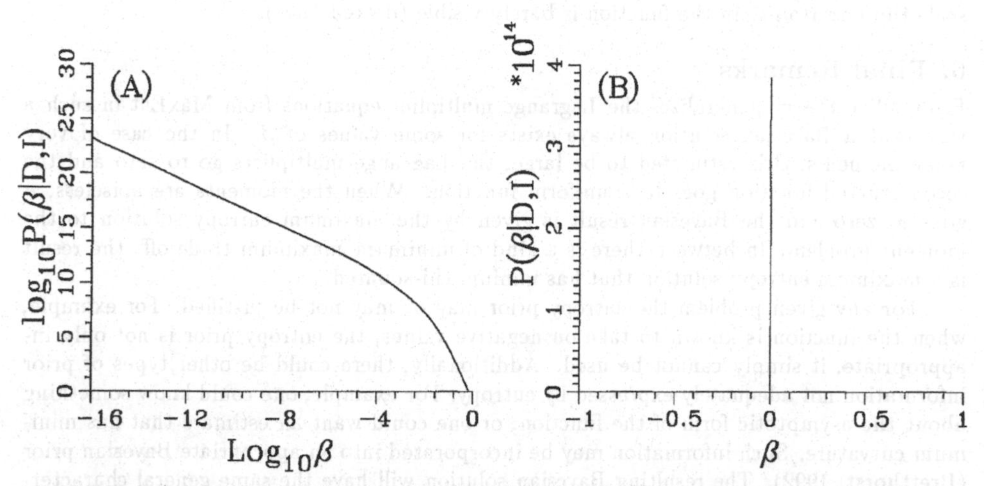

Fig. 1. Panel A is a Log plot of the probability for β. Note that this probability density function drops some 22 orders of magnitude. Panel B is the fully normalized probability density function. For noiseless moments this function should be a delta function. However, computationally we could not compute it for values of β smaller than 10^{-16}, yet the fully normalized density function is still a good approximation to a delta function.

Figure 2: The Expected Function

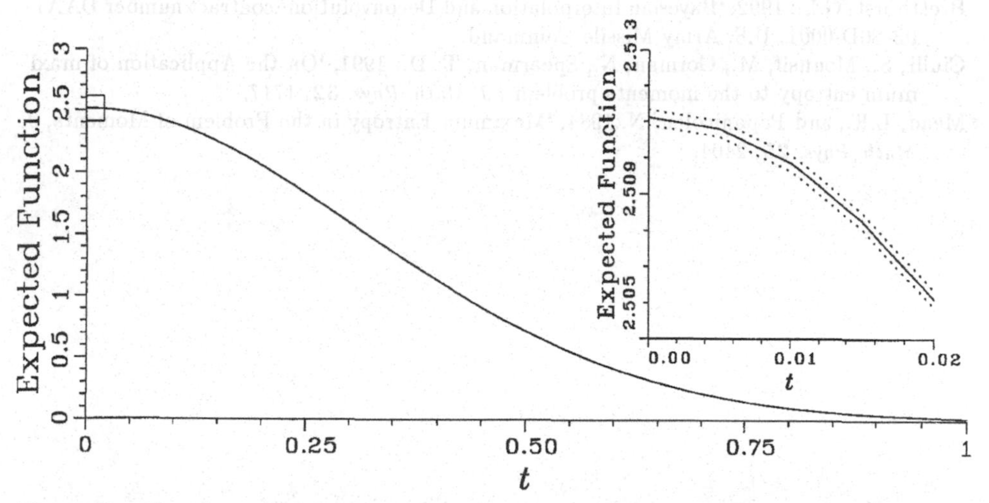

Fig. 2. On this scale no difference is seen between the expected function and the (mean $\pm$ standard deviation) estimates. The large box is an expansion of the region around $t = 0$. The expected function is shown as the solid line, the one standard deviation errors are shown as the dotted lines.

have plotted this in the upper right-hand-corner of Fig. 2. Even on this highly expanded scale the uncertainty in the function is barely visible (dotted lines).

6. Final Remarks

Probability theory generalizes the Lagrange multiplier equations from MaxEnt in such a way that a Bayesian solution always exists for some values of β. In the case of very noisy moments, β is estimated to be large, the Lagrange multipliers go to zero and the reconstructed function goes to a uniform function. When the moments are noiseless, σ goes to zero and the Bayesian result is given by the maximum entropy solution to the moment problem. In between there is a kind of minimum–maximum trade off: the result is a maximum entropy solution that has minium chi-squared.

For any given problem the entropy prior may or may not be justified. For example, when the function is known to take on negative values, the entropy prior is not only inappropriate, it simply cannot be used. Additionally, there could be other types of prior information not adequately expressed by entropy. For example, one could know something about the asymptotic form of the function, or one could want an estimate that has minimum curvature. Such information may be incorporated into an appropriate Bayesian prior (Bretthorst, 1992). The resulting Bayesian solution will have the same general characteristics as the one exhibited here: the uncertainty in the function will be a well-behaved quantity, and in the noiseless limit the Bayesian solution will be equal to the solution of some constrained optimization problem.

ACKNOWLEDGMENTS. SK would like to thank the Kosciuszko Foundation for financial support.

REFERENCES

Bretthorst, G.L.: 1992, 'Bayesian Interpolation and Deconvolution',contract number DAAL-03-86D-0001, U.S. Army Missile Command.

Ciulli, S., Mounsif, M., Gorman, N., Spearman, T. D.: 1991, 'On the Application of maximum entropy to the moments problem', *J. Math. Phys.* **32**, 1717.

Mead, L.R., and Papanicolau, N.:1984, 'Maximum Entropy in the Problem of Moments, *J. Math. Phys.* **25**, 2404.

BAYES-OPTIMAL VISUALIZATION OF VECTOR FIELDS

Robert W. Boesel and Yoram Bresler
Coordinated Science Laboratory and the Beckman Institute
University of Illinois
1308 West Main St., Urbana, IL 61801, USA

ABSTRACT. This paper addresses the effective display of noisy vector-valued image data. A new game-theoretic model of a human observer in a visualization task is developed, and used to derive optimal algorithms for the display of a maximally informative fused monochrome image.

1. Introduction

We address the effective display of noisy vector-valued image data for analysis by a human observer. Such data, where a vector value, rather than a scalar, is associated with each pixel, arises in various applications in science, engineering, medicine, and biology, when a multichannel sensor is used. Examples include Magnetic Resonance Imaging (MRI), multispectral remote sensing, and multi-sensor fusion, e.g., of optical, radar, and IR sources. Each component of the vector corresponds to a different imaging modality, or a different combination of imaging parameters, and may provide different levels of contrast sensitivity between different regions. The question therefore arises, as to how to present this wealth of data to a human observer.

Our goal is to present the observer with an image that will maximize *his* chances to correctly segment the image. Our premise is that the observer is able to bring to bear all his knowledge and experience, which are difficult to capture in a computer program, on the final analysis process. This is often key to the detection of subtle and otherwise elusive features in a noisy image. We therefore rule out the generation of an automatically segmented image, which not only fails to include this knowledge, but also would deprive the observer of the opportunity to exercise it, by presenting him with hard labels for different regions in the image. Instead we concentrate on the fusion of the vector data into a single most informative scalar grey-scale image.

2. Model for Visualization of Vector Fields

The region of interest $\Psi \subset I\!R^2$ is composed of p spatially disjoint subregions, $\Psi = \cup_{i=1}^{p} \Psi_i$, belonging to q ($q \leq p$) distinct classes ω_j, $j = 1, \ldots, q$. Different regions can belong to the same class. For example, these classes could be tissue types present in a particular part of the body under consideration. The partition of the region into subregions, along with their labels, is called the *underlying image*, and denoted by UI. The region is probed by a d-parameter imaging modality (equivalently, the images produced by different modalities are registered and stacked), producing a vector field $\boldsymbol{X} = \{\boldsymbol{x}_i\}_{i=1}^{N}$, $\boldsymbol{x}_i \in I\!R^d$ of N pixels. We assume that the d-dimensional feature vector $\boldsymbol{x}_i$ for a particular pixel is a random vector, whose distribution is dependent on the class of the pixel. The randomness is introduced by

A. Mohammad-Djafari and G. Demoments (eds.), Maximum Entropy and Bayesian Methods, 31–38.

measurement noise, and by natural within-class variability.

The visualization problem is to find a mapping from the vector field $\boldsymbol{X}$ to a real grey-scale image $\boldsymbol{Y} = \{y_i\}_{i=1}^N$, $y_i \in I\!R$, such that the performance of a *human observer* in recovering the underlying image UI from $\boldsymbol{Y}$ is optimized. Hence, the performance is measured by success in an unsupervised segmentation task of the displayed image.

The human observer is modeled as an optimal unsupervised Bayesian decision maker, who has only access to the scalar image $\boldsymbol{Y}$, but is equipped with an *a-priori* distribution on the UI, and a cost matrix for weighing the different segmentation errors. The class conditional densities of the scalar image are unknown to the observer, but are estimated by him from the image. His goal is to minimize the Bayes risk in performing the segmentation. The *a-priori* distribution represents his prior knowledge about the number of classes (tissue types) present and their relative frequency and abundance, the relative contrast between various classes provided by the different modalities, and the morphology of the region of interest. The cost matrix represents the observer's bias regarding the relative costs of misclassification (e.g., costs of false positives versus false negatives).

In contrast, the visualization algorithm has access to the vector field $\boldsymbol{X}$, but not to the cost matrix, or the *a priori* distribution, which reside with the human observer. The visualization algorithm (the mapping from the vector field to the scalar image) is to be designed to optimize the performance of the human observer in his segmentation task.

Note that the evaluation of the observer's performance requires expressions for the probability of error in unsupervised classification, which can not be obtained in closed form. However, through a series of numerical studies, we are able to conclude that such models do not lead to significantly different results than the Bayes error criterion for supervised classification. This allows us to ignore the unsupervised nature of the problem and model the observer as if he designs a supervised classifier, i.e. has labeled training samples. Assuming further that many samples from each class are available, the observer's estimates of the unknown means approach the actual means, and the error rate simplifies to the Bayes error rate with known statistics. Hence we assume that the observer is well-modeled as performing a Bayesian classifier design with known statistics.

3. Problem Formulation

We assume that the feature vectors for the different classes are distributed as multivariate normals with distinct means, but a common and known covariance[1], which is taken as a scaled identity, without loss of generality. Hence, a pixel of class ω_j is distributed as $p(\boldsymbol{x}|\omega_j) = \mathcal{N}_{\boldsymbol{x}}(\boldsymbol{\mu}_j, \sigma^2 \boldsymbol{I})$ for $j = 1, ..., q$, with $\boldsymbol{\mu}_i \neq \boldsymbol{\mu}_j$, $\forall i \neq j$. Furthermore, we assume that the feature vector for different pixels are independent. Finally, we assume that the noise level σ^2 is known, both to the algorithm and the human observer, but the class means $\boldsymbol{\mu}_j$, $j = 1, c$ are unknown to the human observer. The visualization algorithm obtains good estimates of the class means and the UI by automatic segmentation of the vector-field image (to which it has access) and uses them in the choice of mapping.

In this paper we restrict our attention to adaptive linear space-invariant mappings $y_i = \boldsymbol{w}'\boldsymbol{x}_i$ of the vector field to the scalar image, which are specified by one vector $\boldsymbol{w} \in I\!R^d$

[1]the case of different, unknown covariance, which is of interest when natural within-class variability dominates the measurement noise can be similarly treated

for all N image pixels. In turn, w is determined by optimization of a criterion depending on μ_i and hence, on the data. Several such criteria, of different levels of realism and complexity, are used to define the optimality of the weight vector w.

To effectively model the fact that our projection algorithm will not know the observer's a-priori probabilities nor the cost matrix, we propose a game-theoretic formulation. The resulting model will lead to several criteria designed to effectively quantify the image quality resulting from the transformation $\mathbf{w}$.

Let $\boldsymbol{\pi}$, $\mathbf{C}$, and $\boldsymbol{\delta}$ denote the prior probabilities, cost matrix, and decision rule that the observer applies to the scalar-valued image, respectively. Then his risk is given by

$$r(\boldsymbol{\delta}, \boldsymbol{\pi}, \mathbf{C}, \mathbf{w}) = \sum_{j=1}^{q} \pi_j R_j(\boldsymbol{\delta}) \tag{1}$$

where $R_j(\boldsymbol{\delta})$ is the class-conditional risk of the decision rule $\boldsymbol{\delta}$,

$$R_j(\boldsymbol{\delta}) = \sum_{i=1}^{q} c_{ij} P_{\boldsymbol{\delta}}(d_i/\omega_j), \tag{2}$$

and $P_{\boldsymbol{\delta}}(d_i/\omega_j)$ is the probability that the decision rule $\boldsymbol{\delta}$ will make decision d_i (i.e. decide that a particular pixel is of class ω_i), given that the pixel is actually of class ω_j. For a general randomized decision rule in a one-dimensional classification problem, $P_{\boldsymbol{\delta}}(d_i/\omega_j)$ is given by

$$P_{\boldsymbol{\delta}}(d_i/\omega_j) = \int_{-\infty}^{\infty} \delta_i(y) p(y/\omega_j) dy \tag{3}$$

where $p(y/\omega_j)$ are the class-conditional density functions of a pixel value after the transformation $y = w'x$.

We assume a game in which there are three players:

(1) A task ($\mathcal{T}$), which picks $\boldsymbol{\pi}$ and $\mathbf{C} \in \Xi$, where Ξ is a set of suitably constrained costs to prevent arbitrarily large risk, (e.g. $\Xi = \{\mathbf{C} : \sum_{ij} c_{ij} = 1\}$).

(2) The observer ($\mathcal{O}$), who picks $\boldsymbol{\delta} \in \Delta$ to minimize $r(\boldsymbol{\delta}, \boldsymbol{\pi}, \mathbf{C}, \mathbf{w})$ for given $\boldsymbol{\pi}$, $\mathbf{C}$, and $\mathbf{w}$. Initially, we will assume that the observer may choose from the space of all possible decision rules, denoted as Δ. (The observer is assumed to know $\boldsymbol{\pi}$ and $\mathbf{C}$, which are determined by the task, of which he is aware.)

(3) The projection algorithm ($\mathcal{P}$), which picks $\mathbf{w}$ to minimize $r(\boldsymbol{\delta}, \boldsymbol{\pi}, \mathbf{C}, \mathbf{w})$ for *unknown* $\boldsymbol{\delta}$, $\boldsymbol{\pi}$, and $\mathbf{C}$. It is assumed that $\mathcal{P}$ does have knowledge of the constraint set Ξ.

Players $\mathcal{O}$ and $\mathcal{P}$ are on a team which has the objective of minimizing $r(\boldsymbol{\delta}, \boldsymbol{\pi}, \mathbf{C}, \mathbf{w})$. However, while $\mathcal{O}$ picks $\boldsymbol{\delta}$ for given $\boldsymbol{\pi}$, $\mathbf{C}$, and $\mathbf{w}$, (determined by $\mathcal{T}$ and $\mathcal{P}$, respectively), $\mathcal{P}$ must choose $\mathbf{w}$ without knowledge of $\boldsymbol{\pi}$, $\mathbf{C}$, or $\boldsymbol{\delta}$. The task, $\mathcal{T}$, determines the priors and costs arbitrarily, so in choosing $\mathbf{w}$, $\mathcal{P}$ must consider a worst case $\boldsymbol{\pi}$ and $\mathcal{C}$ and protect against it. Hence $\mathcal{T}$ is modeled as an adversary to the team $(\mathcal{O}, \mathcal{P})$, picking the worst costs and priors for any choice of $\mathbf{w}$.

This game theoretic formulation leads to the optimality criterion

$$J(\mathbf{w}) = \max_{\boldsymbol{\pi}, \mathbf{C} \in \Xi} \min_{\boldsymbol{\delta}} r(\boldsymbol{\delta}, \boldsymbol{\pi}, \mathbf{C}, \mathbf{w}) \tag{4}$$

which the player $\mathcal{P}$ attempts to minimize, i.e. find $\mathbf{w}_{opt} = \arg\min_{\mathbf{w}} J(\mathbf{w})$. Thus $J(\mathbf{w})$ becomes the relevant measure of the quality of an image formed by the transformation $y = w'x$. This *maximin* criterion is very general in that the priors and costs are unknown to $\mathcal{P}$, and the form of the decision rule used by $\mathcal{O}$ is unconstrained. We now introduce some criteria that are special cases of (4). Some of these criteria will assume knowledge, either complete or partial, of the cost matrix, and others will constrain the space of decision rules from which the observer may choose. The result is a number of criteria that differ in their computational requirements and performance, which we will explore.

Combining Equations (1), (2), and (3), the general risk for a given decision rule, priors, and costs is

$$r(\delta, \pi, \mathbf{C}, \mathbf{w}) = \sum_{j=1}^{q} \sum_{i=1}^{q} \pi_j c_{ij} P_\delta(d_i/\omega_j) . \tag{5}$$

The following is a list of the special cases we consider:

Unknown Priors, Equal Costs. Assume that the costs are known and equal, i.e. $c_{ij} = 1\ i \neq j$, $c_{ii} = 0$. The resulting criterion is

$$J_{upe}(\mathbf{w}) = \max_{\pi} \min_{\delta} r(\delta, \pi, \mathbf{w}), \tag{6}$$

where the dependence of the risk on $\mathbf{C}$ has been suppressed, since it is known and fixed.

Unknown Priors, Arbitrary Costs. Assume that the costs are known, but arbitrary. We denote this criterion

$$J_{upa}(\mathbf{w}) = \max_{\pi} \min_{\delta} r(\delta, \pi, \mathbf{w}). \tag{7}$$

The motivation for introducing different notation for J_{upe} and J_{upa} is that they are computed in completely different fashions.

Unknown and Constrained Priors and Costs. Assume both priors and costs are unknown, and choose the following form of constraint on the costs and priors to prevent arbitrarily large risk. Let $\alpha_{ij} = \pi_j c_{ij}$ in (5), then define the set

$$\mathcal{A} = \left\{ \alpha : \sum_{ij} \alpha_{ij} = 1, \alpha_{ij} \geq 0, \alpha_{ii} = 0,\ i, j = 1, 2, \cdots, c \right\}.$$

With this notation, the maximin risk for unknown priors and costs, $J_{uc}(\mathbf{w})$, is defined

$$J_{uc}(\mathbf{w}) = \max_{\alpha \in \mathcal{A}} \min_{\delta} \sum_{i=1}^{q} \sum_{j=1}^{q} \alpha_{ij} \int_{-\infty}^{\infty} \delta_i(y) p(y/\omega_j) dy = \max_{\alpha \in \mathcal{A}} \min_{\delta} r(\delta, \alpha, \mathbf{w}) \tag{8}$$

Unknown Priors and Costs, Constrained Costs. Again assume both priors and costs are unknown, but restrict some of the α_{ij}'s to be zero. Specifically fix $\alpha_{ij} = 0$ if classes i and j are not adjacent in the physical image. The reasoning is that physically non-adjacent regions will not be confused by the human observer even if they are assigned the same grey level. As such let $\mathcal{A}_c$ be the set

$$\mathcal{A}_c = \left\{ \alpha : \sum_{ij} \alpha_{ij} = 1, \alpha_{ij} \geq 0, \alpha_{ii} = 0,\ i, j = 1, 2, \cdots, q,\ \alpha_{kl} = 0,\ kl \in AR \right\}$$

where the set AR consists of all pairs of adjacent regions. Then the criterion J_{ucc} is defined as

$$J_{ucc}(\mathbf{w}) = \max_{\alpha \in \mathcal{A}_c} \min_{\delta} \sum_{i=1}^{q} \sum_{j=1}^{q} \alpha_{ij} \int_{-\infty}^{\infty} \delta_i(y) p(y/\omega_j) dy = \max_{\alpha \in \mathcal{A}_c} \min_{\delta} r(\delta, \alpha, \mathbf{w}) \tag{9}$$

The criteria, (6) - (9), assume the observer is capable of implementing an arbitrary, possibly randomized decision rule, and thus the criterion J_{upa} is implicitly $J_{upa}(\mathbf{w}) = \max_{\pi} \min_{\delta \in \Delta} r(\delta, \pi)$, where Δ is the space of all randomized decision rules. It is not clear that an observer will be able to implement such complex rules in general. This, and the fact that the computation of J_{upa}, J_{uc}, and J_{ucc} can be quite intensive, motivate the following alternative criteria.

Simply Conneceted Rules. Assume that the space of decision rules the observer may consider is restricted to the class of rules that are deterministic and consist of a simply connected partition of the observation space, $\mathbb{R}$ for this case, into A_i, $i = 1, 2, \cdots, q$, such that if if an observation y falls in A_i, the decision that y is of class ω_i is made. Furthermore, the regions A_i are forced to be simply connected. Denote this space of decision rules by Δ_s. Then we define the following criteria as alternatives to J_{upa}, J_{uc}, and J_{ucc} respectively:

$$J_{sp}(\mathbf{w}) = \max_{\pi} \min_{\delta \in \Delta_s} r(\delta, \pi, \mathbf{w}) \tag{10}$$

$$J_{sc}(\mathbf{w}) = \max_{\alpha \in \mathcal{A}} \min_{\delta \in \Delta_s} r(\delta, \alpha, \mathbf{w}) \tag{11}$$

$$J_{scc}(\mathbf{w}) = \max_{\alpha \in \mathcal{A}_c} \min_{\delta \in \Delta_s} r(\delta, \alpha, \mathbf{w}) \tag{12}$$

Fixed Rules. Assume the space of decision rules is even further restricted, to contain only the Bayes decision rule for equal priors and equal costs, δ^{epc}. We define the following criteria:

$$J_{fp} = \max_{\pi} r(\delta^{epc}, \pi, \mathbf{w}) \tag{13}$$

$$J_{fc} = \max_{\alpha \in \mathcal{A}_c} r(\delta^{epc}, \alpha) \tag{14}$$

We present results of simulations to compare these various criteria in Section 5..

4. Optimization Algorithms

The choice of a $\mathbf{w}$ optimizing one of the criteria (6)-(14) is complicated by their multimodal nature and the required minimax optimizations. We have developed a efficient techniques [4] for finding the global optima in these problems, relying on the minimax theorem and on a special structure of the multimodal cost funstion. For resons of space, these are described elsewhere [4].

5. Simulations

We consider the example for which the class-conditional means and image geometry are as shown in Figure 1. The resulting images found by optimizing the various criteria are shown in Figure 2. Note that the criteria, J_a and J_{upe} both assume equal costs while the images shown for the remaining criteria, (first two rows), use a cost matrix with zero cost between non-adjacent regions (or the constraint set $\mathcal{A}_c$, for those criteria that are undefined for unknown costs), and a cost of one between adjacent regions. (Note: images

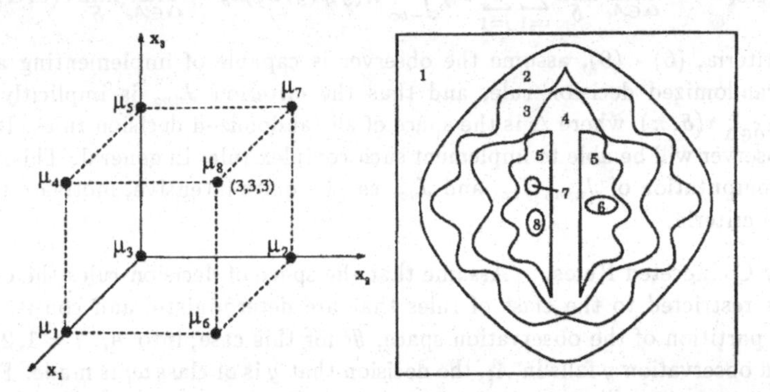

Figure 1: Class-conditional means and image geometry for simulations.

for J_{uc}, J_{ucc}, and J_{upa} are left out, due to computational difficulty of finding the optimum $\mathbf{w}$ for this example with a large number of classes ($q = 8$).) In row three we show three different images, each of which is an approximate global minimum of J_{fc}. The quality of these "equivalent" images is obviously not equal to an observer. Though none causes the loss of a region, the images corresponding to $\mathbf{w}_1$ and $\mathbf{w}_2$ appear superior to the image corresponding to $\mathbf{w}_3$. This is a situation that can occur in general, and indicates that in addition to the global minimum of the criterion, several of the local minima nearest to it should be considered and evaluated for providing usueful candidate images.

Returning to the images in rows one and two, we can see that the images corresponding to J_{sp} and J_{sc} are significantly poorer than those for J_{fp} and J_{fc}. In particular, the poorer images both result in the loss of contrast between regions #4 and #5. Why should the similar criteria J_{fp} and J_{fc} perform so much better ? One possible explanation for this counter-intuitive example is that the the space of decision rules over which the observer can optimize is not sufficiently constrained; e.g. J_{sp}, computed as

$$J_{sp} = \min_{\delta \in \Delta,} \max_{\pi} r(\delta, \pi)$$

can, for a particular projection vector, result in apparently unreasonable decision rules when a cost matrix with zero cost for non-adjacent regions is used. For example, consider the projection vector optimizing J_{sp}, $\mathbf{w}_{opt} \approx (-.0051, -.7008, .7134)^T$. The projected class-conditional means for this vector are shown in Figure 3, along with the decision boundaries $\{b_i\}_{i=1}^{q-1}$ that result in the minimum risk. Note that of the classes projected together, only $m_4 \approx m_5$, causes a problem, the other classes with equal class-conditional means are non-adjacent. Near $m_4 \approx m_5$, the optimum decision rule makes the decision d_1, a reasonable choice since neither ω_4, nor ω_5 has a cost for being confused with ω_1. The corresponding class-conditional risks, R_i, for this case are

$$R_1 \approx .016,\ R_2 \approx .016,\ R_3 \approx .016,\ R_4 \approx 0,\ R_5 \approx .016,\ R_6 \approx 0,\ R_7 \approx 0,\ R_8 \approx 0,$$

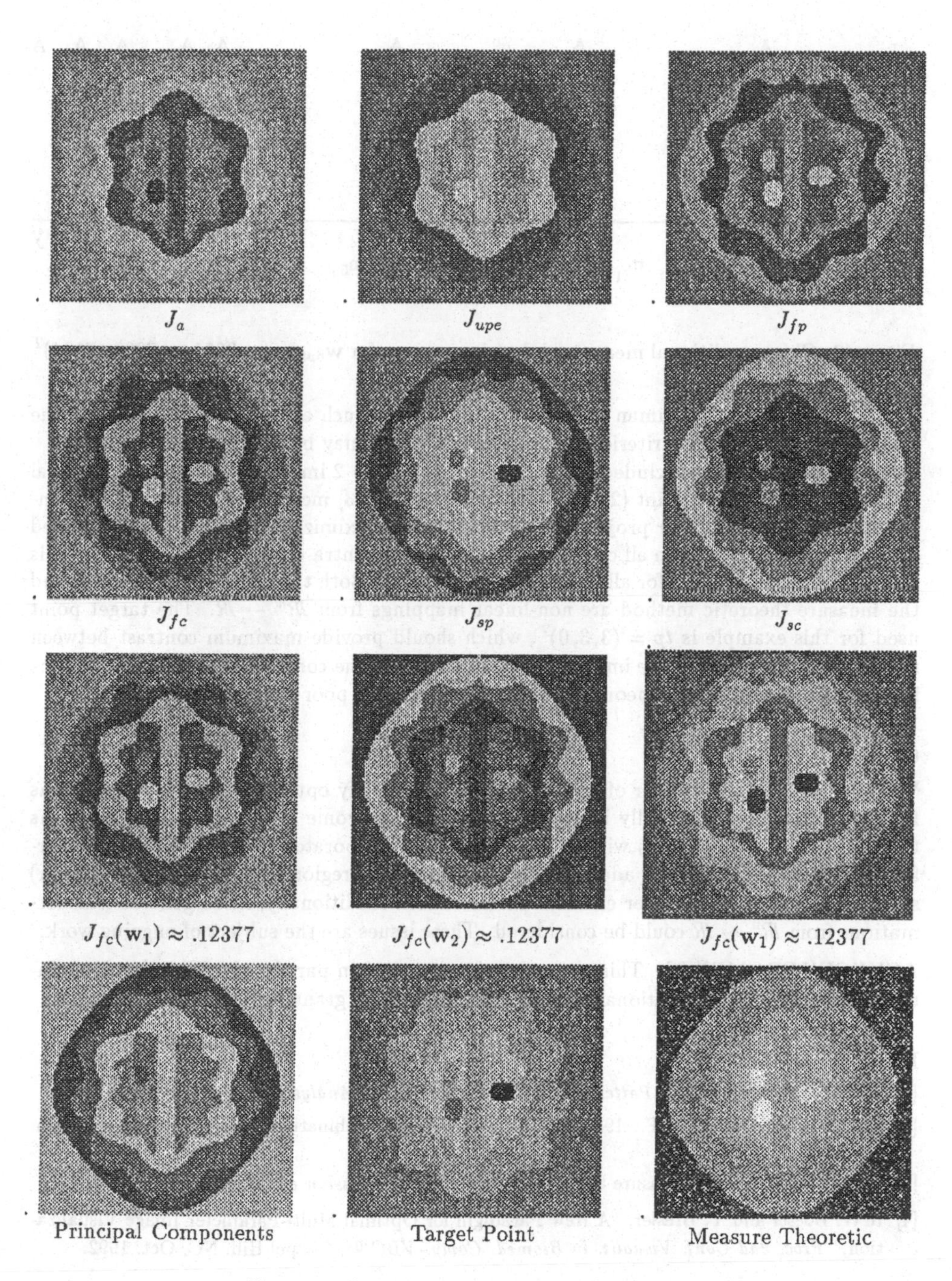

Figure 2: Optimal Images for simulations.

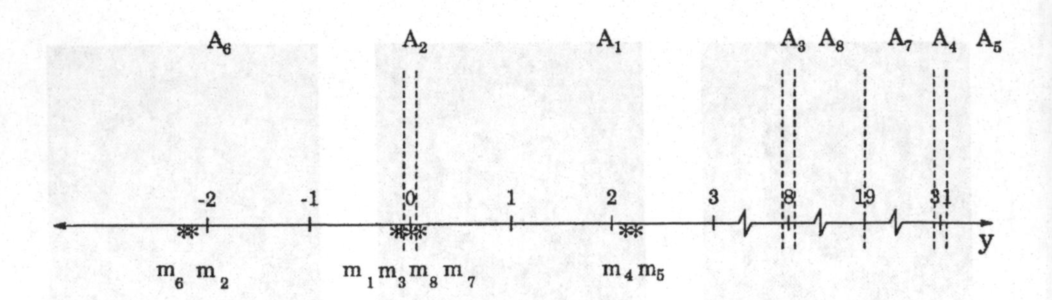

Figure 3: Class-conditional means and decision regions for $\mathbf{w}_{opt} \approx (-.0051, -.7008, .7134)^T$.

It is unlikely that a human observer will implement such a decision rule, and thus the images formed using the criterion J_{sp} (and likewise J_{sc}) may be unpredictable.

For comparison, we included in the last row of Figure 2 images formed using principal components [1], target point [2], and measure theoretic [3] methods . The principal components method is a linear projection that finds $\mathbf{w}$ to maximize the variance of the pooled data, (i.e. data points from all classes) normalized by the intra-class variance, which for this problem is assumed equal for all classes and fixed $\sigma = 1$. Both the target point method, and the measure theoretic method are non-linear mappings from $\mathbb{R}^n \rightarrow \mathbb{R}$. The target point used for this example is $tp = (3,3,0)^T$, which should provide maximum contrast between region #6 and the rest of the image. Note however, that the contrast between other regions is not good. The measure theoretic method also provides poor overall contrast.

6. Conclusions

We have presented a number of criteria designed to specify optimal linear transformations for the display of a maximally informative fused monochrome image. In general, it seems that the criteria J_{fp} and J_{sp} with region adjacency incorporated into the cost matrix perform the best. In the future, additional information, (e.g. region size and boundary length) should be included to further enhance performance. Additionally, more general transformations from $\mathbb{R}^n \rightarrow \mathbb{R}$ could be considered. These issues are the subject of ongoing work.

ACKNOWLEDGMENTS. This research was supported in part by the Whitaker Foundation and in part by the National Science Foundation PYI grant No. MIP 9157377.

References

[1] Duda, R, Hart, P.: 1973, *Pattern Classification and Scene Analysis*, Wiley, New York.

[2] Buxton, R.B, Greensite, F.: 1989, "Target-point image combination," Abst., *Soc. Magn. Resn. Imag.*,, **7**, 41.

[3] Greensite, F.: 1988, "Measure theoretic imaging," *Mach. Vision and Appl.*, **1**, 169.

[4] R.W. Boesel and Y. Bresler, "A New Paradigm for Optimal Multi-Parameter Image Visualization," *Proc. 2nd Conf. Visualiz. in Biomed. Comp.-VBC'92*, Chapel Hill, NC, Oct. 1992.

A SURVEY OF CONVERGENCE RESULTS FOR MAXIMUM ENTROPY METHODS

J.M. Borwein and A.S. Lewis
Department of Combinatorics and Optimization
University of Waterloo, Waterloo, Ontario, Canada
N2L 3G1

ABSTRACT. Maximum entropy methods seek to estimate an unknown density function, typically nonnegative, on the basis of some known moments – integrals of the function with respect to given weights. The estimate is chosen to minimize some measure of entropy, a convex integral functional of the density, subject to the given moment constraints. A desirable feature of such a method is that the estimates should converge to the unknown density as the number of known moments grows. We survey recent results demonstrating how various types of convergence (weak-star, weak and norm in L_1 and L_p, measure, and uniform) result from various properties of the entropy (strict convexity, smoothness, and growth conditions). This investigation may be seen as an extension of the classical study of the convergence of Fourier series to the framework of maximum entropy problems. Rather than present the most general theorems known, the unified pattern of the results is illustrated on a single fairly general model problem. References are given for the more general results and their proofs.

1. Introduction and notation

The problem of estimating an unknown density function on the basis of some known moments is an underdetermined inverse problem which occurs frequently in the physical sciences. We consider the following model. We assume the underlying set S is a compact metric space with an associated nonnegative regular Borel measure μ with full support, and $(a_i)_1^\infty$ is a sequence of (real-valued) continuous functions whose linear span is dense in $C(S)$, the Banach space of continuous functions with supremum norm. We denote the unknown function by $\bar{x} \in L_1(S)$, where for $1 \leq p \leq +\infty$ and measurable x,

$$\begin{aligned} \|x\|_p &= \left(\int_S |x|^p \, d\mu\right)^{1/p}, \quad (1 \leq p < +\infty) \\ \|x\|_\infty &= \text{essential supremum of } x, \text{ and} \\ L_p(S) &= \{x \mid \|x\|_p < +\infty\} \end{aligned}$$

(see for example [36] for terminology).

The problem then is to estimate $\bar{x}$ on the basis of the known moments $\int a_i \bar{x} \, d\mu$, $i = 1, \ldots, n$, and other prior information such as, typically, nonnegativity of $\bar{x}$. Two very typical examples are the 'Hausdorff' case,

$$a_i(s) = s^{i-1}, \tag{1}$$

and the 'Fourier' case,

$$a_i(s) = \begin{cases} \cos \pi(i-1)s, & i \text{ odd}, \\ \sin \pi i s, & i \text{ even}, \end{cases} \tag{2}$$

A. Mohammad-Djafari and G. Demoments (eds.), Maximum Entropy and Bayesian Methods, 39–48.

both on $S = [0,1]$ with μ Lebesgue measure.

The maximum entropy approach to estimating $\bar{x}$ has achieved quite wide popularity in recent years: [14] is a collection of recent survey papers. Various formulations give rise to optimization problems of the following form:

$$(P_n) \quad \begin{cases} \text{minimize} & I_\phi(x) + \int z_0 x \\ \text{subject to} & \int a_i x = \int a_i \bar{x}, \quad (i = 1, \ldots, n) \\ & x \in L_1, \end{cases}$$

where $I_\phi(x) = \int_S \phi(x(s))\, d\mu$, loosely minus the 'entropy', is a 'normal convex integral' [34]. The extended-valued function $\phi : \Re \to (-\infty, +\infty]$ is a closed, convex, proper function in the sense of Rockafellar [35], whose terminology we use throughout, and $z_0 \in C(S)$ (frequently in practice $z_0 \equiv 0$). This integrand ϕ is chosen on the basis of prior information regarding $\bar{x}$, so for example requiring nonnegativity of our estimates may be accomplished by setting $\phi(u) = +\infty$ for $u < 0$. We assume that ϕ is not affine. We say $x \in L_1$ is *feasible* for (P_n) if it satisfies the moment constraints and $I_\phi(x) < +\infty$.

The classical Shannon entropy suggested by Jaynes in this context [26] corresponds to

$$\phi(u) = \begin{cases} u \log u, & u > 0, \\ 0, & u = 0, \\ +\infty, & u < 0, \end{cases} \tag{3}$$

and the Burg entropy suggested in [15] for the time series case corresponds to

$$\phi(u) = \begin{cases} -\log u, & u > 0, \\ +\infty, & u \le 0. \end{cases} \tag{4}$$

We could also consider an 'L_p entropy', $1 < p < +\infty$ (see [24]),

$$\phi(u) = \begin{cases} u^p/p, & u \ge 0, \\ +\infty, & u < 0, \end{cases} \tag{5}$$

and if the range of $\bar{x}$ was known to lie in $[0,1]$ then the Fermi- Dirac entropy

$$\phi(u) = \begin{cases} u \log u + (1-u)\log(1-u), & 0 < u < 1, \\ 0, & u = 0, 1, \\ +\infty, & \text{otherwise}, \end{cases} \tag{6}$$

might be appropriate (see also [11]). An elegant systematic way of deriving ϕ from certain prior knowledge is described in [17], based on earlier ideas from [32].

An important concept in what follows is the Fenchel conjugate of ϕ, the function $\phi^* : \Re \to (-\infty, +\infty]$ defined by $\phi^*(v) = \sup_u \{uv - \phi(u)\}$. In the four cases above we obtain respectively

$$\begin{aligned} \phi^*(v) &= e^{v-1}, \\ \phi^*(v) &= \begin{cases} -1 - \log(-v), & v < 0, \\ +\infty, & v \ge 0, \end{cases} \\ \phi^*(v) &= \max\{0, v^q/q\}, \quad \text{where } 1/p + 1/q = 1, \text{ and} \\ \phi^*(v) &= \log(1 + e^v). \end{aligned}$$

We define $\text{dom}(\phi) = \{u \mid \phi(u) < +\infty\}$, which we assume is not a singleton, and denote the interior and boundary of this set by $\text{ri}(\text{dom}(\phi))$ and $\text{rb}(\text{dom}(\phi))$ respectively. Our investigation will centre on the following three properties of ϕ:

Strict convexity: $\phi(\alpha u + (1-\alpha)v) < \alpha\phi(u) + (1-\alpha)\phi(v)$, for $0 < \alpha < 1$ and $u \neq v$ in $\text{dom}(\phi)$.

Essential smoothness: ϕ differentiable on $\text{ri}(\text{dom}(\phi))$ with $|\phi'(u)| \to +\infty$ as u approaches a point in $\text{rb}(\text{dom}(\phi))$.

Coercivity: $\phi(u)/|u| \to +\infty$ as $|u| \to +\infty$.

The first two properties are dual to each other in the sense that ϕ is strictly convex if and only if ϕ^* is essentially smooth (and vice versa, since $\phi^{**} = \phi$). When ϕ is both strictly convex and essentially smooth, ϕ^* is simply the classical Legendre conjugate. Coercivity of ϕ corresponds to ϕ^* being everywhere finite. The four examples above have all three properties, with the exceptions that the Burg entropy is not coercive and the L_p entropy is not essentially smooth.

2. Convergence

Since $(a_i)_1^\infty$ is densely spanning in $C(S)$ it follows that $\bar{x}$ is uniquely determined by the moment sequence $(\int a_i\bar{x})_1^\infty$. It is therefore natural to expect that if x_n is feasible, or in particular optimal, for the problem (P_n) then x_n should approach $\bar{x}$ in some sense. Indeed, the strength of this convergence may be regarded as one measure of the success of a maximum entropy method. For example, as has been observed in the literature, when $\bar{x}$ vanishes on regions of positive measure, Burg entropy reconstructions tend to be very 'spiky' [37].

A number of authors have therefore considered this question of convergence of x_n to $\bar{x}$ [31], [23], [22], [21], [17], [38]. The aim of this survey is to demonstrate in a systematic way how the three properties of the entropy introduced previously – strict convexity, essential smoothness and coercivity (and finally smoothness of $\bar{x}$) – lead to various distinct types of convergence. Rather than presenting results in their greatest generality we will confine ourselves to the model of the previous section to maintain a unified presentation and emphasize patterns. In each case we give references for the more general results and detailed demonstrations, but we will give a brief sketch of the ideas behind each proof: we believe the theory underlying this problem is a beautiful illustration of the power of a variety of areas of mathematical analysis, in particular convex optimization and analysis, functional analysis, classical real analysis and approximation theory.

We study the following types of convergence for a sequence $(x_n)_1^\infty$ in L_1.

Weak-star: $\int(x_n - \bar{x})w \to 0$, for all w in $C(S)$.

Weak: $\int(x_n - \bar{x})w \to 0$, for all w in $L_\infty(S)$.

Measure: $\mu\{s \mid |(x_n - \bar{x})(s)| > \epsilon\} \to 0$, for all $\epsilon > 0$.

Norm: $\|x_n - \bar{x}\|_1 \to 0$.

When $(x_n)_1^\infty$ and $\bar{x}$ lie in L_p, $1 < p < +\infty$, we also have

L_p weak: $\int(x_n - \bar{x})w \to 0$, for all w in $L_q(S)$.

L_p norm: $\|x_n - \bar{x}\|_p \to 0$.

Finally, if $(x_n)_1^\infty$ and $\bar{x}$ are in L_∞ we can also consider:

Uniform: $\|x_n - \bar{x}\|_\infty \to 0$.

Weak and measure convergence in L_1 are independent. For example, the characteristic functions $n\chi_{[0,1/n]} \to 0$ in measure but not weakly, whereas $\cos 2n\pi s \to 0$ weakly (by the Riemann-Lebesgue lemma) but not in measure. Weak and measure convergence both holding is equivalent to norm convergence [20].

In the Fourier case (2) with $\phi(u) = u^2/2$ for all u in $\Re$, the optimal solution of (P_n) is just the n'th partial sum of the Fourier series for $\bar{x}$. This investigation can therefore be seen as a generalization to maximum entropy problems of the classical study of the convergence of Fourier series.

3. Duality

The maximum entropy problem (P_n) is a convex optimization problem, and is best analyzed from this perspective. A fundamental tool is the corresponding dual problem

$$\sup_\lambda \inf_x L(x;\lambda),$$

where the Lagrangian $L(x;\lambda) = I_\phi(x) + \sum_{i=1}^n \lambda_i(\int a_i\bar{x} - \int a_i x)$. The dual therefore becomes [5]:

$$(P_n^*) \qquad \begin{cases} \text{maximize} & \sum_{i=1}^n \lambda_i \int a_i\bar{x} - \int \phi^*(\sum_{i=1}^n \lambda_i a_i - z_0) \\ \text{with} & \lambda \in \Re^n, \end{cases}$$

or equivalently

$$\begin{cases} \text{maximize} & \int(\bar{x}q - \phi^*(q - z_0)) = h(q) \\ \text{where} & q = \sum_{i=1}^n \lambda_i a_i, \quad \lambda \in \Re^n, \end{cases} \tag{7}$$

which, in the Hausdorff or Fourier case, is an unconstrained maximization problem over polynomials q of fixed degree. In any case, the problem (P_n^*) is an unconstrained, finite-dimensional, concave maximization problem, typically smooth, and is thus relatively easy to solve numerically by standard algorithms [18]. That this is typically the most natural computational approach to solving the maximum entropy problem (P_n) is a consequence of the following 'strong duality' theorem.

We will assume throughout this paper that the following 'constraint qualification' holds for all large n.

Constraint Qualification There exists $\hat{x}$ in L_1 with $\hat{x}(s) \in \text{ri}(\text{dom}(\phi))$ a.e. and $\int a_i\hat{x} = \int a_i\bar{x}$, for $i = 1, \ldots, n$.

This assumption is a 'quasi-interior' condition [7]. For example, it holds in the Hausdorff and Fourier cases with Shannon, Burg or L_p entropies providing $\bar{x} \geq 0$ a.e. and $\bar{x} \neq 0$ [8], [5], [27]. It is important to understand that without such a condition Lagrange multipliers (which are exactly optimal solutions of (P_n^*)) may fail to exist [12], and furthermore that the standard tools from infinite-dimensional optimization fail since the positive cone $\{x \in L_1 \,|\, x \geq 0 \text{ a.e.}\}$ has empty interior. When this constraint qualification fails, reduction techniques must be used [30].

Theorem 1 *[7], [8], [9] The primal and dual optimal values, $V(P_n)$ and $V(P_n^*)$, are equal, with attainment in the dual if finite. If ϕ is strictly convex, λ^n is optimal for the dual (P_n^*)* **and (P_n) has an optimal solution,** *then it is given uniquely by*

$$x_n(s) = (\phi^*)'\left(\left(\sum_{i=1}^{n} \lambda_i^n a_i - z_0\right)(s)\right). \tag{8}$$

The assumption that (P_n) has an optimal solution is crucial for (8). For example, in the Burg case, with $n = 2$, $z_0 \equiv 0$, $a_1 \equiv 1$, $a_2(s) = s^{1/2}$, and $\bar{x} = 4\chi_{[0,1/4]}$, existence fails. To solve this problem we need to extend the space of densities L_1 to allow measures: in this setting the optimal solution is in fact $x(s) = (s^{-1/2} + \delta_0(s))/3$, where δ_0 indicates a delta-function at 0. For a rigorous analysis, see [9]. Theorem 1 is a special case of a very general family of duality results [7], [8].

4. Consequences of the constraint structure

The basic tool in many of the results we present is compactness in various topologies: loosely speaking, if optimal solutions x_n of P_n can be confined to a compact set then in that topology x_n must approach $\bar{x}$. Our first result uses the Alaoglu theorem (eg [25]) to deduce relative weak-star compactness of the sequence of optimal solutions $(x_n)_1^\infty$.

Theorem 2 *[29] As $n \uparrow \infty$, the optimal values $V(P_n) \uparrow I_\phi(\bar{x}) + \int z_0 \bar{x}$ (finite or infinite).*

This 'convergence in value' result has natural extensions to inequality constrained problems.

Theorem 3 *[6] If x_n is optimal (or even just feasible) for (P_n) then $x_n \to \bar{x}$ weak-star.*

The special case of the Hausdorff moment problem was studied in [31]. If x_n is optimal we have $V(P_n) = I_\phi(x_n) + \int z_0 x_n$. Weak-star convergence shows $\int z_0 x_n \to \int z_0 \bar{x}$, so the above results give convergence of the entropies: $I_\phi(x_n) \to I_\phi(\bar{x})$.

It is in fact possible to extend weak star convergence slightly, but not in general to weak convergence. The following example is illustrative. It considers the Hausdorff case with $\bar{x} \equiv 1$.

Theorem 4 *[6] Suppose $w \in L_\infty[0,1]$. Then $\int (x_n - 1)w \to 0$ for every sequence $0 \leq x_n \in L_1[0,1]$ satisfying $\int (x_n(s) - 1)s^{i-1} = 0$ $(i = 1, \ldots, n)$ if and only if w is* **essentially Riemann integrable***: that is $w = \hat{w}$ a.e. for some $\hat{w}$ which is continuous a.e. In particular, there exists such a feasible sequence (x_n) such that $x_n \not\to 1$ weakly.*

Thus to guarantee weak convergence we need to impose conditions on the entropy integrand ϕ. For more general results, see [6].

5. Strict convexity

Given the convergence in value result, Theorem 2, an elementary geometric argument suggested by [38] implies convergence in measure when ϕ is strictly convex.

Theorem 5 *[29], [3] Suppose ϕ is strictly convex, $I_\phi(\bar{x}) < +\infty$ and x_n is optimal for (P_n). Then $x_n \to \bar{x}$ in measure.*

Convergence in measure is quite robust under, for instance, compositions. For example, if $F : \Re \to \Re$ is continuous with bounded range then $x_n \to \bar{x}$ in measure implies $\int F(x_n(s)) \to \int F(\bar{x}(s))$.

When ϕ is not strictly convex it is a consequence of the Lyapunov theorem on the convexity of the range of vector measures [25] that convergence in measure can fail [3].

6. Essential smoothness

Let us denote an optimal solution of the dual problem (P_n^*) by λ_n, and the corresponding generalized polynomial in (7) by $q_n = \sum_{i=1}^n \lambda_i^n a_i$. Again using the convergence in value result, Theorem 2, we can deduce that (q_n) is a maximizing sequence for the dual function h. A result in [2] guarantees that when ϕ is essentially smooth such sequences must converge to the unconstrained maximizer of h, due to the Fréchet differentiability of $(-h)^*$. This gives the following 'dual' convergence result, which may be extended to consider inexact dual solutions. A weak precursor appeared in [28].

Theorem 6 *[29] Suppose ϕ is essentially smooth and the essential range of $\bar{x}$ is a bounded subset of ri(dom(ϕ)). Then if q_n is optimal for (7) we have $q_n \to \phi'(\bar{x}(\cdot)) + z_0$ in norm.*

Notice that if x_n is optimal for (P_n) then we must have $q_n = \phi'(x_n(\cdot)) + z_0$, so the above result becomes $\phi'(x_n(\cdot)) \to \phi'(\bar{x}(\cdot))$ in norm.

In the Shannon case for the Hausdorff problem it was observed heuristically in [31] that when the polynomials q_n were expanded in orthogonal polynomials each resulting coefficient converged in n. This follows immediately from the above result.

7. Coercivity and strong convexity

When ϕ is coercive a result of de la Vallée Poussin states that the level sets $\{x \in L_1 \,|\, I_\phi(x) + \int z_0 x \leq \alpha\}$ are weakly compact for any α in $\Re$ (see eg [34]). This guarantees the existence of optimal solutions for (P_n), and we obtain the following result.

Theorem 7 *[6] Suppose ϕ is coercive and $I_\phi(\bar{x}) < +\infty$. Then (P_n) has an optimal solution x_n, and any sequence of optimal solutions $x_n \to \bar{x}$ weakly.*

Since weak and measure convergence together imply norm convergence we can immediately strengthen this result.

Theorem 8 *Suppose ϕ is coercive and strictly convex, and $I_\phi(\bar{x}) < +\infty$. Then (P_n) has a unique optimal solution x_n, and $x_n \to \bar{x}$ in norm.*

This result was first proved directly in the Shannon case in [4]. In the more general case it was demonstrated independently in [38], using the above approach, and in [10], using the ideas below. Various more abstract set-theoretic constraints can be considered.

If, rather than coercivity, $\phi(u)$ actually grows as fast as $|u|^p$, for some $1 < p < \infty$ then by applying weak compactnees in L_p rather than L_1 we get weak convergence in L_p in Theorem 7 [6]. This holds for example with the L_p entropy.

The L_p norms $(1 < p < \infty)$ are sometimes called 'strongly convex', to indicate that the function $f(x) = \|x\|_p^p$ is strictly convex, has weakly compact level sets (due to the reflexivity of L_p), and has the *Kadec* property: $x_n \to \bar{x}$ weakly in L_p and $f(x_n) \to f(\bar{x}) < +\infty$ implies $x_n \to \bar{x}$ in L_p norm (Clarkson's theorem – see [19]). In consequence, if $\bar{x} \in L_p$ and we use the L_p entropy $(1 < p < +\infty)$ then the optimal solution of (P_n), $x_n \to \bar{x}$ in L_p norm.

By analogy, it can be shown that I_ϕ is strongly convex as a function on L_1 if and only if ϕ is strictly convex and coercive, as was first observed in [39] (see also [10]). This gives an alternative approach to Theorem 8.

When ϕ fails to be coercive, as in the Burg case, the question of existence of optimal solutions of (P_n) becomes much more delicate. The example of Section 3 is a case in point. It is known for the Burg case that the optimal solutions must exist (in L_1) in the Hausdorff and Fourier cases, and in the two-dimensional trigonometric case, but not necessarily for the three-dimensional case [33]. For a complete analysis see [10].

8. Uniform convergence

Arguments leading to uniform convergence of optimal soutions x_n to $\bar{x}$ are approximation-theoretic in nature. For example, the following result uses a simple interpolation argument. Versions also exist in the Fourier case. Assume for simplicity that $z_0 \equiv 0$ throughout this section.

Theorem 9 *[13] Suppose x_n is optimal for (P_n) in the Hausdorff case, ϕ is strictly convex and essentially smooth, and $\phi'(\bar{x}(\cdot))$ is analytic on the 'racetrack' region of the complex plane of points no greater than distance d from the line segment* $[0,1]$, *for some $d > 1$. Then $x_n \to \bar{x}$ uniformly.*

More sophisticated arguments use the Jackson theorems on the rate of best uniform approximation by algebraic and trigonometric polynomials (see eg [16]) to find a lower bound on $\|q_n - \phi'(\bar{x}(\cdot))\|_1$, where q_n is the optimal polynomial in (7). Using some results in [1] and [40] bounding the growth rate of $\|r_n\|_1/\|r_n\|_\infty$ for algebraic or trigonometric polynomials r_n of degree n, we can then translate this into a bound on $\|q_n - \phi'(\bar{x}(\cdot))\|_\infty$. The following are sample results. The notation C^k means k times continuously differentiable.

Theorem 10 *[4] Using the Shannon entropy, suppose that $\bar{x}$ is strictly positive and either C^1 (in the Fourier case) or C^2 (in the Hausdorff case). Then $x_n \to \bar{x}$ uniformly.*

For a more general entropy we deduce a similar result if $\bar{x}$ is smoother. Notice that ϕ is not required to be coercive.

Theorem 11 *[3] Suppose that ϕ is strictly convex and essentially smooth, and that on $ri(dom(\phi^*))$, ϕ^* is C^2 with $(\phi^*)''$ strictly positive. Suppose also that $\bar{x}(s) \in ri(dom(\phi))$ for all s, and $\phi'(\bar{x}(\cdot))$ is C^2 (in the Fourier case) or C^4 (in the Hausdorff case). Then $x_n \to \bar{x}$ uniformly.*

In each case error bounds can be computed, and multi-dimensional versions may also be derived.

References

[1] D. Amir and Z. Ziegler. Polynomials of extremal L_p-norm on the L_∞-Unit sphere. *Jounal of Approximation Theory*, 18:86–98, 1976.

[2] E. Asplund. Fréchet differentiability of convex functions. *Acta Mathematica*, 121:31–47, 1968.

[3] J.M. Borwein and A.S. Lewis. Convergence in measure, mean and max norm for sequences in L_1. Forthcoming.

[4] J.M. Borwein and A.S. Lewis. Convergence of best entropy estimates. *SIAM Journal on Optimization*, 1:191–205, 1991.

[5] J.M. Borwein and A.S. Lewis. Duality relationships for entropy-like minimization problems. *SIAM Journal on Control and Optimization*, 29:325–338, 1991.

[6] J.M. Borwein and A.S. Lewis. On the convergence of moment problems. *Transactions of the American Mathematical Society*, 325:249–271, 1991.

[7] J.M. Borwein and A.S. Lewis. Partially finite convex programming, Part I, Duality theory. *Mathematical Programming B*, pages 15–48, 1992.

[8] J.M. Borwein and A.S. Lewis. Partially finite convex programming, Part II, Explicit lattice models. *Mathematical Programming B*, pages 49–84, 1992.

[9] J.M. Borwein and A.S. Lewis. Partially-finite programming in L_1 and the existence of maximum entropy estimates. *SIAM Journal on Optimization*, 1992. To appear. CORR 91-05, University of Waterloo.

[10] J.M. Borwein and A.S. Lewis. Strong convexity and optimization. *SIAM Journal on Optimization*, 1993. To appear.

[11] J.M. Borwein, A.S. Lewis, and M.A. Limber. Entropy minimization with lattice bounds. Technical Report CORR 92-05, University of Waterloo, 1992. Submitted to Journal of Approximation Theory.

[12] J.M. Borwein and H. Wolkowicz. A simple constraint qualification in infinite dimensional programming. *Mathematical Programming*, 35:83–96, 1986.

[13] Peter Borwein and A.S. Lewis. Moment-matching and best entropy estimation. Technical Report CORR 91-03, University of Waterloo, 1991. Submitted to Journal of Mathematical Analysis and Applications.

[14] B. Buck and V.A. Macaulay, editors. *Maximum entropy in action.* Oxford University Press, Oxford, 1991.

[15] J.P. Burg. Maximum entropy spectral analysis. Paper presented at 37th meeting of the Society of Exploration Geophysicists, Oklahoma City, 1967.

[16] E.W. Cheney. *Introduction to approximation theory.* McGraw-Hill, New York, 1966.

[17] D. Dacunha-Castelle and F. Gamboa. Maximum d'entropie et problème des moments. *Annales de l'Institut Henri Poincaré*, 26:567–596, 1990.

[18] J.E. Dennis and R.B. Schnabel. *Numerical methods for unconstrained optimization and nonlinear equations.* Prentice-Hall, New Jersey, 1983.

[19] J. Diestel. *Sequences and Series in Banach Spaces.* Springer-Verlag, New York, 1984.

[20] N. Dunford and J.T. Schwartz. *Linear operators*, volume 1. Interscience, New York, 1958.

[21] B. Forte, W. Hughes, and Z. Pales. Maximum entropy estimators and the problem of moments. *Rendiconti di Matematica, Serie VII*, 9:689–699, 1989.

[22] F. Gamboa. *Methode du Maximum d'Entropie sur la Moyenne et Applications.* PhD thesis, Universite Paris Sud, Centre d'Orsay, 1989.

[23] E. Gassiat. Probléme sommatoire par maximum d'entropie. *Comptes Rendues de l'Académie des Sciences de Paris, Série I*, 303:675–680, 1986.

[24] R.K. Goodrich and A. Steinhardt. L_2 spectral estimation. *SIAM Journal on Applied Mathematics*, 46:417–428, 1986.

[25] R.B. Holmes. *Geometric functional analysis and its applications.* Springer-Verlag, New York, 1975.

[26] E.T. Jaynes. Prior probabilites. *IEEE Transactions*, SSC-4:227–241, 1968.

[27] A.S. Lewis. Pseudo-Haar functions and partially-finite programming. Manuscript.

[28] A.S. Lewis. The convergence of entropic estimates for moment problems. In S. Fitzpatrick and J. Giles, editors, *Workshop/Miniconference on Functional Analysis/Optimization*, pages 100–115, Canberra, 1989. Centre for Mathematical Analysis, Australian National University.

[29] A.S. Lewis. The convergence of Burg and other entropy estimates. Technical Report CORR 92-08, University of Waterloo, 1992.

[30] A.S. Lewis. Facial reduction in partially finite convex programming. Technical Report CORR 92-07, University of Waterloo, 1992.

[31] L.R. Mead and N. Papanicolaou. Maximum entropy in the problem of moments. *Journal of Mathematical Physics*, 25:2404–2417, 1984.

[32] J. Navaza. The use of non-local constraints in maximum-entropy electron density reconstruction. *Acta Crystallographica*, A42:212–223, 1986.

[33] R. Nityananda and R. Narayan. Maximum entropy image reconstruction – a practical non-information-theoretic approach. *Journal of Astrophysics and Astronomy*, 3:419–450, 1982.

[34] R.T. Rockafellar. Integrals which are convex functionals. *Pacific Journal of Mathematics*, 24:525–539, 1968.

[35] R.T. Rockafellar. *Convex Analysis.* Princeton University Press, Princeton, N.J., 1970.

[36] W. Rudin. *Real and complex analysis.* McGraw-Hill, New York, 1966.

[37] J. Skilling and S.F. Gull. The entropy of an image. *SIAM-AMS Proceedings*, 14:167–189, 1984.

[38] M. Teboulle and I. Vajda. Convergence of best ϕ-entropy estimates. *IEEE Transactions on Information Theory*, 1992. To appear.

[39] A. Visintin. Strong convergence results related to strict convexity. *Communications on partial differential equations*, 9:439–466, 1984.

[40] Z. Ziegler. Minimizing the $L_{p,\infty}$-distortion of trigonometric polynomials. *Journal of Mathematical Analysis and Applications*, 61:426–431, 1977.

MAXIMUM ENTROPY METHOD AND DIGITAL FILTER DESIGN

Rabinder N. Madan
Office of Naval Research
800 N. Quincy St.
Arlington, Virginia 22217, U.S.A.

ABSTRACT. A new procedure that makes use of the maximum entropy method (MEM) for the design of linear phase FIR digital filters is described here. It is shown here that by applying MEM to the inverse of the desired gain function or its square root function, it is possible to generate linear phase FIR filters that match the given gain function to any desired degree of accuracy. Moreover, an iterative algorithm makes the design procedure very efficient, since the higher order filters can be recursively generated from the lower order ones. To minimize the effect of any residual passband distortion, a final averaging scheme on the lower order filters generates a class of passband distortion-free linear phase FIR filters. Simulation results that compare the present procedure with other well known methods are also presented here.

1. Introduction

The Maximum Entropy Method (MEM) is one of the techniques for estimating the power spectrum of a wide sense stationary stochastic process from partial information available about itself. In the ideal case, this information is in the form of knowledge about a portion of its autocorrelation sequence $r_k = E[x(nT)\,x^*((n+k)T)]$, $k = 0 \to \infty$, where $x(nT)$ represents a discrete-time wide sense stationary stochastic process. Then $r_k = r^*_{-k}$ and the power spectral density $S(\omega)$ associated with this process is given by

$$S(\omega) = \sum_{k=-\infty}^{+\infty} r_k\, e^{-jk\omega} \geq 0 \,. \tag{1}$$

Clearly, $r_k = \frac{1}{2\pi}\int_{-\pi}^{\pi} S(\omega) e^{jk\omega} d\omega$, $|k| \geq 0$. Given a finite set of autocorrelations $r_0, r_1, \ldots, r_n$ of a process, the spectrum estimation problem refers to evaluating a nonnegative function $S(\omega)$ such that its first $(n+1)$ Fourier coefficients matches with the given autocorrelation sequence (correlation matching property). It is well known that a necessary and sufficient condition for the existance of a nonnegative function as in (1) is the nonnegativity property of every Hermitian Toeplitz matrix $\mathbf{T}_k$ generated from the correlations $r_0, r_1 \ldots, r_k$, i.e., $S(\omega) \geq 0 \Longleftrightarrow \mathbf{T}_k \geq 0$, $k = 0 \to \infty$ [1].

Given a finite set of correlations, the maximum entropy method(MEM) eliminates the remaining ambiguity about the underlying process by choosing that process with maximum entropy. Since maximization of entropy subject to given values of $r_0, r_1, \ldots, r_n$, leads to a Gaussian process, the desired MEM-solution can be obtained by maximizing the entropy H of the Gaussian process given by

$$H = \frac{1}{2\pi}\int_{-\pi}^{\pi} ln\, S(\omega) d\omega \,. \tag{2}$$

A. Mohammad-Djafari and G. Demoments (eds.), Maximum Entropy and Bayesian Methods, 49–54.

It is well known that, given $r_0, r_1, \ldots, r_n$, maximization of (2) leads to a stable autoregressive (all-pole) model of order n, i.e., [3],

$$S(\omega) = 1/|H_n(e^{j\omega})|^2 \tag{3}$$

where

$$H_n(z) = a_0^{(n)} + a_1^{(n)} z^{-1} + a_2^{(n)} z^{-2} + \cdots + a_n^{(n)} z^{-n}. \tag{4}$$

The coefficients of the above polynomial satisfy the standard Yule-Walker equations and can be generated through a recursive procedure known as the Levinson recursion algorithm [4]

$$\sqrt{1 - s_n^2}\, H_n(z) = H_{n-1}(z) - s_n z^{-n} H_{n-1}(1/z), \quad n = 1 \to \infty. \tag{5}$$

The above update rule (5) starts with $H_0(z) = \frac{1}{\sqrt{r_0}}$, $s_1 = \frac{r_1}{r_0}$. Here, s_n, $n = 1 \to \infty$, are known as the reflection coefficients and at stage n it can be recursively computed from

$$s_n = \left(\sum_{k=0}^{n-1} a_k^{(n-1)} r_{n-k} \right) a_0^{(n-1)}. \tag{6}$$

Clearly, the single-step algorithm in (5)–(6) brings in the new information entirely through the reflection coefficients. The unique extension so obtained in (3) has other interesting properties as well. To examine this, notice that for an *arbitrary* stochastic process with integrable power spectal density $S(\omega)$, i.e., $\int_{-\pi}^{\pi} S(\omega) d\omega < \infty$, the functional in (2) is referred to as the causality criterion, since the condition $H > -\infty$, (Paley-Wiener criterion), guarantees the factorization of $S(\omega)$ in terms of a unique function $B(z)$ that is analytic together with its inverse in $|z| > 1$ and has square summable impulse response, i.e.,

$$S(\omega) = |B(e^{j\omega})|^2\,; \quad B(z) = \sum_{k=0}^{\infty} b_k z^{-k}, \quad b_0 > 0, \tag{7}$$

and $\sum_{k=0}^{\infty} |b_k|^2 < \infty$. Clearly, this minimum phase function $B(z)$, known as the Wiener factor of the process, represents a causal (one-sided) digital filter with square summable impulse response. For Gaussian processes, the Paley-Wiener criterion refers to its entropy and in general we will refer to H as the entropy functional of that process.

Thus, subject to the additional constraint $H > -\infty$, the given autocorrelations $r_0 \to r_n$ will generate a positive-definite Toeplitz matrix and in that case the spectrum estimation problem has an infinite number of solutions [4]. To see this, it is best to refer to the well known geometric interpretation for the class of all admissible solutions [4]. Letting Δ_k denote the determinant of $\mathbf{T}_k$, $k = 0 \to \infty$ the given covariances satisfy $\Delta_n > 0$ and clearly, any admissible extension r_k, $k = n+1 \to \infty$ must satisfy $\Delta_k > 0$, $k = n+1 \to \infty$. Consequently, at the first step r_{n+1}, must be chosen so that $\Delta_{n+1} > 0$. In general, this gives the well known relation [4]

$$|r_{k+1} - \xi_k|^2 \le R_k\,, \quad where \quad R_k = \frac{\Delta_k}{\Delta_{k-1}} > 0\,, \quad k = n \to \infty, \tag{8}$$

which in turn implies that r_{k+1} must be inside the above admissible circle. Thus, at every stage $k = n \to \infty$, the unknown correlation can be chosen in an infinite number of ways satisfying (8), and since at each stage the present value of the parameters ξ_k and R_k depend on the choice at the previous stages, clearly, there are an infinite number of solutions to the spectrum extention problem.

To identify the maximum entropy extension described before among the class of all such extensions, it is necessary to connect the entropy H with the extension parameters in (8). Towards this, notice that for any $k \geq n$, we can write [4]

$$\frac{\Delta_{k+1}}{\Delta_k} = \frac{\Delta_k}{\Delta_{k-1}} \left[1 - \left(\frac{\Delta_{k-1} |r_{k+1} - \xi_k|}{\Delta_k} \right)^2 \right] \tag{9}$$

and this gives the inequality, $R_{k+1} \leq R_k \leq R_n\,, \quad k = n+1 \to \infty$, i.e., the radii of the admissible circles in (8) form a monotone nonincreasing positive sequence of numbers bounded by R_n. Letting R_∞ denote the limit of this sequence, we have

$$R_\infty \leq R_k \leq R_n = \frac{\Delta_n}{\Delta_{n-1}}\,, \quad k = n+1 \to \infty\,. \tag{10}$$

Interestingly, this 'final radius' R_∞ is related to the power spectral density function $S(\omega)$ through the relation [1, 4] $R_\infty = exp\left[\frac{1}{2\pi}\int_{-\pi}^{\pi} ln\, S(\omega) d\omega\right] = exp[H]$. It follows that maximization of entropy is equivalent to choosing the extension in (8) such that the radius of the 'final circle' has its maximum possible value R_n. Using (10), clearly R_∞ takes its maximum possible value Δ_n/Δ_{n-1} if and only if *every* $R_k = \Delta_n/\Delta_{n-1}$, $k = n+1 \to \infty$, i.e.,

$$(R_\infty)_{\max} = \frac{\Delta_n}{\Delta_{n-1}} \ \text{ iff } \ R_k = \frac{\Delta_n}{\Delta_{n-1}}\,, \; k = n+1 \to \infty\,.$$

Then, from (9), we have [4] $r_{k+1} = \xi_k\,, \quad k = n \to \infty$, and every extension r_{k+1} must be at the centers of the respective admissible circles in the case of maximum entropy method. Since the center of the circle is maximally away from the circumference of the circle, such an extension is also most robust and maximally noncommittal. From previous remarks, maximum entropy extension always leads to an autoregressive solution, and to explore its application in linear phase FIR filter design, we next examine a possible approach.

2. Linear Phase FIR Filter Design

If a given gain function $G(\theta)$ can be written as

$$G(\theta) = |H(e^{j\theta})|^2\,; \quad H(z) = h_0 + h_1 z^{-1} + \cdots + h_n z^{-n}\,, \tag{11}$$

then $G(\theta)$ represents a finite impulse response (FIR) filter. If $H(e^{j\theta})$ also has linear phase, i.e., $H(e^{j\theta}) = A(\theta)e^{\psi(\theta)}$ and $\psi(\theta) = k\theta$, then $H(z)$ is a linear phase FIR filter. It is well known that a sufficient condition for a polynomial to have linear phase is that it be reciprocal, i.e., $H(z) = z^{-n}H^*(1/z^*) \stackrel{\Delta}{=} \tilde{H}(z)$. For any rational function, define $H_*(z) = H^*(1/z^*)$, then (11) can be rewritten as $G(\theta) = H(z)H_*(z)|_{z=e^{j\theta}} = H(e^{j\theta})H^*(e^{j\theta})$.

If the given $G(\theta)$ does not represent the gain funtion associated with an FIR filter, nevertheless, it may be possible to obtain an FIR filter $H(z)$, such that (11) is satisfied approximately. More accurately, if the deviation between $G(\theta)$ and $|H(e^{j\theta})|^2$ is negligible for all θ, then $H(z)$ represents a reasonable FIR approximation to the gain funtion $G(\theta)$. In that sense, the quantity

$$\eta \triangleq \sup_{0\leq\theta\leq2\pi} \left| G(\theta) - |H(e^{j\theta})|^2 \right| \tag{12}$$

which represents the supremum error norm can be used as a meaningful measure in selecting $H(z)$.

Since a power spectral density function $S(\theta)$ and a gain function have a one-to-one correspondence, comparing (7) and (11), the filter design problem is conceptually similar to the spectral factorization problem. As a result, given a gain function $G(\theta)$ that satisfies the integrability condition and the causality criterion, let

$$G(\theta) = \sum_{k=-\infty}^{+\infty} r_k e^{-jk\theta} \geq 0 \tag{13}$$

represent its Fourier representation. By treating $G(\theta)$ as a spectral density and applying the maximum entropy method to $(n+1)$ of its autocorrelations $r_0, r_1, \ldots, r_n$, one can generate the corresponding Levinson polynomials $H_n(z)$ as in (5)–(6) and from (3), for any $n \geq 1$, we also have

$$\frac{1}{|H_n(e^{j\theta})|^2} = \sum_{k=-n}^{n} r_k e^{jk\theta} + o(e^{j(n+1)\theta}). \tag{14}$$

Clearly, as n increases, the difference $G(\theta) - \sum_{k=-n}^{n} r_k e^{jk\theta} \longrightarrow 0$. Since from (13), $r_k \to 0$ as $k \to \infty$, and from (14), $H_n(z)$ represents a reasonable IIR approximation to $G(\theta)$.

Following the above arguments, an FIR approximation is straightforward. Instead of starting with the given gain function $G(\theta)$, define

$$K(\theta) = 1/G(\theta). \tag{15}$$

Naturally for $K(\theta)$ to satisfy the integrability condition and the Paley-Wiener criterion, it is sufficient that $G(\theta) > 0$ for all θ. Evidently, applying maximum entropy method to $K(\theta)$ allows one to approximate it by $|1/P_n(e^{j\theta})|^2$, where $P_n(z)$ represents the Levinson polynomial associated with $1/G(\theta)$. As n increases, since the sup. error in (12) for $1/G(\theta)$ decreases to zero, we can write

$$K(\theta) = 1/G(\theta) \doteq 1/|P_n(e^{j\theta})|^2 \tag{16}$$

or $G(\theta) \doteq |P_n(e^{j\theta})|^2$. Thus, $P_n(z)$ is an nth order FIR filter that approximates the given gain function $G(\theta)$. Even though the filter $P_n(z)$ in (16) represents an FIR approximation to $G(\theta)$, its response does not possess linear phase. However, as shown below, the remedy is near by and this can be achieved by beginning with the function

$$K_1(\theta) = 1/\sqrt{G(\theta)}. \tag{17}$$

Clearly, once again it is possible to generate the Levinson polynomials $H_n(z)$ of $K_1(\theta)$ from its Fourier coefficients and for large n, we have

$$K_1(\theta) \doteq 1/|H_n(e^{j\theta})|^2$$

and using (17), $G(\theta) \doteq \left|A(e^{j\theta})\right|^2$, where $A(z) = H_n(z)\tilde{H}_n(z) = \tilde{A}(z)$ represents a symmetric (reciprocal) polynomial of degree $2n$.

Thus, any given power gain function $G(\theta)$ can be approximated by a linear phase FIR filter. The approximation can always be improved by increasing n and to accomplish that one can make use of the recursions in (5)–(6). The sup. error norm in (12) can be utilized to decide upon a stopping criterion.

A further refinement can be employed to minimize any passband distortion in the FIR approximations $P_n(z)$ and $A_n(z)$ so obtained. To illustrate this , refer to (15)–(17) and consider a new polynomial approximation of the form

$$G(\theta) \doteq |\sum_{k=0}^{n} \alpha_k P_k(e^{j\theta})|^2 = |Q_n(e^{j\theta})|^2 \tag{18}$$

such that the error $\sqrt{G(\theta)} - \sum_{k=0}^{n} \alpha_k P_k(e^{j\theta})$ is minimized by selecting the coefficients α_k, $k = 0 \rightarrow n$ in some optimal fashion. Here $P_k(z)$ represents the Levinson polynomials of $1/G(\theta)$ as given in (16). Minimization of the square error $|\sqrt{G(\theta)} - \sum_{k=0}^{n} \alpha_k P_k(e^{j\theta})|^2$ *that has been weighted* with respect to $1/G(\theta)$ gives a set of $n+1$ linear equations in $n+1$ unknowns and these equations can be solved for α_k, $k = 0 \rightarrow n$. However, since the off diagonal elements $\sqrt{R_k/R_m} < 1$, the approximate solution $\alpha_k = 1$, $k = 0 \rightarrow n$ turns out to be an excellent choice here in terms of minimizing the ripples in the passband as well as the stopband. In that case, with $P_k(z) = b_0^{(k)} + b_1^{(k)} z^{-1} + \cdots + b_k^{(k)} z^{-k}$, the desired FIR filter of order n is given by [5]

$$Q_n(z) = \sum_{k=0}^{n} P_k(z). \tag{19}$$

The orthogonal property of the reciprocal polynomials in (18) has effectively decoupled the n linear equations and hence no matrix operations are involved in the above approximation. These filters, which in a sense combine the maximum entropy as well as the minimization of the mean square error criteria, have been found to possess uniformly low distortion in the passband irrespective of the transition band size as well as the stop band size.

Notice that as before (see (17)– (18)), in this case also, linear phase can be maintained by starting with $\sqrt{G(\theta)}$ rather than $G(\theta)$ in (18). In that case with $H_n(z)$ representing the Levinson polynomials associated with $1/\sqrt{G(\theta)}$, the desired linear phase FIR filter is given by

$$B(z) = Q_n(z)\tilde{Q}_n(z). \tag{20}$$

and $Q(z) = \sum_{k=0}^{n} H_k(z)$. Clearly $B(z)$ in (20) possesses linear phase.

To illustrate these ideas, Fig. 1 shows the design of a typical lowpass and a bandpass filter with details as indicated there.

3. Conclusions

This paper describes a new technique to design linear phase FIR digital filters with/without linear phase. The design procedure makes use of the inverse of the desired magnitude gain functions (or its square root) and is able to achieve any desired degree of accuracy by incorporating a higher order model. An iterative procedure makes the implementation of the

higher order filters very efficient. To minimize the effect of the residual passband distortion, a final averaging scheme generates a class of distortion free linear phase FIR filters. Since the proposed method is based entirely on 'digital' techniques, it becomes possible to include features such as transition band characterization into the design procedure itself. The new set of filters are seen to be more robust to changes in transition bandwidth as well as stop band level. This is quite valuable from a practical filter design point of view.

Acknowledgments

The author wishes to acknowledge Prof. S. U. Pillai of Polytechnic University,Brooklyn, New York, for several useful discussions on this topic.

References

[1] U. Grenander and G. Szegö, *Toeplitz Forms and Their Applications*, New York: Chelsea, 1984.

[2] G. Szegö, *Orthogonal Polynomials*, Amer. Math. Soc., Colloquium Publications, vol. 23, pp. 287–292, 1985.

[3] J. P. Burg, "Maximum entropy analysis," presented at *the 37th Annual Meeting, Soc. Explor. Geophysics*, Oklahoma City, Oklahoma, 1967.

[4] D. C. Youla, "The FEE: A new tunable high-resolution spectral estimator, part I," Tech. Report, no. 3, Polytechnic Inst. of New York, Brooklyn, New York, 1980. Also, RADC Report, RADC-TR-81-397, AD A114996, Feb. 1982.

[5] S. U. Pillai and R. N. Madan, "Design of Linear Phase FIR Filters with uniformly Flat Band-Response," *Proc. of twenty-fifth Asilomar Conference on Signals, Systems and Computers, Pacific Grove, CA*, pp. 1015–1020, Nov. 4–6, 1991.

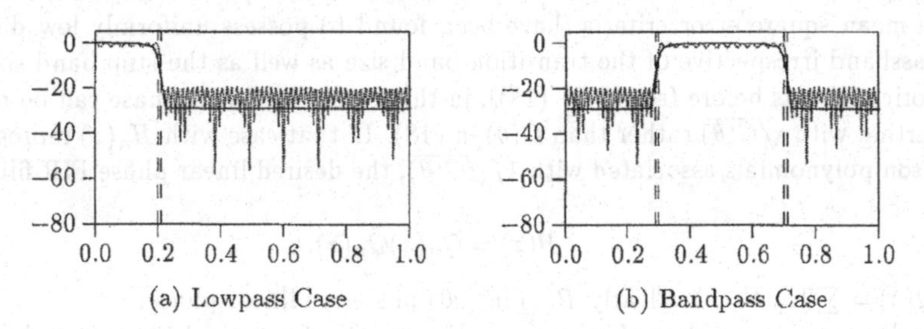

Fig. 1. Linear phase FIR filer design (a) Lowpass case with pass bandwidth BW = 0.2. (b) Bandpass case with pass bandwidth BW = 0.4. Common transition bandwidth TB = 0.01. Solid curves represent the present technique based on MEM. Dotted curves represent the Parks-McClellan technique. In each case, filter order N = 128.

A THEORETICAL PERFORMANCE ANALYSIS OF BAYESIAN AND INFORMATION THEORETIC IMAGE SEGMENTATION CRITERIA

Ian B. Kerfoot and Yoram Bresler
Beckman Institute
University of Illinois
405 N. Mathews · Urbana, IL 61801, USA

ABSTRACT. This paper presents a theoretical analysis of the performance characteristics of image segmentation objective functions that model the image as a Markov random field corrupted by additive white Gaussian noise, or equivalently, use Rissanen's Minimum Description Length criterion. The analysis is decoupled into the problems of false alarm region detection, parameter selection, detection threshold, and a priori boundary structure analysis. The various aspects of the problem are analyzed by standard methods of signal detection and estimation.

1. Introduction

Numerous image segmentation techniques have been proposed in the literature, but little formal performance analysis has been presented. The recent introduction of model-based criteria that define the noise and uncorrupted image (UI) as stochastic processes implicitly defines the segmented image (SI) as a stochastic process. We present a theoretical approach, based on standard methods of signal detection and estimation, to the performance analysis of Leclerc's [1] minimum description length (MDL) based segmentation objective function. Since this objective function is equivalent to a maximum *a posteriori* (MAP) criterion for a Markov random field (MRF) image model, our analysis also applies to this popular class of models. The analysis focuses on the performance innately possible with the given objective function, assuming its perfect global optimization.

Leclerc models the noisy image (NI), with pixels $\{z_{i,j}\}$, as a piecewise constant UI, $\{\mu_{i,j}\}$, corrupted by additive white Gaussian noise (AWGN) of known covariance, σ^2. An MDL-based derivation yields the objective function (codelength),

$$L(\mathrm{NI}, \mathrm{UI}) = \sum_{i,j \in I} a \left(\frac{z_{i,j} - \mu_{i,j}}{\sigma} \right)^2 + b n_b, \tag{1}$$

$$a = \frac{1}{2 \ln (2)}, \tag{2}$$

$$b = \lg (3), \tag{3}$$

where n_b is the number of boundaries between pixels in the UI, and $\lg(x) = \log_2(x)$. Leclerc's objective function omits several needed terms. To compensate for this, Leclerc proposes the use of $b = 8$. The segmented image is obtained as the minimizer of Equation (1), i.e.,

$$\mathrm{SI} = \arg\min_{UI} L(\mathrm{NI}, \mathrm{UI}). \tag{4}$$

A. Mohammad-Djafari and G. Demoments (eds.), Maximum Entropy and Bayesian Methods, 55–61.

Owing to the duality of the MDL and maximum a posteriori (MAP) probability principles, Equation (1) equals the negative logarithm of the posterior probability of an MRF, $\{\mu_{i,j}\}$, corrupted by AWGN. The quadratic term corresponds to the AWGN likelihood, and bn_b is the negative logarithm of the a priori probability of an MRF with a Gibbs prior defined by the energy function

$$\mathcal{E}(\mathrm{UI}) = b\sum_{i,j} \delta(\mu_{i,j} - \mu_{i-1,j}) + \delta(\mu_{i,j} - \mu_{i,j-1}), \tag{5}$$

with respect to the nearest neighbor system and the Kroenecker delta. In this interpretation, b represents a parameter of the clique potential.

We decouple the theoretical performance analysis into four distinct aspects: false alarm region detection, parameter selection, detection threshold, and *a priori* boundary structure analysis. Error probabilities are based on the probability that the codelength (equivalently, the a posteriori probability) of an error scenario is smaller (equivalently, larger) than assignment to the UI. The false alarm region density is shown to be independent of the noise level and separation of means in the region, a constant false alarm rate (CFAR) system. A Neyman-Pearson method of selecting the parameter b is presented. Region detection thresholds are determined. A method of analyzing the a priori boundary structure is proposed. An experimental study yields performance characteristics similar to our theoretical results. We briefly describe the application of these techniques to the analysis of vector-field segmentation.

2. False Alarm Region Density

The false alarm region density is the expected fraction of pixels that will be misassigned to spurious regions—SI regions that do not correspond to UI regions. In general, spurious regions could contain an arbitrary number of pixels, and there are a multitude of mutually exclusive, partially overlapping, candidate spurious regions with coupled probabilities of detection. We have been unable to adequately analyze the general problem, so we consider only the case of spurious single pixel regions. Thus, the false alarm region density equals the false alarm probability for a single pixel region.

The false alarm problem can be posed in terms of a chi-square random variable, yielding a false alarm density, $P_F(\Delta_b)$, of

$$P_F(\Delta_b) = 2\Phi\left(-\sqrt{\frac{b\Delta_b}{a}}\right) \approx \sqrt{\frac{2a}{\pi b\Delta_b}}\exp\left(\frac{-b\Delta_b}{2a}\right), \tag{6}$$

where $\Phi\{\tau\}$ is the standard normal cumulative density function, and $\Delta_b \in \{1,2,3,4\}$ is the increase in boundary length for the four possible single-pixel false alarm configurations. A homogeneously surrounded single-pixel false alarm has $\Delta_b = 4$ since a boundary segment is created for each neighbor. A single-pixel false alarm lying on a straight horizontal boundary creates three boundary segments, so $\Delta_b = 3$. Similarly, a diagonal boundary requires only two boundary segments for a false alarm, as illustrated in Figure 2, with $\Delta_b = 2$. Figure 2 also illustrates the $\Delta_b = 1$ scenario where only a single boundary segment must be added for a false alarm. The false alarm density is rather high for $b = \lg(3)$ and decreases exponentially with b. The false alarm density is independent of the distance between means of UI regions or noise level σ; a constant false alarm rate (CFAR) system.

3. Parameter Selection

The Neyman-Pearson method of parameter selection is particularly well suited for parameter selection for CFAR systems. We constrain the false alarm densities to satisfy

$$\begin{aligned} P_F(1),\, P_F(2) &\leq 1.0 \times 10^{-3}, \\ P_F(4) &\leq 1.0 \times 10^{-5}. \end{aligned} \tag{7}$$

The expected false alarm density is roughly ten pixels of each type in a 1024×1024-pixel image, since the $\Delta_b = 1$ and $\Delta_b = 2$ occur only along boundaries and $\Delta_b = 4$ occurs in the interior of regions. The minimum b satisfying the constraints is $b = 7.82$. This is equivalent to Leclerc's $b = 8$, although Leclerc does not specify whether his selection was made on a theoretical or experimental basis. We have also developed a receiver operating characteristic (ROC) approach to parameter selection [2], which we omit for brevity.

4. Detection Thresholds

The detection of legitimate regions and boundary bumps (protrusions from an otherwise straight boundary) involves the binary decision of assigning a set of pixels to either its correct mean, μ_H, or to the mean, μ_M, of an adjacent region to which it will be misassigned. Assignment to μ_H or μ_M shall be referred to as a *hit* or a *miss*, respectively. Regions are characterized by their boundary complexity parameter f. For an N_R-pixel region of Manhattan perimeter p, $f = p/\sqrt{N_R}$, which satisfies $0 < f \leq 2(N_R+1)/\sqrt{N_R}$. Two boundary bumps are considered: the horizontal, which consists of a height h, N_B pixel protrusion from an otherwise straight boundary aligned with the horizontal (or vertical) axis, and the diagonal, an N_B pixel protrusion from an otherwise straight boundary aligned with a diagonal boundary.

The miss probabilities for both regions and bumps are monotone decreasing functions of the contrast to noise ratio, $\mathrm{CNR} = |\mu_M - \mu_H|/\sigma$, the Mahalanobis distance between μ_M and μ_H. Therefore, we determine detection thresholds $\mathcal{D}_R$ and $\mathcal{D}_B$ for an N_R-pixel region or an N_B-pixel bump, respectively, such that for all $\mathrm{CNR} \geq \mathcal{D}_R$ or $\mathrm{CNR} \geq \mathcal{D}_B$, the respective miss probabilities do not exceed specified levels $\bar{P}_{MR}$ and $\bar{P}_{MB}$,

$$\mathcal{D}_R = \frac{1}{\sqrt{N_R}}\left(\tau_R + \sqrt{\tau_R^2 + \frac{bf\sqrt{N_R}}{a}}\right) \approx \sqrt{\frac{bf}{a}}\, N_R^{-0.25}, \tag{8}$$

$$\mathcal{D}_B = \begin{cases} \dfrac{1}{\sqrt{N_B}}\left(\tau_B + \sqrt{\tau_B^2 + \dfrac{2bh}{a}}\right) \approx \sqrt{\dfrac{2bh}{a}}\, N_B^{-0.5}, & \text{horizontal} \\ 2\tau_B N_B^{-0.5}, & \text{diagonal,} \end{cases} \tag{9}$$

$$\tau_R = \Phi^{-1}\left(1 - \bar{P}_{MR}\right), \tag{10}$$

$$\tau_B = \Phi^{-1}\left(1 - \bar{P}_{MB}\right). \tag{11}$$

We use $\bar{P}_{MR} = 10^{-3}$ and $\bar{P}_{MB} = 10^{-1}$, corresponding to $\tau_R = 3.09$ and $\tau_B = 1.28$, respectively, because we consider a boundary bump to be less important than a region.

It is clear from Equation (9) that single pixel horizontal bumps have the highest detection threshold ($\mathcal{D}_B = 6.11$ for $b = 7.82$) of any bump, making them the limiting factor for

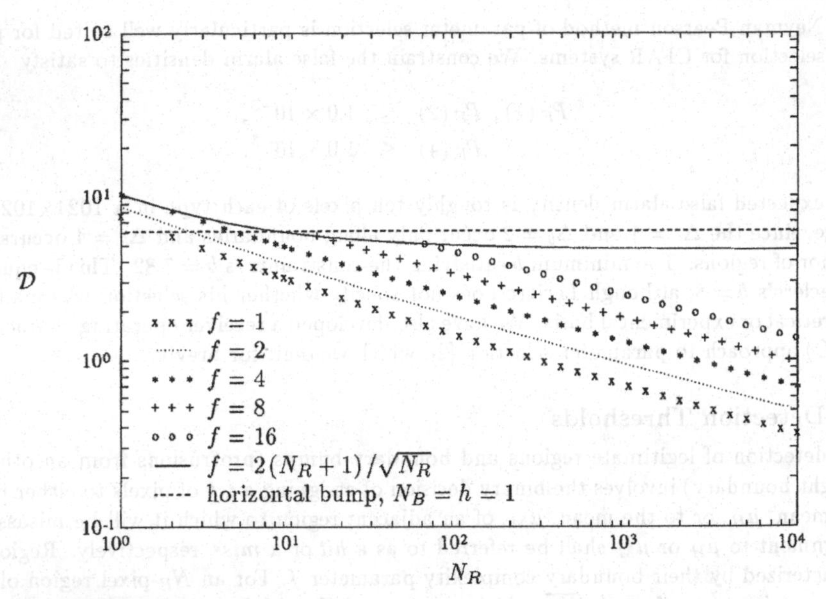

Figure 1: Detection thresholds for $b = 7.82$, $P_{MR} = 10^{-3}$, $P_{MB} = 10^{-1}$.

boundary reconstruction. The region detection thresholds are plotted in Figure 1 for various f as a function of N_R. The single pixel horizontal bump $\mathcal{D}_B$ is included as a reference. The $\mathcal{D}_B$ for a single pixel bump significantly exceeds the $\mathcal{D}_R$ for all regions except those with either very few pixels or very high boundary complexity.

It is often possible to detect regions with high f at lower CNR than Figure 1 would suggest. Consider a region of small f inscribed in a UI region of high f, such that both regions contain roughly equal numbers of pixels. The inscribed region has a lower $\mathcal{D}_R$ than the UI region. Thus, the UI region could be detected in the distorted form at the lower $\mathcal{D}_R$. Other configurations with still lower $\mathcal{D}_R$ may exist, so this remains an upper bound on the true $\mathcal{D}_R$. The boundaries will be accurately reconstructed only above the boundary bump detection threshold.

5. Expected A Priori Boundary Roughness

To quantify the significance of b, we also consider the geometrical characteristics of an image generated by the MRF model with a given value of b, or equivalently, by the a priori distribution induced by the boundary codelength bn_b. As shown by our analysis, the large scale structure of the image can be accurately determined at low CNR. Therefore, we feel that an analysis of the a priori probability of small scale fluctuations from a specified large scale framework adequately characterizes the a priori boundary structure. If only those configurations consisting of a straight horizontal (or vertical) boundary with bumps a single pixel high, of arbitrary length, along either side of the boundary are considered, then the expected fraction γ of increments along the straight line containing vertical transitions is

$$\gamma = \frac{1}{1+2^{(b-0.5)}}. \tag{12}$$

The MDL choice, $b = \lg(3)$, corresponds to $\gamma = 0.3204$ (fairly rough), while the Neyman-Pearson choice, $b = 7.82$, corresponds to $\gamma = 6.219 \times 10^{-3}$ (unreasonably straight). From this result, and from the previous sections, it appears that the single degree of freedom available in the choice of b is insufficient to ensure both low false alarm densities and good boundary and region detection at low CNR. This problem is aggravated in the segmentation of vector fields as discussed below.

6. Vector Field Segmentation

In [2], we present several MDL-based objective functions for vector field segmentation and use the analytic techniques presented in this paper to compare their performance characteristics. The objective functions are based on those of Leclerc [1] and Keeler [3] for scalar images and the vector clustering objective function of Rissanen and Ristad [4], with the addition of the codelength to specify the set of means, $L(\{\bar{\mu}_k\})$.

If $L(\{\bar{\mu}_k\})$ is omitted, then (for Leclerc's model) b must be increased with dimension, d, to keep the false alarm density low. This causes both $\mathcal{D}_B$ and $\mathcal{D}_R$ to increase with d, thus requiring higher CNR at larger dimensionality. If $L(\{\bar{\mu}_k\})$ is used, $\mathcal{D}_R$ still increases with d, but $\mathcal{D}_B$ is independent of d and equals the scalar $\mathcal{D}_B$ discussed in this paper. Leclerc's boundary model is shown to be somewhat superior to Keeler's under certain circumstances and significantly superior to Rissanen and Ristad's clustering model almost always. Since $\mathcal{D}_B$ is independent of d for objective functions including $L(\{\bar{\mu}_k\})$ and penalizing the spatial structure of the image (Leclerc's and Keeler's), it is unlikely that any boundary model could have significantly better performance than Leclerc's with $L(\{\bar{\mu}_k\})$ when d is high. However, alternate boundary models would be useful for low or moderate d.

Our current research is focused on the development of optimization algorithms that include $L(\{\bar{\mu}_k\})$ and are suitable for implementation on SIMD parallel architectures (e.g., the Connection Machine). The ability to analyze the merit of an objective function before developing an optimization algorithm is a major benefit of our theoretical performance analysis.

7. Experimental Study

In this section, we present an experimental study of the performance characteristics of Leclerc's scalar objective function. The segmentations are performed with Leclerc's [1] optimization algorithm.

The false alarm region densities are determined by segmenting a single test image containing a large number of candidate false alarm regions of the desired class. Figure 2

contains small portions of the experimental UI, NI, and SI for $\Delta_b = 1$ and $\Delta_b = 2$. The experimental CNR $= 10.0$ is well above $\mathcal{D}_R$ and $\mathcal{D}_B$, so the only relevant error scenarios are false alarms. Experimental false alarm densities are tabulated in Table 1.

The experimental false alarm rates are low by a factor of two, and the segmented images contain a few large false alarm regions.

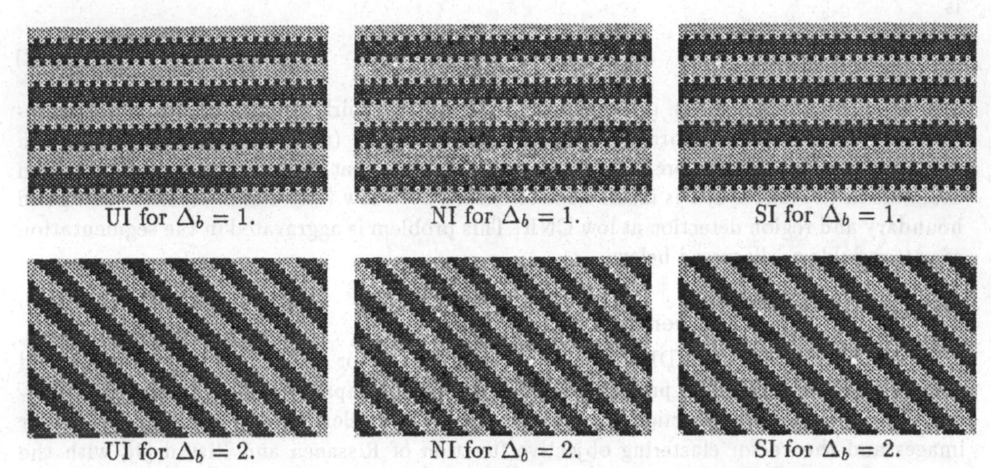

Figure 2: Experimental false alarm regions with $b = \lg(3)$ and CNR = 10.0.

Table 1: False Alarm Densities for $b = \lg(3)$.

Δ_b	$P_F(\Delta_b)$ Theoretical	$P_F(\Delta_b)$ Experimental	Maximum Number of FA Possible
1	1.38×10^{-1}	6.40×10^{-2}	1000
2	3.61×10^{-2}	1.26×10^{-2}	2699

The region detection threshold has been analyzed by segmenting a two-region test image at a series of CNR's, as illustrated in Figure 3.

The results of a series of 100 segmentations at each CNR are plotted in Figure 4.

The experimental $\mathcal{D}_R$ curve is similar to the theoretical one, but somewhat higher.

The theoretical and experimental results are in rough agreement, but have significant differences between them. It is unclear whether the differences are due to inaccuracy of our theoretical analysis, or imperfections in the optimization algorithm. The experimental analysis demonstrates that while our theoretical analysis is useful for designing an objective function, there is still a need for experimental analysis of the collective performance of the objective function and the optimization algorithm. It would be interesting to incorporate characteristics of optimization algorithms in the theoretical analysis, although we have not done so.

ACKNOWLEDGMENTS. This research was supported in part by the Whitaker Foundation and in part by the National Science Foundation PYI grant No. MIP 9157377.

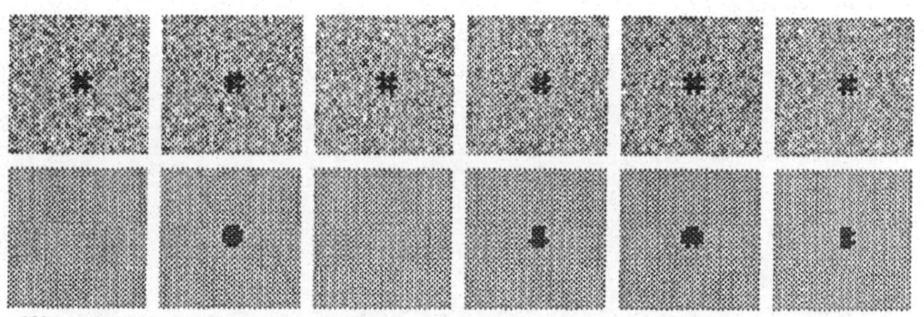

CNR = 5.25 CNR = 5.50 CNR = 5.75 CNR = 6.00 CNR = 6.25 CNR = 6.50

Figure 3: Experimental region detection. NI on top, SI on bottom.

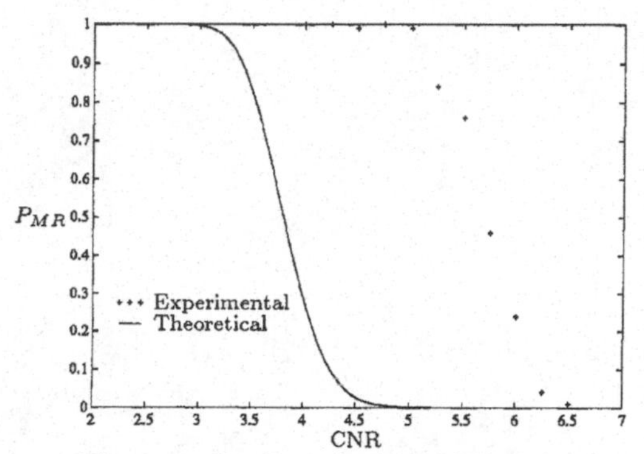

Figure 4: Region miss probabilities.

References

[1] Leclerc, Y. G.: 1989, "Constructing simple stable descriptions for image partitioning," *International Journal of Computer Vision*, **3**, 73.

[2] Kerfoot, I. B.: 1992, "Information-theoretic segmentation criteria for vector fields: analysis and design," Master's thesis, University of Illinois, Urbana, Illinois, USA.

[3] Keeler, K.: 1990, "Minimum-length encoding of planar subdivision topologies with application to image segmentation," *The Theory and Application of Minimal-Length Encoding Wokshop Notes*, (Stanford University), American Association for Artificial Intelligence.

[4] Rissanen, J. and Ristad, E.: 1991, "Unsupervised classification using stochastic complexity," Tech. Rep., IBM.

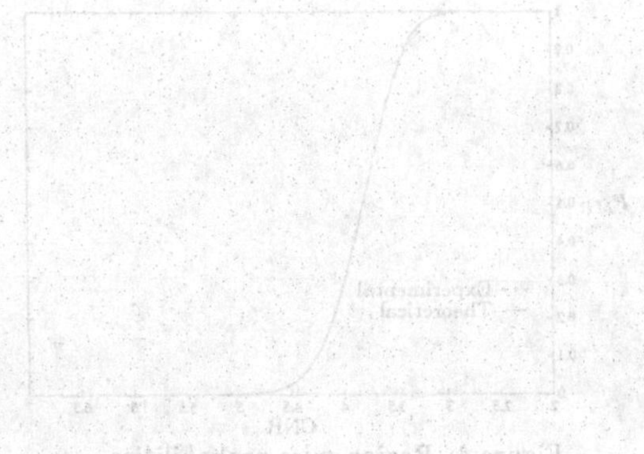

ON APPLICATION OF MAXENT TO SOLVING FREDHOLM INTEGRAL EQUATIONS

Sławomir Kopeć
Jagellonian University
Institute of Physics
Reymonta 4
30-059 Kraków, Poland.

ABSTRACT. We discuss the properties of MaxEnt approximate solutions to nonhomogenous linear equations with a compact kernel. The MaxEnt approximations exist and weakly converge to the exact solution of he problem. We find an estimate of an *a posteriori* upper error bound for the MaxEnt solution and illustrate the analysis on the numerical example.

1. Introduction

There have been several succesful attempts to approximate solving differential and integral equations by means of the MaxEnt method. In fact, these approaches can be clasified into two distinct categories. In the first one the expectation value of the solution is determined using the probability distribution found by MaxEnt (eg (Baker-Jarvis, 1989)). This approach allows the solution to be of any sign. The analysis, including computation of error bounds, can be carried out within the probability theory. In the second approach (eg (Mead, 1986)) the argument of the entropy functional is the function itself. Therefore, the solution cannot be negative. The main advantage of this technique is its computational simplicity. The natural formalism for study this variant of MaxEnt is the operator theory.

In this paper we discuss the properties of the approximate solutions to nonhomogenous linear integral equations, obtained by the second variant of MaxEnt. In particular, we will show that MaxEnt leads to convergent results for a wide class of equations as well as demonstrate the way of finding error bounds for solutions.

We shall be concerned with a nonhomogenous linear equation for $x(s)$

$$y(s) = x(s) - \lambda \int_0^1 k(s,t)x(t)dt, \quad 0 \leq s \leq 1, \tag{1}$$

where $x, y \in L_2(0,1)$, the space of square-integrable functions on $(0,1)$ with an inner product $(a,b) = \int_0^1 a(t)b(t)dt$, $Kx(s) = \int_0^1 k(s,t)x(t)dt$ is a compact linear operator on $L_2(0,1)$ with a nonnegative and measurable kernel $k(s,t)$. The sufficient condition for K to be compact is $\int_0^1 ds \int_0^1 dt|k(s,t)|^2 < \infty$. We assume that λ is not a characteristic value of K. Then Eq. (1) possesses a unique solution and, by the Fredholm alternative, an associate homogenous equation has only a trivial solution for every $y \in L_2(0,1)$. We further assume that the solution of Eq. (1) is nonnegative. For simplicity we assume that the kernel is

A. Mohammad-Djafari and G. Demoments (eds.), Maximum Entropy and Bayesian Methods, 63–66.

symmetric. Eq. (1) can be written as $y = Ax$, with $Ax := x - \lambda Kx$. We introduce in $L_2(0,1)$ an orthonormal basis of vectors ω_i for $i = 1, \ldots, \infty$ and take an inner product of (1) with each of basis vectors ω_i. As a result we obtain a system of equations

$$(\Omega_i, x) = \mu_i, \quad i = 1, \ldots, \infty, \tag{2}$$

with $\mu_i = (\omega_i, y)$, $\Omega_i = A\omega_i$.

In order to find an approximate solution of (2) we take only first n moment equations

$$(\Omega_i, x) = \mu_i, \quad i = 0, 1, \ldots, n. \tag{3}$$

Since the system (3) has infinitely many solutions in $L_2(0,1)$, an additional condition should be posed to make the procedure definite. In MaxEnt one claims that the functional $S(x) = \int_0^1 dt x(t) \log x(t)$ takes on its minimal value subject to constraints given by (3). The MaxEnt solution $x^{(n)}$ of (1) can be found by solving the system of equations

$$x^{(n)} = Bz^{(n)}, \quad z^{(n)} = \sum_{i=1}^{n} \alpha_i A\omega_i, \quad (\omega_i, Rz^{(n)}) = 0, \, i = 1, \ldots, n, \tag{4}$$

where $Bz = \exp z$ and $Rz := ABx - y$.

It can be shown that $x^{(n)}$ exist and are unique for every n (Kopeć, 1991a).

2. Convergence of Sequences of MaxEnt Solutions

Let $\tilde{x}$ denote the exact solution of (1). We have $S(x^{(1)}) \leq S(x^{(2)}) \leq \ldots \leq S(x^{(n)}) \leq \ldots \leq S(\tilde{x})$, so that, in particular, $S(x^{(n)}) \leq S(\tilde{x})$. This inequality and the property $S(x) \to \infty$ as $\|x\| \to \infty$ show that $\{x^{(n)}\}$ forms a bounded set. Hence, one can find a function $\hat{x} \in L_2(0,1)$ and a sequence $n_1 < n_2 < \ldots$ such that

$$\lim_{i \to \infty} (x^{n_i}, \beta) = (\hat{x}, \beta), \quad \forall \beta \in L_2(0,1).$$

As $(\Omega_i, \tilde{x} - x^{(n)}) = 0$ for $i = 1, 2, \ldots, N$, we have

$$\lim_{n \to \infty} (\Omega_i, \tilde{x} - x^{(n)}) = (\Omega_i, \tilde{x} - \hat{x}) = (\omega_i, A(\tilde{x} - \hat{x})) = 0, \qquad i = 1, 2, \ldots, \infty,$$

which is equivalent to $A(\tilde{x} - \hat{x}) = 0$. Since λ is not a characteristic value of K, this equation implies that $\tilde{x} = \hat{x}$. Moreover, by uniqueness of $\tilde{x}$, the entire sequence $\{x^{(n)}\}$ tends to $\tilde{x}$.

Thus *the sequence* $\{x^{(n)}\}$ *of MaxEnt approximations weakly converges to the exact solution* $\tilde{x}$ *of (1).*

3. Error Bounds for MaxEnt Solutions

The correspondence between MaxEnt and the Galerkin–Petrov method (Krasnoselskii, 1972) makes it possible to prove (Kopeć, 1991b) that

$$\|\tilde{x} - x^{(n)}\| \leq \|x^{(n)}\| \frac{\chi_n r_n}{1 - (q_n + r_n \chi_n)} \tag{5}$$

if the following conditions are fulfilled:
(i) there exists $\chi_n > 0$ such that $\|[R'(z^{(n)})]^{-1}\| \leq \chi_n$;
(ii) there exist constants $0 < \delta_n < 1$, $0 \leq q_n < 1$ such that

$$\sup_{\|z-z^{(n)}\| \leq \delta_n} \|[B'(z^{(n)})]^{-1}[B'(z) - B'(z^{(n)})]\| \leq q_n;$$

(iii) $r_n := \|Rz^{(n)}\| \leq \delta_n(1-q_n)/\chi_n$.

Let us define $\alpha = \inf|1 - \lambda/\lambda_k|$, where λ_k is a characteristic value of the kernel $k(s,t)$. Since $\|[R'(z^{(n)})]^{-1}\| \leq e_n/\alpha$, with $e_n = (\int_0^1 [x^{(n)}(t)]^{-2}dt)^{1/2}$, we take $\chi_n = e_n/\alpha$. Condition (ii) is satisfied when $q_n = \delta_n/(1-\delta_n)$ and $q_n < 1$.
Condition (iii) is then equivalent to the inequality $q_n^2 + q_n(\chi_n r_n - 1) + \chi_n r_n \leq 0$, which has a solution for $q_n \in [q_-, q_+]$. Since the error is an increasing function of q_n, we choose q_n as small as possible (but greater than q_-).

Thus, in order to find the error bound from inequality (5), we calculate:
(1) $x^{(n)}$ and $\|x^{(n)}\|$;
(2) $e_n = (\int_0^1 [x^{(n)}(t)]^{-2}dt)^{1/2}$;
(3) $\alpha = \inf|1 - \lambda/\lambda_k|$;
(4) $\chi_n = e_n/\alpha$;
(5) $r_n = \|Rz^{(n)}\| = \|Ax^{(n)} - y\|$;
(6) $q_n > q_-$.

4. Sample Calculation

Let us consider the equation

$$x(t) - \frac{3}{2}\int_0^1 \exp(|t-s|)x(s)ds = -1, \tag{6}$$

of which the five-moment MaxEnt solution has been found by Mead (Mead, 1986).
The norms of $x^{(5)}$ and $1/x^{(5)}$ are $\|x^{(5)}\| \approx 0.85480$ and $e_5 \approx 1.23854$, respectively.
In order to find α we compute the least characteristic value λ_1 of the kernel $\exp(|t-s|)$. Using the Kellog method (Mikhlin, 1949), we find that $\lambda_1 \geq 0.69105$. By the kernel traces method (Mikhlin, 1949) we have $0.69068 \leq \lambda_1 \leq 0.70669$. Therefore, we take $\lambda_1 = 0.69105$. The second characteristic value λ_2 is approximately given by $\lambda_2 = 1/\lambda_1(2/(A_2^2 - A_4))^{1/2}$, with $A_{2m} = 2\int_0^1 ds \int_0^s k_m^2(s,t)dt$, where $k_1(s,t) = k(s,t)$ and $k_2(s,t) = \int_0^1 k(s,u)k(u,t)du$. In our case $\lambda_2 \approx 3.15134$. Thus $\alpha = \inf|1 - \lambda/\lambda_k| \approx 0.5240$ and $\chi_5 \approx 2.3637$.
Now we make use of the values of

$$w^{(5)}(t) = (3/2)\int_0^1 \exp(|t-s|)x^{(5)}(s)ds - 1,$$

also calculated in (Mead, 1986), to find $r_5 = \|x^{(5)} - \lambda K x^{(5)} - y\| = \|x^{(5)} - w^{(5)}\| \approx 0.00082$. Inequality in condition (iii) has a solution for $q_5 > 0.002069$. We safely choose $q_5 = 0.00207$. The desired error bound is

$$\|\tilde{x} - x^{(5)}\| \leq 0.0017.$$

Since the exact solution of (6) is known (Mead, 1986), we can compare the above result to the 'exact' norm

$$\|\tilde{x} - x^{(5)}\| \approx 0.00093.$$

We see that the method presented above gives quite a restrictive estimate of the *a posteriori* error bound for the solution of (6).

5. Final Remarks

As we have shown in Sec.4, our method gives means to obtain realistic bounds for solutions of Fredholm equations. The error bounds can be, however, made more restrictive. In particular, one can try to find better estimate of $\|\exp(\tilde{z} - z^{(n)})\|$ and $\|B'(z^{(n)})\|$.

The assumption that $k(s,t)$ is symmetric is essential, because it enables us to estimate $\|[R'(z^{(n)})]^{-1}\|$. Therefore, once the kernel in (1) is not symmetric, one should first reduce the equation to the one with the symmetric kernel, which is always possible (Tricomi, 1957).

Our approach implies nonnegativity of the solution. Quite frequently, however, a change of variable suffices to convert the equation we want to solve to the equation with a nonnegative solution (Mead, 1986).

The analysis presented here can be generalized to solving other types of equations. For example, in order to apply the formalism to solving differential equations, one can transform the original equation into its weak form and look for the solution belonging to the appropriate Sobolev space.

ACKNOWLEDGMENTS. This work has been supported by the Polish Government Grant No 1772/2/91.

REFERENCES

Baker-Jarvis, J.: 1989, 'Solution to boundary value problems using the method of maximum entropy', *J. Math. Phys.* **30**, 302.

Mead, L.R.: 1986, 'Approximate solution of Fredholm integral equations by the maximum-entropy method', *J. Math. Phys.* **27**, 2903.

Kopeć, S.: 1991a, 'Properties of Maximum Entropy Approximate Solutions to Fredholm Integral Equations', *J. Math. Phys.* **32**,1269.

Krasnoselskii, M.A., Vainikko, G.M., Zabreyko, P.P., Rutitskii, Y.B., and Stetsenko, V.Y.: 1972, *Approximate Solutions of Operator Equations*, Wolters–Noordhoft, Groeningen.

Kopeć, S.:1991b, 'Error Bounds for Maximum Entropy Approximate Solutions to Fredholm Integral Equations', *J. Math. Phys.* **32**, 3312.

Mikhlin, S.G.: 1949, *Integral Equations and Their Applications*, OGIZ, Moscow.

Tricomi, F.G.: 1957, *Integral Equations*, Interscience, New York.

AN ALGORITHM TO PROCESS SIGNALS CONVEYING STATISTICAL INFORMATION

L. Rebollo Neira
Department of Electrical and Electronic Engineering
Imperial College
Exhibition Road, London SW7 2BT, England

A. Plastino
Departamento de Física, Universidad Nacional de La Plata, c.c
67, 1900 La Plata, Argentina

ABSTRACT. A method for processing signals containing information about the state distribution of a physical system is presented. The concomitant algorithm is specifically devised to suitably adapt lineal restrictions so as to take into account the presence of noise due to experimental errors.

1. Introduction

We call Statistical Signals the ones which convey information about systems that consist of subsystems of known properties whose relative proportions we want to find. We shall adopt a vectorial representation denoting a signal f as a vector $|f\rangle$ and a measurement as a mapping that assigns to it a real number.

For the sake of definiteness we assume that the system S we are interested in consists of a number M of subsystems S_n. Our purpose is that of finding out the relative population of S, assuming that the one corresponding to S_n is $C_n \geq 0$ (unknown). We take the view [1] that in order to study S one interacts with it by means of an input signal $|I\rangle$, the interaction between the signal $|I\rangle$ and S resulting in a response signal $|f\rangle$. The corresponding process is represented according to

$$\hat{W}|I\rangle = |f\rangle, \tag{1}$$

where the linear operator $\hat{W}$ portrays the effect that the system produces upon the input signal and can be decomposed in the following fashion

$$\hat{W} = \sum_{n=1}^{M} C_n \hat{W}_n, \tag{2}$$

where $\hat{W}_n|I\rangle = |n\rangle$. We work under the hypothesis that we know the response $|n\rangle$ evoked by S_n and that this set of vectors gives rise to a linear space U_M of dimension M. From

A. Mohammad-Djafari and G. Demoments (eds.), Maximum Entropy and Bayesian Methods, 67–72.

(1) and (2) it is clear that the response $|f\rangle$ is contained whiting U_M and carries information concerning the numbers C_n we are tying to find out. In order to accomplish such a goal one needs to perform observations upon $|f\rangle$. The corresponding measurement procedure provides numbers $\{f_1, \ldots, f_N\}$ out of $|f\rangle$ which can be regarded as the numerical representation of the signal.

2. Treatment of a Numerical Representation

Let's suppose that the numerical representation of $|f\rangle$ is obtained in such a way that measurements are performed as a function of a parameter x which adopts the values x_i with $(i = 1, \ldots, N)$. If the measurements are performed independently, we can regard the x_i as defining an (orthogonal) set of vectors $|\hat{x}_i\rangle$ that span an N-dimensional linear space E. We associate this space with the measurement instrument [2].

Let us $|f\rangle_p$ be the projection of $|f\rangle$ in E, i.e.

$$|f\rangle_p = \sum_{i=1}^{N} \langle \hat{x}_i | f \rangle | \hat{x}_i \rangle. \tag{3}$$

The expressions $\langle \hat{x}_i | f \rangle$ in a general case represent bilinear forms [2] and they are supposed to be given by experimental observations, so what we really have are numbers f_i^o affected by uncertainties Δf_i^o. Thus, instead of (3) we have, for the representation of $|f\rangle$ in E

$$|f^o\rangle_p = \sum_{i=1}^{N} f_i^o | \hat{x}_i \rangle. \tag{4}$$

The problem we face is that of building up a vector $|f^\star\rangle \in U_M$

$$|f^\star\rangle = \sum_{n=1}^{M} C_n^\star |n\rangle \tag{5}$$

out of the $\{f_i^o, i = 1, \ldots, N\}$-set, such that the $C_n^\star$ constitute a good approximation to the "true" C_n. For this purpose we construct the representatives in E of $|n\rangle$ and $|f^\star\rangle$

$$|n\rangle_p = \sum_{i=1}^{N} \langle \hat{x}_i | n \rangle | \hat{x}_i \rangle \quad ; \quad |f^\star\rangle_p = \sum_{n=1}^{M} C_n^\star |n\rangle_p. \tag{6}$$

The nearest vector $|f^\star\rangle_p$ to $|f^o\rangle_p$ that can be built is the one that fulfills the least distance equations [2]. These equations can be written in the form

$$F_n = \sum_{k=1}^{M} C_k^\star a_{n,k} \quad ; \; n = 1, \ldots, M. \tag{7}$$

where the $a_{n,k}$ are constructed out of the projections of vector $|n\rangle$ in E while the F_n contains the experimental data

$$a_{n,k} = \sum_{i=1}^{N} \langle n | \hat{x}_i \rangle \langle \hat{x}_i | k \rangle \quad ; \quad F_n = \sum_{i=1}^{N} \langle \hat{x}_i | n \rangle f_i^o. \tag{8}$$

Of course, as the f_i^o are affected by the experimental uncertainties Δf_i^o so will the F_n be subjected to corresponding uncertainties ΔF_n. Furthermore, the set of conditions (7) do not restrict the C_n^* to the domain of the non-negative real numbers, so we will adopt an algorithm to obtain a non-negative set of C_n that fulfills the set of equations (7), *within the margin allowed by the uncertainties* ΔF_n.

3. A Maximum Entropy Algorithm

We start by writing the equations (7) in the form

$$F_k = A \sum_{n=1}^{M} p_n a_{k,n} \quad ; k = 1, \ldots, M, \tag{9}$$

where $A \geq 0$ is a constant such that $\sum_{n=1}^{M} p_n = 1$. We can now think of the weights p_n as defining a probability space over a discrete set of M events whose informational content is given by

$$H = -\sum_{n=1}^{M} p_n \ln p_n. \tag{10}$$

We regard each F_k in (9) as proportional to the mean value of a random variable that adopts the values $a_{k,n}$; $(n = 1, \ldots, M)$ with a probability distribution given by the $\{p_n\}$-set. As A is an unknown constant, we employ one of the equations , say the l-th one, to determine it and are now in a position to solve the set of equations in an iterative fashion. We start our iterative process, by employing the Maximum Entropy Principle [3] in each step in order to construct an "optimal conjecture", that improves upon the results obtained in the previous step .

The zeroth-order approximation (first step) is devised by requiring that the zeroth-order weights p_n maximize H. This entails $p_n^{(0)} = 1/M$ so that we *predict* a zeroth-order value for the F_k. The quality of our conjecture can be measured by defining the "predictive error" ϵ_k as

$$\epsilon_k = \frac{|F_k - F_k^{(0)}|}{|F_k|} \quad ; k = 1, \ldots, M. \tag{11}$$

In order to construct our first order approximation we select, among the ϵ_k, the largest one, let us call it ϵ_{l1}. We shall then obtain the first-order weights $p_n^{(1)}$ by requiring that they maximize H with a constraint that ensures that the $l1$-th equation in (9) be fulfilled. We evaluate now the $F_k^{(1)}$ and the concomitant (new) set of ϵ_k. After selecting the largest one, ϵ_{l2}, say, we obtain the $p_n^{(2)}$ by maximizing H with the constraint that *both* the equations (9) for $k = l1$ and $k = l2$ be fulfilled, etc ,etc. The j-th order approximation is given by

$$p_n^{(j)} = \frac{exp(-\sum_{i=1}^{j} \lambda_i [F_l a_{li,n} - F_{li} a_{l,n}])}{\sum_{s=1}^{M} exp(-\sum_{i=1}^{j} \lambda_i [F_l a_{li,s} - F_{li} a_{l,s}])}, \tag{12}$$

where the j Lagrange multipliers λ_i are obtained by solving the j equations

$$\sum_{n=1}^{M} p_n^{(j)} [F_l a_{li,n} - F_{li} a_{l,n}] = 0 \quad ; i = 1, \ldots, j. \tag{13}$$

The iterative process is to be stopped when

$$\epsilon_k \leq \frac{\Delta F_k}{|F_k|} \quad ; \; \Delta F_k = \sum_{i=1}^{N} \Delta f_i^o |\langle k|\hat{x}_i\rangle| \quad ; \; k = 1\ldots, M. \tag{14}$$

Let us assume that the "convergence" (14) is attained at the L-th iteration. With this solution we can evaluate the numerical values

$$\langle \hat{x}_i | f^{(L)} \rangle = \sum_{n=1}^{M} C_n^{(L)} \langle \hat{x}_i | n \rangle \quad ; \; i = 1, \ldots, N. \tag{15}$$

If these conjectures are such that

$$|\langle \hat{x}_i | f^{(L)} \rangle - f_i^o| \geq \Delta f_i^o \quad \textit{for some } i, \tag{16}$$

the number of iterations can be augmented until the direction of the inequality is reversed. However, there is no guarantee that this type of convergence will *always* be achieved. Even more, in the realistic cases where we only can guess same estimations for the errors, to require that the direction of the inequality be reversed for all i it becomes a too stringent requirement. Although in the application we will discuss this type of convergence can be achieved, we wish to keep the discussion open so as to suitably adapt the "stop" point to the errors concomitant to any given particular model.

4. Numerical Text

Consider that we have a mixture of M = 11 different rare earth elements which satisfies a simple paramgnetic model [4], their respective proportions in the mixture being denoted by p_n. For any given n we list the corresponding quantum number S_n, L_n and J_n in Table I and set $|n\rangle \equiv |S_n L_n J_n\rangle$. We take a series of N =40 values of the magnetic field at the temperature T, which generates the parameters $x_i = H_i/T$;$(i = 1, \ldots, 40)$ The projection of vector $|n\rangle$ for a given value x_i is given by the magnetization of the ion n in Table I

$$\langle \hat{x}_i | n \rangle = g_n J_n \mu_B B_n(x_i), \tag{17}$$

where μ_B is the Bohr magneton, g_n is the spectral factor for the ion n, and $B_n(x_i)$ the appropriate Brillouin function[4]

$$g_n = 1 + \frac{J_n(J_n+1) + S_n(S_n+1) - L_n(L_n+1)}{2J_n(J_n+1)}, \tag{18}$$

$$B_n(x_i) = \frac{2J_n+1}{2J_n} cotgh[\frac{2J_n+1}{2J_n} x_i] - \frac{1}{2J_n} cotgh[\frac{x_i}{2J_n}]. \tag{19}$$

Two sets of "weights" $\{p_n\}$ and $\{p'_n\}$ are listed in Table I which correspond to two hypothetical mixtures S and S'. By recourse to these sets we have numerically simulated a series of measurements of the magnetization and have randomly distorted them within a 3% range (bars in Fig 1) where a) correspond to S and b) to S'.

n	Ion	S_n	L_n	J_n	p_n	$p_n^{\star}$	$p_n^{(1)}$	p'_n	$p'^{\star}_n$	$p'^{(2)}_n$
1	Ce^{3+}	$\frac{1}{2}$	3	$\frac{5}{2}$	0.003	2661	0.006	0.549	535	0.457
2	Pr^{3+}	1	5	4	0.004	-5966	0.003	0.329	-1246	0.296
3	Nd^{3+}	$\frac{1}{2}$	6	$\frac{4}{2}$	0.005	6886	0.003	0.076	1432	0.110
4	Pm^{3+}	2	6	4	0.006	-2979	0.005	0.042	-598	0.137
5	Gd^{3+}	$\frac{7}{2}$	0	$\frac{7}{2}$	0.055	496	0.600	0.000	93	0.000
6	Tb^{3+}	3	3	6	0.156	1606	0.163	0.000	310	0.000
7	Dy^{3+}	$\frac{5}{2}$	5	$\frac{15}{2}$	0.338	-8984	0.334	0.000	-1653	0.000
8	Ho^{3+}	2	6	8	0.301	9517	0.296	0.000	1751	0.000
9	Er^{3+}	$\frac{3}{2}$	6	$\frac{15}{2}$	0.104	-2510	0.106	0.000	-477	0.000
10	Tm^{3+}	1	5	6	0.019	-34	0.019	0.000	-4	0.000
11	Yb^{3+}	$\frac{1}{2}$	3	$\frac{7}{2}$	0.005	-6124	0.004	0.000	-124	0.000

Table I. For each rare-earth ion the pertinent quantum numbers are given. The relative properties in the rare-earth mixture are denoted by p_n (system S) and p'_n (system S').$p_n^{(1)}$ and $p_n^{\star}$ are theoretical results for system S that allow for a 3% error in the input data (see fig $1a$)). The $p_n^{\star}$ correspond to a least-square treatment and the $p_n^{(1)}$ to a first-order version of the present approach. $p'^{\star}_n$ and $p'^{(2)}_n$ are theoretical results for system S'.

The algorithm of section 3 gives the weights $p_n^{(1)}$ (up to first order) for the mixture S and the weights $p'^{(2)}_n$ (up to second order) for mixture S'. If we employ a Marquardt's algorithm [5] in order to get a real least-distance solution (it is known as a least-square approximation), we obtain the $p_n^{\star}$ and coefficients for the systems S and S' respectively. (All the pertinent figures are listed in Table I)

6. Conclusions

We have discussed a method, based upon the optimum conjecture derived from the Maximum Entropy Principle, that on the basis of measurements performed on the signal the system produces, allows one to find out its state distribution. The main idea is to build-up a signal belonging to a subspace determined by the system whose projection in a subspace, determined by the measurements instrument, is close to a vector constructed by experimental observations. The algorithm presented has been specifically devised so as to deal with lineal restrictions. It was achieved by a Maximum Entropy criterion and the requirement that the distance between observed measurements and predicted measurements be bounded by experimental errors. In order to compare the present approach with a least square approximation a low dimension numerical test has been performed. A larger dimension realistic situation involving X ray diffraction has also been tackled but is not included in this paper [2]. In both cases the algorithm has reached a rapid convergence, and few parameters are needed.

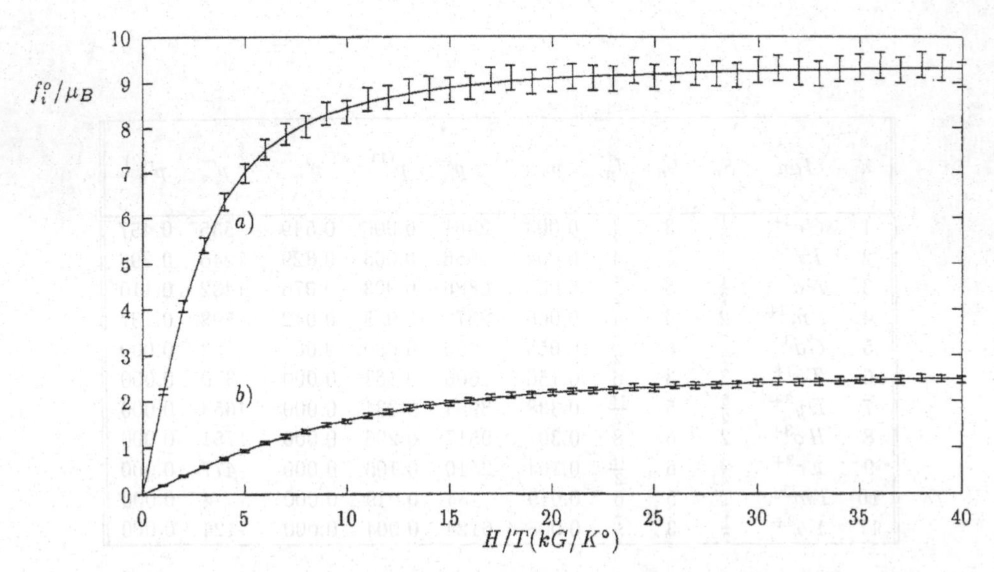

Fig. 1. Magnetization vs external applied field at the temperature T. The error bars are the input data of the numerical test and allow for a 3% distortion. The continuous curve represents both the predictions obtained with the present approach and with the least-squares approximation. Curve $a)$ correspond to system S and curve $b)$ to system S'

ACKNOWLEDGMENTS. L. Rebollo Neira acknowledges support from CONICET (Beca Externa) and is a member of CIC (Argentina). A. Plastino is a member of CONICET Argentina.

REFERENCES

[1] A. Plastino, L. Rebollo Neira, A. Alvarez, *Phys. Rev.* **A40** (1989) 1644.
[2] A. Plastino, L. Rebollo Neira, F. Zyserman, A. Alvares, R. Bonetto, H. Viturro.(To be published) (1992).
[3] E.T. Jaynes, *Phys. Rev.* **160** (1957) 620.
[4] C. Kittel.'Introduction to Solid State Physics'. John Wiley and sons, Inc, New York (1959).
[5] D. W. Marquardt, *J. Soc. Indust. Appl. Math.* **11** (1963) 2.

GENERALIZATION OF THE MAXIMUM ENTROPY CONCEPT AND CERTAIN ARMA PROCESSES

S. Unnikrishna Pillai and Theodore I. Shim
Department of Electrical Engineering
Polytechnic University
333 Jay Street
Brooklyn, New York 11201, U.S.A.

ABSTRACT. Given $(n+1)$ consecutive autocorrelations of a stationary discrete-time stochastic process, one interesting question is how to extend this finite sequence so that the power spectral density associated with the resulting infinite sequence of autocorrelations is nonnegative everywhere. It is well known that when the Hermitian Toeplitz matrix generated from the given autocorrelations is positive-definite, the problem has an infinite number of stable solutions and the particular solution that maximizes the entropy functional results in a stable all-pole model of order n. Since maximization of the entropy functional is equivalent to maximization of the minimum mean-square error associated with one-step predictors, in this paper the problem of obtaining admissible extensions that maximize the minimum mean-square error associated with k-step $(k \leq n)$ predictors, that are compatible with the given autocorrelations, is studied. It is shown here that the resulting spectrum corresponds to that of a stable ARMA$(n, k-1)$ process. This true generalization of the maximum entropy extensions for a two-step predictor turns out to be a unique ARMA$(n,1)$ extension, the details of which are presented here.

1. Introduction

An interesting problem in the study of autocorrelation functions and their associated power spectral densities is that of estimating the spectrum from a finite extent of its autocorrelation function. Towards this, consider a discrete-time zero mean wide sense stationary stochastic process $x(nT)$ with autocorrelation function $\{r_k\}_{k=-\infty}^{\infty}$ as

$$E[x((n+k)T)\, x^*(nT)] = r_k = r^*_{-k}\,, \quad k = -\infty \to \infty\,. \tag{1}$$

Its power spectral density function $S(\theta)$ is given by

$$S(\theta) = \sum_{k=-\infty}^{\infty} r_k e^{jk\theta} \geq 0\,, \quad -\pi < \theta \leq \pi \tag{2}$$

and assume that the process has finite power and satisfies the causality criterion, i.e.,

$$\frac{1}{2\pi}\int_{-\pi}^{\pi} S(\theta)\, d\theta = r_0 < \infty\,, \tag{3}$$

and

$$\mathcal{H} = \frac{1}{2\pi}\int_{-\pi}^{\pi} ln\, S(\theta)\, d\theta > -\infty\,. \tag{4}$$

A. Mohammad-Djafari and G. Demoments (eds.), Maximum Entropy and Bayesian Methods, 73–78.

In that case, every Hermitian Toeplitz matrix $\mathbf{T}_k$, formed from the autocorrelations $r_0 \to r_k$, $k = 0 \to \infty$ is positive definite, and moreover, there exists a unique function [1]

$$B(z) = \sum_{k=0}^{\infty} b_k z^k \tag{5}$$

that is analytic together with its inverse in $|z| < 1$ such that $\sum_{k=0}^{\infty} |b_k|^2 < \infty$, and

$$S(\theta) = |B(e^{j\theta})|^2 , \quad a.e. \tag{6}$$

In addition, $B(z)$ is free of poles on $|z| = 1$ and it represents the minimum-phase Wiener factor associated with the power spectrum $S(\theta)$.

The trigonometric moment problem, also known as the interpolation problem, can be stated as follows: Given $r_0, r_1, \ldots, r_n$ from a stationary regular stochastic process that satisfies (3) and (4), determine all solutions for the power spectral density $S(\theta)$ that are compatible with the given data; i.e., such an admissible solution $S(\theta)$ must satisfy

$$S(\theta) \geq 0 \quad \text{and} \quad \frac{1}{2\pi} \int_{-\pi}^{\pi} S(\theta) e^{-jk\theta} \, d\theta = r_k , \quad k = 0 \to n , \tag{7}$$

in addition to satisfying (3) and (4). From (2) and (4), for the existence of such a solution, the positivity of $\mathbf{T}_n$ is necessary, and interestingly that condition is also sufficient! It is well known that an infinite number of solutions exist to the above problem and various approaches can be used to obtain these solutions. Before examining the extension associated with maximization of the multi-step minimum mean-square prediction error, we begin with the general parametrization of the class of all extensions.

2. Parametrization of Admissible Extensions

By making contact with network theory, Youla has given an interesting physical characterization of this problem, and in this context he has derived explicit formulas that express the admissible spectra in terms of arbitrary bounded functions[1] [2]. Youla's parametrization given by

$$H_\rho(z) = \frac{\Gamma(z)}{A_n(z) - z\rho(z)\tilde{A}_n(z)} \tag{8}$$

is minimum-phase and the corresponding spectrum matches the first $(n+1)$ autocorrelations for *every* choice of the arbitrary bounded function $\rho(z)$ in (8); i.e.,

$$|H_\rho(e^{j\theta})|^2 = \sum_{k=-n}^{n} r_k e^{jk\theta} + O(e^{j(n+1)\theta}) \tag{9}$$

irrespective of the choice of $\rho(z)$. In (8), $\Gamma(z)$ represents the unique solution to the factorization

$$1 - |\rho(e^{j\theta})|^2 = |\Gamma(e^{j\theta})|^2 \tag{10}$$

[1] A function $\rho(z)$ is said to be bounded if (i) it is analytic in $|z| < 1$ and (ii) $|\rho(z)| \leq 1$ in $|z| < 1$.

that is analytic and free of zeros in $|z| < 1$. The above factorization is possible iff $\frac{1}{2\pi}\int_{-\pi}^{\pi} ln(1 - |\rho(e^{j\theta})|^2)d\theta > -\infty$. Further, $A_n(z)$ represents the Levinson polynomial associated with the given autocorrelations $r_0, r_1, \ldots, r_n$, i.e.,

$$A_n(z) = \frac{1}{\sqrt{\Delta_n \Delta_{n-1}}} \begin{vmatrix} r_0 & r_1 & \cdots & r_n \\ r_1^* & r_0 & \cdots & r_{n-1} \\ \vdots & \vdots & \vdots & \vdots \\ r_{n-1}^* & r_{n-2}^* & \cdots & r_1 \\ z^n & z^{n-1} & \cdots & 1 \end{vmatrix} = \sum_{k=0}^{n} a_k z^k \tag{11}$$

and $\tilde{A}_n(z) \triangleq z^n A_n^*(1/z^*)$. Here Δ_n represents the determinant of the Hermitian Toeplitz matrix generated from $r_0, r_1 \cdots r_n$. The Levinson polynomials $A_n(z)$, $n \geq 1$, are free of zeros in $|z| \leq 1$ and they can be computed using the recursion

$$\sqrt{1 - |s_k^2|}\, A_k(z) = A_{k-1}(z) - z s_k \tilde{A}_{k-1}(z), \quad k \geq 1. \tag{12}$$

that starts with $A_0(z) = 1/\sqrt{r_0}$, where the reflection coefficients s_k satisfy

$$s_k = \left\{ A_{k-1}(z) \sum_{i=1}^{k} r_i z^i \right\}_k A_{k-1}(0), \quad k \geq 1 \tag{13}$$

with $\{\ \}_k$ representing the coefficient of z^k in $\{\ \}$.

The maximum entropy extension can be identified easily by considering the expression for the entropy functional $\mathcal{H}_\rho$ associated with an otherwise arbitrary extension in (8). From (5) and (8), since for any process, the entropy $\mathcal{H} = ln\,|B(0)|^2$, we have from (8)

$$\mathcal{H}_\rho = \frac{1}{2\pi}\int_{-\pi}^{\pi} ln\, S(\theta)d\theta = \frac{1}{2\pi}\int_{-\pi}^{\pi} ln\,|B_\rho(e^{j\theta})|^2 d\theta = ln\,|B_\rho(0)|^2 \tag{14}$$

$$= ln\,|\Gamma(0)|^2 - ln\,|A_n(0)|^2 = ln\,(|1/A_n(0)|^2) - ln\,(|1/\Gamma(0)|^2) \tag{15}$$

Since $A_n(0) = \sqrt{\Delta_{n-1}/\Delta_n}$ does not depend on $\rho(z)$ and since $|\Gamma(0)| \leq 1$, clearly the maximum value for $\mathcal{H}_\rho$ is achieved by selecting $\Gamma(0) = 1$. In that case, from maximum modulus theorem [3], $\Gamma(z) \equiv 1$ and from (10), $\rho(z) \equiv 0$. Thus the maximum entropy extension corresponds to

$$B_{ME}(z) = \frac{1}{A_n(z)} \sim \mathrm{AR}(n), \tag{16}$$

and

$$S_{ME}(\theta) = \frac{1}{|A_n(e^{j\theta})|^2} = \sum_{k=-n}^{n} r_k e^{jk\theta} + O(e^{j(n+1)\theta}) \tag{17}$$

which is the same as the all-pole model derived by Burg [4]. From (14), since maximization of the entropy is equivalent to maximization of the one-step minimum mean-square prediction error $|B(0)|^2$, the natural generalization of maximizing the multi-step prediction error is interesting. We start with the two-step predictor case, where a complete answer can be given to the problem.

3. Two-Step Predictor

Given the autocorrelations $r_0, r_1, \ldots, r_n$, with $\mathbf{T}_n > 0$, of all the admissible completions given by (8), the problem here is to determine the particular extension that maximizes the k-step minimum mean-square prediction error. Since maximization of the one-step prediction error leads to an all-pole model, next we examine the two-step predictor case. The two-step minimum mean-square prediction error P_2 is given by [5]

$$P_2 = |b_0|^2 + |b_1|^2 = (1 + |\alpha_1|^2)\, exp\left(\frac{1}{2\pi}\int_{-\pi}^{\pi} ln\, S(\theta)d\theta\right) \tag{18}$$

where b_0 and b_1 are as given in (5) and $\alpha_1 = \frac{1}{2\pi}\int_{-\pi}^{\pi} ln\, S(\theta)e^{-j\theta}d\theta$. Naturally, maximization of P_2 is with respect to the unknown autocorrelations $r_{n+1}, r_{n+2}, \ldots$ and this leads to

$$\partial\alpha_1/\partial r_k = (1/2\pi)\int_{-\pi}^{\pi} e^{j(k-1)\theta}/S(\theta)d\theta\,, \quad |k| \geq n+1$$

and hence

$$\begin{aligned}\frac{\partial P_2}{\partial r_k} &= \frac{|b_0|^2}{2\pi}\int_{-\pi}^{\pi}\left(\frac{1+|\alpha_1|^2+\alpha_1 e^{j\theta}+\alpha_1^* e^{-j\theta}}{S(\theta)}\right)e^{jk\theta}d\theta \\ &= \frac{|b_0|^2}{2\pi}\int_{-\pi}^{\pi}\frac{|1+\alpha_1 e^{j\theta}|^2}{S(\theta)}e^{jk\theta}d\theta = 0, \quad |k| \geq n+1\,.\end{aligned} \tag{19}$$

Since $|b_0|^2 > 0$, (19) implies that the Fourier series expansion for the periodic nonnegative function $(|1+\alpha_1 e^{j\theta}|^2)/S(\theta)$ truncates after the nth term and hence it must have the form

$$\frac{|1+\alpha_1 e^{j\theta}|^2}{S(\theta)} = \sum_{k=-n}^{n} c_k e^{jk\theta} = |g(e^{j\theta})|^2 \tag{20}$$

where $g(z) = g_0 + g_1 z + \cdots + g_n z^n$ represents the strict Hurwitz polynomial (free of zeros in $|z| \leq 1$) associated with the factorization in (20). (Notice that since $S(\theta)$ is free of poles on the unit circle, $g(z)$ cannot have any zeros on $|z| = 1$.) Then,

$$S(\theta) = \frac{|1+\alpha_1 e^{j\theta}|^2}{|g(e^{j\theta})|^2} = |B_2(e^{j\theta})|^2 \tag{21}$$

where

$$B_2(z) = a(z)/g(z) \tag{22}$$

represents the minimum-phase factor and $a(z)$ represents the strict Hurwitz polynomial of degree 1 associated with the factorization $|1+\alpha_1 e^{j\theta}|^2 = |a(e^{j\theta})|^2$. Depending on the value of α_1, this gives rise to two choices, i.e., $a(z) = (1+\alpha_1 z)$ *or* $(\alpha_1 + z)$. Thus, given $r_0, r_1, \ldots, r_n$, the Wiener factor $B_2(z)$ in (22) that maximizes the two-step minimum mean-square prediction error, if it exists, is of the type ARMA$(n,1)$. To complete the argument, we must demonstrate the existence of such a factor that is analytic together with its inverse in $|z| < 1$.

Towards this purpose, notice that in the case of real autocorrelations, this specific extension, if admissible, should follow from (8) for a certain choice of the rational bounded function $\rho(z)$ that is also real for real z, and on comparing (22) and (8), because of degree restrictions, $\rho(z)$ must have the form [6]

$$\rho(z) = \frac{1}{a + bz}. \tag{23}$$

For (23) to be bounded-real, it is necessary that there exist no poles in $|z| \leq 1$, i.e., $|a/b| > 1$ and $|1/(a+be^{j\theta})| \leq 1 \iff (a \pm b)^2 \geq 1$, in addition to a and b being real. Interestingly, there is always one and only one bounded function $\rho(z)$ that maximizes the two-step minimum mean-square prediction error [6].

It can be shown that, given $r_0, r_1, \ldots, r_n$, always there exists a *unique* Wiener factor that maximizes the minimum mean-square error associated with the two-step predictors that are compatible with the given autocorrelations. This Wiener factor turns out to be an ARMA$(n, 1)$ filter given by [5,6]

$$B_2(z) = \frac{\alpha + \beta z}{g_0 + g_1 z + \cdots + g_n z^n} \tag{24}$$

where

$$g_k = a_{k-1} b + a_k a - a_{n-k+1}, \quad k = 0 \to n,$$

$$\alpha = \pm\frac{1}{2}\left(\sqrt{(a+b)^2 - 1} + \sqrt{(a-b)^2 - 1}\right),$$

$$\beta = \pm\frac{1}{2}\left(\sqrt{(a+b)^2 - 1} - \sqrt{(a-b)^2 - 1}\right),$$

with

$$a = \begin{cases} -2bR\cos(\phi/3 + 2\pi/3) \\ (a_n^2 - a_0^2)/a_1 a_n \end{cases} \left(\begin{array}{c} \textit{whichever is} \\ \textit{largest in} \\ \textit{magnitude} \end{array}\right) \tag{25}$$

and

$$b = a_0/a_n. \tag{26}$$

Here, a_k, $k = 0 \to n$, represent the coefficients of the degree n Levinson polynomial in (11)–(12). The signs of α and β should be selected so as to satisfy the identity $\alpha\beta = ab$. Further, R and ϕ satisfy

$$R = sgn(a_1/a_0)\sqrt{(2/3)(1 + 1/b^2 + a_1^2/(2a_0^2))}$$

and

$$\cos\phi = |a_1/a_0|(1 - 1/b^2)/2|R|^3.$$

Since $1 > \cos\phi \geq 0$ here, without loss of generality, $0 < \phi < \pi/2$, and this gives $\frac{1}{2} < -\cos(\phi/3 + 2\pi/3) \leq \sqrt{3}/2$. As remarked earlier, the largest quantity (in magnitude) between the two choices for the parameter a in (25) gives rise to a *unique* bounded-real function $\rho(z)$ in (23) and an admissible Wiener factor that maximizes the two-step minimum mean-square prediction error in (18). The corresponding spectrum satisfies (3), (4) and the interpolation property (7). Since maximization of the entropy functional is equivalent to maximization of the minimum mean-square error associated with one-step predictors, this is a true generalization of Burg's maximum entropy extension which results in a stable AR(n) filter.

The uniqueness of the above ARMA(n, 1) extension should not be confused with other admissible ARMA(n, 1) extensions. In fact, from (23), with b as given by (26) and, for example, letting $|a| > |b| + 1$ results in a bounded-real function $\rho(z)$ that generates an admissible ARMA(n, 1) extension. Thus, there exist an infinite number of admissible ARMA(n, 1) extensions that match the first $(n+1)$ autocorrelations, and the particular one described above is distinguished by the fact that it maximizes the two-step minimum mean-square prediction error among all such admissible extensions.

The natural generalization to k-step predictors ($k > 2$) that maximize the corresponding minimum mean-square error, given r_0, r_1, ..., r_n has been shown to result in a well-structured stable ARMA($n, k-1$) filter [5,6]. Since the k-step minimum mean-square prediction error is given by $P_k = \sum_{i=0}^{k-1} |b_i|^2$, maximization of P_k with respect to the free parameters $r_{n+1}, r_{n+2}, \ldots$ leads to $\partial P_k / \partial r_m = 0$, $|m| > n$ and this gives [6, 7]

$$H_k(z) \triangleq \frac{\alpha_0 + \alpha_1 z + \cdots + \alpha_{k-1} z^{k-1}}{g_0 + g_1 z + g_2 z^2 + \cdots + g_n z^n} \sim \text{ARMA}(n, k-1), \tag{27}$$

i.e., given r_0, r_1, ..., r_n, the transfer function that maximizes the k-step minimum mean-square prediction error is a stable ARMA($n, k-1$).

References

[1] N. Wiener and P. Masani, "The prediction theory of multivariate processes, part I: the regularity condition," *Acta Math.*, vol. 98, 1957.

[2] D. C. Youla, "The FEE: A new tunable high-resolution spectral estimator," Part I, Technical note, no. 3, Department of Electrical Engineering, Polytechnic Institute of New York, Brooklyn, New York, 1980: also RADC Rep. RADC-TR-81-397, AD A114996, February 1982.

[3] P. Dienes, *The Taylor Series*, New York: Dover Publications, 1957.

[4] J. P. Burg, *Maximum Entropy Spectral Analysis*, Ph.D. dissertation, Stanford University, Stanford, CA, May 1975.

[5] U. Grenander and G. Szegö, *Toeplitz Forms and Their Applications*, New York: Chelsea, 1984.

[6] S. U. Pillai, T. I. Shim, and H. Benteftifa, "A new spectrum extension method that maximizes the multistep minimum prediction error — Generalization of the maximum entropy concept," *IEEE Trans. Signal Processing*, vol. 40, no. 1, pp. 142–158, January 1992.

[7] T. I. Shim, *New Techniques for ARMA-System Identification and Rational Approximation*, Ph.D. Dissertation, Polytechnic University, Brooklyn, New York, June 1992.

MAKING SENSE OF QUANTUM MECHANICS: WHY YOU SHOULD BELIEVE IN HIDDEN VARIABLES

Anthony J.M. Garrett
Inference Consultants
Byron's Lodge, 63 High Street, Grantchester
Cambridge CB3 9NF, England

ABSTRACT. The nondeterministic character of quantum measurement can, and should, be taken to imply a deeper 'hidden variable' description of a system, which reproduces quantum theory when the unknown values of the variables are marginalised over. Differences in measurements on identically prepared systems then represent differences in the hidden variables of the systems. Not to seek the hidden variables, as the Copenhagen interpretation of quantum mechanics arbitrarily instructs, is to give up all hope of improvement in advance, and is contrary to the purposes of science. Correspondingly, it can never be proven that hidden variables don't exist; the most that can be done is to place hypothetical restrictions on them and prove that the resulting theory cannot be marginalised into quantum theory – implying that the hidden variables violate the restriction. The best known example is Bell's theorem, which rules out all local hidden-variable theories and explains why the variables are so coy: it is not easy to isolate and influence something nonlocal. A promising way forward is to proceed from the Dirac theory of the electron in its natural, geometric language of Clifford algebra. This approach, due to Hestenes, suggests that the free electron is in fact a massless point particle executing zitterbewegung at the speed of light, with radius the electron Compton wavelength. This proposal improves the correspondence of the physics with the formalism and resolves many paradoxes; it also hints that the hidden variables are concerned with those parameters needed to define the electron orbit completely. These ideas all represent a demystification of quantum theory.

In the 19th century, microscopes advanced to the point at which a new phenomenon, today called Brownian motion after its chronicler, could be seen. The small particles, newly made visible ('dust' particles), were observed to jiggle around in a manner the scientists could not predict, for no apparent reason. Motion of any individual dust particle could not be predicted; however, statistical details of the motion, such as the average (drift) velocity or the mean square velocity, were reproducible.

The scientists of the day, probably without realising it, made an assumption: that there was a reason for the jiggling, and that if they could find this, they could predict the precise motion of individual dust particles. They never said "Maybe the dust particles just behave like that and there isn't a why; maybe we can go no further and this is the end of the line".

A. Mohammad-Djafari and G. Demoments (eds.), Maximum Entropy and Bayesian Methods, 79–83.

As we know, their persistence paid off: they eventually found that the motion was due to particles, too small to be seen even by their microscopes, continually bombarding the specks of dust and jolting them. These smaller particles are today called atoms (or molecules, a few atoms joined together); the atmosphere, in which the dust particles move, primarily consists of molecules of nitrogen and oxygen. This bombardment is undergone by all objects surrounded by a liquid or a gas, even a ball bearing resting on a smooth table (though gravity keeps the motion two-dimensional, on the table). We cannot see the jiggling of the ball bearing simply because it is too massive: a dust particle is light enough to be thrown around appreciably by the bombardment, but a ball bearing moves only imperceptibly. (On such large objects, we call the bombardment *air pressure.*)

Surely we would count this discovery as an advance. An underlying reason was postulated, searched for, and found, with the result that precise prediction could (in principle) be made of the motion of individual dust particles. Improvement is made formal in the (Bayesian) theory of comparative hypothesis testing.

In the 1920s, much the same thing happened again, at a deeper level. The new theory of quantum mechanics – one of the great creations of the human intellect – was worked out which successfully predicted many things beyond the old, Newtonian theory. It could, however, predict only the statistics of experimental measurements on observable quantities and not (with certainty) the outcome of any one measurement. This time (perhaps because of a loss of confidence within the civilisations in which science reached maturity), it came to be generally accepted that this was the end of the line; that beyond this point, prediction could not be improved. To ask what was the underlying reason, and why apparently identical systems behave differently when measurements are made, became taboo. This prevailing view is the centrepiece of the Copenhagen interpretation of quantum mechanics, named after a conference there. No consensus exists on the precise meaning of 'Copenhagen interpretation', but underlying explanations are rejected by all its adherents.

If this is so, then it is indeed the end of the line. But we can never be certain that this is so, for failure to find any underlying reason may reflect only our technological incapability, not the systematic absence of a reason; we cannot distinguish. By contrast, success in finding a reason obviously confirms its presence. But, if you accept the positivistic *non sequitur* of Copenhagen – I can't predict it, so it can't be predicted – then you will never go out and see if you can do any better. You are stuck; for how do you know you can't find something if you won't look?

Again: if you do as the 19th century physicists did, you will do your utmost – experimentally and theoretically – to do better; to arrange and predict the outcomes of individual experiments. Please note that I am not making the unprovable claim that we can 'know everything'; merely that if we begin by assuming we can't, then we surely won't. Invariably, systematic seekers stand the best chance of finding. If you succeed, you will have made a predictive advance of the kind science exists to make. So the Copenhagen interpretation, which argues against this, is, literally, anti-scientific; and in fact a startling amount of spurious material has been written recently concerning mysticism and quantum mechanics.

The 'many worlds' interpretation, in which every conceivable outcome of the measurement is held actually to arise, in a multiplicity of mutually unobservable universes branching from the point of measurement, is an attempt to to explain *why* observers could never do any better. If it were true, observers would indeed be unable to improve prediction. But the branching premise is untestable; we can never know whether it is true. Consequently it simply defers the arbitrariness of the Copenhagen veto one step back.

As a further example of Copenhagen logic, suppose two aeroplanes are built to identical specifications. One suffers engine failure; the other doesn't. Copenhagen asserts that there is no meaning in asking why this should be, and no point in trying to develop better diagnostics which could look more closely at the aeroplanes beforehand and try to distinguish differences which lead one of them to fail. Pilots would not be content with this situation; nor should physicists.

It should be emphasised that in all these cases the logic is precisely the same; the results, in yes/no binary form, can be passed on by an intermediary who does not state whether they concern aeroplanes crashing (or not) or electrons in either spin configuration. The context should be irrelevant to the logical reasoning. Nor can one escape by asserting that the planes encountered differing environments, flying through different airspace at different times; for the same applies equally to the electrons.

Invoking 'quantum logic' for the case of the electron does not resolve the problem, for our minds reason by ordinary, binary true-false logic, so that any explanation must, by definition of 'explanation' (to a human mind, at least) use true-false logic. I know of no explanation of what it means to say that an aeroplane both crashes and lands safely; that a cat is both alive and dead; or that, equally, an electron spin state is both up and down.

A further analogy is with the Cardinals who, it is said, refused to look through Galileo's telescope because their ideology told them that moons of Jupiter could not exist. Those moons would never have been found by people holding such views. Equally (had it been the case) the absence of any moon brighter than the sensitivity threshold of the telescope would never have been confirmed; and, today, there is no systematic experimental program to find hidden variables or to rule out specific types of them.

There is no relief from the problem in going from quantum mechanics to quantum field theory, which is conceptually just a reformulation to allow for variable numbers of particles: the cause of the trouble is the non-commutation of operators in the quantum formalism, and this persists.

What is the next level, knowledge of which would enable us to make precise predictions of quantum measurements – just as knowledge of atoms specifies the Brownian motion – and would restore a one-to-one correspondence between the mathematical formalism and observable quantities? This we do not yet know: the relevant variables are as yet hidden from us, just as atoms were hidden 100 years ago, or parts of aeroplane engines are hidden from someone without a tool kit. The variables are called – temporarily, we hope – hidden variables, and it is differences in the hidden variables which cause apparently identical systems having identical quantum descriptions – each set up according to an identical procedure, which obliterates previous quantum descriptions – to behave differently under subsequent measurement. It is not that we cannot see the effects of the 'hidden' variables; rather that we cannot influence them. There have been attempts, starting with John von Neumann, to prove by logic that hidden variables cannot reproduce the predictions of quantum theory when statistically averaged ('marginalised') over, but these have always been found to contain errors. Logically, such attempts are doomed: a hidden-variable theory can always be taken to underlie any probability distribution, and can conflict only with intuitive notions already violated by quantum mechanics.

To see which of these intuitive notions are mistaken, we use them to place restrictions on the hidden-variable theory and see whether quantum theory still emerges under statistical averaging. In a famous theorem, John Bell [1] proved that no *local* hidden-variable theory could reproduce the (experimentally confirmed) quantum predictions of experiments on spinning particle pairs. A local theory is one which does not admit a physical connection between spatially separated places. Since only a single example of nonlocality is needed in order to confirm it in general, it follows that the hidden variables, which we should as scientists seek, are nonlocal. It is therefore remarkable that they lead to *any* coherent prediction! It also partly explains why we have not yet been successful in influencing ('finding') them: the traditional strategy of science is to study something by isolating it, and this is impossible by definition with something nonlocal. Science has in fact faced this issue before, with the law of gravity 300 years ago, when the issue was called action-at-a-distance, and 100 years ago with electromagnetism. It is reasonable to suppose that we can do the same again, with just as great an increase in our understanding.

In Bell's theorem, measurement of the spin of one particle gives information about its hidden variable, which is assumed internal to it (locality), and, through the particle correlation, tells us about the hidden variable of the second particle; just as learning the speed of a bullet tells us the recoil of the gun. But observed measurements on the spin of the second particle cannot be accounted for in this way. Measuring the spin of the first particle therefore actually alters the second, no matter how distant, and the hidden variables are nonlocal. (A tutorial exposition is given by the present author [2].) More surprising than nonlocality is acausality: the connections between space and time which become apparent close to the speed of light, described by Einstein's theory of special relativity, imply that an observer passing rapidly enough through the laboratory could see the measurements on the two particles take place in the reverse order from the laboratory experimenter. So it is not possible to state which particle signals to which. Changes in the hidden variables are dictated not only by the past, but by the future, and in fact an interpretation of quantum mechanics based on this idea has proved helpful in several contexts [3]. This is less surprising once it is seen that acausality is the natural equivalent, with respect to time, of nonlocality with respect to space, and that space and time are connected relativistically. For an observer moving at an intermediate speed, the measurements take place at the same instant; this does not imply a faster-than-light connection, since the signal might propagate subluminally forward in time from one electron to a third body, and backwards to the second. It is required only that the laws governing the hidden variables satisfy the relativity principle of invariance in every inertial frame. We hope that a hidden-variable theory would unify special and general (gravitational) relativity with quantum principles more readily than today's investigations, whose extreme abstraction is, by all precedent, a flag that something more is needed.

Since paradoxes are associated with directly observable acausality ("I sent a message back to my parents never to meet each other"), it is possible that probabilistic prediction provides the shield from the paradox, so that, wherever acausality follows upon comparing the outputs of both apparati, quantum predictions are the best possible. Even then, a hidden-variable theory might generate predictions of those individual events which do not demand an acausal explanation – giving it the potential to beat quantum mechanics in the arena of comparative hypothesis testing.

Whether hidden variables exist is a problem distinct from that of 'wavefunction collapse': the reconciliation of a quantum treatment of the system-and-measuring-device with the quantum treatment of the system by a classically treated measuring device (an 'observer'). This clearly depends on what is meant by 'classical', and a satisfactory criterion (based on the density of the spectrum of quantum states) has reconciled the two descriptions in a simple, but representative, model problem [4]. The hidden variables pertaining to the system (observation guarantees that they can be so labelled, even though nonlocal) would specify the result of the measurement to the classically treated observer. At present we can state only that the probability of observing any one eigenvalue of the operator representing the measured observable is the square modulus of the contraction of the system wavefunction with the corresponding eigenstate.

I propose that a promising lead already exists towards hidden variables. If you do not find plausible what I am about to say, please do not be put off the parts you have already read; the arguments as to why hidden variables should be sought, and what they might be, are quite unrelated. Over the past three decades, David Hestenes, of Arizona State University, has reformulated Dirac's special-relativistic quantum theory of the electron – of which ordinary, non-relativistic quantum mechanics is a limiting case – in the natural, geometrical language of Clifford algebra [5]. At this point there is no change in the physical content of the theory, but its features are revealed far more clearly; many have not been recognised before. This is one of the few modern quantum ideas of which Dirac himself latterly approved. In particular, there is no mysterious factor of $i \equiv \sqrt{-1}$: everything is real and interpretable. The resulting picture strongly suggests that the electron is a point particle of zero rest mass and zero intrinsic spin executing gyro-rotation, due to interaction with its own electromagnetic field, at the speed of light ('zitterbewegung'). The radius of gyration is the electron Compton wavelength; complete details of the trajectory would involve the hidden variables. Quantum effects, in particular the Pauli exclusion principle (which is central to nonlocality) are explained as zitterbewegung resonance phenomena. The new 'moving' picture of the electron overcomes old problems (the 4/3 factor) associated with taking all the mass as electromagnetic in character, and explains immediately the (leading order) value of the magnetic moment. This natural corollary of the Dirac theory is not inconsistent with any experiment to date and is testable in relatively inexpensive quantum optics and scattering experiments which work close to the values of the zitterbewegung parameters. Further, the interpretation placed on this relativistic theory generates a corresponding interpretation of non-relativistic quantum theory. It may also suggest a consistent extension of the Dirac theory to two or more charged particles without encountering the infinities of field theory; it is hardly coincidental that there is no classical relativistic Hamiltonian for two or more charged particles. I regard the development and the testing of these ideas as the most important task in fundamental physics.

REFERENCES

1. Bell, J.S.: 1964. 'On the Einstein Podolsky Rosen Paradox', *Physics*, **1**, 195-200.
2. Garrett, A.J.M.: 1990. 'Bell's Theorem and Bayes' Theorem'. *Foundations of Physics*, **20**, 1475-1512 & 1991, **21**, 753-755 (correction).
3. Cramer, J.G.: 1986. 'The Transactional Interpretation of Quantum Mechanics', *Reviews of Modern Physics*, **58**, 647-687.
4. van Kampen, N.G.: 1988. 'Ten Theorems About Quantum Mechanical Measurements', *Physica A*, **153**, 97-113.
5. Hestenes, D.: 1990. 'The Zitterbewegung Interpretation of Quantum Mechanics'. *Foundations of Physics*, **20**, 1213-1232; and references therein.

BAYESIAN METHODS AND COURT DECISION-MAKING

G.A. Vignaux and Bernard Robertson
Victoria University
PO Box 600
Wellington, New Zealand

ABSTRACT. There has been vigorous debate in the legal literature about whether the axioms of probability apply to decisions on facts in legal cases. It is seriously considered that *legal probability* is different from *mathematical probability*. Analysis of these objections reveals that they are actually objections to frequentist statistics. A grasp of Probability as Logic solves these problems.

> "more and more quickly, inference problems in all sorts of disciplines are being brought within the purview of Bayesian/MaxEnt analysis."
> (Skilling, 1989)

1. The legal debate

As a preliminary to making a decision, courts have to 'find facts' which requires them to reason under uncertainty. In some cases the reasoning process itself is examined in an appeal. The result may be a statement by the court about how facts ought to be thought about. Alternatively the way facts are thought about in a particular case may be seized upon as a precedent for future cases. For example, courts have argued about whether inferences can be drawn 'beyond reasonable doubt' from evidence which is itself not proved beyond reasonable doubt (Robertson and Vignaux, 1991a). A logical approach to probability offers an obvious standard by which to judge such precedents. The interest in probability in the legal literature however, does not lie in puzzling out the contribution that probability theory has to make. Rather it lies in a vigorous debate about whether probability theory is appropriate to court decisions at all.

Writers have denied that the axioms of probability apply in court cases, or that court cases ought not to be thought about in this way even if they do apply (Tribe, 1971). An alternative line is to argue that some special kind of probability applies in legal cases, with its own axioms and rules. An entire book has been written to argue this with the result that conventional probability has become known in the jurisprudential world as Pascalian. (Cohen, 1977). This has all been at a theoretical level and this paper will consider some of the arguments in detail below. But in practice one commonly finds statements such as:

> "The concept of 'probability' in the legal sense is certainly different from the mathematical concept; indeed, it is rare to find a situation in which these two usages co-exist, although when they do, the mathematical probability has to be taken into the assessment of probability in the legal sense and given its appropriate weight" (Case: *Re J.S. (a minor)* p 1066)

A. Mohammad-Djafari and G. Demoments (eds.), Maximum Entropy and Bayesian Methods, 85–92.

In the Common Law system a comment made by a superior court is regarded, subject to certain conditions, as authority for that proposition. Thus if one wishes to describe the law, as opposed to discuss what the law should be, it may be sufficient to quote such a case. Although such a statement is technically only binding on the lower courts in the same legal system, the Common Law world is used to citing cases from other jurisdictions, regarding them as being of persuasive rather than binding authority. The fact that this statement is made by the English Court of Appeal therefore leads to it being cited in textbooks as authority that legal and mathematical probability are different.

2. Objections to probability in the legal literature

Many objections to probabilistic analysis of legal decision making stem from an insufficient understanding of probability by legal commentators. These have been extensively dealt with elsewhere by a number of authors. Many practical problems in court stem from the evidence of scientific witnesses who fail to acknowledge the logical nature of probability (Robertson and Vignaux, 1992). Some of the more cogent objections however are both deserving of serious consideration and revealing as to the commentators' underlying assumptions.

2.1. The Legal Literature

These objections are shown below together with illustrative quotations from the legal literature.

a. Things either happen or they don't; they don't probably happen.

"Propositions are true or false; they are not 'probable'."
(Jaffee, 1985 p 934)

b. Courts are concerned with single instances not long runs.

Thus, descriptively:

"Trials do not typically involve matters analogous to flipping coins. They involve unique events, and thus there is no relative frequency to measure." (Allen, 1991a p 376).

And normatively:

"Application of substantive legal principles relies on, and due process considerations require, that triers must make individualistic judgments about how they think a particular event (or series of events) occurred." (Bergman and Moore, 1991 p 591).

c. Frequency approaches hide relevant considerations.

Hypothetical situations involving readily quantifiable evidence are often constructed to illustrate the application of probability theory. These are frequently met with the claim that there are numerous factors relevant to the particular case which either have not been or cannot be expressed in probabilistic terms. For an extended example of this argument see Ligertwood (1990 p 14).

d. The legal system is not supposed to be subjective

Allen (1991a p 379) refers to: "the desire to have disputes settled by reference to reality rather than the subjective state of mind of the decision maker".

e. Evidence must be interpreted

"The implicit conception [in the probability debate] of 'evidence' is that which is plopped down on the factfinder at trial. ...the evidence must bear its own inferences...each bit of evidence manifests explicitly its characteristics. This assumption is false. Evidence takes on meaning for trials only through the process of being considered by a human being. ...the underlying experiences of each deliberator become part of the process, yet the probability debates proceed as though this were not so." (Allen, 1991b)

f. People actually compare hypotheses.

"Meaning is assigned to trial evidence through the incorporation of that evidence into one or more plausible stories which describe 'what happened' during events testified to at trial.I The level of acceptance will be determined by the coverage, coherence and uniqueness of the 'best' story"
(Pennington and Hastie, 1991 p 529).

g. Assessment of a prior probability "appears to fly in the face of the presumption of innocence" (Ligertwood, 1990 p 14)

2.2. Corresponding Bayesian Arguments

Many of these comments are sound in themselves but are not objections to a Bayesian or logical approach to probability. To illustrate this it is possible to quote parallel comments in the scientific literature made by writers who are advocating a probabilistic approach to fact finding but reject the frequentist approach to probability.

a. Things either happen or they don't; they don't probably happen.

"Nature does not prepare distributions, . . .she prepares states."
(Jaynes, 1990 p10).

The problem of course is that we have to make decisions on the basis of necessarily imperfect information. We are therefore concerned to measure our state of knowledge and belief rather than to measure any quality of the real world. That Jaffee could believe that the matters which statistics measure 'probably happen' indicates how powerful is the Mind Projection Fallacy.

b. Courts are concerned with single instances not long runs.

"The gamma-ray astronomer does not want to know how an observation of a gamma ray burst would compare with thousands of other observations of that burst; the burst is a unique event which can be observed only once, and the astronomer wants to know what confidence should be placed in conclusions drawn from the one data set that actually exists."
(Loredo, 1990).

In the real world, as in the court-room, decisions constantly have to be made about unique events. Probability viewed as an extension of logic tells us how to do this.

c. Frequency approaches hide relevant considerations.

"One who views human diseases or machine failures as 'stochastic processes' as described in some orthodox textbooks, would be led thereby to think that in gathering statistics about them he is measuring the one controlling factor; the physically real 'propensity' of a person to get a disease or a machine to fail; and that is the end of it."(Jaynes, Draft)

Bayesian analysis does not cause the thinker to ignore information which it is difficult to quantify, indeed it directs the mind to all the relevant issues.

d. Subjectivity?

This objection in the legal literature arises from the saddling of Bayesian probability with the label 'subjective'. Examination reveals that the labels 'objective' and 'subjective' are not useful.

"Operational difficulties with frequentist theory clearly indicate that it is as subjective as Bayesian probability theory, and in some contexts even more subjective.... In many problems, [the notion of randomness] is substantially more subjective [than Bayesian probability], since the identification of random variables and their probability assignments can depend in disturbing ways on the thoughts of the experimenter" (Loredo, 1990 pp 92-3).

All probabilities are subjective in the sense that they are assessments made by unique individuals with unique combinations of knowledge and experience. It is to be hoped however that assessments are objective in two senses: first, they are arrived at by universally acknowledged rules and second, all individuals, given identical relevant information, should come to the same conclusions.

e. Evidence must be interpreted

"one of the main practical messages of this work is the great effect of prior information on the conclusions that one should draw from a given set of facts. Currently much discussed issues such as environmental hazards or the toxicity of a food additive, cannot be judged rationally if one looks only at the current data and ignores the prior information that scientists have about the mechanism at work" Jaynes (Draft, p v). In court as elsewhere, the data cannot 'speak for itself'. It has to be interpreted in the light of the competing hypotheses put forward and against a background of knowledge and experience about the world. Bayesian Probability provides a mechanism for doing this.

f. People actually compare hypotheses.

"There is not the slightest use in rejecting any hypothesis H_0 unless we can do it in favour of some definite alternative H_1 which better fits the facts. Bayes theorem tells us much more than this: *Unless the observed facts are absolutely impossible on hypothesis* H_0, *it is meaningless to ask how much those facts tend 'in themselves' to confirm or refute* H_0. Not only the mathematics but also our innate common sense ... tells us that we have not asked any definite, well-posed question until we specify the possible alternatives to H_0." Jaynes (Draft p 104-105). Discussion of

Bayesian Analysis in both legal and scientific literature is usually in terms of prior odds and likelihood ratios. In reality, the task is to compare hypotheses which are not exhaustive, in which case one is talking about relative probabilites rather than about odds. A court often has to compare multiple hypotheses and debate has arisen in the legal literature as to whether the task of the court is to compare the case of the prosecutor/plaintiff with all possible alternatives (i.e. determine posterior odds) or to compare the two stories told by the two parties. Objectors to probability analysis such as Cohen and doubters such as Allen say that the latter is the court's role and that this does not conform to the requirements of probability theory.

It is of course impossible to arrive at true posterior odds since it is impossible to consider the infinite number of alternative hypotheses available. Furthermore the law recognises this by having rules which cast an evidential burden for certain issues upon the defence in a criminal case. If the defence wish to have matters such as self-defence, automatism, or duress considered there must be some evidence introduced which makes them worth considering. Defence counsel cannot just say to the jury in closing, "It might have been self-defence, have you considered that?" Many hypotheses are thus excluded from consideration; the alternatives are clearly not exhaustive.

On this analysis the aspiration of the legal system is to approach an assessment of odds. The means by which this is done in the vast majority of cases is to consider the two parties' respective hypotheses. Jaynes shows that a good approximation of the probability of an hypothesis can usually be attained by comparing it with the next most likely hypothesis (Jaynes, Draft). On the assumption that the two most likely hypotheses in a legal case are those advanced by the respective parties, this is what the legal system does. The cases which cause problems in the legal system are precisely the cases in which this principle would not hold true in scientific inference either, namely where there are three or more roughly equally plausible hypotheses or where a party's hypothesis is extremely vague owing to lack of information.

Consideration of evidence is not merely a process of applying likelihood ratios to the prior probability ratios of fixed hypotheses. As each piece of evidence is considered in turn it may lead to the formation of a new and more refined hypotheses which will become the alternative hypotheses against which the next piece of evidence is evaluated. An easy example is provided by race and DNA analysis. If good evidence is given that an offence was committed by someone of Vietnamese appearance we now have 3 hypotheses: the perpetrator was the accused (who is Vietnamese), the perpetrator was some other Vietnamese, and the perpetrator was of some other race. The third hypothesis, provisionally at least, drops out of consideration. The DNA evidence will then be considered in the light of the hypotheses that it was either the accused or some other Vietnamese. This analysis solves a number of the doubts felt about probabilistic analysis. In particular, the problem of the very large number of interdependencies between items of evidence falls away. This is first because many possible hypotheses are not actively considered and second because, rather than discuss the dependence of race and DNA, we discuss the probability of a particular analysis given that a person is of a particular race. This is much easier to grasp.

g. Assessment of a prior conflicts with the presumption of innocence

> "Any probability P(A|X) that is conditional on X alone is called a prior probability. But we caution that the term prior is another of those terms from the distant past that can be inappropriate and misleading today. In the first place, it does not necessarily mean 'earlier in time'. ... The distinction is a purely logical one; any additional information beyond the immediate data D of the current problem is by definition 'prior information'." Jaynes (Draft p 68).

> In fact, legal text-books agree that the presumption of innocence is a restatement of the burden and standard of proof and therefore something which matters at the end of a trial not at the beginning. This alleged objection is simply a matter of semantic confusion. For a full discussion of the meaning of the 'presumption of innocence' and a demonstration that it corresponds with a logical approach to probability see Robertson and Vignaux (1991b).

3. Different kinds of probability?

It is common to find references to different kinds of probability in both legal and traditional statistical literature. Thus William Twining (1980) and Glanville Williams (1979) identify classical and empirical probabilities as separate concepts. Dawid (1991) divides these further. Unhelpfully these are frequently known by different names and and so it is necessary to define them.

Classical probabilities refer to cases where there are a finite number of possible outcomes each of which is assumed to be equally probable. Empirical probability refers to some observation that has been carried out that in series Y event X occurs in a certain proportion of cases. Subjective probability refers to a judgment as to the chances of some event occurring based upon evidence. Unfortunately Twining treats any judgment a person might choose to express in terms of 'probability' as a 'subjective probability'. This leads him to say that subjective probabilities 'may or may not be Pascalian'.

L Jonathan Cohen (1977) invented "inductive" or "Baconian" probabilities, claiming that they represent common sense reasoning and scientific method. Unlike "Pascalian" probabilities these do not conform to the traditional axioms (although Twining has fudged this by his conflation of different concepts under subjective probabilities). They obey their own set of axioms which correspond essentially to the fallacies detected in untutored thinking. Cohen justifies the introduction of a new system of probability by analogy with the different systems of geometry. "[No one] would suppose today that the only kind of geometry we can have is a Euclidean one, even if Euclid still monopolises the elementary text-books".

The idea of different kinds of probability is both unhelpful and based upon false models and analogies. It is unhelpful in that the description of different kinds of probability is an open invitation to invent a new kind – and never mind that it happens to be radically different in not conforming to the axioms. It is based on false analogies because the belief that there are different kinds of probability governing different events stems from the Mind Projection Fallacy (Jaynes, 1989). This is the error into which Cohen falls with his analogy with different kinds of geometry. Different types of geometry are required to measure objects on planes on the one hand or curved surfaces such as the surface of the globe on the other. Probability, on the other hand, relates to thought processes. In order to argue for a different type of probability one must show why facts ought to be thought about in a distinctive way in legal discourse. Cohen does not do so; in fact he argues for a distinction between 'quantifiable' (i.e. some scientific) and 'non-quantifiable' evidence and claims that his system of axioms represents 'common sense reasoning'. The distinctive features of legal proceedings relate to the way decisions are to be made once probabilities have been assessed. There seems to be no good reason why facts should be thought about differently.

If I am a football captain considering the probability a coin will land heads, I will only reply 0.5 (classical probability) if I have no information relating to the past performance of the referee (empirical probability) or the way the referee intends to toss the coin this time (subjective probability). If I have such information I will take it into account in assessing the probability and yet according to the above scheme each of these pieces of evidence relates to a different kind of probability.

Clearly Bayesian probability encompasses all these considerations. The most helpful view is that there is one and only one kind of probability, the measure of the strength of belief in a proposition. Such belief must be based upon evidence and there are several different *kinds of evidence* on which one might rationally draw in coming to one's judgment.

4. Legal probability?

In the legal literature discussing the use of probability there is little consideration of what the requirements of a rational system for thinking about facts are. It is never explained why the axioms of probability represent desirable ways of thinking, they are simply stated and examples of their operation are given. Examples are invariably in terms of coin tossing or bags of coloured beads. This leads to the obvious riposte: "In actual trials, unfortunately, few cases involve sacks containing known quantities of different colored marbles" (Bergman and Moore, 1991).

Our experience is that when the basic desiderata (Jaynes, Draft; Loredo, 1990) are put to lawyers there is no dissent. The only exception is that evidence may be excluded for ideological reasons. Thus illegally obtained evidence may be excluded; evidence may be 'privileged' in order to protect certain relationships such as lawyer/client or husband/wife; and evidence may be 'immune' to protect the security of the state. These exclusions need cause no problems for our analysis provided two conditions apply. First it must be explicitly recognised that the search for truth is being sacrificed in order to protect some other value. Secondly the rules determining the exclusion of such evidence must be reasonably clear and capable of consistent application. In this way the system can still maintain consistency by consistently excluding certain evidence. Problems will arise when a rule of exclusion is justified as an aid to accurate decision making. This is one of the justifications for the hearsay rule, for example, and it is clear from cases from a variety of jurisdictions that the judges are increasingly impatient with this claim.

Thus these objections are not objections to probability but to the frequentist definition of probability which the objectors have been exposed to. That this is so is revealed even in asides such as references to "fashionable . . .mathematical analysis" (Tapper, 1991). Once one is aware of the Mind Projection Fallacy and regards probability as a generalisation of logic (Smith and Erickson, 1989) these objections evaporate. The most perceptive comment in Twining's review of the debate (1980) is his closing remark that "the life of the lawyer of the future will need to include logic as well as statistics."

References

[1] Allen, R.J. (1991a) 'The Nature of Juridical Proof' *Cardozo Law Review* 13, 373.

[2] Allen, R.J. (1991b) , 'On the significance of batting averages and strikeout totals: a clarification of the "naked statistical evidence" debate, the meaning of "evidence", and the requirement of proof beyond reasonable doubt ' *Tulane Law Review*, 65, 1093

[3] Bergman, P. and Moore, A., (1991), 'Mistrial by likelihood ratio; Bayesian analysis meets the f–word' *Cardozo Law Review* , 13, 589

[4] Cohen, L.J., (1977) "The probable and the provable", Oxford University Press

[5] Cohen, L.J., (1980) 'The logic of proof', *Crim Law Review*, 91-107

[6] Dawid, P., (1991) 'Appendix on Probability and Proof' in Anderson, T. and Twining, W.(1991)

[7] Jaffee, L.R. (1985) 'Of Probativity and Probability : statistics, scientific evidence, and the calculus of chance', *University of Pittsburgh Law Review*, 46, 924-1082.

[8] Jaynes, E.T. (1989) 'Clearing up mysteries – the original goal' in Skilling, J. (ed) Maximum Entropy and Bayesian Methods, 1-27

[9] Jaynes E.T. (1990) 'Probability Theory as Logic' in Maximum Entropy and Bayesian Methods (1989) ed Fougere

[10] Jaynes, E. T. (Draft) Probability theory – the logic of science, (forthcoming)

[11] Lempert, R., (1986) 'The new evidence scholarship – analyzing the process of proof' *Boston University Law Review*, 66, n 3-4, 439-477

[12] Ligertwood, A.L.C., (1989) 'Australian Evidence', Butterworths, Sydney

[13] Loredo, T.J., (1990) 'From Laplace to Supernova SN1987A: Bayesian Inference in Astrophysics' 81-142 in Fougere, P.F. (ed) Maximum Entropy and Bayesian Methods.

[14] Pennington N and Hastie R.A. ,(1991) Cognitive Theory of Juror Decision Making: The Story Model. *Cardozo Law Review*, 13, 519 - 557

[15] Robertson, B.W.N. and Vignaux, G.A., (1991a), 'Inferring Beyond Reasonable Doubt' *Oxford Journal of Legal Studies*, 11, 431-438.

[16] Robertson, B. and Vignaux, G.A., (1991b), 'Extending the conversation about Bayes' *Cardozo Law Review*, 13, 629.

[17] Robertson, B. and Vignaux, G.A., (1992), 'Expert evidence : law, practice and probability' *Oxford Journal of Legal Studies*, 12, 392.

[18] Skilling, J (1989) Preface to Skilling, J. (ed) Maximum Entropy and Bayesian Methods, vii.

[19] Smith, C.R. and Erickson, G. (1989) 'From rationality and consistency to Bayesian probability' in Skilling, J. (ed) Maximum Entropy and Bayesian Methods, 29-44

[20] Tapper, C. F. H. (1991) book review *Law Quarterly Review*, 107, 347

[21] Thompson, W.C. and Schumann, E.L., (1987) 'Interpretation of statistical evidence in criminal trials / The prosecutor's fallacy and the defense attorney's fallacy' *Law and Human Behaviour*, 11, 167-187

[22] Tribe L., (1971), 'Trial by mathematics' *Harvard Law Review*, 84, 1329

[23] Twining, W. (1980) 'Debating Probabilities' *Liverpool Law Review*, 2, 51

[24] Williams, G., (1979) 'The mathematics of proof', *Crim Law Review*, 297-354

[25] case: *J S, a minor, Re* [1980] 1 All England Reports 1061 (Court of Appeal)

WIGMORE DIAGRAMS AND BAYES NETWORKS

G. A. Vignaux and Bernard Robertson
Victoria University
PO Box 600
Wellington, New Zealand

ABSTRACT. Formal techniqes for the analysis of decision problems such as Bayes networks and Influence Diagrams were anticipated by the famous lawyer John Henry Wigmore (1913) when he devised charts for examining the structure of evidence in legal cases. We examine the relationship between these independently developed tools.

1. Introduction

In a court case one party has to prove a charge (in criminal cases) or a claim (in civil cases) which will contain a number of elements. Thus in a murder case it will be necessary to prove (i) that the accused performed an act, (ii) which caused death (iii) intentionally (or with whatever mental state is required by the local law). In a civil case it may be necessary to prove (i) the defendant performed an act (ii) which breached an obligation and (iii) which caused loss or damage.

To prove these points the parties will have available a mass of evidence some of which will be relevant only to one element of the proceeding and some of which may be relevant to more than one. The sheer quantity of such evidence is a data handling problem on its own, far exceeding what is usually met in classical decision analysis problems.

Daily, ordinary people are having to make decisions of vital importance to the parties under these sorts of conditions. There is a view that the sheer complexity of these problems makes decision analysis inapplicable and that jurors actually make their decisions "holistically" (whatever this may mean). On the other hand "It is hard to imagine how we can imbibe the evidence we 'see' without performing some sort of mental analysis, which by definition seems to involve some sort of dissection" (Tillers, 1989 p1252).

The challenge for decision analysis therefore is to provide a systematic way of considering such complex problems. Such an approach should both assist fact finders in their reasoning and provide standards by which to judge their reasoning processes.

2. Wigmore Diagrams

This was realised as far back as 1913 by John Henry Wigmore. In that year he published *The Science* (later *The Principles) of Judicial Proof*. This proposed a "*novum organum*" for the examination of evidence in legal cases. This was an inference chart designed for lawyers preparing for trial. A case was to be broken down into propositions, each stating a single fact. A chart was then drawn showing the structure of the argument by which the matters to be proved were inferred from the evidence. In this way the mind could work

A. Mohammad-Djafari and G. Demoments (eds.), Maximum Entropy and Bayesian Methods, 93–98.

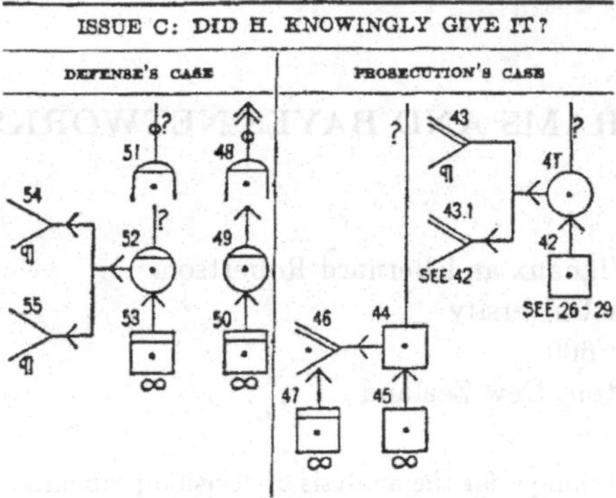

Figure 1: A Wigmore Diagram

its way through a mass of facts considering only a few matters at a time and building up inference hierarchically.

Although Wigmore is well known for other work in Evidence and Comparative Law his "*novum organum*" sank "like a lead balloon" (Twining, 1984). Its refloating is due to a small number of legal scholars, working independently in some cases, of whom Twining and Anderson have been the most active evangelists. In the last ten years Wigmore's methods have been developed, publicised and integrated into academic and professional legal training (Carter, 1988; Robertson, 1990; Anderson and Twining, 1991; Schum and Tillers, 1991.) The common aim is to develop Wigmore's proposal to make his method more useful to practising lawyers and investigators. This includes attempts to overcome the greatest weakness in the system – that an ultimate hypothesis must be formulated before analysis can begin. Attention is now being devoted to analysing the mess of facts confronting an investigator or lawyer who has to identify the appropriate hypotheses. (Schum and Tillers, 1991; Robertson, 1990).

3. Bayes Networks and Influence Diagrams

In recent years graphical methods have been developed that combine the fact analysis approach with Bayesian probability theory. These are Bayes Networks and Influence Diagrams (Oliver and Smith, 1990) These are quantitative in that they assist the analyst, not only in describing the problem and communicating information about structure but also in calculating the effect of the truth of one proposition or piece of evidence on the plausibility of others.

Bayes Networks set out, in an acyclic network of nodes and directed arcs, the propositions to be proved, the intermediate propositions that go to support that proof, and the evidence (in the form of propositions) on which it is based. Thus in Figure 2, nodes A, B, C, D, and E represent propositions. A is the hypothesis we seek to establish; B, C, and E are pieces of evidence and D is an intermediate proposition. Directed arcs between the nodes indicate conditional dependence (or, more precisely, the absence of an arc joining two nodes asserts conditional independence between them). Thus nodes B and C are conditionally dependent upon A but conditionally independent of each other.

Conditional probabilities are associated with each of the nodes, the conditions being the states of the nodes linked to it by incoming arcs. Prior probability distributions are

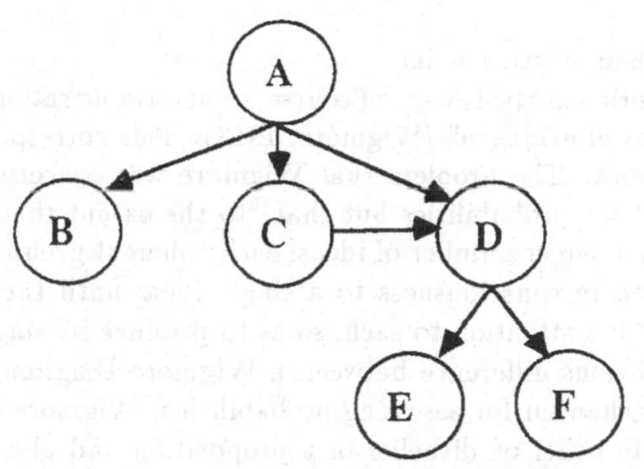

Figure 2: An Example of a Bayes Network

associated with source nodes (those with no entering arcs, such as A). Before the truth of evidential propositions is asserted the network represents the complete probability structure of the model. The effect of asserting the truth of a piece of evidence is discovered by carrying out standard Bayesian manipulations. These correspond to operations on the network together with the appropriate probabilistic calculations.

For example using Bayes Rule corresponds to reversing an arc in the network and generating any additional arcs required to carry over conditional dependence. Thus, suppose C represents a piece of evidence which is either true or not. Reversing the arc AC and stating that C is true corresponds to using Bayes Rule to find $P(A \mid C)$ from $P(A)$ and $P(C \mid A)$. Of course B, though conditionally independent of C, is not independent since knowing C tells us something about B through the effect of A. Similarly, marginalisation corresponds to eliminating a node as we might wish to do for node D once we have established the truth or otherwise of proposition E.

Even in the simplest cases, there may be dozens or hundreds of nodes. Computer programs are now available to carry out the calculations required by these complicated networks.

Influence Diagrams extend Bayes Networks to include decision nodes, reflecting choice and optimization, deterministic nodes to add straightforward calculations (of profit or loss, for example) and value nodes that measure utilities. Decision nodes are evaluated by maximising the expectation of following value nodes. These diagrams are used to represent and solve decision problems and are a useful improvement on Decision Trees where the latter become too complex. Since courts have to make decisions, influence diagrams may prove to be a valuable way of describing and assisting with their deliberations. There are many hurdles to be overcome before this is feasible, not the least of which is deciding what utility functions should be used. Edwards (1991) makes some valuable comments on this.

4. Wigmore and Decision Analysis

Wigmore developed his system working alone and years ahead of the publication of any similar methods in the scientific literature (Wright, 1921). Naturally, it seems very old-fashioned in style and modern developments have simplified the notation. Bearing in mind the antiquity of Wigmore Diagrams it is useful to consider their relationship to modern-day decision analysis, particularly as lawyers are increasingly considering fact analysis and taking Wigmore as their starting point.

The aim of Wigmorian analysis was "of course, to determine rationally the net persuasive effect of a mixed mass of evidence" (Wigmore, 1913). This corresponds precisely with the aim of a Bayes Network. The problem that Wigmore was concerned with, however, was not the calculation of the probabilities but that "to the extent that the mind is unable to juxtapose consciously a larger number of ideas, each coherent group of detailed constituent ideas must be reduced in consciousness to a single idea; until the mind can consciously juxtapose them with due attention to each, so as to produce its single final idea".

Thus the most obvious difference between a Wigmore Diagram and a Bayes Network is that there is no mechanism for assessing probabilities. Wigmore did provide a symbolic vocabulary to indicate belief or disbelief in a proposition and also to indicate whether a proposition supported or undermined the next higher proposition in the chain of inference. These symbols could show strength of belief in a crude way – one dot meant a proposition was believed, two dots meant it was strongly believed. He gave no instructions as to how to formulate these strengths of belief or how to combine them. Likewise propositions which affected one another's value as evidence could be linked by horizontal lines but how these influences were to be incorporated was not explained. Such interactions are handled using conditional probability arrays in modern Bayes Networks.

As a case is examined a Wigmore diagram tends to expand as more steps in the reasoning are teased out. This typically involves making explicit the generalisations on which steps in reasoning are based. These generalisations (as opposed to the facts asserted) may turn out to be the weakest points in an opponent's argument and one of the most valuable uses of the Wigmore diagram is to expose them. The tendency in drawing a Bayes network is to simplify as much as possible; the factors affecting probability assessments disappear into the nodes. Alternative explanations for the same event will usually appear as separate propositions in a Wigmore Diagram but in a single node in a Bayes Network.

Wigmore viewed his system as "only an attempt at a working method, which may suffice for lack of any other yet accessible". Other methods are now accessible. Wigmore's system is perfectly consistent with modern methods of decision analysis, but incomplete as he himself realised. He is an unrecognised founder of decision analysis.

5. Fact analysis and probability

In the New Evidence Scholarship, probability and Fact Analysis are treated as separate issues. Whether or how they are connected is often not made clear. For example Anderson and Twining (1991) is primarily about Fact Analysis but contains an appendix on probability with limited discussion of its relationship to the main theme of the book (Dawid, 1991).

If probability is a matter of logic then clearly Wigmorian analysis and Probabilistic analysis of a case are the same operation. To assess probabilities one must dissect the case into simple verifiable propositions and represent their logical relationships in some way. Conversely the decision to connect one symbol to another is a decision that the one is relevant to the other which can only mean that one has assessed the degree of association (that is, essentially, a likelihood ratio) and decided that it is high enough to compensate for the additional complication to the chart.

Consideration of more detailed matters also reveals a close connection between a Wig-

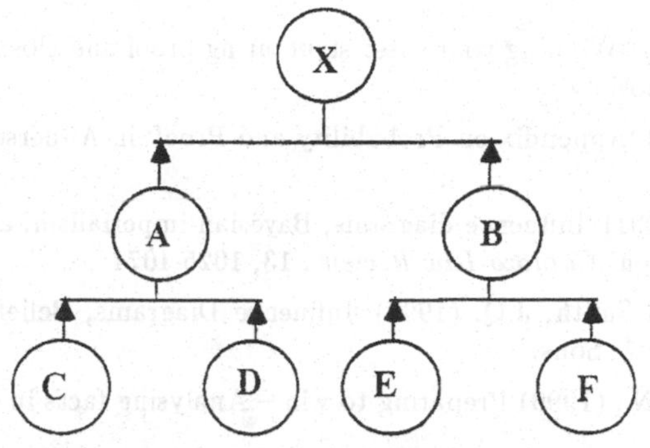

Figure 3: A Simple 2-level Wigmore Diagram

more Diagram and probabilistic analysis. The "extension of the conversation" is easily visualised in terms of a Wigmore diagram by a descending tree structure expanding at each level.

$$P(X) = \sum_{AB} P(X|AB)P(AB)$$

and, at the next level, $P(AB) = P(A)P(B|A) = P(A)P(B)$ (where we have assumed that A and B are conditionally independent),

$$P(A) = \sum_{CD} P(A|CD)P(CD).$$

Traditionally in a Wigmore Diagram evidence is regarded as supporting or undermining a particular hypothesis which is of interest. A problem in charting has always been (at least for us) how to represent alternative explanations for a particular event. The solution we have adopted is an expanded node which contains all the alternative hypotheses rather than having a separate node for each. Relevant evidence is then connected to this expanded node rather than to any particular hypothesis. Once this is done, one's point of view of the evidence is changed. Rather than seeing particular pieces of evidence as supporting particular hypotheses one now sees more clearly that the role of each item of evidence is to help in discriminating between the alternatives. If this is the case, one is immediately forced to considering likelihood ratios as the mechanism for assessing how well the evidence succeeds in discriminating between pairs of hypotheses. One is then forced to consider the probability of the evidence given each hypothesis which means that one is drawing arrows *downwards* from the hypothesis to the evidence rather than *upwards* as the arrows in a Wigmore diagram normally flow. The diagram more and more closely resembles a Bayes network.

Probability as logic provides the connection between fact analysis and probability which, until now, has not been noticed owing to the concentration on a statistical model of probability.

References

[1] Anderson, T. and Twining, W.,(1991) "Analysis of Evidence", Weidenfeld and Nicholson

[2] Carter, N., (1988) "Winning your case: structuring proof and closing evidentiary loopholes", NZ Law Society

[3] Dawid, P., (1991) 'Appendix on Probability and Proof' in Anderson, T. and Twining, W.

[4] Edwards, W., (1991) 'Influence diagrams, Bayesian imperialism, and the Collins case: an appeal to reason' *Cardozo Law Review* , 13, 1025-1074

[5] Oliver, R.M. and Smith, J.Q. (1990) 'Influence Diagrams, Belief Nets and Decision Analysis' J Wiley & Sons.

[6] Robertson, B.W.N., (1990) Preparing to win – Analysing facts in legal cases, NZ Law Society

[7] Schum, D. and Tillers, P., (1991) 'Marshalling evidence for adversary litigation', *Cardozo Law Review*, 13, 657-704

[8] Tillers, P., (1989) 'Webs of things in the mind: a new science of evidence' *Michigan Law Review*, 87, 1225

[9] Twining, W.,'Taking facts seriously', (1984) *Journal of Legal Education*, 34, 22.

[10] Wigmore, J.H., (1913) The Science of judicial proof, Little, Brown and Co

[11] Wright, S., (1921) 'Correlation and causation' *Journal of Agricultural Research*, 20, 557-585

AN INVARIANCE PRINCIPLE FOR UPDATING PROBABILITIES

Vicente Solana
Instituto de Matemáticas y Física Fundamental
National Research Council of Spain
Serrano 123, E-28006 Madrid, Spain

ABSTRACT. This paper considers the search for an inductive inference method based on the observed data of an uncertain quantity, using the same rationale as used in axiomatic derivations of fundamental probability rules. The basic inference case deals with the assignment of probabilities to a future observation of the random quantity given only a set of observed data and the order of these data in a reference set. A principle of invariance under different equivalent ways of updating is stated as a particular case of the strong invariance principle. A major consequence of the new updating principle is the derivation of the data based inference procedure from only this principle. It is concluded that there is one solution defining completely the mathematical form of the inference procedure, that it corresponds to a linear transformation of probabilities of the reference distribution on each elementary interval bounded on the observed data, and that it is identical to the solution obtained by the relative entropy version using fractile constraints.

1. INTRODUCTION

The modelling of an uncertain quantity on a domain based on observational data can be made using a plausible logical language (Polya 1954, Carnap 1950). In this framework, logical probabilities given as numerical values are assigned to the plausibilities, or degrees of certainty, of an inferred sentence on the given evidence of observed data and any evidential context. Probabilities for uncertain quantities are usually expressed as continous probability distribution functions on the given domain.

The fundamental rules of probability theory, i.e. the sum and the multiplicative rules of logical probabilities, were derived by Cox (1946, 1961), from only two axioms. Jaynes (1957, 1988) extended the derivation from a rationale that includes three desiderata.

However, apparently no data-based inference method has been derived from only the axioms and the rationale by Cox and Jaynes, or from any other axiomatic derivation of probability. The known available methods for data-based inference employ some rules or principles *ad hoc* that have not been justified precisely on the same rationale, such as the use of the likelihood function of observed data associated with the Bayes' theorem, or the use of a variational principle extremizing an entropy functional.

A main problem remains, therefore, in logical probability theory. The question is whether a data-based inference method can be derived from the same rationale as used

A. Mohammad-Djafari and G. Demoments (eds.), Maximum Entropy and Bayesian Methods, 99–112.

in logical probabilities by Cox and Jaynes. If such a method exist, it could be denominated *"The " data-based inference procedure* and should be preferred to any other inference method.

The paper presents a derivation of such an inference procedure. It is solelly a consequence of the invariance principle for probabilistic updates also stated in the paper, which forms part of the Jaynes-Cox rationale.

Updating, i.e. changing of probability assignments in view of additional observed data, is usually interpreted as the change from a prior distribution of an observable quantity encoding the initial data, to a posterior distribution; the change is made, for instance, in Bayesian methods through the distributions of non-observable parameters. This paper considers a more general updating process, in which the succesive updates are referred to a common probability distribution, namely the reference distribution.

Probabilistic inference methods used for uncertain quantity modelling should satisfy certain desiderata or requirements. Among other rules for inductive inference Jeffreys (1948) stated that *"a theory of inference must be consistent, i.e. it must not be possible to derive contradictory conclusions from the postulates and the data"*. An application of this rule determines the invariance of probabilistic assignments under regular transformations of uncertain quantities and parameters. This rule was used by Cox and was also proposed by Jaynes as one of the fundamental desiderata for inductive inference. Shore and Johnson (1980) and Skilling (1988) considered invariance principles derived from this rule in the axiomatic derivations of relative entropy methods using mathematical expectations as constraints. A new consistency principle of invariance under conditioning of the uncertain quantity domain has been stated (Solana 1990). A new version of relative entropy methods using probability fractile constraints has been developed (Lind and Solana 1990, Solana and Lind 1992). The relative entropy version with fractile constraints satisfies this invariance principle.

Jeffreys' consistency rule can be regarded as a general principle that prescribes fundamental requirements for consistency of any inference procedure. Such a principle has been restated as the following *strong invariance principle* (Solana 1990):

> *"Alternative ways of using the same observational data and assumptions in probability assignments should give the same result"*.

This principle is called *strong* to distinguish it from the *"weak" invariance principle* in which invariance of probability assignments is only required under regular transformations of the uncertain quantities and the non-observable parameters, as employed in the invariance theory by Jeffreys, which is applied to Bayesian and entropy methods.

2. THE INVARIANCE PRINCIPLE FOR PROBABILITY UPDATES

The search for a data-based inference procedure has been oriented in the past mainly to find an appropriate method for the simplest inference case, namely estimating the probabilities of a random quantity given only a set of observed data, without taking any possible order of the observed data into account. Main rules and principles *ad hoc* of inductive inference refer to this standard problem.

Usual frequentist methods, such as the maximum likelihood method and the method of moments, and data-based methods utilizing logical probabilities, such as the method that combines Bayes' formula with likelihood functions, were developed specifically for this

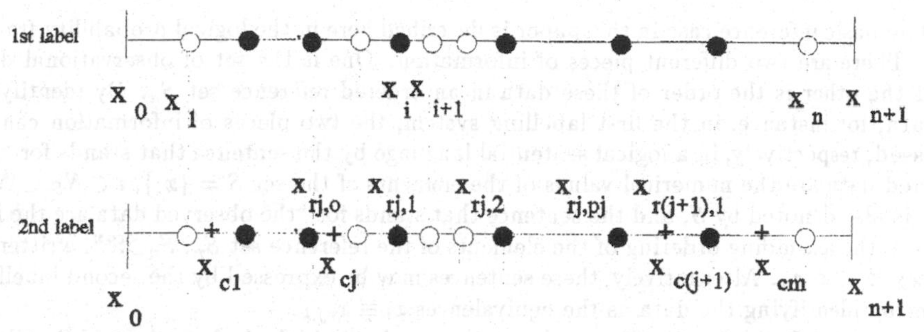

Fig. 1. Data Labelling.

standard problem of estimation. These methods have been extended since to other more complicated cases of inference, as when the probability estimates refer to ordered random variables or some related functions termed order statistics.

This paper, however, considers a different basic inference problem: The more general case in which the probabilities for a future observation of a random entity are assigned given only a set of observed data and the order of these data in an ordered reference set. Application of the strong invariance principle to this case leads to the invariance principle for probabilistic updates formulated in this Section, from which one data-based inference procedure is derived in Section 3.

NOTATION AND LABELLING OF DATA

Let X be a real-valued random quantity, defined on a given domain E, let S be a set of observed data of X and let S_n be any ordered reference set of n elements such that $S \subset S_n$. Let X be a future observation of X, and denote by x any possible value of X, $x \in E$. For the sake of simplicity, a simple domain $E = (x_o, x_{n+1})$ given by the extreme value data x_o and x_{n+1} is considered here.

In addition to the data, consider the partition of E by an arbitrary set, denoted S_c, of m possible values x_{cj} of X, indexed $cj = c1, c2, .., cm$, and arranged in one ascending order written as $x_{c1} < x_{c2} < ... < x_{cm}$. The domain E is partitioned by the set $\{x_{cj}\}$ into $m+1$ simple subdomains $E_j = (x_{cj}, x_{c(j+1)})$, $j = 0, 1, ..., m$ such that $E_o = (x_o, x_{c1})$ and $E_m = (x_{cm}, x_{n+1})$.

As the elements of S are arranged in ascending order two labelling systems are possible, Figure 1. First, the observations denoted as $S = \{x_i\}$ are simply labelled using the index i, such that $i \in N_S \subset N$, $N = 1, 2..., n$, in the same ascending order as the elements of the ordered reference set S_n written as $x_1 < x_2 < ... < x_n$. Second, the identically ordered observations denoted $S = \{x_{rj,k}\}$ are labelled using the double index (rj, k) by regarding the partition of E into the intervals E_j. Let the index pj, such that $pj = p0, p1, ..., pm$, $pj \in N$, denotes the number of data of S within the subdomain E_j. The index rj, such that $rj = r0, r1, ..., rm$, which is in general different from x_{cj}, expresses the interval E_j the observation belongs to; the index k such that $k = 1, 2, ..., pj$, $k \in N$, indicates the order of an observation in the subset of data lying on that interval. In this labelling the observations are arranged in ascending order written as $x_{rj,1} < x_{rj,2} < ... < x_{rj,pj} < x_{r(j+1),1}$, $j = 0, 1, 2, .., m$.

BASIC INFERENCE PROBLEM

The basic inference case in this paper is described here in the logical probability framework. There are two different pieces of information. One is the set of observational data S and the other is the order of these data in an ordered reference set S_n. By identifying the data, for instance, in the first labelling system, the two pieces of information can be expressed, respectively, in a logical sentential language by the sentence that stands for "the observed data are the numerical values of the elements of the set $S = \{x_i\}$, $i \in N_S \subset N$ ", which is also denoted by S, and the sentence that stands for "the observed data are the i-th values in the ascending ordering of the elements of the reference set S_n, $S_n \supset S$ ", written as $x_1 < x_2 < ... < x_n$. Alternatively, these sentences may be expressed by the second labelling system by identifying the data as the equivalences $x_i \equiv x_{rj,k}$.

Denote by D the joint evidence that corresponds to the logical product of the above pair of sentences. This joint evidence, is represented identically by any of the following two expressions:

$$D \equiv (S, x_1 < x_2 < ... < x_n) = (\{x_i\}, x_1 < x_2 < ... < x_n), i \in N_S \subset N; \quad (1)$$

$$D \equiv (S, x_1 < x_2 < ... < x_n) = (\{x_{rj,k}\}, x_1 < x_2 < ... < x_n), x_{rj,k} \equiv x_i, i \in N_S \subset N. \quad (2)$$

In these expressions, $\{x_i\}$ includes some of the i-th ordered values of the reference set S_n such that $i \in N_S \subset N$ and $\{x_{rj,k}\}$ includes all the corresponding (rj,k)-th values throughout the equivalences $x_{rj,k} \equiv x_i$. The logical product of sentences is represented here and in the following by the symbol "(,)", as used in (1) and (2).

In particular, the joint evidence D_n that refers to the observations of S when the reference ordering is the ascending order of the elements of the set S itself such that $S_n \equiv S$, is represented as the expression

$$D_n \equiv (S_n, x_1 < x_2 < ... < x_n) = (\{x_i\}, x_1 < x_2 < ... < x_n), i \in N. \quad (3)$$

The evidence of the ordered data of S lying in the interval E_j can be also given as the logical product of the joint evidence D and the sentence denoted by e_{xj}: $x_{rj} \leq \mathrm{X} \leq x_{r(j+1)}$, that stands for "the next observation X is contained in E_j", represented by the expression

$$(D, e_{xj}) \equiv (\{x_{rj,k}\}, x_1 < x_2 < ... < x_n, x_{rj} \leq \mathrm{X} \leq x_{r(j+1)}), x_{rj,k} \equiv x_i, i \in N_S \subset N. \quad (4)$$

The elements of S lying in the interval E_j can be also ordered with regard to the subset S_{nj} of the nj elements of the reference set S_n within this interval. Let (rj, o) be the index of the element before the first data of the reference subset S_{nj}. Then the elements $\{x_{rj,k}\}$ are the $[(rj,k) - (rj,o)]$-th values in the ascending order of the elements of S_{nj}. The joint evidence (4) may be represented as

$$(D, e_{xj}) \equiv (\{x_{rj,k}\}, x_{rj,o} < x_{(rj,o)+1} < x_{(rj,o)+2} < ... < x_{(rj,o)+nj}), \ k = 1, 2, ..., pj$$

$$= (\{x_{rj,k}\}, x_{(rj,1)-(rj,o)} < x_{(rj,2)-(rj,o)} < ... < x_{(rj,nj)-(rj,o)}), \ k = 1, 2, ..., pj. \quad (5)$$

ALTERNATIVE INFERENCE WAYS

Logical probabilities are always conditioned by the evidence. In the basic inference case of this Section, probabilities are the certainty degrees for the inferred sentence $X \leq x$ that stands for "the next observation X is less than or equal to x", on the given joint evidence D and the sentence describing the evidential context denoted by $e_x : x_o \leq X \leq x_{n+1}$, that stands for "the next observation X is contained into the domain E". They are represented by the conditional distribution function $Q_X(x \mid D, e_x) \equiv Prob(X \leq x \mid D, e_x)$, given by the following expressions

$$Q_X(x \mid D, e_x) = Prob[X \leq x \mid \{x_i\}, x_1 < x_2 < ... < x_n, x_o \leq X \leq x_{n+1}], i \in N_S \subset N$$

$$= Prob[X \leq x \mid \{x_{rj,k}\}, x_1 < x_2 < ... < x_n, x_o \leq X \leq x_{n+1}], x_{rj,k} \equiv x_i, i \in N_S \subset N. \quad (6)$$

In general, given a reference distribution $P_X(x) \equiv Prob(X \leq x \mid e)$, it is assumed that the inference procedure of assigning a conditional probability distribution on the basis of the given evidence D and an evidential context e, can be formulated as the operator "$g[.]$" such that

$$Q_X(x \mid D, e) = g[P_X(x \mid e), I(D, e)], \forall x \in E. \quad (7)$$

The existence of such an operator for evidential contexts e_x and e_{xj} is postulated in this Section. In (7) the expression $I(D, e)$ represents any information function encoding all the available evidence.

There are two different ways of applying this procedure to the basic inference case:

- (I) In the *first way*, the probability distribution $Q^I_X(x \mid D, e_x)$ is directly assigned on the basis of available information, on the whole domain E, by specializing the expression (7) to this case as

$$Q^I_X(x \mid D, e_x) = g[P_X(x \mid e_x), I(D, e_x)], \forall x \in E. \quad (8)$$

- (II) In the *second way*, the probability distribution $Q^{II}_X(x \mid D, e_x)$ is assigned regarding an arbitrary partition of E given by the set $S_c = \{x_{cj}\}$, by the following three steps:
- (II.1) First, the probabilities of the distribution function $Q^{II}_X(x \mid D, e_x)$ at the data values $x = x_{cj}$ are directly assigned as in way (I), using the expression (7) on the basis of D and the whole domain E, specialized to the values x_{cj} such that

$$Q^{II}_X(x_{cj} \mid D, e_x) =$$
$$= Q^I_X(x_{cj} \mid D, e_x) \equiv Prob[X \leq x_{cj} \mid D, e_x] = g[P_X(x), I(D, e_x)] \mid_{X=x_{cj}}, \forall j = 1, 2, ..., m. \quad (9)$$

- (II.2) Second, the conditional probability distribution $Q^{II}_X(x \mid D, e_{xj})$ is directly assigned on every subdomain E_j, by taking into account the data contained in the subdomain and the order of these data in the reference subset S_{nj}, given by the joint evidence represented by (4) and (5), and the conditional reference distribution $P_X(x \mid e_{xj}) \equiv Prob(X \leq x \mid x_{cj} \leq X \leq x_{c(j+1)})$ as the expressions

$$Q_X^{II}(x \mid D, e_{xj}) \equiv Prob[X \leq x \mid D, e_{xj}] = g[P_X(x \mid e_{xj}), I(D, e_{xj})], \forall x \in E_j; j = 0, 1, ..., m. \quad (10)$$

- (II.3) Third, the distribution $Q_X^{II}(x \mid D, e_x)$ is obtained from the probability assignments (II.1) and (II.2) given by (9) and (10). The application of the associative (multiplicative) rule to these logical probabilities, by taking the logic product $(e_x, e_{xj}) \equiv e_{xj}$ into account, gives

$$Prob(X \leq x, x_{cj} \leq X \leq x_{c(j+1)} \mid D, e_x) =$$
$$Prob[X \leq x \mid D, e_{xj}] \cdot Prob[x_{cj} \leq X \leq x_{c(j+1)} \mid D, e_x], \forall x \in E_j;\ j = 0, 1, .., m. \quad (11)$$

The substitution of (10) and the expressions

$$Prob(X \leq x, x_{cj} \leq X \leq x_{c(j+1)} \mid D, e_x) = Q_X^{II}(x \mid D, e_x) - Q_X^{II}(x_{cj} \mid D, e_x)$$

$$Prob(x_{cj} \leq X \leq x_{c(j+1)} \mid D, e_x) = Q_X^{II}(x_{c(j+1)} \mid D, e_x) - Q_X^{II}(x_{cj} \mid D, e_x),$$

into (11), gives the distribution $Q_X^{II}(x \mid D, e_x)$ as a function of the probabilities assigned by the operator $g[.]$ in (II.1) and (II.2), as the following final expressions in the way (II)

$$Q_X^{II}(x \mid D, e_x) = Q_X^{II}(x_{cj} \mid D, e_x) +$$
$$+ Q_X^{II}(x \mid D, e_{xj})[Q_X^{II}(x_{c(j+1)} \mid D, e_x) - Q_X^{II}(x_{cj} \mid D, e_x)], \forall x \in E_j; j = 0, 1, ..., m. \quad (12)$$

There are also two different ways of applying the inference procedure (7) in the case of any context given by the logical sum of a set of evidences e_{xj}. Then, probabilities in the step II.1 are assigned on the subdomain equivalent to the union of a set of simple subdomains E_j. Expressions (12) are easily generalized in this case (Solana, 1992).

INVARIANCE PRINCIPLE

When the domain E is divided into $m+1$ subdomains E_j such that every element of the set $S_c = \{x_{cj}\}$ equals to any of the elements of the set of observed data S, then $\{x_{cj}\} = \{x_{rj,o}\}$. In this case no new data, i.e. no other data than those in $S = \{x_{rj,k}\}$, are introduced for probability assignments through the way (II). Then, since the data are the same in the ways (I) and (II) the strong invariance principle in Section 1 must be invoked.

In this particular case in which $x_{cj} = x_{rj,o}$, the probability assignments in the ways (I) and (II) constitute in fact two different forms for updating probabilities with respect to a reference distribution. This updating corresponds to the change from the probability distribution given the initial observed data $\{x_{rj,o}\}$, to the probability distribution given the initial and additional observed data $S = \{x_{rj,k}\}$.

The application of the strong invariance principle to this case, therefore, determines the principle of invariance under the different forms (I) and (II) of updating probabilities, which is termed *the invariance principle for probability updates*. This principle states that:

The identical updated probability distribution inferred with regard to any reference distribution should be obtained when:

- *(a) The probabilities are assigned on the whole domain E, on the basis of the set $S = \{x_{rj,k}\}$ of observations that complete with additional data one set of initially observed data $\{x_{rj,o}\}$, the order of these data into an ordered reference set S_n, and the given reference distribution on this domain;*

- *(b) The probabilities are obtained from the two following assignments: (b.1) the probabilities at the initial data points $\{x_{rj,o}\}$ of the distributions assigned in the way (a), and (b.2) the probability distributions on every initial data-bounded subdomain E_j, on the basis of the additional observed data, and the order of these data into the ordered reference subset of S_{nj} and the conditional reference distribution for every subdomain.*

For any partition by observed data the invariance principle may be formally expressed as the following invariance requirements:

$$Q_X^I(x \mid D, e_x) = Q_X^{II}(x_{rj,o} \mid D, e_x) +$$

$$+ Q_X^{II}(x \mid D, e_{xj}) \cdot [Q_X^{II}(x_{r(j+1),o} \mid D, e_x) - Q_X^{II}(x_{rj,o} \mid D, e_x)], \forall x \in E_j,\ j = 0, 1, ..., m;$$

$$\text{for any} \{x_{rj,o}\} \subset S. \tag{13}$$

These requirements can be generalized in the case of any context given by the logical sum of a set of evidences e_{xj} (Solana, 1992).

3. DERIVATION OF THE INFERENCE PROCEDURE

The updating principle has been formalized as the application of the strong invariance principle stated in Section 1 to the basic inference case proposed in this paper. A major consequence of the strong invariance principle when applied as the updating principle is just the derivation of one data-based inference procedure directly from only this principle (Solana 1992).

The procedure, which is necessary for doing inference based on observed data in a self-consistent way, is presented in this Section. The derivation is developed in three steps by identifying the mathematical form of the inference operator g(.) postulated in Section 2. It considers, first, the particular updating case in which there are not new observed data into any of the subdomains bounded on the initial data; the application of the updating principle to this particular case leads to the statement of the mathematical form of the inference operator. Next, the general case of updating is considered using the previous resulting mathematical form; then again, the application of the updating principle to this case determines the basic functional equations for the coefficients of the mathematical form. Finally, one solution of these functional equations is found out defining completely the data-based inference procedure.

MATHEMATICAL FORM OF THE INFERENCE OPERATOR.

Consider the partition of E as the set of initially observed data $\{x_{rj,o}\}$ into the subdomains $E_j = (x_{rj,o}, x_{r(j+1),o})$, $j = 0, 1, 2, ..., m$, and let $S = \{x_{rj,k}\}$ be an updated set of observed data. In addition, consider the particular updating situation in which *no new observed data* fall into any specific subdomain E_j, such that $pj = 0$. Then, the only information that can be associated to this subdomain is simply the evidential context sentence e_{xj}: $x_{rj,o} \leq X \leq x_{r(j+1),o}$.

First of all, it is necessary to disregard the possibility of an inference operator that does not infer a distribution in the way II if there are not data in a specific subdomain. Otherwise , it should be incoherent with the use of the same operator in the way I on the whole domain, and the posterior conditioning that provides a conditional distribution on the subdomain.

In this updating case, therefore, given the conditional reference distribution $P(x \mid e_{xj})$, one probability distribution $Q_X^{II}(x \mid e_{xj})$ should be assigned on the whole subdomain E_j, directly in the way II. If there are no data on a subdomain E_j, then the inferred distribution must be identical everywhere to the conditional reference distribution on this subdomain, in the manner expressed as

$$Q_X^{II}(x \mid e_{xj}) = g[P_X(x \mid e_{xj}), I(e_{xj})] = P_X(x \mid e_{xj}), \forall x \in E_j. \tag{14}$$

Otherwise, if the inferred distribution differs from the conditional reference distribution, then either some piece of information other than e_{xj} has been also considered through the operator g(.), which is contrary to the assumption of having only the information e_{xj}, or the conditional reference distribution does not correspond in fact to the assumed reference distribution for this case.

It must be noticed that the statement expressed as (14) is one of simple implication. Consequently the identity of the conditional inferred distribution and the reference distribution could be also possible when there are some data in a subdomain.

On the other hand, by taking the logical product $(e_x, e_{xj}) \equiv e_{xj}$ into account, application of the multiplicative rule for logical probabilities to the reference distribution gives

$$Prob(X \leq x, x_{rj,o} \leq X \leq x_{r(j+1),o} \mid e_x) =$$
$$Prob(X \leq x \mid e_{xj}) \cdot [Prob(x_{rj,o} \leq X \leq x_{r(j+1),o} \mid e_x)], \; \forall x \in E_j. \tag{15}$$

Hence, the conditional reference distribution is

$$P_X(x \mid e_{xj}) = [P_X(x \mid e_x) - P_X(x_{rj,o} \mid e_x)][P_X(x_{r(j+1),o} \mid e_x) - P_X(x_{rj,o} \mid e_x)]^{-1}, \; \forall x \in E_j. \tag{16}$$

Next, by substituting the distribution $Q_X^{II}(x \mid e_{xj})$ calculated by (14) and (16) into the expression of *the updating principle* (13) specialized to this particular updating case, the following is obtained

$$Q_X^{I}(x \mid D, e_x) = Q_X^{II}(x_{rj,o} \mid D, e_x) + \mu_j[P_X(x \mid e_x) - P_X(x_{rj,o} \mid e_x)], \; \forall x \in E_j, \tag{17}$$

where

$$\mu_j = [Q_X^{II}(x_{r(j+1),o} \mid D, e_x) - Q_X^{II}(x_{rj,o} \mid D, e_x)][P_X(x_{r(j+1),o} \mid e_x) - P_X(x_{rj,o} \mid e_x)]^{-1}. \tag{18}$$

According to (17) the assigned distribution for the next observation of X on the whole domain E, should be a linear function of the reference distribution $P_X(x \mid e_x)$ on the interval E_j containing no additional observed data.

Finally, as no other data exist for every subdomain limited by initial observed data, the expression (17) specialized for the joint evidence $D \equiv (\{x_{rj,o}\}, x_1 < x_2 <, \cdots, < x_n)$ applies to every subdomain E_j. This case just corresponds to the probability assignment on the basis of only the initial data and the order of these data in a reference set. Therefore the expression (17) specialized for this joint evidence determines the *linear mathematical form* of the inference operator

$$Q_X(x \mid D, e_x) =$$
$$g[P_X(x \mid e_x), I(\{x_{rj,o}\}, x_1 < x_2 < ... < x_n)] = a_j + b_j P_X(x \mid e_x), \forall x \in E_j,\ j = 0, 1, \cdots, m. \tag{19}$$

The coefficients of the linear function (19) are the functions of both the unknown assigned probabilities and the reference distribution probabilities at the observed data points $x = x_{rj,o}$, given by

$$a_j = Q_X(x_{rj,o} \mid D, e_x) - \mu_j P_X(x_{rj,o} \mid e_x),$$
$$b_j = \mu_j \quad j = 0, 1, ..., m. \tag{20}$$

The linear form (19) of the inference operator is *unique* in the sense that any inference method using other mathematical form violates the updating principle expressed as the invariance requirement (13).

NECESSARY CONDITIONS FOR INVARIANCE.

Now, in the case of the data set $S = \{x_{rj,k}\}$ that includes the *additional observed data*, the linear mathematical form of the inference procedure is employed to assign the inferred distribution before applying the invariance principle.

First, the inference procedure given by (19) and (20) applies directly, in the *way* (I), to the joint evidence in this case D represented by (1) and (2) on the intervals $E_{j,k} = (x_{rj,k}, x_{rj,(k+1)})$ as follows

$$Q_X^I(x \mid D, e_x) = Q_X^I(x_{rj,k} \mid D, e_x) +$$
$$+\mu_{rj,k}^I [P_X(x \mid e_x) - P_X(x_{rj,k} \mid e_x)], \forall x \in E_{j,k};\ k = 0, 1, ..., pj\ ,\ j = 0, 1, ..., m. \tag{21}$$

The coefficients of the linear form (21) for different intervals $E_{j,k}$ correspond to the scale parameters

$$\mu_{rj,k}^I = [Q_X^I(x_{rj,k+1} \mid D, e_x) - Q_X^I(x_{rj,k} \mid D, e_x)][P_X(x_{rj,k+1} \mid e_x) - P_X(x_{rj,k} \mid e_x)]^{-1}. \tag{22}$$

Second, the same inference operator given by (19) and (20) is applied in the *way* II, by taking into account the partition of E as the initially observed data set $\{x_{rj,o}\}$, to the joint evidence (D, e_{xj}) represented by expressions (4) and (5)

$$Q_X^{II}(x \mid D, e_{xj}) = Q_X^{II}(x_{rj,k} \mid D, e_{xj}) +$$

$$+\mu^{II}_{rj,k}[P_X(x \mid e_{xj}) - P_X(x_{rj,k} \mid e_{xj})],\ \forall x \in E_{j,k};\ k = 0,1,...,pj\ ,\ j = 0,1,...,m. \quad (23)$$

The coefficients of the linear forms (23) for different intervals $E_{j,k}$ correspond to the scale parameters

$$\mu^{II}_{rj,k} = [Q^{II}_X(x_{rj,k+1} \mid D, e_{xj}) - Q^{II}_X(x_{rj,k} \mid D, e_{xj})][P_X(x_{rj,k+1} \mid e_{xj}) - P_X(x_{rj,k} \mid e_{xj})]^{-1}. \quad (24)$$

The substitution of (23) into the second term of the expression (13) of the updating principle gives

$$Q^{II}_X(x \mid D, e_x) = Q^{II}_X(x_{rj,o} \mid D, e_x) + [Q^{II}_X(x_{r(j+1),o} \mid D, e_x) - Q^{II}_X(x_{rj,o} \mid D, e_x)]\cdot$$
$$\cdot [Q^{II}_X(x_{rj,k} \mid D, e_{xj}) + \mu^{II}_{rj,k}[P_X(x \mid e_{xj}) - P_X(x_{rj,k} \mid e_{xj}])]\ ,\ \forall x \in E_{j,k};$$
$$k = 0,1,...,pj,\ j = 0,1,...,m. \quad (25)$$

Then, by taking again the logic product $(e_x, e_{xj}) \equiv e_{xj}$ into account, the multiplicative rule of probabilities gives

$$[P_X(x \mid e_{xj}) - P_X(x_{rj,k} \mid e_{xj})][P_X(x_{rj,k+1} \mid e_{xj}) - P_X(x_{rj,k} \mid e_{xj})]^{-1} =$$
$$= [P_X(x \mid e_x) - P_X(x_{rj,k} \mid e_x)][P_X(x_{rj,k+1} \mid e_x) - P_X(x_{rj,k} \mid e_x)]^{-1}. \quad (26)$$

Finally, by (24) and (26) the distribution assigned in *way* (II) may be written as the following linear function of $P_X(x \mid e_x)$

$$Q^{II}_X(x \mid D, e_x) = Q^{II}_X(r_{rj,o} \mid D, e_x) + [Q^{II}_X(x_{r(j+1),o} \mid D, e_x) - Q^{II}_X(x_{rj,o} \mid D, e_x)]\cdot$$
$$\cdot [Q^{II}_X(x_{rj,k} \mid D, e_{xj}) + [Q^{II}_X(x_{rj,k+1} \mid D, e_{xj}) - Q^{II}_X(x_{rj,k} \mid D, e_{xj})][P_X(x \mid e_x) - P_X(x_{rj,k} \mid e_x)]\cdot$$
$$[P_X(x_{rj,k+1} \mid e_x) - P_X(x_{rj,k} \mid e_x)]^{-1}],\ \forall x \in E_{j,k};\ k = 0,1,...,pj,\ j = 0,1,...,m. \quad (27)$$

The *invariance principle* for probabilistic updates states the equivalence of the linear mathematical forms given by the expressions (21) and (27). The identification of the coefficients of $P_X(x \mid e_x)$ in both linear forms and the equivalence of probabilities in the ways (I) and (II) at the initial data points given by (9), results in the following conditions.

$$Q^{II}_X(x_{rj,k+1} \mid D, e_{xj}) - Q^{II}_X(x_{rj,k} \mid D, e_{xj}) =$$
$$= [Q^{I}_X(x_{rj,k+1} \mid D, e_x) - Q^{I}_X(x_{rj,k} \mid D, e_x)][Q^{I}_X(x_{r(j+1),o} \mid D, e_x) - Q^{I}_X(x_{rj,o} \mid D, e_x)]^{-1},$$
$$k = 0,1,...,pj,\ j = 0,1,...,m. \quad (28)$$

On the other hand, the same equations (28) are obtained from the identification of the complementary terms associated to the reference probabilities $P(x_{rj,k} \mid e_x)$. The identification of the remainder terms in the linear mathematical expressions (21) and (27), gives the following equation

$$Q_X^{II}(x_{rj,k} \mid D, e_{xj}) =$$
$$= [Q_X^I(x_{rj,k} \mid D, e_x) - Q_X^I(x_{rj,o} \mid D, e_x)][Q_X^I(x_{r(j+1),o} \mid D, e_x) - Q_X^I(x_{rj,o} \mid D, e_x)]^{-1},$$
$$k = 0, 1, ..., pj, \; j = 0, 1, ..., m. \tag{29}$$

Subtraction of the equations (29) for succesive orders k and $k+1$ leads exactly to the equations (28). Moreover, the expressions (29) lead exactly to the necessary expressions that generalize (28) in the case of any context given by the logical sum of a set of evidences e_{xj}, by using the Bayes' formula for conditioned probabilities on this set at data points $x_{rj,k}$ (Solana 1992).

Consequently the expressions (29), are the *necessary and sufficient conditions* for invariance under probabilistic updates. These conditions state: The equivalence among the probabilities of the assigned distribution on the whole domain conditioned on the initial data-bounded subdomains E_j, specialized at data points set $S = \{x_{rj,k}\}$, and the probabilities of the conditional distributions assigned on every subdomain E_j at the same data points.

FUNCTIONAL EQUATIONS AND SOLUTION.

The linear parametric form of the inference operator written by (19) and the expressions (29) of the unkown values of probability distributions at the observed data points are the necessary and sufficient conditions for invariance under probabilistic updates.

The last necessary conditions (29) can be analysed separately as functional equations of the general class of functions $F[x_i, D, e]$ that express the probability values at any observed data point in the updating cases (I) and (II).

There exists one simple solution for these functional equations. Given the set of observed data S and the i-th orders of the observed data into an ordered reference set S_n of n elements, the solution of the functional equations (29) is the particular class of functions such that their values for the observed data x_i equal exactly to the (i/n) probability fractiles, written by

$$F[x_i, D, e] = i/n, \; \forall i \in N_S \subset N. \tag{30}$$

Indeed, the probability values obtained by specialization of (30) to the observed data and the different joint evidences and domains (D, e) in the ways (I) and (II) fulfil the necessary and sufficient conditions (29), as is showed in the following.

First, the specialization of the class of functions (30) to the updating case (I) by using the joint evidence (2), gives

$$Q_X^I[x_{rj,k} \mid \{x_{rj,k}\}, x_1 < x_2 < ... < x_n, e_x] = (rj,k)/n, \forall (rj,k) \equiv i, \; i \in N_S \subset N;$$
$$k = 0, 1, ..., pj, \; j = 0, 1, ..., m. \tag{31}$$

where the (rj, k)-th order of the observed data are defined through the equivalences $x_{rj,k} \equiv x_i$.

Similar, the specialization of (30) to the updating case (II) by employing the joint evidence (6), gives

$$Q_X^{II}[x_{rj,k} \mid \{x_{rj,k}\}, x_{(rj,o)} < x_{(rj,o)+1} < ... < x_{(rj,o)+nj}] = [(rj,k) - (rj,o)]/nj,$$

$$\forall k = 1,2,...,pj;\; j = 0,1,...,m. \tag{32}$$

where the index $[(rj,k) - (rj,o)]$ means the order of each observed datum lying on E_j, into the ordered reference subset S_{nj} having the number of elements nj.

Next,by substituting the probabilities calculated by (31) into the second term of the expression (29)

$$[Q_X^I[x_{rj,k} \mid D, e_x] - Q_X^I[x_{rj,o} \mid D, e_x]]\,[Q_X^I(x_{r(j+1),o} \mid D, e_x) - Q_X^I(x_{rj,o} \mid D, e_x)]^{-1} =$$

$$[(rj,k)/n - (rj,o)/n][(r(j+1),o)/n - (rj,o)/n]^{-1}. \tag{33}$$

Since the variable n cancels in this term and the number of elements nj equals to $[(r(j+1),o) - (rj,o)]$, the expression (33) equals also to (32) and, therefore, the functional equations (29) are exactly satisfied by the class of functions (30).

The *uniqueness* of the solution (30) of functional equations derived from expressions (29) has been also showed (Solana 1992).

THE INFERENCE PROCEDURE.

Consider again the basic inference case in this paper in which the probability distribution is assigned on the domain E, given a reference distribution, on the only basis of the evidence D that stands for the initial data set $S = \{x_{rj,o}\}$ and the order of these data into the ascending order of the elements of a reference set S_n, such that $x_{rj,o} = x_i$, $i \in N_S \subset N = 1,2,...,n$.

By substituting the probability values given by the class of functions (30) into the expressions (18) and (20) and both of them into the linear parametric form (19), the inference procedure is completely determined on every initial data-bounded subdomain E_j, as the following operator

$$Q_X(x \mid D, e_x) = g[P_X(x \mid e_x), I(\{x_{rj,o}\}, x_1 < x_2 < ... < x_n)] =$$

$$= (rj,o)/n + \mu_j[P_X(x \mid e_x) - P_X(x_{rj,o} \mid e_x)], \forall x \in E_j;\; j = 0,1,2,...,m, \tag{34}$$

where the scale parameters on every subdomain E_j are

$$\mu_j = (1/n)[(r(j+1),o) - (rj,o)][P_X(x_{r(j+1),o} \mid e_x) - P_X(x_{rj,o} \mid e_x)]^{-1}. \tag{35}$$

This result states that:

The inference procedure is a linear transformation of the probabilities of the reference distribution as different scale parameters on each elementary data-bounded subdomain, such that the assigned probabilities at the observed data points , $x_i = x_{rj,o}$, equal exactly to the fractiles $(rj,o)/n$ of the order of the observed data on the number of elements of the ordered reference set.

In the particular case in which the reference ordering of observed data is the ascending order of the elements of the set S itself, such that $S = S_n$, $i \in N = 1, 2, ..., n$, then the inference procedure (34) simplifies as follows

$$Q_X(x \mid D, e_x) = g[P_X(x \mid e_x), I(\{x_i\}, x_1 < x_2 < ... < x_n)] =$$
$$= i/n + \mu_i[P_X(x \mid e_x) - P_X(x_i \mid e_x)], \forall x \in E_i;\ i = 0, 1, ..., n, \tag{36}$$

where the scale parameters on every subdomain E_i are given as

$$\mu_i = (1/n)[P_X(x_{i+1} \mid e_x) - P_X(x_i \mid e_x)]^{-1} \tag{37}$$

Since a set of observed data may be ordered in an ascending order through the inference procedure without considering additional information, the particular case in which $S = S_n$ just corresponds to the standard inference problem considered by the usual inference methods. Consequently the inference procedure (36) can be applied directly in the case of assigning probabilities for a random quantity given only a set of observed data.

4. CONCLUSIONS

(1) To be self-consistent any data-based inference method should lead necessarily to the linear mathematical form of a reference distribution given by (19), and the coefficients of this form should satisfy the basic conditions (29). Otherwise, such an inference method violates the invariance principle for probabilistic updates formulated in Section 2.

(2) The inference procedure given in Section 3 by (34) and (36) is self-consistent. The procedure has been derived from only the invariance principle for probabilistic updates. Uniqueness of the mathematical form (19) and the solution (30) of functional equations (29) have been demonstrated. Hence, the uniqueness of the linear procedure and the coefficients given by (34) and (36) can be assured.

(3) There is one method called *the relative entropy method with fractile constraints*, or the REF method for short, that provides exactly the same results as the inference procedure derived in this paper. The REF method uses logical probabilities and satisfies the fundamental desideratum for consistency given in Section 2. Using relative entropy, the REF method is a variant of the entropy-based inference formalism developed by Jaynes and others. The data are introduced in the form of mathematical constraints as prescribed probability fractiles. The Kullback-Leibler entropy functional is minimized subject to these fractile constraints, in contrast with the well known classical version of the relative entropy method by Shore and Johnson that uses moment constraints.

(4) The prescribed fractiles in the REF method were obtained as a sample rule based on symmetry arguments as used in the Laplace's rule. They are identical to the solution (30) of the functional equations (29), which are logically derived in this paper from only the invariance principle under probabilistic updates.

(5) It can be shown that many of the *usual inference methods*, including the method that combines the Bayes' formula with the likelihood function of observations, do not provide the required linear mathematical form (36). They lack therefore consistency.

ACKNOWLEDGMENTS. The author wish to thank to Niels Lind, Emilio Rosenblueth and José L. Cervera the useful comments to this paper. The work reported in this paper received financial support from the Direccion General de Investigación Científica y Técnica of Spain as a part of the research project PB 89-0077.

REFERENCES

Carnap, R.: 1950, *Logical Foundations of Probability*, University of Chicago Press.

Cox, R.T.: 1946, Probability, Frequency and Reasonable Expectation, *Am. Journal of Physics*, 17, 1-13.

Cox, R.T.: 1961, *The Algebra of Probable Inference*, John Hopkins University Press, Baltimore MD.

Heckerman, D.E.: 1988, An Axiomatic Framework for Belief Updates, in *Uncertainty in Artificial Intelligence 2*, Lemmer J.F and Kanal L.N. (eds.), Elsevier Science Publishers.

Jaynes, E.T.: 1957, *"How Does the Brain Do Plausible Reasoning"* Microwave Laboraty Report, No. 421. University of Stanford; reprinted in G.J. Ericson and C.R. Smith (eds), *Maximum Entropy and Bayesian Methods in Science and Engineering*, Vol.I, 1-24, Kluwer Academic Publishers, 1988.

Jeffreys, H.: 1939, 2nd and 3th. ed. 1948, 1961, *Theory of Probability*, Oxford University Press, London.

Lind, N.C. and Solana V.: 1990, Fractile Constrained Entropy Estimation of Distributions based on Scarce Data. *Civil Engineering Systems*, Vol. 7, No.2, 87-93.

Polya, G.: 1954, *Mathematics and Plausible Reasoning*, Vol. 2, Princenton University Press.

Shore, J.E. and Johnson, R.W.: 1980, Axiomatic Derivation of the Principle of Maximum Entropy and the Principle of Minimum Cross Entropy. *IEEE Transactions on Information Theory* IT-26 Vol.1, 26-37.

Skilling, J.: 1988, The Axioms of Maximum Entropy, in Maximum Entropy and Bayesian Methods, Erickson, G.J. and Smith C.R. (eds.) Kluwer Academic Press.

Solana, V.: 1990, Consistency Principle for data based Probabilistic Inference, in *Maximum Entropy and Bayesian Methods in Science an Engineering*. Paul Fouguere(ed.) Kluwer Academic Publishers.

Solana, V., and Lind, N.C.: 1992, Relative Entropy Method using Tail Fractiles, *Reliability Engineering and System Safety* (to appear).

Solana, V.: 1992, The Derivation of a Data-based Inference Procedure using Logical Probabilities. *Research Report*. Instituto Matemécticas y Física Fundamental, Madrid.

Tribus, M.: 1988, An Engineer Look at Bayes. In *Maximum Entropy and Bayesian Methods*, Erikson G.J. and Smith C.R. (eds.) Kluwer Academic Publishers.

ALPHA, EVIDENCE, AND THE ENTROPIC PRIOR

C. E. M. Strauss, D. H. Wolpert*, and D. R. Wolf
Los Alamos National Laboratory
Los Alamos NM 87545
USA

ABSTRACT. First, the correct entropic prior is computed by marginalization of alpha. This is followed by a discussion of improvements to the "evidence" approximation. Surprisingly, it appears that the approximations used to restore the famous "Susie" image may have questionable aspects.

1. Introduction

In the Classic Maxent formalism, the conditional prior $P(\vec{f}|\alpha)$ over an r-element positive additive distribution $\vec{f}$ contains a regularization parameter α and a "default model" $\vec{m}$ (Skilling 1989):

$$P(\vec{f}|\alpha) = \frac{\exp[\alpha S(\vec{f})]}{Z_S(\alpha)\prod_{i=1}^{r}\sqrt{f_i}}, \text{ where } S(\vec{f}) \equiv \sum_{i=1}^{r} f_i - m_i - f_i \ln[\frac{f_i}{m_i}] \qquad [1]$$

and $Z_S(\alpha)$ is a normalization constant. While some discussion at recent MAXENT conferences has addressed the question of "setting alpha" in [1], this is something of a misdirection brought on by the legacy of regularization. Formally, alpha does not have a single value and should be marginalized in order to obtain $P(\vec{f})$:

$$P(\vec{f}) = \int_0^\infty d\alpha \, P(\vec{f}|\alpha) P(\alpha). \qquad [2]$$

The posterior, $P(\vec{f}|D = Data)$, has no alpha dependence and is proportional to the product of the likelihood function, $P(D|\vec{f})$, and $P(\vec{f})$.

Various approximation methods have been developed to make calculation of the posterior for the entropic prior tractable (Bryan 1990, Gull 1989). In particular, under the popular "evidence approximation", the posterior distribution is taken to be proportional to the product of the likelihood function, $P(D|f)$, and $P(\vec{f}|\alpha')$, where α' is the special value of alpha which maximizes its own posterior, $P(\alpha|D)$ (Gull 1989, MacKay 1991).

It turns out that straightforward marginalization over alpha to find $P(\vec{f})$ is not as intractable as one might think. In the next section of this article we present a method for marginalization of alpha which bypasses the need for approximations such as setting α

A. Mohammad-Djafari and G. Demoments (eds.), Maximum Entropy and Bayesian Methods, 113–120.

to a single value. We consider both the case where $\vec{f}$ is a probability vector and the case where it is an unconstrained distribution. The last section discusses the implications of the first section for the evidence method of setting alpha; important conditions of validity are discussed.

2. The Entropic Prior

When the f_i Are Independent.

The unconstrained case (i.e., independent f_i) is the usual one considered in MAXENT conferences. To address this case, first one must compute the normalization term, $Z_S(\alpha)$, in [1]:

$$Z_S(\alpha) = \int_0^\infty df_1\, df_2 \cdots df_r \frac{\exp[\alpha \sum_{i=1}^r f_i - m_i - f_i \ln[\frac{f_i}{m_i}]]}{\prod_{i=1}^r \sqrt{f_i}}. \qquad [3]$$

Since the f_i are independent, this factors into the product of r one– dimensional integrals. Making the change of variable $f_i = g_i m_i e$, we can rewrite this in a compact manner as

$$Z_S(\alpha) = \prod_{i=1}^r \sqrt{em_i}\, Z_1(em_i\alpha) \quad \text{where } Z_1(\beta) \equiv \int_0^\infty dg \frac{\exp[-\beta(g\ln(g) + \frac{1}{e})]}{\sqrt{g}}. \qquad [4]$$

This reduces calculation of $Z_S(\alpha)$ to the evaluation of a 1-dimensional definite integral, Z_1, parameterized by the dimensionless variable $\beta = \alpha m_i e$. While an analytic result is unavailable, the expression is amenable to numerical integration; the integrand decays more rapidly than $\exp(-\beta g)$ at large g, and the singularity at the origin is integrable.

We have computed Z_1 for values of β ranging over twelve orders of magnitude and stored them in a lookup table. Figure 1 displays the ratio of $Z_1(\beta)$ to its asymptotic form (for large β, $Z_1(\beta) \approx \sqrt{\frac{2\pi}{\beta}}$). We pause to note that all problems using an entropic prior will use the *same* lookup table; once computed, this table will be used for any estimation problem (i.e., it will be used regardless of changes in the model, likelihood function, data, prior over alpha, or dimension r).

The second task is to perform the integral in [2], which can be expressed as

$$P(\vec{f}) = I[S(\vec{f})] \prod_{i=1}^r \sqrt{\frac{1}{f_i}} \quad \text{, where } I[x] \equiv \int_0^\infty d\alpha \frac{\exp(\alpha x)}{Z_S(\alpha)} P(\alpha). \qquad [5]$$

We have thus reduced the calculation of $P(\vec{f})$ to the tabulation of $I(x)$. Like Z_S, this integral is also a one dimensional definite integral and is parameterized by a scalar, $x = S$. To evaluate this integral we now need $P(\alpha)$. Barring a peculiar $P(\alpha)$, it is straight-forward to carry out the numerical integration. Note that $I(S)$ is data–independent; it will be necessary to recompute $I(S)$ only when either the prior over alpha or the model changes. For illustration, in Figure 1 right we plot $I(S)$ for a flat model ($m_i = 1/e$) and a $P(\alpha)$ flat over the range 10^{-3} to 10^6, for dimensions r $= 16, 256, 4096$, and 16384. In general, $I(S)$ decreases monotonically from its peak at $S = 0$. (As defined in [1], $S \leq 0$.) Note that *if* one believes, as has been widely claimed, that the posterior is extremely insensitive to the

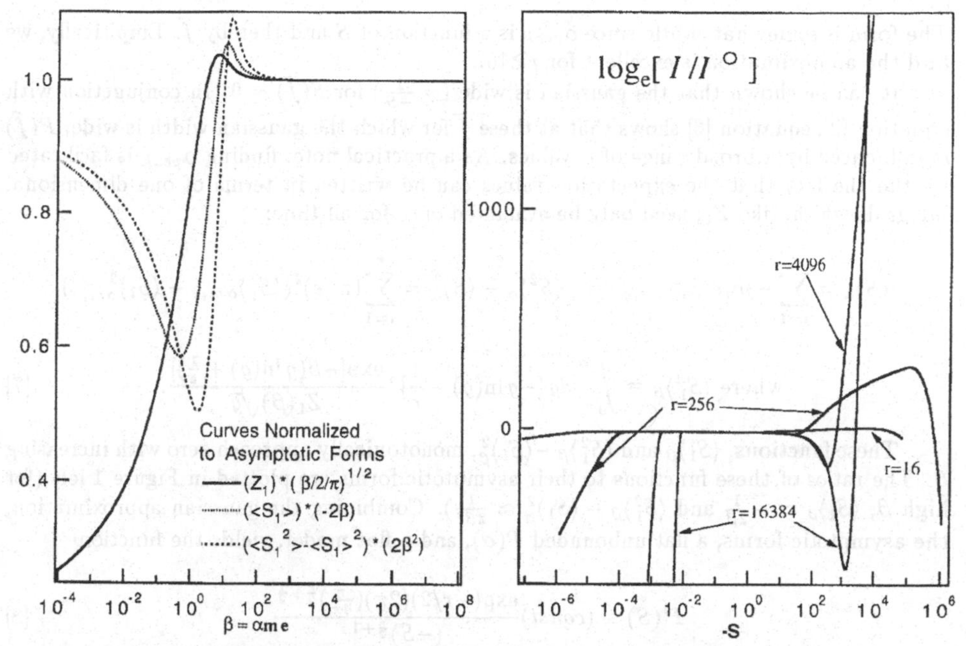

Fig. 1 **left** : The three dimensionless functions $Z_1(\beta)$, $\langle S_1\rangle_\beta$, and $(\langle S_1^2\rangle_\beta - \langle S_1\rangle_\beta^2)$, each divided by its high-β asymptotic form. **right**: The ratio of $I(S)$ to $I^\circ(S)$ for various dimensions, r, in the case of a tophat-shaped $P(\alpha)$ and a flat model, $\vec{m}$.

choice of $P(\alpha)$ for all "reasonable" $P(\alpha)$, then this specific example is sufficient to compute the posterior in all problems with "reasonable" $P(\alpha)$ (and a flat model).

Three things should be noted about these results. First, there are just two single dimensional integrals to be computed, regardless of the $\vec{f}$–space dimension, r. The underlying reason for this is the separability of the integrals for $Z_s(\alpha)$. Second, each of the two integrals is parameterized by a single variable. The underlying reason for this is that the probability of a given r-dimensional point $\vec{f}$ can be written as the function I of the *scalar* $S(\vec{f})$. Third, the computation time depends only weakly on the dimension r. Each of the plotted curves was calculated on a Macintosh II on a time scale of minutes.

High Dimensional Cases

Although it is not difficult to compute $I(S)$ numerically, even in high dimensions, an approximate form clarifies several important characterisitics of the prior. Taylor expanding $\ln[P(\vec{f}|\alpha)]$ to second order in α gives the gaussian expression

$$P(\vec{f}|\alpha) \approx \frac{\exp[\alpha_{peak} S(\vec{f})]}{Z_S(\alpha_{peak}) \prod_{i=1}^{r} \sqrt{f_i}} \exp[-(\langle S^2\rangle_{\alpha_{peak}} - \langle S\rangle^2_{\alpha_{peak}})(\alpha - \alpha_{peak})^2] \qquad [6]$$

where $\langle S^n\rangle_\alpha \equiv \int_0^\infty d\vec{f}\, S(\vec{f})^n P(\vec{f}|\alpha)$ and α_{peak} is the solution to $S(\vec{f}) = \langle S\rangle_\alpha$.

The form is somewhat subtle since α_{peak} is a function of S and thereby $\vec{f}$. Empirically, we find the approximation is excellent for $r \gtrsim 16$.

It can be shown that the gaussian is wide ($\sim \frac{1}{-S}$) for $S(\vec{f}) \sim 0$. In conjunction with equation [2], equation [6] shows that at these $\vec{f}$ for which the gaussian width is wide, $P(\vec{f})$ is influenced by a broad range of α values. As a practical note, finding α_{peak} is facilitated by the the fact that the expectation values can be written in terms of one dimensional integrals which, like Z_1, need only be evaluated once for all time:

$$\langle S\rangle_\alpha = \sum_{i=1}^{r} -m_i e \langle S_1\rangle_{\alpha m_i e} \quad ; \quad \langle S^2\rangle_\alpha - \langle S\rangle_\alpha^2 = \sum_{i=1}^{r} (m_i e)^2 (\langle S_1^2\rangle_{\alpha m_i e} - \langle S_1\rangle_{\alpha m_i e}^2)$$

$$\text{where } \langle S_1^n\rangle_\beta \equiv \int_0^\infty dg\,[-g\ln(g) - \frac{1}{e}]^n \frac{\exp[-\beta(g\ln(g) + \frac{1}{e})]}{Z_1(\beta)\sqrt{g}}. \qquad [7]$$

These functions, $\langle S_1\rangle_\beta$ and $\langle S_1^2\rangle_\beta - \langle S_1\rangle_\beta^2$, monotonically approach zero with increasing β. The ratios of these functions to their asymptotic forms are plotted in Figure 1 left (for high β, $\langle S_1\rangle_\beta \approx -\frac{1}{2\beta}$ and $\langle S_1^2\rangle_\beta - \langle S_1\rangle_\beta^2 \approx \frac{1}{2\beta^2}$). Combining the gaussian approximation, the asymptotic forms, a flat unbounded $P(\alpha)$, and a flat model, yields the function

$$I^\circ(S) \equiv (const)\frac{\exp(-r/2)(2\pi)(\frac{r}{4\pi})^{\frac{r}{2}+\frac{1}{2}}}{(-S)^{\frac{r}{2}+1}}. \qquad [8]$$

As $S \to 0$, for a flat, unbounded $P(\alpha)$, $I(S) \to I^\circ(S)$. Figure 1 right shows $\frac{I(S)}{I^\circ(S)}$ when $I(S)$ is instead calculated with a bounded (top-hat shaped) $P(\alpha)$. The drops in the ratio at very high and low $|S|$ are a direct result of the lower and upper bounds on alpha, respectively. (The drops are seen most readily for r=256.) The flat regions show where $I(S)$ behaves like $I^\circ(S)$. As $|S|$ increases beyond the flat region, the ratio dips and then rises, due to the breakdown of the asypmtotic forms used in I° (the gaussian approximation is still quite good).

WHEN $\vec{f}$ IS A PROBABILITY.

In many problems, it is natural to constrain the elements of $\vec{f}$ and $\vec{m}$ to sum to a fixed value, which, after a change of variables, can be taken to be one. In such a case $\vec{f}$ is a probability. This constrained entropic prior has received little attention, perhaps due to the 'technical infelicity' of having that constraint in the integrals (Skilling 1989). Here we describe a means of circumventing this difficulty.

To begin, expand the exponential in the integrand in [3], $\exp(\alpha S)$, in a Taylor series about $\alpha = 0$, and integrate each term. This gives

$$Z_S(\alpha) = (Z_0)\sum_{i=0}^{\infty} \frac{\langle S^i\rangle_{\alpha=0}\alpha^i}{i!}\text{, where } Z_0 \equiv \int d\vec{f}\,\delta(1 - \sum_{j=1}^{r} f_j)\prod_{i=1}^{r}\sqrt{\frac{1}{f_i}} = \frac{\pi^{r/2}}{\Gamma(r/2)} \qquad [9]$$

$$\text{and, for this constrained case, } \langle S^n\rangle_{\alpha=0} = \int d\vec{f}\,\frac{S(\vec{f})^n}{Z_0}\delta(1 - \sum_{j=1}^{r} f_j)\prod_{i=1}^{r}\sqrt{\frac{1}{f_i}}.$$

It is proven in (Wolpert 1993b) that 1) this expansion converges for all positive alpha, and 2) the error introduced in $I(S)$ by truncating this expansion vanishes with increasing terms.

The moment integrals, $\langle S^i \rangle_{\alpha=0}$, all have a product convolution form and can be solved in closed form by Laplace transform methods. The first moment is given by

$$\langle S^1 \rangle_{\alpha=0} = \Psi(\frac{r+2}{2}) - \Psi(\frac{3}{2}) + \frac{1}{r}\sum_{i=1}^{r} \ln(m_i) \text{ , where } \Psi(x) = \frac{\partial \ln \Gamma(x)}{\partial x}. \quad [10]$$

The higher moments, while more complex, are also available in closed form as sums of higher order polygamma functions $\Psi^{(n)}(x)$ (Wolpert 1993a). $I(S)$ can now be computed numerically using a truncated series approximation for $Z_S(\alpha)$. (As an aside we note that it is the choice of α=0 for the expansion point which allows us to evaluate the moments.)

3. The Posterior

The posterior $P(\vec{f}|D)$ is proportional to the product of $P(\vec{f})$ (calculated above) and the likelihood function $P(D|\vec{f})$:

$$P(\vec{f}|D) = \frac{P(D|\vec{f})}{P(D)} \int_0^\infty d\alpha \, P(\vec{f}|\alpha)P(\alpha) = \frac{P(\vec{f})P(D|\vec{f})}{P(D)}. \quad [11]$$

A popular approximation to the posterior is the "evidence approximation":

$$P(\vec{f}|D) = \int_0^\infty d\alpha \, P(\vec{f}|\alpha, D)P(\alpha|D) \approx (const)\frac{P(\vec{f}|\alpha = \alpha')P(D|\vec{f})}{P(D)} \quad [12]$$

where α' is the α which maximizes $P(\alpha|D)$. (The constant fixes the overall normalization.) The following three conditions are sufficient (thought not necessary) for this to be a valid approximation:

i) $P(\alpha|D)$ is sharply peaked about some $\alpha = \alpha'$.

ii) $P(\vec{f}|\alpha, D)$, viewed as a function of α, varies little across the peak of $P(\alpha|D)$ (for those $\vec{f}$ with non-negligible posterior).

iii) $\int d\alpha \, P(\vec{f}|\alpha, D)P(\alpha|D)$, where the integral ranges over those alpha away from the evidence peak, is much less than $P(\vec{f}|D)$ (for those $\vec{f}$ with non-negligible posterior).

We note that these conditions are general and not restricted to the entropic prior; for example, they also apply when evidence is used to set a noise level. We will defer discussion of the validity of these conditions to a subsequent section. For now, we assume that (i) through (iii) hold, so that the evidence procedure works, and show how our results can improve upon that procedure.

Improving the Evidence Approximation

First, we emphasize that the calculations of section 2 give a more accurate estimate of $P(\vec{f}|D)$ than is given by the evidence procedure. Indeed, comparing [11] and [12], we see that in essence the evidence procedure approximates $P(\vec{f})$ by using a delta-function approximation to the integrand in [11]. In contrast, here we estimate this integral either by arbitrarily accurate numerical techniques (or series approximations) or by the gaussian approximation described previously. Recall that for $S \sim 0$ the width of the gaussian can be large ($\sim \frac{-1}{S}$), making a delta–function fit a relatively poor approximation.

Second, for some applications, in addition to being able to evaluate $P(\vec{f}|D)$, it is important to know the region of $\vec{f}$–space for which $P(\vec{f}|D)$ is non-negligible. For example, gradient ascent to find the peak in the posterior is facilitated by starting at an $\vec{f}$ in that region; an unfacilitated search over $P(\vec{f}|D)$ may be difficult when it has multiple maxima (Bryan 1990). In these applications, the evidence procedure might provide an efficient algorithm for estimating the location of this region. Once this region is found however, one may perform the subsequent analysis (e.g., the subsequent gradient ascent) using the more accurate approximation to the posterior surface we provide.

Third, one difficulty with the common algorithmic implementation of the evidence procedure (see Gull 1989, Bryan 1990) is that it relies on an approximation to $Z_S(\alpha)$. In this implementation α' is the solution to the "stopping condition", $-2\alpha' S(\hat{f}) = \sum_{j=1}^{n} \frac{\lambda_j}{(\lambda_j+\alpha')}$ (where the λ_j are the eigenvalues of the local hessian of the likelihood function evaluated at the peak, $\hat{f}$, of $P(f|\alpha', D)$). This equation follows, in part, from the high–β asymptotic approximation to $Z_1(\beta)$. As can be seen in Figure 1 left, this approximation goes awry below $\beta = \alpha m e \sim 100$. In this regime, the resultant error in $\alpha' S$ can lead to significant 'overfitting' of the data. One can speculate that some of the 'overfitting' Gull reported (1989) in the deblurring of the famous "Susie" images may have been due to this phenomenon: although the text is ambiguous, it seems that Susie was deconvolved using $\alpha' \sim 10^{-4.7}$ and $m_i < 10^3$, which is well outside the range of validity of the asymptotic approximation.

Fortunately, using the $Z_S(\alpha)$ calculated in section 2, we can compute a numerical correction to the stopping condition:

$$-2\alpha' S(\hat{f}) = \sum_{j=1}^{n} \frac{\lambda_j}{\lambda_j + \alpha'} - 2\alpha' \left. \frac{\partial \ln((\frac{\alpha}{2\pi})^{r/2} Z_S(\alpha))}{\partial \alpha} \right|_{\alpha=\alpha'} \quad [13]$$

The reader may visually estimate the importance and sign of the correction term by noting that the argument of the logarithm is, for a flat model, the Z_1 ratio plotted in Figure 1 left raised to the power r. For the case of the "Susie" image, it appears the correction term is actually larger than other terms in [13]. (The change the correction term induces in α' depends upon the λ_j and $\frac{\partial \lambda_j}{\partial \alpha}$, which were not available; there are also other confounding approximations we have not discussed.)

The Validity Of The Evidence Approximation

We now return to the sufficiency conditions on which the evidence approximation is almost always based. Let us examine the second condition more closely. That condition would fail if $P(\vec{f}|\alpha, D)$ changed rapidly around, α', the peak of $P(\alpha|D)$. However, Bayes theorem yields

$$P(\vec{f}|\alpha, D) = \left(\frac{1}{P(D)}\right) \frac{P(\vec{f}, \alpha, D)}{P(\alpha|D)}. \quad [14]$$

Since the highly peaked $P(\alpha|D)$ appears in the denominator, $P(\vec{f}|\alpha, D)$ is highly dipped at α' unless $P(\vec{f}, \alpha, D)$ is coincidentaly peaked there as well. In other words, unless for all

$\vec{f}$ with non-negligible posterior the joint probability $P(\vec{f}, \alpha, D)$ is peaked close (in terms of the width of the alpha–peak of the evidence) to α', condition (ii) is violated.

The authors know of no *general* reason to suppose that $P(\vec{f}, \alpha, D)$ should be peaked within the width of the peak of $P(\alpha|D)$, for the $\vec{f}$ with non-negligible posterior. Accordingly, to formally justify one's use of the evidence procedure, condition (ii) needs to be explicitly checked. Unfortunately, it appears that this is often a non-trivial task (Wolpert 1993b). However we can offer a test which is necessary for condition (ii) to hold (though not sufficient): for all $\vec{f}$ with non-negligible posterior, test whether $P(\vec{f}, \alpha, D)$ is peaked near α'. Solving for the maxima (α_{peak}) in $P(\vec{f}, \alpha, D)$ yields the condition

$$\langle S \rangle_{\alpha_{peak}} - S(\vec{f}) = \frac{\partial}{\partial \alpha} \ln[P(\alpha)]_{\alpha = \alpha_{peak}}. \qquad [15]$$

Therefore unless for all $\vec{f}$ which the evidence posterior considers to have non-negligible posterior α_{peak} lies close to α' (on the scale of the width of the peak in $P(\alpha|D)$), condition (ii) fails to hold. The value of α_{peak} may be computed using the expressions in [7]. In addition to checking condition (ii), this test also serves as a check on algorithmic error in the calculation of α' in the case that (ii) does hold. Indeed, if one knows –by other means– that condition (ii) holds, then α_{peak} may be a better choice of α than the α' estimated by the stopping condition [13].

Is This Needless Worry?

We have noticed that it is common practice to check only the first of conditions (i) through (iii). However, it was recently found that in the case of a different prior (gaussian as opposed to entropic) for which the evidence procedure was believed to work well (because $P(\alpha|D)$ is peaked), the evidence procedure actually works poorly; the other sufficiency conditions are violated (Wolpert 1992).

We can apply the $\{\langle S \rangle_{\alpha_{peak}} - S(\vec{f}) = \frac{\partial}{\partial \alpha} \ln[P(\alpha)]\}$ test to the "Susie" image in (Gull 1989). Although the text is ambiguous, we infer that Susie was reconstructed using the values of $\alpha' \sim 10^{-4.75}$, $-2\alpha' S \sim 790$, and $m \sim 155$, and we considered both the case where $P(\alpha)$ is the Jefferies prior and the case where it is flat. Assuming these values, we find $\alpha_{peak} \sim 10^{-3.5}$, which lies well outside the evidence peak. Therefore this reconstruction appears to be inconsistent with sufficiency condition (ii). (It interesting to note that this value of α_{peak} will cause less 'overfitting' and thereby produce a more pleasing image of Susie in Gull's reconstruction procedure than his inferred value of α' produced.)

Given the preceding arguments concerning (i) through (iii) it is reasonable to wonder if there might be some other basis for why the evidence procedure "works", in the sense that it gives "reasonable" looking results. Perhaps there exist other sufficiency conditions, not yet found, that are always satisfied. Alternatively, it is of interest to note that, under the evidence approximation, finding the MAP $\vec{f}$ is equivalent to regularized maximum likelihood; the MAP $\vec{f}$ is the one which minimizes $\chi^2 - \alpha' S$. As it turns out, most techniques which choose the value of the regularization parameter to give a χ^2 which is a sizeable fraction of the number of data points will tend to produce visually pleasing results. (For those primarily interested in such visually pleasing results, (Donoho 1992) examines the entropic prior in the context of regularization.) Accordingly, the fact that the evidence procedure produces a "reasonable" looking MAP estimate might merely reflect its choice of an alpha with this regularization property, and not reflect any accuracy in the evidence procedure's estimate of the Bayesian posterior.

4. Summary

When there are no constraints on the $\vec{f}$-vectors, it is feasible to numerically marginalize α from the entropic prior distribution. When a summation constraint is imposed on the $\vec{f}$-vectors it is possible to give a series representation for $Z_S(\alpha)$. The validity of the evidence approximation should be checked in any given application (peaked $P(\alpha|D)$ is *not* sufficient). Our methods allow the posterior to be computed without the evidence approximation.

ACKNOWLEDGMENTS. Los Alamos National Laboratory is operated by the University of California for the U.S. DOE under contract W-7405-ENG-36. * - D.H. Wolpert is supported by NLM grant F37-LM00011 and by the Santa Fe Institute (1660 Old Pecos Trail, Sante Fe, NM 87501).

REFERENCES

Bryan, R.K. (1990), 'Maximum Entropy Analysis of Oversampled Data Problems', Eur. Biophys. J., **18**, 165.

Donoho et al. (1992), 'Maximum Entropy and The Nearly Black Object', J. Roy. Soc. B, **54**, 41.

Gull, S.F. (1989), 'Developments in Maximum Entropy Data Analysis', in J. Skilling (ed.), *Maximum Entropy and Bayesian Methods*, Kluwer, The Netherlands.

MacKay, D. (1991), 'Bayesian Interpolation' presented at Neural Networks for Computing conference, Snowbird, Utah.

Skilling, J. (1989), 'Classic Maximum Entropy', in J. Skilling (ed.), *Maximum Entropy and Bayesian Methods*, Kluwer, The Netherlands.

Wolpert, D.H., and Wolf, D.R. (1993a), 'Estimating Functions of Probability Distributions from a Finite Set of Samples: Part I', in preparation.

Wolpert, D.H., Strauss, C.E.M., and Wolf, D.R. (1993b), 'On Evidence and the Marginalization of Alpha in the Entropic Prior', in preparation.

Wolpert, D.H. (1992), 'A Rigorous Investigation of Evidence and Occam Factors in Bayesian reasoning', Santa Fe Institute TR 92-03-013.

UNIFORMLY DISTRIBUTED SEQUENCES
AND THE MAXIMUM ENTROPY PRINCIPLE

D. Giuliani and J. L. Gruver
Universität Kaiserslautern
Fachbereich Mathematik
D-6750 Kaiserslautern, Germany

ABSTRACT. A new definition of uniformly distributed sequences based on the entropy is presented. A description of a set of points via density functions using the Maximum Entropy Principle is given and an entropic measure of uniformity is proposed.

1. Introduction

In mathematics the concept of uniformity of sets and sequences of points in an interval was in its origin based on common sense [1-3]. On the other hand in physics it is known that the state of equilibrium of an ideal monoatomic gas at constant temperature in a container is reached when the spatial density of the atoms is uniform [4]. Now thinking the points of a finite subset of a sequence as the positions of the atoms of the gas, we might assume a uniform distribution of points as a state of maximum entropy.

In analogy with statistical mechanics we want to assigne a density function which describes the points. The maximum entropy principle (MEP) and information theory (IT) [5-7] provide us with a powerful tool to solve this problem without involving arbitrary additional assumptions. In this frame, we construct a "natural density function" (see definition 1) associated with a set of points. The formalism for case where the points are in the open unit interval $(0,1) \subset \Re$ is developed.

A definition of uniformity in the entropic sense is first discussed. In the third section a natural density function is constructed and some simple examples are shown. Then an entropic measure of uniformity is proposed. At the end we present some concluding remarks.

2. Uniformity as a Maximum Entropy State

Let us analyze the meaning of uniform distribution of a single point in $(0,1)$. A point $p \in (0,1)$ will not be thought as a mathematical point but as the average position of an atom. With this interpretation in mind we want to construct a density function $f^{(p)}$ that represents the point p. Therefore $f^{(p)}$ must satisfy the relations

$$\int_0^1 f^{(p)}(x)\, dx = 1 \,, \tag{1}$$

A. Mohammad-Djafari and G. Demoments (eds.), Maximum Entropy and Bayesian Methods, 121–125.

$$\int_0^1 x f^{(p)}(x)\, dx = p\,. \tag{2}$$

It is known that given $p \in (0,1)$ one can find many density functions that satisfy eqs. (1-2). This lack of unicity leads to consider what the best density $f^{(p)}$ that represents the point p might be, with the only information of p. The method of maximum entropy provides a functional form, namely the entropy

$$S(f) = -\int_0^1 f(x)\, ln f(x)\, dx\,, \tag{3}$$

to be optimized with the only constraints (1-2) [6]. Using the usual procedure of introducing an auxiliary functional with undetermined Lagrange multipliers to include the constraints and functionally differentiating with respect to f yield the maximum-entropy solution of the problem:

$$f^{(p)}(x) = exp(-\lambda_0 - \lambda_1 x)\,,$$

where the values of λ_0 and λ_1 are uniquely determined by the conditions (1-2).

We can now pose the question: For which $p \in (0,1)$ is $S(f^{(p)})$ maximum? The answer is: $p = \frac{1}{2}$.

What happens when there is more than one point? Keeping in mind the analogy between the points and the positions of the atoms of the ideal gas, consider two one-dimensional containers having each one atom. Without loss of generality suppose that the length of each container is equal to $\frac{1}{2}$. The state of equilibrium in each container is reached when the entropy is equal to zero and that means that the atom must be in the middle of the container. Both containers being identical, the entropy of the total system will also be equal to zero. Therefore the points $\frac{1}{4}$ and $\frac{3}{4}$ maximize the entropy in the whole interval of length 1 and are uniformly distributed in the entropic sense. It is clear that for more than two points the same reasoning can be used and leads to the following assertion:

The set $\omega_N^u := \left\{\frac{i-\frac{1}{2}}{N}\right\}_{i=1}^{N}$ is uniformly distributed in the entropic sense.

We recall that the set of points ω_N^u is also the best uniformly distributed from the number theoretical point of view [3].

Having the points we still have the problem of their characterization via a density function. The following definition associates a density function to a set of points in a very general way. The definition is mathematically vague, but since we do not want to restrict the defined object, we rather prefer to say it is an *informal* definition:

Definition 1 *A "natural density function" f_N associated with a set of points ω_N is a density function that maximizes the Boltzmann entropy subject to constraints exclusively derived from the positions of the points.*

In the case of the set ω_N^u we call the density function u_N.

We need now to transform the points into a set of constraints which can be used in practice. In the next section we present a mapping T that performs such transformation. When u_N is identically 1 we will say that this mapping T preserves uniformity.

In the case of sequences the uniformity is an asymptotic property. This leads to the following definition:

Definition 2 *A sequence of points $\omega = \{x_i\}_{i=1}^{\infty} \subset (0,1)$ is "asymptotically uniformly distributed" in the entropic sense if and only if*

$$\lim_{N\to\infty} S_{u_N}(f_N) := \lim_{N\to\infty} \int_0^1 f_N(x)\, ln\left(\frac{f_N(x)}{u_N(x)}\right) dx = 0\,.$$

3. Construction of a Natural Density Function

In this section a way of constructing a natural density function is proposed.

We choose the mapping T defined by

$$T : D \subset \Re^N \longrightarrow T(D) \subset D$$

$$T(\mathbf{x}) := \frac{1}{N}\Big(\sum_{i=1}^{N} x_i, \sum_{i=1}^{N} x_i^2, \ldots, \sum_{i=1}^{N} x_i^N\Big)$$

where

$$D := \{\mathbf{x} = (x_1, x_2, \ldots, x_N) \in (0,1)^N : x_1 < x_2 < \ldots < x_N\},$$

This mapping transforms a set of N points into the set of their first N moments and is bijective, and therefore provides a set of moments which can be used as constraints. We make use of the method of moments [8-10] to construct a natural density function associated with the set of points. The moment problem seeks for a positive density function $f(x)$ from knowledge of its power moments

$$\int_0^1 x^k\, f(x)\, dx \;=\; \mu_k\,,\quad k = 0, 1, 2, \ldots\,.$$

In practice, only a finite number $N+1$ of moments (without loss of generality we assume that the zero order moment $\mu_0 = 1$) are available, and there are many different density functions whose first $N+1$ moments coincide. As before, the maximum entropy approach offers a criterium to choose one of them [9].

Maximizing the entropy functional defined by (3) subject to the moment constraints

$$\int_0^1 x^k\, f(x)\, dx \;=\; \mu_k\,,\quad k = 0, 1, \ldots, N, \tag{4}$$

and introducing Lagrange multipliers to include the $N+1$ constraints (4) results in the entropy functional form

$$-\int_0^1 f(x)\, ln f(x)\, dx + \sum_{k=0}^{N} \lambda_k \left(\int_0^1 x^k f(x)\, dx \;-\; \mu_k\right). \tag{5}$$

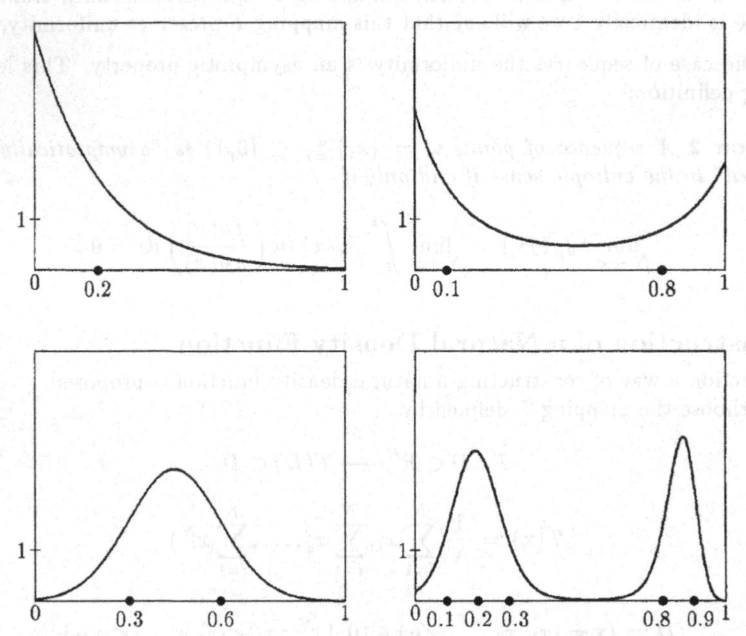

Fig. 1. Natural density functions for $N = 1, 2$ and 5 and for different locations of the points $x_1, \ldots, x_N$.

The maximum-entropy solution exists and has the form

$$f_N(x) = exp\left(-\lambda_0 - \sum_{k=1}^{N} \lambda_k x^k\right), \tag{6}$$

where the Lagrange multipliers $\lambda_0, \lambda_1, \ldots, \lambda_N$ are uniquely determined by the constraints (4). For details we refer e.g. to [9,10].

To give the reader a flavour of the f_N's, some figures (fig. 1) for $N = 1, 2$ and 5 are included. The numerical simulations were performed using a Newton minimization procedure.

4. The Entropic Measure

A measure of uniformity, the discrepancy, was introduced by Bergström [11] and has been widely used. Although easy to calculate in one dimension, it becomes more and more complicated in higher dimensions. To overcome this problem a new measure of uniformity, the range, was proposed by Sobol [2]. Although from the computational point of view the range is much easier to calculate than the discrepancy, it lacks sensitivity.

We propose as a measure of uniformity of a set $\omega_N = \{x_i\}_{i=1}^N \subset (0, 1)$ the following:

$$S_{u_N}(f_N) = \int_0^1 f_N(x)\, ln\left(\frac{f_N(x)}{u_N(x)}\right) dx\ .$$

Even if estimates were not yet developed, the great advantage of this measure is that it gives information about the uniformity of a sequence taking into account the information that every point carries.

5. Conclusions

A new point of view to describe uniformity of sequences was presented. The MEP guarantees that in the continuous representation of sets of points via natural density functions no arbitrary information is added. In the construction of the natural density functions we have chosen the moments of the set of points as constraints, but do not discard other constructions.

We intend to find a discretized version of our model which may considerably simplify the numerical aspect and will be probably based on a discretized version of the Boltzmann's entropy, namely the Shannon's entropy.

In the future, generalizations to more than one dimension will be investigated.

ACKNOWLEDGMENTS. D. G. thanks the Graduiertenkolleg Technomathematik of the University of Kaiserslautern for its financial support. J. L. G. acknowledges support from the German Academic Exchange Service (DAAD). Both want to thank Professor Dr. R. Illner and Dr. J. J. Valdes for a carefully reading of the work and useful suggestions. Both are glad that this work could be done in the frame of the A.G. Technomathematik directed by Professor Dr. H. Neunzert.

REFERENCES

[1] Weyl H., 'Über die Gleichverteilung von Zahlen mod. Eins', *Math. Ann.* **77**, 313-352, (1916).

[2] Sobol I.M., 'Punkte, die einen mehrdimensionalen Würfel gleichmässig ausfüllen', Bericht 51, A.G.T. University of Kaiserslautern, (1991).

[3] Kuipers L., Niederreiter H., '*Uniform Distribution of Sequences*', Wiley & Sons, (1974).

[4] Reif F., '*Fundamentals of Statistical and Thermal Physics*', Mc Graw-Hill, (1965).

[5] Jaynes E.T., 'Information Theory and Statistical Mechanics', *Phys. Rev.* **106** (4), 620-630, (1957).

[6] Drabld D.A., Carlsson A.E. and Fedders P.A., 'Applications of Maximum Entropy to Condensed Matter Physics', in J. Skilling (ed.), *Maximum Entropy and Bayesian Methods*, Kluwer, Dordrecht, (1989).

[7] Aliaga J., Gruver J.L. and Proto A.N., 'Generalized Bloch equations for time-dependent N-level systems', *Phys. Lett. A* **153**, 317-320, (1991).

[8] Widder D.V., '*The Laplace Transform*', Princeton U. P., Princeton, (1946).

[9] Mead L.R. and Papanicolaou N., 'Maximum entropy in the problem of moments', *J. Math. Phys.* **25** (8), 2404-2417, (1984).

[10] Mohammad-Djafari A., 'A Matlab Program to calculate the Maximum Entropy Distributions', *Proc of the 10th Int MaxEnt Workshop, Laramie, Wyoming*, published in Maximum-entropy and Bayesian Methods, T.W. Grandy ed., (1990).

[11] Bergström V., 'Einige Bemerkungen zur Theorie der diophantischen Approximationen', *Fysiogr. Sälsk. Lund. Förh.* **6**, 1-19, (1936).

DUALITY WITHOUT TEARS

Claude Lemaréchal
INRIA – Domaine de Voluceau
BP 105 – Rocquencourt
78153 Le Chesnay Cédex – France

ABSTRACT. Duality is a theoretical tool to transform an optimization problem with constraints into another optimization problem, without constraints. This technique is by no means new, but the aim of these notes is to explain its essentials, in a language as intelligible as possible.

1. Introduction

Let X be a set, and let be given $m+1$ real-valued function $f, c_1, \ldots, c_m$ defined on X.

Consider the optimization problem:

$$\begin{cases} \min_{x \in X} f(x) \\ c_j(x) = 0 \quad j = 1, \ldots, m; \end{cases} \tag{1}$$

f is the **objective function,** the c_j are the **constraints.**

The aim of duality is to eliminate the constraints to obtain another problem, which in some cases is simpler to study or to solve numerically. Its principle can be attributed to Legendre but it was fully developed some 40 years ago (W. Fenchel, J.J. Moreau, ...).

For each given real numbers $u_1, u_2, \ldots, u_m$, consider the real number

$$L(x, u) = f(x) + \sum_{j=1}^{m} u_j c_j(x). \tag{2}$$

Thus, L is a real valued function defined on $X \times R^m$, the **Lagrange function.** Suppose that, for fixed $u \in R^m$, one minimizes the Lagrange function with respect to x varying in the whole set X. This gives the real number:

$$q(u) = \min\{L(x, u) \quad / \quad x \in X\}. \tag{3}$$

Therefore q is a real-valued function defined on R^m. The u_j are called the **dual variables,** and q is the **dual function.**

We remark that:

- Replacing a constraint $c_j(x) = 0$ by a linear term $u_j c_j(x)$ in the objective function can be interpreted as adding to f a linear cost, with marginal price u_j, which has to be paid whenever the constraint is violated.

A. Mohammad-Djafari and G. Demoments (eds.), Maximum Entropy and Bayesian Methods, 127–132.

- Instead of solving (1), we are now faced with the unconstrained problem (3), which is supposedly preferable in some sense (otherwise, the approach is useless). We will see in the next sections how (3) can help solving (1).
- The approach is purely abstract in the sense that we have made absolutely no assumption on X (it must be a set ...) nor on f or c_j (they can be discontinuous ...even continuity is an irrelevant concept, since X can be a finite set).
- The writing in (3) is a bit unformal because the minimum may not exist. Consider for example $X = R$, $m = 1$, $f(x) \equiv 0$, $c(x) = x$. Then, $L(x, u) = ux$ and $q(u)$ does not exist when $u \neq 0$; however, in such cases one adopts the convention $q(u) = -\infty$.

2. Basic properties

Theorem 1 *For any x feasible in (1) (i.e. in X and such that $c_j(x) = 0$, $j = 1, \ldots, m$) and any $u \in R^m$, there holds*

$$f(x) \geq q(u).$$

Proof: By definition of L and q, we have for any $x \in X$ and $u \in R^m$,

$$q(u) \leq L(x, u) = f(x) + \sum u_j c_j(x).$$

In particular, if x is feasible, then $L(x, u) = f(x)$. □

Theorem 2 *Let $\bar{x} \in X$ and $u \in R^m$ be such that $L(\bar{x}, u) = q(u)$. Set $a_j = c_j(\bar{x})$, $j = 1, \ldots, m$. Then $\bar{x}$ is a solution of:*

$$\begin{cases} \min_{x \in X} f(x) \\ c_j(x) = a_j \quad j = 1, \ldots, m. \end{cases} \tag{4}$$

Proof: By definition of a, $\bar{x}$ is feasible in (4). By definition of $\bar{x}$ and u, we have for any $x \in X$:

$$f(\bar{x}) + \sum u_j a_j = L(\bar{x}, u) = q(u) \leq L(x, u) = f(x) + \sum u_j c_j(x).$$

In particular, if x is feasible in (4), then $u_j a_j = u_j c_j(x)$; hence $\bar{x}$ is optimal in (4). □

These two simple results clearly show how (3) can be useful in terms of (1): suppose that, for some $u \in R^m$, one solves (3) to obtain some solution $\bar{x}$. This $\bar{x}$ has the following properties:

i) $L(\bar{x}, u)$ is a lower bound on the optimal cost in (1).

ii) If, by any chance, $\bar{x}$ is feasible in (1), it is also optimal in (1).

iii) If $\bar{x}$ is "almost" feasible in (1), i.e. if all the $c_j(\bar{x})$ are close to 0, then $\bar{x}$ is "almost" a solution (because, then, (4) can be considered as a perturbation of (1)).

It remains to see how u can be found so that the corresponding $\bar{x}$ is a solution of (1). Actually, it is easy to guess:

Corollary 3 *If u is such that there exists $\tilde{x}$ satisfying*

$$L(\tilde{x}, u) = q(u) \text{ and } c_j(\tilde{x}) = 0 \qquad j = 1, \ldots, m$$

then u maximizes q over R^m.

Proof: In the stated circumstances, we have:

$$f(\tilde{x}) = L(\tilde{x}, u) = q(u).$$

Therefore Theorem 1 implies that, for all v in R^m:

$$q(v) \leq f(\tilde{x}) = q(u). \qquad \square$$

Thus maximizing q not only gives the tightest possible lower bound for the optimal cost in (1); it is also the only chance to obtain an optimal x.

However, it should be noted that the converse of Corollary 3 is not true: if u maximizes q, then the corresponding $\tilde{x}$ may not be feasible for (1).

3. Maximizing the dual

Duality consists in replacing (1) by two nested optimization problems without constraints; the inner one (minimizing the Lagrange function) depends only on the original problem, but the outer one (maximizing the dual function) has some general properties which do not depend on the particular X, f and c_j.

Theorem 4 *q is concave.*

Proof: We have to show that, for any $u \in R^m$, $v \in R^m$ and $t \in [0,1]$

$$tq(u) + (1-t)q(v) \leq q(tu + (1-t)v).$$

Take arbitrary $u \in R^m$, $v \in R^m$ and $x \in X$. From the definition of q and from the fact that t and $(1-t)$ are nonnegative:

$$\begin{aligned} tq(u) + (1-t)q(v) &\leq tL(x,u) + (1-t)L(x,v) \\ &= (t + 1 - t)f(x) + t\sum u_j c_j(x) + (1-t)\sum v_j c_j(x) \\ &= f(x) + \sum [tu_j + (1-t)v_j] c_j(x) = L(x, tu + (1-t)v). \end{aligned}$$

Now suppose that q is not concave. Then, we would have:

$$q(tu + (1-t)v) < tq(u) + (1-t)q(v) \leq L(x, tu + (1-t)v) \quad \forall x \in X$$

which contradicts the definition of q because $q(tu + (1-t)v)$ could not be reached by L values. $\square$

This result means that the dual problem is well-posed, in the sense that its solutions are defined without ambiguity.

Theorem 5 *Suppose that a certain technical assumption is satisfied, concerning continuity in (1). If, for all $u \in R^m$, L is minimized at a unique $\tilde{x}(u)$, then*

$$\frac{\partial q}{\partial u_j}(u) = c_j(\tilde{x}(u)).$$

Proof: We cannot prove this statement in its full generality (it requires a profound knowledge of convex analysis). Therefore we will restrict ourselves to the simple situation when $X = R^n$, f and c_j are differentiable, and $\tilde{x}(u)$, a smooth function of u. Then $q(u) = L(\tilde{x}(u), u)$, hence

$$\frac{\partial q}{\partial u_j}(u) = \frac{\partial L}{\partial u_j}(\tilde{x}(u), u) + \sum_{i=1}^{n} \frac{\partial L}{\partial x_i}(\tilde{x}(u), u) \frac{\partial \tilde{x}_i}{\partial u_j}(u).$$

But since $\tilde{x}(u)$ is minimal, it is stationary hence the last sum is zero and the only remaining term is $\dfrac{\partial L}{\partial u_j} = c_j(\tilde{x})$. □

These two results enlight Corollary 3 of the previous section: to say that a given $\tilde{u}$ corresponds to a feasible $\tilde{x}$ is to say that the gradient of $q(\tilde{u})$ is 0, i.e., that $\tilde{u}$ maximizes the concave function q. However, Theorem 5 does not say that q is always differentiable, and the following example shows that duality may not work: $X = [-1, 1]$, $m = 1$, $f(x) \equiv -x^2$, $c(x) \equiv x$. Therefore the problem is:

$$\min -x^2 \qquad -1 \leq x \leq +1 \qquad x = 0$$

whose solution is $x = 0$, the only feasible point, for which the optimal value is $f(0) = 0$. The Lagrange function is:

$$L(x, u) = -x^2 + ux$$

and the dual function is obtained by solving the parametrized problem

$$\min_{-1 \leq x \leq +1} \quad -x^2 + ux.$$

It is easy to see that:

for $u > 0$ the solution is $\tilde{x}(u) = -1$ and $q(u) = -1 - u$
for $u > 0$ the solution is $\tilde{x}(u) = 1$ and $q(u) = -1 + u$

Unfortunately, q is maximized at $\tilde{u} = 0$, the only u for which q is not differentiable (because both $\tilde{x} = -1$ and $\tilde{x} = 1$ minimize $L(x, 0)$).

To conclude this section, we can say that duality will safely solve (1) if the Lagrange function has a unique minimizer when the dual variables are at their dual optimal value; and this, in practice, requires X to be convex and L to be strictly convex with respect to x.

4. Two illustrative examples in crystallography

Example 1: Consider the problem

$$\begin{cases} \min\limits_{p \geq 0} \sum\limits_{r \in D} p_r \log p_r \\ F_h(p) = y_h, \qquad h \in H \end{cases} \tag{5}$$

where p is a (real, periodical) unknown function defined on a given (discretized) domain D, H is a given set of integers, and for each $h \in H$, y_h is a given complex number, while $F_h(p) = \sum\limits_{r \in D} p_r e^{ihr}$ is the hth Fourier coefficient of p.

To adapt the results of sections 1 to 3, we take X as the set of positive functions p defined on D; m is twice the number of elements in H; to each complex constraint, we associate a complex dual variable u_h. We have to place ourselves in the real field, i.e. to split the real and imaginary parts and we observe that, if u and z are two complex numbers,

$$Re(u).Re(z) + Im(u).Im(z) = \frac{1}{2}(uz^* + u^*z).$$

This relation can be conveniently used to write the Lagrange function in the form:

$$L(p,u) = \sum_{r\in D} p_r \log p_r + \frac{1}{2}\sum_{h\in H}[u_h^*(F_h(p) - y_h) + u_h(F_h(p) - y_h)^*].$$

Setting

$$a_r(u) = \frac{1}{2}\sum_{h\in H}[u_h^* e^{ihr} + u_h e^{-ihr}]$$

we obtain

$$L(p,u) = \sum_{r\in D}[p_r \log p_r + a_r(u)p_r] - \frac{1}{2}\sum_{h\in H}[u_h y_h^* + u_h^* y_h]. \tag{6}$$

This is a strictly convex function of p; equating its derivatives to zero gives:

$$p_r(u) = e^{-1-a_r(u)} \quad \text{for each } r \text{ in } D. \tag{7}$$

Plugging this value into (6), readily gives $q(u)$ and its partial derivatives.

Thus (5) can easily be solved by any relevant unconstrained optimization algorithm applied to maximize q, such as gradient method, or even Newton method (differentiating $F_h[p(u)] - y_h$ with respect to u and u^* readily gives the hessian matrix of q).

Example 2:

$$\begin{cases} \min \sum_{r\in D} p_r \log p_r \\ F_h(p) = y_h & h \in H \\ |y_h| = I_h & h \in H. \end{cases} \tag{8}$$

The notations are as in (5), but now the y_h's are also unknown complex variables subject to modulus-constraints.

A possibility to form a Lagrange function is to take

$$X = \{p, y \quad / \quad p_r \geq 0, \quad |y_h| = I_h\}$$

i.e. to keep the nonlinear constraints explicitly, and to dualize only the linear constraints. Thus, we obtain the expression (6) as before but now, it must be minimized with respect to p and y.

Minimizing with respect to p is as before and still gives formula (7). To compute the optimal y, we see that we must solve for each $h \in H$:

$$\begin{cases} \max u_h^* y_h + u_h y_h^* \\ y_h y_h^* = I_h^2 \end{cases}$$

whose solution is $y_h(u_h) = \frac{u_h}{|u_h|} I_h$ (observe that the function to be maximized is the length of the projection of y_h onto u_h, considered as vectors in R^2; this length is maximal when the vectors are colinear).

We conclude that the conditions of Theorem 5 are fulfilled if $u_h \neq 0$ for any $h \in H$. If some $u_h = 0$, then every y_h with $|y_h| = I_h$ is optimal, and the gradient of q disappears. Unfortunately, the dual function is precisely maximized at $u = 0$. To see this, observe that:

$$q(u) = \bar{q}(u) - \sum_{h \in H} |u_h| I_h \tag{9}$$

where $\bar{q}(u)$ is the dual function of Example 1 with $y = 0$; it is maximal at $u = 0$. Therefore, when u deviates from 0, $\bar{q}$ increases with an order $|u|^2$ while the second sum in (9) decreases with an order $|u|$. Globally q decreases.

5. Some bibliographical references

A more complete exposition of duality is given in [5].

In the present paper, we have intentionally left out some aspects of duality, which require a rather sound culture in a branch of mathematics: convex analysis. Along these lines, [8] is the best book, although it is difficult to read for a non-mathematician. A pedagogical introduction to the basic concepts in convex analysis is done in the Appendices of [7] (a good and instructive text making [6] readable), which gives a minimal presentation of duality, similar to the present paper. On the other hand [1] contains most of the necessary material, is easier to read, and gives the theory of *augmented Lagrangians*, which is useful when duality does not work.

Finally, we mention [2] for a way to motivate the introduction of the Lagrange function, and [4] for a few words on problems of the type of Example 1 in section 4. A proposal to solve Example 2 is given in [3].

References

[1] D.P. Bertsekas, "*Constrained optimization and Lagrange multiplier methods*", Academic Press, (1982).

[2] E. Cansado, "Dual programming problems as hemi-games", Management Science, 15, 9, (1969), 539-549.

[3] A. Decarreau, D. Hilhorst, C. Lemaréchal, J. Navaza, "Dual methods in entropy maximization. Application to some problems in crystallography", SIAM Journal on Optimization, 2, 2(1992), 173-197.

[4] J. Eriksson, "A note on solution of large sparse maximum entropy problems with linear equality constraints", Mathematical Programming 18, 2, (1980), 146-154.

[5] A.M. Geoffrion, "Duality in nonlinear programming : a simplified applications oriented development", SIAM Review 13, 1 (1971), 1-37.

[6] R.C. Grinold, "Lagrangian subgradients", Management Science, 17, 3 (1970), 185-188.

[7] L.S. Lasdon, "*Optimization methods for large scale problems*", Mac Millan (1970).

[8] R.T. Rockafellar, "*Convex analysis*", Princeton University Press, (1970).

Chapter 2

Quantum Physics and Quantum Information

ENTROPY, CHAOS, AND QUANTUM MECHANICS

M. Hossein Partovi
Department of Physics and Astronomy
California State University, Sacramento
Sacramento, California 95819 USA

ABSTRACT.
A characteristic feature of quantum theory is the nontrivial manner in which information about a quantum system is inferred from measurements. These measurements are in general incomplete and do not provide an exhaustive determination of the state of the system. Incomplete information is thus an inherent feature of microphysics and naturally calls for the application of entropic methods. Such methods have in recent years been successfully applied to a number of important problems in quantum dynamics. Here, we will concentrate on a recent formulation of chaos using entropic methods. This formalism, which is equally valid for classical and quantum dynamics, provides a new way of extracting Lyapunov exponents from measured or computed data. The basic quantity is the measurement entropy associated with a dynamical variable of the system, the characteristic exponents being related to the asymptotic growth rate of this entropy. Examples of the use of this method in quantum and classical dynamics, as well as a recent demonstration of the absence of sensitivity to initial conditions in quantum mechanics are discussed. Other applications of entropic methods such as the quantum maximum entropy principle, quantum thermodynamics, and decay of correlations in an open system are briefly recalled.

1. Introduction

Understanding natural phenomena at the macroscopic level often requires understanding the statistical behavior of large aggregates of constituents in addition to the dynamics of the constituents at the microscopic level. The classic example is the general program of statistical mechanics, while specific cases include the kinetic theory of dilute systems, many-body systems and condensed matter physics, the measurement problem of quantum theory (where the coupling of the quantum system to a macroscopically large number of degrees of freedom during the act of measurement introduces a stochastic element into the dynamics) and the related problem of *classicality* (i.e., the apparent suspension of the superposition principle at the macroscopic level and the emergence of classical behavior), the Darwinian picture of evolution by natural selection and the associated problem of self-organization and the origin of life, all of which may be referred to as the study of *complex systems*. An essential component of such studies is the dynamics of *information*, the fundamental concept of statistical inference, and the concommitant use of entropic methods. Thus there is a profound universality to the use of entropic methods in all such problems, as is evident in the broad range of topics represented at this meeting. In particular, the formulation of entropic ideas in the quantum context is predicated on the recognition that in dealing with the measurement aspects of quantum theory one is really dealing with a complex system.

Despite its apparent simplicity, the manner in which information about a quantum system is extracted from measurements is highly nontrivial. Strictly speaking, a quantum

A. Mohammad-Djafari and G. Demoments (eds.), Maximum Entropy and Bayesian Methods, 135–144.

measurement is in general incomplete since the resulting information fails to provide an exhaustive determination of the state of the system. By *measurement* we mean not the procedure that produces information on the state of a single quantum system just after the measurement (with the corresponding pre-measurement quantities subject to uncertainty), but rather the process by which a fraction of an ensemble of similarly prepared systems is subjected to interaction with measuring devices, and the collective results of such single measurements are used to assign a state to the ensemble. The fact that such a measurement is in general incomplete follows from the simple observation that a system described by an N-dimensional Hilbert space requires the determination of $N^2 - 1$ elements comprising its density matrix and that N is in general infinite. Thus a quantum measurement in general entails an inverse problem without complete information. This remark should make it clear why the methods of statistical inference and the notion of entropy are important tools in dealing with the problems of quantum dynamics.

The main purpose of this paper is to discuss the application of quantum entropic methods to the characterization of chaos in quantum dynamics. Other applications of these methods will also be briefly mentioned to demonstrate the general power and utility of the entropic ideas. It should also be mentioned here that the presentation is to some degree reflective of the author's perspective on certain issues at the foundation of quantum physics, notably the issues connected with the measurement problem of quantum mechanics and the reversibility problem of statistical mechanics.

2. Measurement and Ensemble Entropies

The primitive notion of entropy in quantum mechanics is that of *measurement entropy* [1,2]. To define it, we need to introduce a bit of notation first. In general, a quantum measurement involves an observable, $\hat{A}$, and a measuring device, D^A. The latter includes a partitioning of the spectrum of $\hat{A}$ into a collection of disjoint sets, $\{\alpha_i\}$, named *bins*, together with a set of projection operators $\{\hat{\pi}_i^A\}$ corresponding to the bins (a caret is used to signify the operator nature of a quantity). Clearly, the union of the bins, $\cup_i \alpha_i^A$, equals the spectrum of $\hat{A}$, the corresponding operator statement being the resolution of identity $\sum_i \hat{\pi}_i^A = \hat{1}$. The results of the measurement are summarized in a set of probabilities $\{\mathcal{P}_i^A\}$, where $\mathcal{P}_i^A$ is the probability that the measured value of $\hat{A}$ turns up in the bin α_i^A. According to the rules of quantum mechanics, $\mathcal{P}_i^A = \mathrm{tr}(\hat{\pi}_i^A \hat{\rho})$, where $\hat{\rho}$ is the density matrix specifying the state of the system.

Measurement entropy is now simply defined be the information entropy associated with the measurement of $\hat{A}$;

$$S(\hat{\rho}|D^A) = -\sum_i \mathcal{P}_i^A ln \mathcal{P}_i^A. \tag{1}$$

Note that $0 \leq S(\hat{\rho}|D^A) \leq \ln(number\ of\ bins)$, and that a refinement of the partition $\{\alpha_i\}$ will in general cause $S(\hat{\rho}|D^A)$ to increase (or remain unchanged). This circumstance corresponds to the fact that a device with a higher resolution is capable of extracting more information. Indeed it is useful to consider a maximal device, D^A_{max}, as the (idealized) limit of a series of devices with ever finer partitions. Clearly, $S(\hat{\rho}|D^A) \leq S(\hat{\rho}|D^A_{\mathrm{max}})$.

For a given state $\hat{\rho}$, there is in general an infinite set of maximal measurement entropies, each associated with a different observable. To identify an overall measure of uncertainty for the ensemble, we should look for the observable that embodies the most information, hence the least entropy. Therefore, we define the *infimum* of the set of maximal measurement entropies to be the overall measure of uncertainty for the ensemble;

$$S(\hat{\rho}) = \inf_{\hat{A}} S(\hat{\rho}|D^{A}_{\max}). \tag{2}$$

It is not difficult to show that the right-hand side of Eq.(2) is equal to $-\mathrm{tr}(\hat{\rho}\ln\hat{\rho})$, the standard formula for *ensemble* or *Boltzmann-Gibbs-Shannon* entropy. It is highly gratifying that the transition from the measurement entropy (observable-specific) to the ensemble entropy (universal) occurs in exact correspondence with the physical meaning of these quantities.

Although we will later recall some interesting applications of the above ideas, our main focus in the following is the entropic formulation of chaos.

3. Chaos

Chaotic phenomena have in recent years attracted a great deal of attention and activity, partly owing to the widespread availability of computers. Chaos is a ubiquitous phenomenon, one that is a general feature of any sufficiently nonlinear, bounded, deterministic dynamical system of three or more phase- space dimensions (two or more for a driven system). A perusal of the papers in Ref. [3] shows the impressive breadth of chaotic phenomena; starting from the first encounter by Poincaré in his studies of celestial mechanics, they are manifested in such diverse systems as atmospheric dynamics, the drip pattern of a leaky faucet, the oscillations of a nonlinear circuit, the dynamics of model economic systems, or the seasonal variations of biological and epidemiological populations. More specifically, however, chaos appears to play an important role in the understanding of turbulent phenomena in fluids and the problem of approach to equilibrium in statistical mechanics, two long-standing problems of classical physics.

Despite the varied nature of phenomena associated with it, the distinguishing characteristic of chaos is quite simple: it is the exponential sensitivity of the motion with respect to initial conditions. To wit, two possible motions of the system, started from nearly identical initial conditions, will in the course of time diverge at a constant average rate such that the distance separating them will grow in an exponential manner. The metric notions in this definition refer to the phase space endowed with the standard Liouville metric. Furthermore, chaos is only relevant to, and defined for, bounded motions, so that all statements concerning diverging trajectories and exponentially increasing distances should be understood to be subject to saturation as trajectories impinge on phase space boundaries.

An immediate consequence of the exponential divergence property is the steady loss of information on the predicated trajectory of a chaotic system. This loss occurs since the inevitable uncertainty in the specification of the initial state of the system (resulting from the finite precision of measured data, the finite number of significant figures in data processing, etc.) will in the course of time amplify exponentially until it is essentially as large as it can be, i.e., comparable to the dimensions of the available phase volume. At this point there is essentially a situation of maximum ignorance with respect to the whereabouts of the system point within the allowed volume in phase space.

The standard measure of chaos is a set of numbers known as the *Lyapunov spectrum*, or *characteristic exponents,* which characterizes the mean, long-time rate of expansions and contractions in various directions as system points evolve in phase space. Indeed the largest of these numbers, the maximum Lyapunov exponent, in general dominates all the others for long times and determines the rate at which the uncertainty in the location of the system point in phase space increases with time.

To state these ideas quantitatively, let us consider a differentiable dynamical system characterized by the variables ξ_i, $i = 1, \cdots, N$, and N first-order equations of motion

$$\dot{\xi}_i = F_i(\xi, t), \tag{3}$$

where ξ represents the N variables collectively. Furthermore, we denote the trajectory that starts from ξ_0 at time $t = 0$ by $R_i(t|\xi_0)$. Thus the trajectory that starts from ξ_0 at $t = 0$ reaches $\xi_i(t) = R_i(t|\xi_0)$ at time t. Next we consider a nearby trajectory starting at $\xi_0' = \xi_0 + \delta\xi_0$ at $t = 0$, and from these find the orbital displacement

$$\delta\xi_i(t) = R_i(t|\xi_0 + \delta\xi_0) - R_i(t|\xi_0). \tag{4}$$

The information we are seeking is embodied in the *sensitivity matrix*, which is simply the matrix of partial derivatives

$$T_{ij}(t|\xi_0) = \lim_{\xi \to \xi'} [\delta\xi_i(t)/\delta\xi_j(0)] = \partial R_i(t|\xi_0)/\partial\xi_{0j}. \tag{5}$$

Now a rather powerful result known as Oseledec's theorem [4] states that if ξ_0 is a point in a bounded, invariant, indecomposable submanifold of phase space, then the limit

$$\Lambda = \lim_{t \to \infty} \frac{1}{2t} \ln[T^\dagger(t|\xi_0) T(t|\xi_0)] \tag{6}$$

exists and is a matrix independent of ξ_0 ("$\dagger$" denotes the adjoint matrix). The eigenvalues of the self-adjoint matrix Λ constitute the Lyapunov spectrum of characteristic exponents; the value zero is always included in the spectrum. Moreover, for a (closed) Hamiltonian system, T is a symplectic matrix, so that when λ is in the spectrum of Λ, so is $-\lambda$; here at least two of the eigenvalues vanish.

As discussed earlier, the phase space distance $d(t) = [\delta\xi_i(t)\delta\xi_i(t)]^{1/2}$ between neighboring trajectories in general grows exponentially with a rate equal to the largest eigenvalue $\lambda_{\max}$. Therefore

$$\lambda_{\max} = \lim_{t \to \infty} \lim_{d(0) \to 0} \frac{1}{t} \ln[d(t)/d(0)]. \tag{7}$$

Eq.(7) is commonly used in numerical studies of chaotic systems.

For Hamiltonian systems, there is a hierarchy of physically significant levels of complexity in the dynamical behavior of the system. The weakest of these is ergodicity, a property that guarantees the equality of time and phase space averages, but not approach to equilibrium. The next level is the *mixing* property, which guarantees that a set of system points in phase space evolves in time in such a manner as to overlap any subset of phase space in proportion to the latter's volume. The mixing property also guarantees the decay of correlations and ensures approach to equilibrium (with no restriction on the speed of this approach) for ensembles of phase points. Furthermore, a sufficient condition for a system to be mixing is sensitive dependence (not necessarily exponential) on initial conditions, that is, an unbounded growth with time of $d(t)/d(0)$. The third important level is chaos, sometimes called *strong mixing*, which is characterized by *exponential* sensitivity to initial conditions as signaled by positive Lyapunov exponents. As discussed earlier, exponential sensitivity implies a steady loss of information and therefore prohibits long-term predictability. For this reason, the (long-time) dynamical output of the system is often characterized as *random*, or *seemingly random*. Actually, in a very real sense, i.e., in the sense of *algorithmic complexity*, the output of a chaotic system *is* random [5]. Briefly, algorithmic complexity of a sequence is the length of the shortest algorithm, or computer code, that will cause a computer to generate the sequence. The complexity of a *random* sequence is essentially the length of the sequence, since there does not exist a more efficient way of computing a random sequence than basically copying it element by element. Insofar as a finite computer will in a finite period of time lose all accuracy in computing the trajectory of a chaotic system, one can see that the long-term behavior of a chaotic trajectory is as complex as that of a random sequence.

Before concluding this section, we should mention an important, general formulation of chaos devised by Kolmogorov [6] and known as the Kolmogorov-Sinai (KS) entropy (or KS invariant). We will not define the KS invariant here — to do that properly requires certain preliminaries — but merely point out that it is essentially an information-theoretic measure of how rapidly the location of a system point in phase space becomes uncertain.

4. Measuring Quantum Chaos

It should be clear from the above discussion that chaotic behavior is characterized by precisely defined statistical features in the dynamical output of a system, and as such ought to be capable of formulation independently of the underlying dynamics. It is therefore reasonable to expect that there exists a commonly accepted definition of quantum chaos on the basis of an appropriately generalized version of one of the measures of stochasticity discussed above. This is not the case [7], however, although there is a rather lively area of research, called "quantum chaos," that deals with certain characteristic features of the spectra and wavefunctions of quantum systems whose classical versions are chaotic [8]. Be that as it may, our interest in examining quantum dynamics for chaotic behavior is to learn about its stochastic properties, and the relevance of these properties to such fundamental issues as approach to equilibrium and irreversibility. For this reason, we shall define quantum chaos in the original sense, i.e., in the sense of increasing uncertainty, or randomness, in the dynamical output of the system.

The standard, classical formulation of chaos in terms of nearby trajectories in phase space is not meaningful in quantum mechanics, so that an alternative approach must be found. Recall that the strong mixing property possessed by a chaotic system gives rise to a continually increasing uncertainty in its dynamical output. For instance, the motion of a tiny drop of food color (composed of a very large number of particles) placed on a ball of dough, while the dough is being kneaded, is such that after a few rounds of stretching and folding it thins out and becomes nearly uniformly spread throughout the body of the dough. Imagine measuring the x-coordinate of the particles of food color after each round, binning the measured data, and calculating the measurement entropy of the probability set associated with the resulting distribution. Clearly, as long as the drop is spreading, this entropy will increase after each round (until the coarse-grained nature of the binning limits the growth). Not unexpectedly, it turns out that chaotic behavior is associated with a linear increase with time of measurement entropy (until saturation sets in), and that the rate of this increase is directly related to the characteristic exponents discussed earlier.

We now proceed to formalize the above scheme. Using the notation introduced in Section 2, let us consider a series of measurements designed to measure the observable $\hat{A}$. These measurements are to be carried out at times t_n, on a quantum system originally in the state $\hat{\rho}$. As before, a device D^A, equipped with a partition $\{\alpha_i\}$, is used to make the measurements. The results are then used to define an *index* according to [9]

$$\lambda_\rho^{D^A} = \lim_{t_n\to\infty} \frac{1}{t_n} S(\hat{\rho}_n|D^A). \tag{8}$$

Here $\hat{\rho}_n$ is the state of the system at time t_n, as evolved from $\hat{\rho}$ according to the Hamiltonian operator $\hat{H}$ that governs the system. Again observe that for any finite partition, $S(\hat{\rho}_n|D^A)$ will eventually saturate, so that effectively the time in Eq.(8) should not run past the saturation time. To remove any possible partition-dependent effects, we may consider a progressive refinement of the partition toward a maximal device to arrive at

$$\lambda_\rho^A = \lim_{D^A\to D^A_{\text{max}}} (\lambda_\rho^{D^A}). \tag{9}$$

The index defined by Eq.(9) is the basis of our definition: *a bounded quantum system is chaotic if it possesses an observable with a positive index for some initial state of the system.* Observe that the index still depends on the observable considered in its definition, as well as on the initial state. For example, if $\hat{A}$ commutes with the Hamiltonian, the index will vanish even if the state is nonstationary, as indeed it must. Similarly, the index always vanishes for a stationary state (i.e., when $\hat{\rho}$ commutes with the Hamiltonian). These statements have strict counterparts in classical dynamics once ensemble distributions (the analogs of mixed states) are considered instead of single trajectories.

It should be clear that the definition given in Eq.(9) can be immediately taken over to the classical domain with no modification. Indeed one can show that classically the index defined in this way is equal to the largest Lyapunov exponent for the system (barring special cases such as those mentioned in the previous paragraph). Indeed we have carried out numerical calculations simulating model classical systems which verify this equality (details of these computations will be published elsewhere). One can go further and establish a correspondence between the Lyapunov spectrum and various multidimensional indices [9].

Returning to quantum systems, we will first explore the workings of our formalism by means of very simple examples. Let us consider the familiar harmonic oscillator in one dimension, defined by the Hamiltonian $\hat{H} = \hat{p}^2/2m + k\hat{x}^2/2$. The initial state is chosen to be $\hat{\rho} = |\psi><\psi|$, a pure state, with $<x|\psi>= \psi(x,0)$ a normalized Gaussian function of width $\sigma/\sqrt{2}$ ($\hbar = 1$ except where noted otherwise). One can now go through the necessary steps in a straightforward manner and find for the measurement entropy of the position operator

$$S(\hat{\rho}|D^x_{\text{max}}) = S_0 + \ln[(\cos\omega t)^2 + (\sin\omega t/m\omega\sigma)^2]^{\frac{1}{2}}, \tag{10}$$

where S_0 is the initial value of the entropy. Clearly here the entropy is a periodic function of time as expected, so that the index (equal to S/t as $t\to\infty$) vanishes, again as expected. Note that we reversed the order of the limits in Eqs.(8) and (9) in this calculation, an operation which is in general not allowed, although it happens to be harmless in this case. Next, consider the special case of $k = 0$ for the above system, corresponding to a body of mass m propagating freely. Eq.(10) then reduces to

$$\lim_{\omega \to 0} [S(\hat{\rho}|D^{x}_{\max})] = S_0 + \ln[1 + (t/m\sigma)^2]^{\frac{1}{2}}, \tag{11}$$

which shows a logarithmic increase in entropy corresponding to the well-known quantum mechanical spreading of a Gaussian wavefunction. Again, the index [$\propto \ln(t)/t$ as $t \to \infty$] vanishes, as would be expected for a freely propagating system.

A second, more interesting limit of the harmonic oscillator is realized if we consider a negative value for k, with ω an imaginary number equal to $i(-k/m)^{1/2}$. This would correspond to inverting the potential well and creating an unstable equilibrium point at $x = 0$ — a system that is sometimes referred to as an *inverted pendulum*. For this case Eq.(10) gives

$$\lim_{\omega \to i\gamma} S(\hat{\rho}|D^{x}_{\max}) = S_0 + \ln[(\cosh \gamma t)^2 + (\sinh \gamma t/m\gamma\sigma)^2]^{\frac{1}{2}}, \tag{12}$$

where $\gamma = (-k/m)^{\frac{1}{2}}$. One can see by inspection that $S \to S_0' + \gamma t$ for large t, exhibiting the linear increase of entropy with time characteristic of chaotic systems, with the index equal to γ. Thus a positive index is found here, as one would expect for a system with a linear instability. However, it should be stated emphatically that this is not an example of a chaotic system since it is not bounded, and in fact it does not even define a proper (i.e., self-adjoint) Hamiltonian. One can devise more sophisticated (albeit contrived) examples with positive indices [9], but invariably some prerequisite of chaos (usually boundedness) is violated.

The fact that one cannot find positive indices in quantum dynamics or otherwise discover an indication of chaotic behavior in quantum systems is not an accident as there is really no chaos in quantum dynamics [10,11]. In fact, there is not *any* sensitivity to initial conditions, however weak, in quantum dynamics, as we will now show. This total lack of sensitivity to initial conditions is rather unexpected, and certainly problematic for statistical mechanics, implying as it does the impossibility of memory loss or approach to equilibrium in Hamiltonian evolution (it is amusing to ponder the stark contrast between this result and the implications of the measurement postulate).

To demonstrate the absence of sensitivity to initial conditions in quantum dynamics [11], we shall first develop the quantum version of the sensitivity matrix discussed in Section 3. Consider an N-dimensional quantum system with canonical observables $\hat{q}_i$ and $\hat{p}_i$, grouped together into a $2N$-dimensional vector $\hat{\xi} = (\hat{q}_i, \hat{p}_i)$. The Hamiltonian is $\hat{H}(\hat{\xi}, t)$, and the commutation relations appear in the form $[\hat{\xi}_i, \hat{\xi}_j] = iZ_{ij}$. Here Z is an antisymmetric $2N \times 2N$ matrix with a simple $N \times N$ block structure whose upper-right block is equal to an $N \times N$ identity matrix. Next consider two "nearby" initial states $\hat{\rho}$ and $\hat{\rho}' = \hat{V}\hat{\rho}\hat{V}^{\dagger}$, where $\hat{V}$ is the unitary operator $\exp(-i\hat{\xi}_i\epsilon_i)$ and ϵ_i constitute a set of $2N$ real, infinitesimal parameters.

The expectation values of the observables $\hat{\xi}_i$ in the states $\hat{\rho}$ and $\hat{\rho}'$ are denoted by $\xi_i(t)$ and $\xi_i'(t)$ respectively. What plays the role of orbital displacement here is $\delta\xi_i(t) = \xi_i'(t) - \xi_i(t)$. Now one can show that to first order in ϵ_i,

$$\delta\xi_i(t) = i\epsilon_j \mathrm{tr}\{[\hat{\xi}_j(0), \hat{\xi}_i(t)]\hat{\rho}\}. \tag{13}$$

Observe that Eq.(13) gives $\delta\xi_i(0) = Z_{ij}\epsilon_j$ for the initial value of the displacement, thereby clarifying the meaning of the parameters ϵ_i.

In strict analogy with the classical case, we now define the sensitivity matrix by

$$\mathcal{T}_{ij}(t) = \lim_{\epsilon\to 0} \frac{\delta\xi_i(t)}{\delta\xi_j(0)}. \tag{14}$$

A comparison of Eqs.(13) and (14) now reveals that the sensitivity matrix is the expectation value of the operator

$$\hat{\mathcal{T}}_{ij}(t) = iZ_{kj}[\hat{\xi}_i(t), \hat{\xi}_k(0)], \tag{15}$$

which may naturally be called the *sensitivity operator*. One can show that this operator possesses the correct correspondence properties in the limit of $\hbar \to 0$.

We can now examine the question of quantum sensitivity to initial conditions by studying the long-time growth properties of $\hat{\mathcal{T}}_{ij}(t)$. To that end, consider the operators $\hat{\zeta}_i(t) = \hat{\xi}_i(t) - \xi_i(t)$, which are so constructed as to have vanishing mean values in the state $\hat{\rho}$. Next notice that $\hat{\mathcal{T}}_{ij}(t)$ is essentially a commutator, so that it can be bounded by an appropriate generalization of the Heisenberg inequality. Indeed one finds

$$|\mathrm{tr}[\hat{\mathcal{T}}_{ij}(t)\hat{\rho}]|^2 \le 4\mathrm{tr}[\hat{\zeta}_i^2(t)\hat{\rho}]\mathrm{tr}\{[Z_{kj}\hat{\zeta}_k(0)]^2\hat{\rho}\}. \tag{16}$$

This inequality basically guarantees that a bounded quantum system cannot have unbounded growth in its sensitivity matrix. To see this, first note that $\mathrm{tr}[\hat{\zeta}_i^2(t)\hat{\rho}]$ is simply the variance $[\Delta\xi_i(t)]^2$ for the observable $\hat{\xi}_i(t)$, so that the above inequality can be written as

$$|\mathcal{T}_{ij}(t)| \le \frac{2}{\hbar}\Delta\xi_i(t)\Delta\xi_{j\pm N}(0), \tag{17}$$

where we have restored $\hbar$ to the inequality. Here $\xi_{j\pm N}$ equals ξ_{j+N} if $j \le N$ and equals ξ_{j-N} if $j \ge N+1$. Observe now that for any bounded system, the dispersions $\Delta\xi_i(t)$ must be bounded. In other words, there must exist time independent positive numbers M_i such that $\Delta\xi_i(t) \le M_i$, since otherwise the system would have to undergo unbounded growth. Consequently, each matrix element $\mathcal{T}_{ij}(t)$ is bounded for all time, implying a total absence of sensitivity to initial conditions for quantum dynamics.

Quantum mechanical motion is thus essentially stable, and places no restriction on long-term predictability. Note also that as $\hbar \to 0$, the inequality in (17) disappears, implying that the inequality does not survive the passage to the classical limit. As mentioned earlier, this remarkable stability of quantum Hamiltonian evolution further complicates the problem of irreversibility and approach to equilibrium. On the other hand, it lends support to the view that irreversible behavior can only occur for open systems. It is worth pointing out in passing that the entire discussion above is about *finite* systems, and that in fact the question of how an infinite system should fit into all of this is far from obvious or settled.

5. Concluding Remarks

Thus far, our discussion has been mainly concerned with the application of entropic ideas to chaos. Here we will briefly mention other examples of the use of entropic methods in quantum mechanics [12]. The first example is the use of measurement entropy as a measure of uncertainty and the entropic formulation of the uncertainty principle [1]. Here one establishes the inequality that, for a fixed state, the sum of the measurement entropies of a pair of incompatible observables cannot be reduced below a certain lower bound. The second example is a formulation of quantum maximum entropy principle [2], including a generalization to the multi-time case where the measured data refer to more than one instant of time. This formulation was used to establish a definitive framework, and a precise meaning, for time-energy uncertainty relations in quantum mechanics[13]. The third example is the formulation of *quantum thermodynamics*, and a dynamical derivation of the zeroth and second laws for ensembles of interacting quantum systems [14]. Here it was shown in detail how molecular collisions drive a system in contact with a heat reservoir toward equilibrium. The final example is a derivation of state reduction during the measurement process [15]. Here a little-known property of entropy known as strong subadditivity is used to demonstrate that the coupling of a quantum system to a macroscopically large number of degrees of freedom during measurement brings about the removal of phase relations.

The measurement problem of quantum mechanics has not yet been resolved. However, there are good reasons to believe that a resolution is forthcoming, that this resolution will be statistical in nature, and that information-theoretical reasoning will play a key role in its formulation. One should therefore expect that quantum entropic methods will continue to play an important rôle in our understanding of the foundations of qunatum mechanics.

References

[1] M. H. Partovi, *Phys. Rev. Lett.* **50**, 1882 (1983).

[2] R. Blankenbecler and M. H. Partovi, *Phys. Rev. Lett.* **54**, 373 (1985).

[3] Hao Bai-Lin, *Chaos II* (World Scientific, Singapore, 1990).

[4] J. P. Eckmann and D. Ruelle, *Rev. Mod. Phys.* **57**, 617 (1985).

[5] J. Ford, *Physics Reports* **213**, 306 (1992).

[6] A. Kolmogorov, *Dokl. Akad. SSSR* **124**, 754 (1959).

[7] J. Ford, in *The New Physics*, edited by P. Davies (Cambridge Univ. Press, Cambridge, England, 1989).

[8] M. Berry, introductory remarks in *Quantum Chaos*, edited by H. A. Cerdeira, R. Ramaswamy, M. C. Gutzwiller, and G. Casati (World Scientific, Singapore, 1990).

[9] M. H. Partovi, *Phys. Lett.* A**151**, 389 (1990).

[10] G. Casati, B. V. Chirikov, F. M. Izrailev, and J. Ford, in *Stochastic Behavior in Classical and Quantum Hamiltonian Systems*, edited by G. Casati and J. Ford (Springer, N.Y., 1979).

[11] M. H. Partovi, *Phys. Rev.* A**45**, R555 (1992).

[12] M. H. Partovi, in *Complexity, Entropy and the Physics of Information,* edited by W. Zurek (Addison-Wesley, Reading, MA, 1990), p. 357.

[13] M. H. Partovi and R. Blankenbecler, *Phys. Rev. Lett.* **57**, 2887 (1986).

[14] M. H. Partovi, *Phys. Lett.* A**137**, 440 (1989).

[15] M. H. Partovi, *Phys. Lett.* A**137**, 445 (1990).

PHASE-SPACE PRIORS IN INFORMATION ENTROPY

R. Englman
Sorec Nuclear Research Center
Yavne 70600, Israel

ABSTRACT. It is possible to demand that in estimating probabilities of events in the real world, one should include physical measures by taking into account the phase-space belonging to each event. In many cases this appears absurd, as in pattern recognition, etc. or in economic events and indeed in most situations that involve personal choices. Still I suggest (in consonance with several previous, unheeded suggestions made by others) that the demand should not be jettisoned, and for the following reasons:

(a) There are macroscopic events for which the physical phase-space can be estimated with greater or lesser difficulty at one or other level of sophistication. Even for these cases predictive procedures (like MaxEnt) are currently practised without bothering about the physical probability measure, as ought to be done.

(b) A consistent formulation of MaxEnt, encompassing both the information and statistical-physical domains, is possible by starting with a Boltzmann-Gibbs entropy in which macroscopic events are included as marginals. The physical measure (estimated on the basis of microscopic considerations) emerges naturally as the prior probability for the macro-events.

The paper attempts to determine the conditions under which phase-space considerations do and do not determine the macro-event probabilities. (In the latter case we are left with the Shannon-Jaynes information-entropy maximization). I shall modelize some simple macro-events, like a biased coin or a roulette, at different levels of sophistication and obtain computed and analytic results that exhibit the relative importance of physical versus informational constraints. A necessary, though probably not sufficient, condition for ignoring the phase-space prior in macroscopic events is to have a constraint on the width of the distribution (and not only on the mean). It is shown that, as a consequence, the "Surprisal" will have macroscopic magnitude.

Data from single fibre fragmentation performed by Roman and Aharonov (1991) in aluminium based composites are analysed by combining phase space and informational inputs to predict fragment-length distributions in a macroscopic system. A single constraint (arising from the physics of the experiments) suffices to yield qualitive agreements; a further constraint brings quantitative fit.

1. Introduction

The conceptual relationship between information entropy (or information deficiency) and physical (Gibbs-von Neumann) entropy has been described from several, not necessarily identical points of view [1-3]. A formal relation has been proposed by Levine and used by him and coworkers (mainly) for molecular and nuclear collision dynamics [3]. In the data-interpretation the degeneracy of states acted as the prior to the changed distribution which was information-laden.

The present work extends the above idea in the following senses: It includes macroscopic systems; it therefore includes macroscopic observables whose values place constraints on

A. Mohammad-Djafari and G. Demoments (eds.), Maximum Entropy and Bayesian Methods, 145–154.

microscopic physical states; it even includes hypotheses as representing groupings of physical states (not excluding hypotheses in the human or social domains, e.g. election outcomes or economic trends, whose physical entropy is imaginable only in an extreme reductionist scenario); it uses priors that involve the physical entropies appropriate to the hypothesis and not just the state-degeneracies; it includes situations in which the difference in the macroscopic observable comes about from different preparation of the macrostate, rather than by a variation of constraints on the microstates.

In spite of its sweeping character the formalism is mature enough to backpedal on its claims, when their extravagance takes it to absurdity. On the positive side it clarifies (at least for the author) the status of information entropy, that features now with a no longer arbitrary multiplying constant, in relation to its 19th Century predecessor; it imposes rules (physical ones) on gambling; it analyses a concrete, classical fragmentation problem from an information theory viewpoint and it says something on algorithmic complexity.

2. Formalism

We envisage a set of hypotheses $H_i (i = 1, \ldots)$, their associated probabilities $p(H_i)$, and physical entropies S_i belonging to H_i. S_i is k (the Boltzmann constant) times the logarithm (to base e) of the phase space volume subject to constraints on the physical states (e.g., fixed energy) consistent with the hypothesis H_i. By the grouping property of entropy, the mean entropy of the distribution of physical states given by

$$S/k = \sum_i p(H_i) \log p(H_i) + \sum_i p(H_i) S_i/k + \sum_r \lambda^r [\sum_i f^r(H_i) p(H_i) - F^r] \tag{1}$$

We have added a set (labelled by r) of constraints involving the hypotheses H_i whose source is observational (e.g., constant observed mean) or otherwise (e.g., informational, of unproven observational content). Upon introducing

$$\Pi_i = e^{S_i/k}/M \tag{2}$$

$$M = \sum_i e^{S_i/k} \tag{3}$$

We can rewrite (1) as

$$S/k = -\sum p(H_i) \log p(H_i)/\Pi_i + \log M + H - \text{constraints} \tag{4}$$

This is an information-entropy (without any arbitrary multiplier) where the prior Π_i is physical, fixed by the phase space and physical constraints. After maximizing S we obtain the probabilities

$$p(H_i) = \Pi_i \exp[\sum_r \lambda^r f^r(H_i)]$$

where the exponential factor adjusts for information on the hypotheses.

It is noted that when the priors are independent of the hypotheses one simply regains the information (or Jaynes) entropy. By extension, we might justify prior-less entropies by our inability to quantify the hypothesis-dependent priors.

It is probably "justification" of a similar kind that has led to the quantification of the hypothesis term in (1), in the form of algorithmic complexity [2]. In many situations the multitude of conceivable hypotheses is prodigious; their enumeration (which is the entropy) is replaced by a length of description.

3. Gamblers' Priors

3.1. BIASED COINS

Coin pieces are borrowed from a mint to establish or to exclude any bias in the manufacture. Each coin is tossed several times, $r = \pm 1$ are registered for H/T, and a distribution of mean values $< r >$ is derived.

To obtain a phase-space prior we postulate the existence of a physical law that relates the mean value $< r >$ to the position x of the centre of mass of the coin across its face and measured from the mid-plane between the faces. We assume that

$$< r >= cx \tag{5}$$

c being a constant. The distribution of $< r >$ is therefore expressible as a probability function $p(x)$, to which the following meaning can be attached. Minting proceeds by alloying two elements, A and B. If this is done uniformly as in an ordered AB alloy the centre of mass is in the mid-plane, $x = 0$. In a random alloy fluctuations yield $x \neq 0$. We have performed a simulation of a random, one dimensional alloying process and found a distribution of x as shown in figure 1 (200,000 simulations were performed on a line having 1000 positions occupied randomly by $N = 500$ A and B atoms). Since the computed distribution is the relative weight of the events in which the centre of mass has the value x, we propose to use it as the prior $\Pi(x)$ for the probability $p(x)$. The following analytical form represents quite well the results of our simulation in which the number of A, B pairs (N) was varied

$$P(x) \propto \exp(-Nx^2/d^2) \tag{6}$$

where d is of the order of the coin width. The distribution has quite reasonably, a width of order $d/\sqrt{N}$.

If the only informational constant in (1) or (4), is a zero average on x, then $p(x) = \Pi(x)$ and the surprisal (Surp) defined by the negative of the first term on the right hand side of (4), is zero. If, on the other hand, the informational constraints is that the mean

$$< x >= \alpha d \qquad (0 < \alpha < 1) \tag{7}$$

then the distribution is, by (4),

$$p(x) \propto \exp[-N(x - \alpha d)^2/d^2] \tag{8}$$

with a macroscopic surprisal of

$$\text{Surp} = N\alpha^2 d^2 \tag{9}$$

An observed mean $< x >$ of the order of $d\sqrt{N}$ will yield a surprisal of order d^2. It is remarkable that the maximum entropy distribution in (8) preserves the <u>shape</u> of the prior

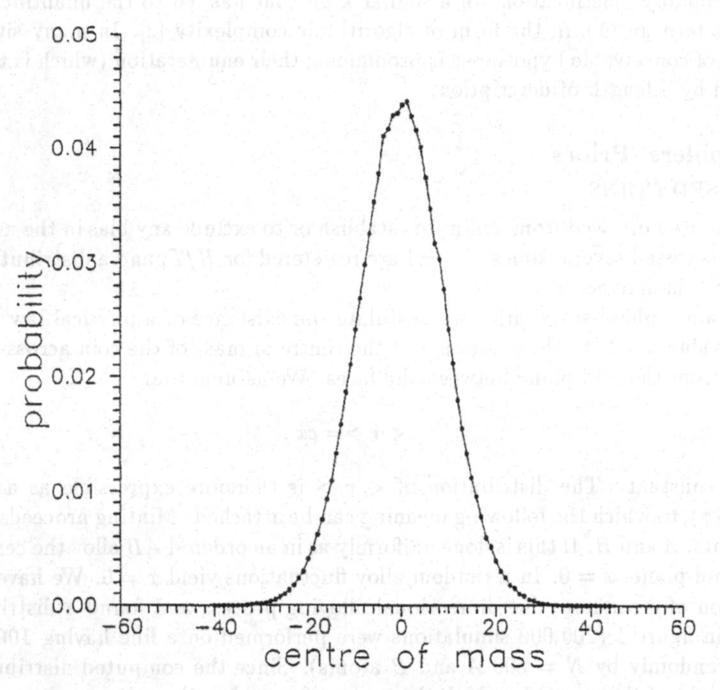

Figure 1: Frequency of centre of mass positions. 200,000 simulations were made on a linear lattice of 1000 points upon which 500 A and 500 B masses were randomly thrown.

and only shifts its peak position. To make the new distribution "forget" the prior, we need to have an information constaint, involving the second moment, $< x >$ to the effect that the width of the distribution is macroscopic.

$$< x^2 >= \beta d^2 \qquad (10)$$

The Lagrange multiplier, λ in (4), will be such as to cancel the N in the exponent of (6) and the distribution will have lost trace of the physical prior. The surprisal will again be macroscopic, of order N as in (9).

The existence of a macroscopic constraint on the spread of the distribution appears to be a general condition to eliminate the effect of the microscopic, phase-space prior on the distribution. (It is not a necessary condition, since the independence of the priors on the hypotheses achieves the same, as pointed out earlier.) It is possible, though does not seem to have been stated before, that the use of the pure information entropy approach rests on the expectation that the width of the distribution is macroscopic.

Here is an example of a table:

Designation	Composite	τ(MPa)	$< s >$ mm	s_M mm	Dev mm^2	$< s^3 >$ mm^3	λ mm^{-1}	σ mm
(a) = Xx	A16061/SCS-2	71 ± 24	2.94	2.6	0.78	3.3	2.1	1
(b) = Xy	A16061/Σ	236 ± 72	0.57	0.55	0.03	0.25	0.5	0.2
(c) = Yx	A11100/SCS-2	41± 12	5.09	5.4	2.0	166	1.2	1.45
(d) = Yy	A11100/Σ	76 ±39	1.73	1.4	0.68	10.7	1.1	1.2

Table 1: Observed, derived and calculated parameters in four metal-composites [5]. (X ≡ A1 6061, Y ≡ A1 1100; x ≡ SCS-2, y ≡ Σ)

3.2. ROULETTE MECHANISMS

The physical prior involved in the operation of a roulette wheel is of a different sort, since the randomness arises not from the manufacture of the roulette but from the mode of its turning. To simplify the analysis we assume that the wheel is not turned manually (since this would require us to quantify phase spaces of biological systems and psychological states), but is operated by a gas gun of a primitive type. The recourse to a mechanical device enables us to model the randomizing mechanism at more than one level of depth. This is also the recommended approach to get level with the philosophical chance-necessity dichotomy.

Suppose that the wheel starts at a preassigned location (you can't beat a quick-witted gambler just by varying the starting location) and that the valve of a gas gun filled with N molecules of mass m at temperature $1/2m\bar{v}^2/k$, is opened. The length travelled by a point on the wheel perimeter will approximately be proportional to the total momentum of the molecules exiting from the gas gun or, equivalently, to the sum of the positive velocities v^+. This sum can be written as an average total momentum plus a stochastic deviation from it, in the form

$$\sum_i v_{i^+} = \frac{1}{2}N\bar{v} + \bar{v}\sqrt{N}\theta \tag{11}$$

where θ is a randomly varying quantity of order unity.

Since both the number of compartments on the wheel and the number of revolutions are of the order unity on the scale of the Avogadro number, or N^0 on the scale the molecular number in the gas gun, the mean $\bar{i}$ of the position of rest i will be of the order of

$$\bar{i} \propto N^0\bar{v} \tag{12}$$

and the deviation from this mean

$$\delta_i \propto \bar{v}\theta/\sqrt{N} \tag{13}$$

Thus we again have a distribution of rest positions with a standard deviation of order $1/\sqrt{N}$. The distribution can be further broadened by allowing the number N of molecules in the gas canister to vary. This causes the mean in (11) to vary; however, the broadening

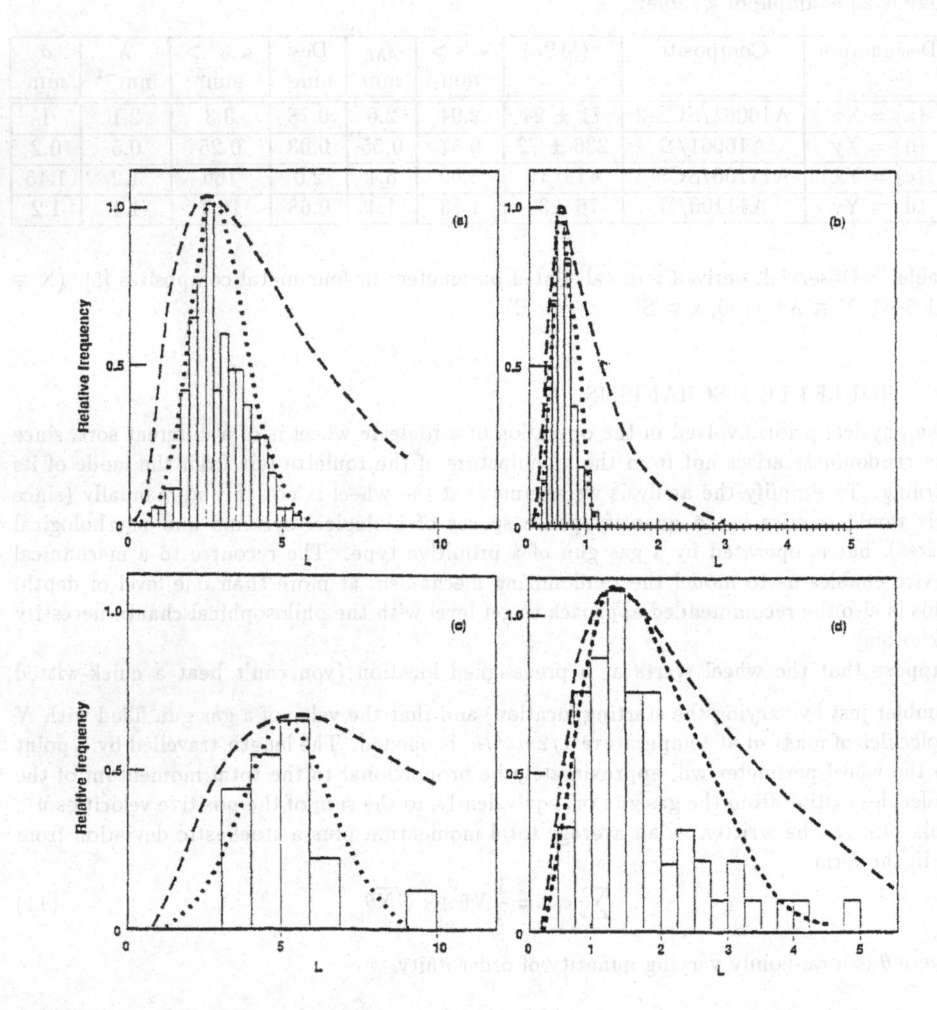

Figure 2: Distribution of fibre fragments (arbitrary scale) against fragment lengths (L) in mm. The histogramms show data by Roman and Aharonov [5]. Broken lines are the result of a prior and one (energy) constraint. Chained lines are plotted with one more constraint. The four composites a-d are designated as in table 1.

is again of order $1/\sqrt{N}$.
Therefore, the prior distribution due to microscopic mechanisms is again very narrow, of the form

$$\Pi(i) \propto exp[-N(i-\bar{i})^2/\Delta i^2 \tag{14}$$

where $\bar{i}$, Δi are of the order unity.

If information arrives that impels us to change the mean $\bar{i}$, the width of the predicted distribution will not change from that in the prior. Only when the information relates to the width's having macroscopic magnitude will the shape of the distribution differ from that of the prior. Without providing a deeper analysis, one may venture that the gambling wheel is set to motion by mechanisms that are beyond a microscopic description and whose effects show up in a broadened distribution.

4. SiC Fragment Data

Tensile fragmentation tests and identation tests were performed by Roman and Aharonov [5] on single fibre composites consisting of two different aluminium-based matrices (a commercial *Al* 1100 alloy and *A*16061) and of two types of SiC reinforcers (labelled SCS-2 and Σ) where the former had an adhesion-promoting 1 μm thick carbon surface layer, lacking in the latter. Fragmentation tests yielded distributions for each combination [Figs. 2(a-d) show experimental histogram from several tests for each combination] and acoustic emission peaks were found to emanate from individual breaks one for one. The fracture profiles, seen in micrographs, differed for the two fibre types: we shall not now consider this difference (significant though it is), nor shall we deal with the identation data.

In table 1, the interfacial shear strength parameter (ISS) τ is derived from the critical fragmentation length [5]. Also shown are the moments, defined by

$$< s^r > \equiv \int_0^\infty s^r f(s) ds / \int f(s) ds$$

where $f(s)$ is the distribution and s the fragment length (in mm)

$$\text{Dev} \equiv < s^2 > - < s >^2$$

s_M is position of maximum; λ and σ are Lagrange parameters to be intoduced later.

4.1. THE INFORMATION LAGRANGEAN

We write $dsP(s)$ for the probability that a site (a monatonic layer?) between $(s+ds, s)$ belongs to a fragment of length s. Evidently the relation to the length-distribution $f(s)$ is given by

$$P(s) = sf(s)/ < s > \qquad [\int P(s)ds = 1]$$

Next maximize the "Lagrangean L" that consists of the entropy and the constraints C

$$L = -\sum_s P(s) \log P(s)/\Pi_s + C$$

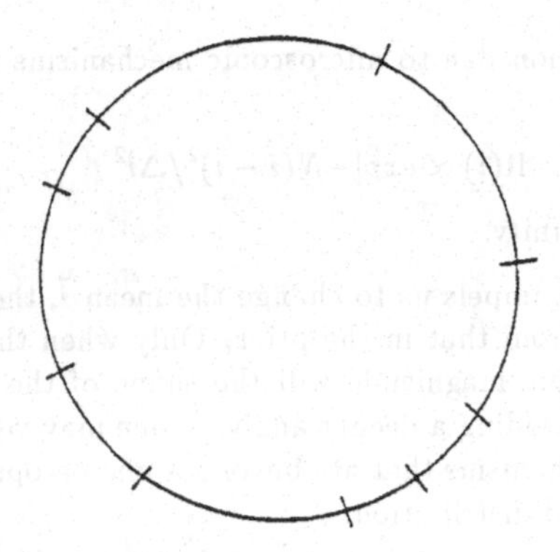

Figure 3: A one-dimensional Poisson process. The ring is broken up into several (9) fragments by randomly placed cuts

where Π_s is the prior for $P(s)$. In the next section we find the prior Φ_s appropriate to $f(s)$ (the length distribution). Analogously to the relation between $P(s)$ and $f(s)$, the relation for the priors is:

$$\Pi_s \propto s\Phi_s.$$

4.2. THE PRIOR DISTRIBUTION

To avoid boundary effects we consider a ring of perimeter length L which is broken up into P pieces by randomly placed cuts (figure 3). The probability of obtaining a piece having a size between s and $s+ds$, anywhere, is equal to the product of probabilities of placing one cut anywhere (=1), of placing a second cut in the interval $(s, s+ds)$ measured from the position of the first cut (probability = ds/L), of not having any of the remaining $P-2$ cuts falling in the interval $L-s$ (probability = $[(L-s)/L]^{P-2}$) and multiplied by the number of cuts ($=P-1$) following the first, Altogether

$$ds\Phi_s(P) = 1\frac{ds}{L}\left(\frac{(L-s)}{L}\right)^{P-2}(P-1).$$

In the limit of L/s and P being large numbers, this reduces to

$$\Phi_s = (1/<s>)\exp(-s/<s>)$$

where $<s>= L/P$ is the mean size.

4.3. QUEST FOR CONSTRAINTS

Closing our eyes to physical theory we attempt to discern some regularity from the data, indicative of constraints on the distribution. Let us calculate the surface formation energy $E_{\Gamma\gamma}$ in a composite designated generally by $\Gamma\gamma$, where $\Gamma = X, Y; \gamma = x, y$ with the latin letters deputizing for the materials as in the legend of Table 1. Denoting by $\gamma_{\Gamma\gamma}$ the surface energy density, A the fibre cross sectional area, L the fibre or matrix length, $N_{\Gamma\gamma}$ the number of fragments we find

$$E_{\Gamma\gamma} = 2\gamma_{\Gamma\gamma} A\, N_{\Gamma\gamma} \int \Pi(s)/s\, ds = 2\gamma_{\Gamma\gamma} A\, N_{\Gamma\gamma} / < s >_{\Gamma\gamma}$$

or

$$= 2\gamma_{\Gamma\gamma} A\, L / < s >^2_{\Gamma\gamma}$$

since the mean fragment size

$$< s >_{\gamma\Gamma} = L/N_{\Gamma\gamma}$$

The quantity

$$\surd(2\gamma_{\Gamma\gamma} A\, L/E_{\Gamma\gamma}) = < s >_{\Gamma\gamma}$$

varies by a factor 10 over the materials in Table 1, but one finds that

$$< s >_{X_x} + < s >_{Y_y} \approx < s >_{X_y} + < s >_{Y_x}$$

indicating a linear dependence of the mean $< s >$ over the matrix and fibre materials. It appears reasonable to convert this law (similar to the law of mixtures) to a regularity over different experiments with the same material. Since $< s >$ and E are related as above, we feel entitled to postulate that $E_{\Gamma\gamma}$ is a constant over different samplings (on the same materials), or that

$$E \propto \int P(s)/s\, ds$$

remains invariant. This gives us a constraint and performing the maximization of the Lagrangean L yields an exponential factor with a Lagrange multiplier λ. We thus obtain for the probability $P(s)$ and the length distribution $f(s)$ the following results (unnormalized):

$$P(s) = \Pi_s e^{-\lambda/s} = \frac{s}{< s >} \exp\left(\frac{s}{< s >} - \frac{\lambda}{s}\right)$$

$$f(s) = \exp(-s/ < s > - \lambda/s)$$

The distribution $f(s)$ in the above expression is compared in Fig. 2 with the experimental histograms for four composites, after choosing Lagrange multipliers (given in Column 8 of table 1) so as to fit the observed mean fragment lengths $< s >$ and normalizing the peak positions. Though the agreement is quantatively poor, the following observations can be made:

The rise and (exponential) drop of the distribution is a direct consequence of having as factors in the distribution both the prior and constraint parts. In the abscence of either, one would get a monotonic behaviour in the distribution. Thus a single constraint already helps to reproduce the trend of the observed data.

A second constraint may be sought to improve the fit quantitatively. This constraint may be rooted in physics or be data-induced. The latter type of constraint implies the existence of some conservation law not superficially obvious, but nevertheless present in the mechanism [3, 4]. Thus one might explore the constancy of the third moment $< s^3 >$. The appropriate factor that is to be added on to the probability $P(s)$ previously written is

$$e^{-(s-<s>)^2/2\sigma^2}$$

where now σ is so chosen (column 9 in table 1) as to yield the observed $< s^3 >$. If the constraint is a true one, some sort of relationship can be presumed to hold between the final distribution and the dynamic equations that guide the time development of the fragment length distribution (e.g., eq. (30) in [5] and eq. (15) in [3])

5. Conclusion

The use of physical or phase-space priors in our information-deficiency formalism is recommended on grounds of consistency and of practicality. Previous works by Levine have demonstrated the utility of this approach in microscopic situations and a claim has been made that frequently only a few (one or two) informational constraints suffice to explain or predict distributions. This paper has treated three macroscopic situations, which are characteristic of many others, but still only represent an infinitesimal portion of the subjects to which MaxEnt can be, has been and will be applied.

Acknowledgements

Dr. Michael Murat is credited with producing computations of which Fig. 1 is an example. Thanks belong to Prof. R.D. Levine for giving me his time (sic).

References

[1] R. Balian, European J. Phys. **10** (1989) 208.

[2] W.H. Zurek, "Complexity, Entropy and the Physics of Information" (Addison-Wesley, New York, 1910).

[3] R. Englman, J. Phys.: Condens. Matter **3** (1991) 1019.

[4] R.D. Levine, "Statistical Dynamics" in *Theory of Chemical Reaction Dynamics*, Ed. M. Baer, Vol. IV. p 1 (C.R.C., Bocca Raton, Florida, 1985).

[5] I. Roman and R. Aharonov, Acta Metall. Mater. **40** (1992) 477.

[6] B. Widom, J. Chem. Phys. **44** (1966) 3888.

[7] W.A. Curtin, J. Materials Sc., 26 (1991) 5239.

MAXIMUM ENTROPY DESCRIPTIONS OF QUANTUM WAVE FUNCTIONS

N. Canosa, R. Rossignoli
Physik Department der Technischen Universität München
D-8046 Garching, Germany

A. Plastino
Departamento de Física, Universidad Nacional de La Plata
1900 La Plata, Argentina

ABSTRACT. A maximum entropy based approximation to the energy eigenstates of a quantum system, suitable for both inference and variational procedures, is presented. Application to many-fermion Hamiltonians and to the single particle Schrödinger equation is discussed.

1. Introduction

The application of information theory and the maximum entropy (ME) principle to statistical physics and quantum mechanics, pioneered by the work of Jaynes [1,2], has attracted great interest in recent years, particularly in problems dealing with non-equilibrium systems [3] and the principles of quantum statistical mechanics and quantum inference [4,5]. In this contribution we discuss a different application of ME based methods, aimed at problems strictly related with the determination and approximation of quantum wave functions. Based on the entropy associated with the probability distribution of a quantum state in a given basis, a method for the approximate description of the ground state of a quantum system in terms of a reduced set of variables associated with relevant observables was recently introduced [6–9]. In particular, it was shown that just with a few parameters, an excellent agreement with the ground state of various many-fermion models could be achieved for all values of the pertinent coupling constant, including transitional regions. The method was also applied to the single particle Schrödinger equation [6], where an accurate inference of the ground state wave function and of the concomitant potential was achieved with just a few expectation values, for various Hamiltonians. In this contribution we briefly review the approach and present the appropriate extension for the description of excited states. A general scheme for generating excitations orthogonal to a ME ground state is developed. Results are presented in a superconducting many-fermion model.

A. Mohammad-Djafari and G. Demoments (eds.), Maximum Entropy and Bayesian Methods, 155–161.

2. Description of Ground States

We shall consider a system described by a Hamiltonian $\hat{H} = \hat{H}_0 + \hat{H}_{int}$, where $\hat{H}_0$ denotes the unperturbed term and $\hat{H}_{int}$ the interaction. Let $\{\hat{O}_\alpha, \ \alpha = 1,\ldots,n\}$ be a set of commuting operators, which are thus diagonal in a common basis $\{|j\rangle\}$. We shall consider a ME based approximation to the ground state of the system, which can be expanded as

$$|0\rangle = \sum_j C_j^{(0)}|j\rangle, \tag{1}$$

of the form [7–10]

$$C_j^{(0)} = \exp\{-\tfrac{1}{2}[\lambda_0 + \sum_\alpha \lambda_\alpha O_\alpha(j)]\}, \tag{2}$$

where $O_\alpha(j) = \langle j|\hat{O}_\alpha|j\rangle$, $\{\lambda_\alpha = \lambda_\alpha^{\rm r} + i\lambda_\alpha^{\rm i}\}$ constitute a set of complex optimizable parameters and $\lambda_0 = \ln\sum_j \exp[-\sum_\alpha \lambda_\alpha^{\rm r} O_\alpha(j)]$ is a normalization constant. The squared moduli of the coefficients (2) have the functional form which maximizes the entropy associated with the probability distribution of the ground state in the common basis,

$$S = -\sum_j |C_j^{(0)}|^2 \ln|C_j^{(0)}|^2, \tag{3}$$

subject to the constraints $\langle\hat{O}_\alpha\rangle_0 \equiv \langle 0|\hat{O}_\alpha|0\rangle = O_\alpha$. An analysis of the information and of the (commuting) observables required for an accurate prediction of the coefficients $|C_j^{(0)}|$ can thus be performed. The operators $\hat{O}_\alpha$ can be chosen for instance such that they commute with $\hat{H}_0$, in which case the states $|j\rangle$ can be taken as eigenstates of $\hat{H}_0$ (unperturbed basis). The entropy (3) will then vanish for the non interacting system and will acquire in general a finite value as the interaction is switched on. The ensuing behaviour of the "quantal" entropy (3) as a function of the pertinent coupling constant may be used to define and infer ground state transitional regions between different regimes and critical points [9].

It turns out that in various many-fermion models, characterized by collective-like two-body interactions, a very accurate prediction of the *ground state* coefficients $|C_j^0|$ can be achieved by including just a few relevant one- and two-body observables $\hat{O}_\alpha$ diagonal in the corresponding unperturbed basis, both for weakly and strongly coupled systems, *including* transitional regions [7]. This fact entails that a very accurate *variational* approximation [7,8] to the ground state can be achieved in these cases with trial wave functions of the form (2). A substantial improvement over both conventional and projected mean field approximations can be obtained in this way, particularly in transitional regions, where mean field methods fail to provide a reliable description.

The formalism is thus able to provide both an inference scheme for $|C_j^{(0)}|$ (and for the full wave function in those cases where phases can be trivially known [6,7]) in which the parameters $\lambda_\alpha^{\rm r}$ are obtained according to the standard equations $-\partial\lambda_0/\lambda_\alpha^{\rm r} = O_\alpha$, in case the O_α's are known, and also a pure variational treatment, in which the parameters λ_α result from the minimization of the ground state energy $\langle\hat{H}\rangle_0 = \langle 0|\hat{H}|0\rangle$. The ensuing equations are

$$-\partial\langle\hat{H}\rangle_0/\partial\lambda_\alpha^{\rm r} = \tfrac{1}{2}\langle\hat{O}_\alpha\hat{H} + \hat{H}\hat{O}_\alpha\rangle_0 - \langle\hat{O}_\alpha\rangle_0\langle\hat{H}\rangle_0 = 0, \tag{4}$$

$$-\partial\langle\hat{H}\rangle_0/\partial\lambda_\alpha^{\rm i} = \tfrac{1}{2}\langle[\hat{H},\hat{O}_\alpha]\rangle_0 = 0. \tag{5}$$

For the exact ground state, (4–5) are obviously satisfied for any operator $\hat{O}_\alpha$. Moreover, the exact coefficients can always be expanded in the form (2) if a complete set of diagonal operators $\hat{O}_\alpha$ is used, so that convergence towards the exact ground state is guaranteed.

If a prior approximate estimate p_j of the coefficients is known, which can either be an approximate starting value or an effective multiplicity factor, the coefficients $C_j^{(0)}$ are to be selected so as to maximize the entropy relative to the "measure" $|p_j|^2$, $S_r = -\sum_j |C_j^{(0)}|^2 \ln |C_j^{(0)}/p_j|^2$. In this case,

$$C_j^{(0)} = p_j \exp\{-\tfrac{1}{2}[\lambda_0 + \sum_\alpha \lambda_\alpha O_\alpha(j)]\}. \tag{6}$$

3. Construction of Excited States

We define the states [10]

$$|\alpha\rangle \equiv (\hat{O}_\alpha - \langle\hat{O}_\alpha\rangle_o)|0\rangle = \sum_j C_j^{(\alpha)} |j\rangle, \tag{7}$$

with $C_j^{(\alpha)} = -2\partial C_j^{(0)}/\partial\lambda_\alpha$, which can formally be regarded as maximum relative entropy coefficients. The states (6) are clearly *orthogonal* to the ME ground state, $\langle 0|\alpha\rangle = 0$, and on account of the stability conditions (4–5), they also verify

$$\langle 0|\hat{H}|\alpha\rangle = \langle\hat{O}_\alpha\hat{H}\rangle_o - \langle\hat{O}_\alpha\rangle_o\langle\hat{H}\rangle_o = 0. \tag{8}$$

Hence, the eqs. (4–5) imply that our approximate ME ground state is indeed stable against the excitations represented by the states $|\alpha\rangle$. A first estimate of the low lying states can thus be obtained by diagonalizing $\hat{H}$ in this reduced non orthogonal space ($\langle\alpha|\alpha'\rangle = \langle\hat{O}_\alpha\hat{O}_{\alpha'}\rangle_o - \langle\hat{O}_\alpha\rangle_o\langle\hat{O}_{\alpha'}\rangle_o$) of dimension n. The resulting approximate eigenstates can be cast as

$$|\beta\rangle \equiv (\hat{Q}_\beta - \langle\hat{Q}_\beta\rangle_o)|0\rangle, \quad \hat{Q}_\beta = \sum_\alpha b_{\alpha\beta}\hat{O}_\alpha, \tag{9}$$

with b the corresponding eigenvector matrix. The orthogonal states (9) can thus be interpreted as *normal* modes.

In order to extend the formalism to higher excited levels, we construct the states $|\gamma\rangle \equiv \prod_\alpha \hat{O}_\alpha^{n_\alpha}|0\rangle$ where $\gamma = (n_1, \ldots, n_n)$, with $0 \le n_\alpha \le k_\alpha$, and diagonalize $\hat{H}$ in the ensuing reduced space. To avoid superposition, we should obviously exclude those operators expressed as products of other $\hat{O}_\alpha$'s. The space spanned in this way is similar to that generated by the states with coefficients

$$C_j'^{(\gamma)} = \partial^k C_j^{(0)} / \prod_\alpha \partial\lambda_\alpha^{n_\alpha}, \tag{10}$$

where $k = \sum_\alpha n_\alpha$. It is expected that a considerable part of the corresponding collective space will be spanned with low values of k_α, so that an accurate description of the low lying energy states can be achieved with a Hamiltonian matrix $\langle\gamma|\hat{H}|\gamma'\rangle$ of small dimension. It is also convenient in some situations to employ instead of $\hat{O}_\alpha$ the normal operators (9) in

(10), in which case an accurate prediction of the excited states can be achieved in many harmonic-like regimes just by orthonormalization of these states, *without* diagonalization of the pertinent Hamiltonian matrix [10].

4. Application to a Many Fermion Model

In order to illustrate our formalism, we shall consider a finite superconducting model, dealing with N fermions distributed among $2m$ 2Ω-fold degenerate single particle levels with unperturbed energies ε_i, coupled by a pairing interaction. The pertinent Hamiltonian is

$$\hat{H} = \sum_i \varepsilon_i \hat{N}_i - G\hat{P}^\dagger \hat{P}, \tag{11}$$

where $\hat{N}_i = \sum_p c^\dagger_{pi} c_{pi} + c^\dagger_{\bar{p}i} c_{\bar{p}i}$, with $p = 1, \ldots, \Omega$, and $\hat{P} = \sum_{pi} c_{pi} c_{\bar{p}i}$ (the bar denotes time reversal and $c, c^\dagger$ are fermion creation and annihilation operators). G is the pairing coupling constant. We shall consider the half filled situation ($N = 2m\Omega$), in which case the ground state belongs to the completely symmetric representation spanned by unperturbed states $|n\rangle \equiv |n_1, \ldots, n_n\rangle$, $\sum_i n_i = N$, n_i even, with n_i denoting the number of particles in the level i. The ground state of (11) can thus be expanded as $|0\rangle = \sum_n C^0_n |n\rangle$, where all coefficients possess the same phase for $G > 0$.

We shall consider now our ME approach using relevant one and two-body diagonal operators in the previous basis, i.e., $\hat{N}_i$, $\hat{N}_i\hat{N}_j$. The ensuing approximate ground state coefficients are

$$C^{(0)}_n = p_n \exp[-\tfrac{1}{2}(\lambda_0 + \sum_i \lambda_i n_i + \sum_{i,j} \lambda_{ij} n_i n_j)], \tag{12}$$

with real λ_i and λ_{ij} determined by the set of equations (4) (equations (5) are in this case trivially satisfied) and $p_n = 1$. An excellent agreement with the exact ground state is obtained in this way for all values of G, both in the variational and inference treatments [8]. For G small (below the critical value) a reliable description is also obtained just with one-body operators (exponential distribution).

Here is interesting to remark that by setting $\lambda_{ij} = 0$ in (12) and employing a weight factor $p^2_n = \prod_i \binom{\Omega}{n_i/2}$, the BCS coefficients are obtained [8]. This factor is just the total number of states (in all relevant representations) with a given value of $n = (n_1, \ldots, n_n)$. Hence, BCS can be rendered in this case as a plain ME approach in the full space, constructed with one-body operators. It is thus apparent that it is possible to go with (12) *beyond* the BCS approximation, both in normal and particle number projected calculations [8]. Nevertheless, including $\lambda_{ij} \neq 0$, results with the previous factor p_n are similar to those with $p_n = 1$.

For excited states, it is verified that a very accurate agreement with the first excited states of the system can be obtained with low values of k_α for all values of G. Moreover, when using the corresponding normal operators (9), a good approximation can be directly obtained (without diagonalization) just by orthonormalization of the states with coefficients (10). In the thermodynamic limit, this procedure leads to the exact eigenstates. It is also easily shown that these orthonormalized states approach the unperturbed states in the limit $G \to 0$ (in which case $\lambda_i \to \pm\infty$). The diagonalization in the reduced basis generated by the states (10), using the normal operators, becomes thus important only in transitional regions (and as N decreases).

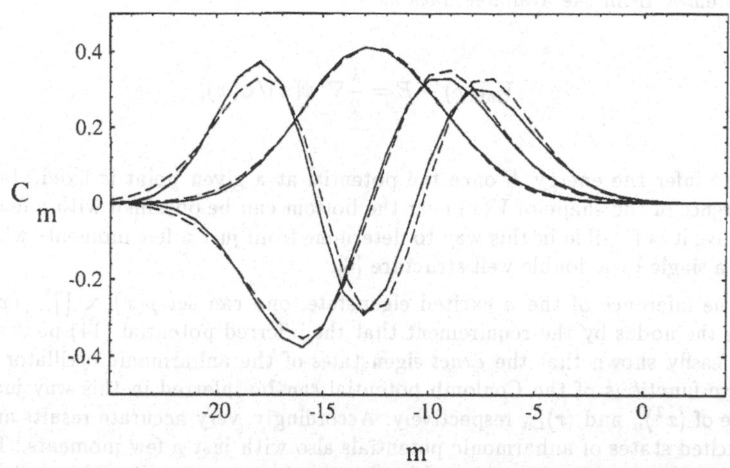

Fig. 1. The coefficients of the first three eigenstates in the two-level case. Solid lines correspond to exact results, dashed lines to the coefficients (10) after orthonormalization, and dashed-dotted lines to the coefficients obtained after a 4×4 diagonalization, which directly overlap with the exact ones in the scale of the figure.

As an example, we depict in Figure 1 variational results for the two level case, widely used in the context of nuclear physics, for $\Omega = 50$ and $g/\varepsilon = 2$ (superconducting regime). We employ the notation $\varepsilon = \varepsilon_2 - \varepsilon_1$, $g = \Omega G$ and $m = \frac{1}{4}(n_2 - n_1)$. There are in this case just two parameters associated with m and m^2. It can be seen that very accurate results are obtained, even just after orthonormalization of the coefficients (10).

5. Application to the Single Particle Schrödinger Equation

Finally, we briefly discuss the application of the foregoing ideas to the problem of approximately determining the wave function $\psi(x)$ corresponding to a bound state which is assumed to be an eigenstate of a Hamiltonian $\hat{H} = \frac{1}{2}\hat{P}^2 + V(\hat{x})$. The maximum entropy ansatz in the coordinate representation, analogous to (2), is

$$\psi(x) = p(x) \exp\{-\frac{1}{2}[\lambda_0 + \sum_{i=1}^{n} \lambda_i O_i(x)]\}, \tag{13}$$

where we can set $p(x) = 1$ for the (nodeless) ground state. The functions $O_i(x)$ can be chosen for instance as moments x^i ($\prod_j x_j^{i_j}$ in more than one dimension). It has been shown in [6] that a very accurate inference of the ground state of various anharmonic oscillators, including single and double well (bistable) potentials, can be obtained just with the data of a few moments $\langle x^i \rangle$. It is also possible to proceed variationally in the same way as in sections 2–3, obtaining an approximate scheme for determining the ground and the first excited states of the system.

In the inference case, having determined the wave function (13), it is then possible to *infer* a potential from the available data as

$$V_{\rm inf}(x) - E = \frac{1}{2}\nabla^2\psi(x)/\psi(x), \tag{14}$$

and also to infer the energy E once the potential at a given point is fixed. In general, a good inference of the shape of $V(x)$ near the bottom can be obtained with a few moments. For instance, it is feasible in this way to determine from just a few moments whether $V(x)$ possesses a single or a double well structure [6].

For the inference of the n excited eigenstate, one can set $p(x) \propto \prod_{i=1}^{n}(x - x_i)$ and determine the nodes by the requirement that the inferred potential (14) possess no poles. It can be easily shown that the *exact* eigenstates of the anharmonic oscillator and of the radial wave functions of the Coulomb potential can be inferred in this way just with the knowledge of $\langle x^2\rangle_n$ and $\langle r\rangle_{l,n}$ respectively. Accordingly, very accurate results are obtained for the excited states of anharmonic potentials also with just a few moments. The present inference scheme provides a useful tool for determining wave functions in particular molecular problems and also in semiclassical methods [11].

6. Conclusions

We have presented a maximum entropy based method for the approximate description of quantum wave functions. The approach is able to provide both an inference scheme for the reconstruction of quantum distributions from particular data, and a variational approximation for determining the ground and the first excited states in terms of a reduced set of variables associated with particular relevant observables. As we have seen, in many situations the amount of information or parameters required for an accurate prediction of the quantum distribution in the present formalism is small, particularly when suitable weight factors are employed for excited states. In many body problems the present formalism is thus able to yield approximate schemes of higher quality than those given by mean field approximations, which can be viewed in some cases as particular one-body ME wave functions. Hence, we consider that the present ideas may enlarge the already extended scope of applications of maximum entropy methods and provide a useful alternative approach in particular quantum mechanical problems.

ACKNOWLEDGMENTS. N.C. and R.R. are supported by an external fellowship of the National Research Council of Argentina (CONICET) and their permanent address is that of A.P. Support from CONICET (N.C. and A.P.) and CICPBA (R.R.) of Argentina for permanent positions is also acknowledged.

REFERENCES

[1] E.T. Jaynes, *Phys. Rev.* **106** (1957) 620; *Phys. Rev.* **108** (1957) 171.

[2] E.T. Jaynes, 'Papers on Probability, Statistics and Statistical Physics', R.D. Rosenkrantz (ed.), *Synthese series* vol. **158**, Reidel, Dordrecht.

[3] R. Balian, Y. Alhassid, H. Reinhardt, *Phys. Rep.* **131** (1986) 1.

[4] R. Balian, N.L. Balazs, *Ann. of Phys. (N.Y.)* **179** (1987) 97.

[5] R. Balian, M. Veneroni, *Ann. of Phys. (N.Y.)* **174** (1987) 229.

[6] N. Canosa, A. Plastino, R. Rossignoli, *Phys. Rev.* **A40** (1989) 519.

[7] N. Canosa, R. Rossignoli, A. Plastino, *Nucl. Phys.* **A512** (1990) 520; *Phys. Rev.* **A43** (1991) 1145.

[8] N. Canosa, R. Rossignoli, A. Plastino, H. Miller, *Phys. Rev.* **C45** (1992) 1162.

[9] L. Arrachea, N. Canosa, A. Plastino, M. Portesi, R. Rossignoli, *Phys. Rev.* **A45** (1992) 7104.

[10] N. Canosa, R. Rossignoli, A. Plastino, *Nucl. Phys.* **A** (in press).

[11] M.Casas, A. Plastino, A. Puente, N. Canosa, R. Rossignoli, to be published.

TOWARDS QUANTUM ϵ–ENTROPY AND VALIDITY OF QUANTUM INFORMATION

V. P. Belavkin
Department of Mathematics
University of Nottingham
Nottingham NG7 2RD, UK

The Bayesian method in quantum signal processing was initiated by C W Helstrom and developed in quantum detection [1], estimation [2, 3] and hypothesis testing [4] theory. The aim of this theory was to find an optimal quantum measurement, minimizing a cost function of quantum state estimation under given probabilities of possible states. The usefulness of entropy restrictions for the finding of the quantum estimation was shown in [2], where a quantum regression problem was solved under the condition of fixed entropy of quantum measurement.

Here we consider a dual maximum entropy problem for the quantum measurements under the condition of a fixed mean error of the quantum estimation. The corresponding classical problem, well-known as the Kolmogorov ϵ–entropy problem, was elaborated in the validity information theory by R. Stratonovich [5]. So we give the formulation and preliminary results for a new branch of quantum measurement and information theory, the quantum ϵ–entropy theory, related with classical problem of optimal discretization [6]. We restrict ourselves to the simplest quantum systems, described by the algebra of all operators in a Hilbert space. The general case can be developed by use of a more general notion of the entropy [7] and relative entropy [8] within a C^*-algebraic approach.

1. Let $\mathcal{H}$ be a Hilbert space, σ be a quantum state, described by a density operator $\sigma \geq 0$ in $\mathcal{H}$ with the trace $\mathrm{Tr}\sigma = 1$, and X be a given observable, described by a non-necessarily bounded operator with in general continuous spectrum (if self-adjoint). Quantum measurement with a discrete spectrum is usually described by an orthogonal resolution $I = \sum E_i$ of identity operator in $\mathcal{H}$, giving the probabilities $\Pr\{i\} = \mathrm{Tr}[\sigma E_i]$. According to the von Neumann projection postulate the state σ after reading of a result $i = 1, 2, \ldots$ normalized on the probability $\Pr\{i\}$ is given by the projection $\rho_i = E_i \sigma E_i$. We replace this very restrictive postulate which is not applicable for continuous spectrum observables, by *non-demolition principle*, according to which the state σ is always supposed to be already reduced: $\sigma = \sum \rho_i$, where ρ_i are *arbitrary* positive operators defining the probabilities $\Pr\{i\} = \mathrm{Tr}\rho_i$ and a posteriori states $\sigma_i = \rho_i / \Pr\{i\}$. One can consider a new representation $\hat{X} = X \otimes \hat{1}$ of the observables X in a dilated Hilbert space $\mathcal{H} \otimes l^2$ of sequences $\eta = \{\eta_1, \eta_2, \ldots\}$ with $\sum \|\eta_i\|^2 < \infty$, in which the measurement is described by the orthogonal identity resolution $\hat{I} = \sum \hat{E}_i$, where

$$\hat{E}_i = I \otimes |i\rangle\langle i|, \qquad |i\rangle = \{\delta_{i_1}, \delta_{i_2}, \ldots\}$$

A. Mohammad-Djafari and G. Demoments (eds.), Maximum Entropy and Bayesian Methods, 163–165.

and the state σ is given by the density matrix

$$\hat{\sigma} = \sum \rho_i \otimes |i\rangle\langle i| : \quad \mathrm{Tr}\{\hat{\sigma}[X \otimes \hat{y}]\} = \sum y_i \mathrm{Tr}[X\rho_i].$$

Here $\hat{y} = \sum y_i |i\rangle\langle i|$ is an observable in l^2, given by a function $i \mapsto y_i$ of the measurement data, represented in $\mathcal{H} \otimes l^2$ by the $\hat{Y} = I \otimes \hat{y}$, which commutes with the systems operators $\hat{X} = X \otimes \hat{1}$. Since this representation is obviously extendable also to continuous observations, the non-demolition principle can be equivalently formulated [8] as a possibility of realization of quantum measurements by observables $\hat{Y}$, commuting with all the operators $\hat{X}$ of the system under the measurement in a proper representation. Such representations we call the non-demolition measurements which indeed are more general than the usual 'demolition' ones, corresponding to the reduced states $\sigma = \sum E_i \sigma E_i$, resoluted by $\rho_i = E_i \sigma E_i$ with the a posteriori states $\sigma_i = \rho_i / \mathrm{Tr}\rho_i$ satisfying *repeatability* condition. Abandoning this condition applicable only for the discrete observations, one can relate with any positive identity resolution $I = \int E(y)\mathrm{d}y$ and an arbitrary state σ a non-repeatable continuous non-demolition measurement $\sigma = \int \rho(y)\mathrm{d}y$ with $\rho(y) = \sigma^{1/2}E(y)\sigma^{1/2}$. In the following we replace the von Neumann repeatability principle by quantum *ϵ-entropy principle*, defining the a posteriori states $\sigma(y) = \rho(y)/\mathrm{Tr}\rho(y)$.

2. Now we formulate the *optimal quantum measurement* problem

$$\epsilon(\mathcal{T}) = \inf_{\rho_i \geq 0} \left\{ \sum_{i=1}^{M} \inf_{y} \mathrm{Tr}[(X-y)^2 \rho_i] \middle| \sum_{i=1}^{M} \rho_i = \sigma \right\} \tag{1}$$

of a given observable X under an informational restriction $\log M \leq \mathcal{T}$. The difference

$$\begin{aligned} v[\rho] &= \inf_y \mathrm{Tr}[(X-y)^2\sigma] - \sum_{i=1}^{M} \inf_y \mathrm{Tr}[(X-y)^2\rho_i] \\ &= \sum_{i=1}^{M} (\mathrm{Tr}[X\sigma_i])^2 \mathrm{Pr}\{i\} - (\mathrm{Tr}[X\sigma])^2 \geq 0 \end{aligned}$$

is called the *validity* of information, $\mathcal{T} = \log M$, given by the discrete measurement $\sum_{i=1}^{M} \rho_i = \sigma$ of the quantum observable X. The optimal quantum measurement realizing the extremum (1) gives the most valuable information on X of the gain $\mathcal{T}$.

The *quantum ϵ-entropy* is defined as the maximum

$$S(\epsilon) = \sup_{\rho(y) \geq 0} \{ -\int \mathrm{Tr}[\rho(y) ln \sigma(y)]\mathrm{d}y \,|\, \int \mathrm{Tr}[(X-y)^2 \rho(y)]\mathrm{d}y \leq \epsilon \} \tag{2}$$

of the a posteriori entropy of $\sigma(y) = \rho(y)/\mathrm{Tr}\rho(y)$ over continuous measurements $\int \rho(y)\mathrm{d}y = \sigma$ corresponding to the mean square error restricted by $\epsilon > 0$. The maximum entropy principle defines the a posteriori state $\sigma(y)$ by the quantum non-demolition measurement corresponding to the error ϵ and the minimal information gain

$$\mathcal{T}[\rho] = \int \mathrm{Tr}[\rho(y) ln\sigma(y)]\mathrm{d}y - \mathrm{Tr}[\sigma\ ln\ \sigma] \geq 0. \tag{3}$$

The following theorem gives an estimate of minimal error (1) in terms of maximal entropy (2).

Theorem. $\epsilon(\mathcal{T}) \geq \epsilon$ if $\mathcal{T} \leq S - S(\epsilon)$, where $S = -\mathrm{Tr}[\sigma\ ln\ \sigma]$ is the quantum a priori entropy.

References

[1] C.W. Helstrom, *Quantum detection and estimation theory*, Academic Press, 1976.

[2] V. P. Belavkin, Optimal quantum randomized filtration, *Problems of Control and Inform Theory*, **3**, 25, 1974.

[3] A. S. Holevo, *Probabilistic aspects of quantum theory*, Kluwer Publisher, 1980.

[4] V. P. Belavkin, Optimal multiple quantum hypothesis testing, *Stochastics*, **3**, 40, 1975.

[5] R. L. Stratonovich, Theory of information, *Sov Radio*, Moscow, 1976.

[6] V. P. Belavkin, Optimal quantization of random vectors, *Isvestia A N USSR, Techn Kibernetika*, **1**, 20, 1970.

[7] F. Hiai, Ohya M, Tsukada M, Sufficiency, KMS condition and relative entropy in von Neumann algebras, *Pacific J Math*, **93**, 99, 1981.

[8] V. P. Belavkin, Staszewski P, C^*-algebraic generalization of relative entropy and entropy, *Ann Inst H Poincaré*, Sect A 37, 51, 1982.

QUANTUM STATISTICAL INFERENCE

R. N. Silver
Theoretical Division
MS B262 Los Alamos National Laboratory
Los Alamos, NM 87545

ABSTRACT. Can quantum probability theory be applied, beyond the microscopic scale of atoms and quarks, to the human problem of reasoning from incomplete and uncertain data? A unified theory of quantum statistical mechanics and Bayesian statistical inference is proposed. QSI is applied to ordinary data analysis problems such as the interpolation and deconvolution of continuous density functions from both exact and noisy data. The information measure has a classical limit of negative entropy and a quantum limit of Fisher information (kinetic energy). A smoothing parameter analogous to a de Broglie wavelength is determined by Bayesian methods. There is no statistical regularization parameter. *A priori* criteria are developed for good and bad measurements in an experimental design. The optimal image is estimated along with statistical and incompleteness errors. QSI yields significantly better images than the maximum entropy method, because it explicitly accounts for image continuity.

1. Introduction

A common data analysis problem is to infer a density function from incomplete and uncertain data. This is an *ill-posed problem*, meaning that there is no unique solution. The best that theory can provide are the probabilities of possible solutions to the problem which are consistent with the data. The most successful approach is to use Bayesian inference, in which the data are systematically combined with the available prior knowledge to restrict the hypothesis space. The most probable density function is presented as the *image* along with estimates of its reliability, which are conditional on the prior knowledge used. Jaynes' hypothesis of *Probability Theory as Logic* postulates the existence of universal principles for inference, although these are incompletely formulated at present.

Quantum statistical mechanics is also a probability theory which provides an empirically flawless description of physical phenomena on the microscopic scale of atoms and nuclei. The goal is to determine the most likely macrostate of a many-body system from incomplete constraints termed *extensive variables* and from prior knowledge of a Hilbert space of possible microstates. Although quantum theory may appear to be very different from Bayesian inference (Jaynes, 1989), perhaps it can provide a big clue toward finding more effecient principles of inference. The present paper demonstrates how an analogy to quantum statistical mechanics can provide a superior approach to data analysis.

A. Mohammad-Djafari and G. Demoments (eds.), Maximum Entropy and Bayesian Methods, 167–182.

The practical motivation for a new approach to data analysis is provided by experience with the maximum entropy (ME) class of Bayesian methods (Jaynes, 1983). It is mathematically analogous to classical statistical mechanics, and it has been successfully applied to numerous problems. The justifications for ME are reasonable (see, e.g., Tikochinsky, 1984), that there should be no prior correlations between points in an image. However, for many data analysis problems ME images exhibit spurious artifacts and overly sharp structure (Skilling, 1989). Many applications of ME are to *continuous* density functions, which means that the first derivatives of the density function are finite. This prior knowledge can only be incorporated by additional image smoothing. The developers of the leading ME code have proposed this smoothing to be a user-chosen *pre-blur* of a *hidden ME image*. While this approach has had some practical success in improving image quality, the choice of pre-blur function is *ad hoc* and the Shannon-Jaynes entropic information measure remains insensitive to the smoothness of an image.

A quantum approach to image smoothing may be motivated by a series of observations. In quantum statistical mechanics a constraint on the average energy will produce a density function which is smoother than the classical density function for the same potential. The smoothing parameter associated with an energy constraint is the *de Broglie wavelength*. Classical statistical mechanics is approached continuously as the de Broglie wavelength is reduced toward zero. The Schrödinger equation of quantum mechanics may be derived from a variational principle on Fisher information (kinetic energy) (Frieden, 1989). Variational principles are preferable to *ad hoc* formulations. Fisher information and entropy are intimately related concepts in information theory (Dembo, 1991).

Hence, a natural way to improve upon ME would be to formulate a Bayesian data analysis method in one-to-one mathematical analogy with quantum statistical mechanics. *Quantum Statistical Inference* (QSI) is the realization of this approach. It introduces a broad new class of prior probabilities incorporating correlations between points in an image. The proposed quantum information measure is sensitive to the smoothness of an image. The QSI images are usually superior to ME images for the same data. Since the classical limit of QSI is ME, QSI preserves ME's successes.

2. The Data Analysis Problem

A typical data analysis problem is to solve a Fredholm equation of the first kind,

$$\hat{D}_k = \int dy \hat{R}_k(y) \hat{f}(y) + \hat{N}_k \ . \tag{1}$$

Here $\hat{D}_k$ are N_d data which are typically incomplete, $\hat{R}_k(y)$ is an integral transform which is often an instrumental resolution function, and $\hat{N}_k$ represents noise. The true $\hat{f}(y)$ which generated the data is usually termed the *object*. Finding a normalized density function, $\hat{f}(y)$, is an ill-posed inverse problem.

In ME, the image is derived from a variational principle: maximize the Shannon-Jaynes entropy,

$$S_c \equiv - \int dy \hat{f}(y) \ln \left(\frac{\hat{f}(y)}{\hat{m}(y)} \right) \quad , \tag{2}$$

subject to the constraints of the data. Here, $\hat{m}(y)$ is a *default model* for $\hat{f}(y)$, so-named because it is the answer ME returns in the absence of data. It is most convenient to remove the explicit reference to the default model by introducing renormalized variables $x \equiv \int_{-\infty}^{y} dy' \hat{m}(y')$, $f(x) \equiv \hat{f}(y)/\hat{m}(y)$, and $R_k(x) \equiv \hat{R}(y)$. Since $\hat{m}(y)$ must also be a normalized density function, clearly $0 \leq x \leq 1$. The data analysis problem is then rewritten in vector notation,

$$\vec{D} = \int_0^1 dx \vec{R}(x) f(x) + \vec{N} \quad . \tag{3}$$

In the special case of noiseless data ($\vec{N} = \vec{0}$), the solution by the method of Lagrange multipliers is

$$f_C(x) = \frac{1}{Z_C} \exp\left(-\vec{\lambda}_C \cdot \vec{R}(x)\right) \qquad Z_C \equiv \int_0^1 dx \exp\left(-\vec{\lambda}_C \cdot \vec{R}(x)\right) \quad . \tag{4}$$

The N_d Lagrange multipliers, $\vec{\lambda}_C$, are to be determined by the fits to the data. This is formally analogous to a density function in *classical* statistical mechanics (subscript "C") with the identifications $\vec{\lambda}_C \cdot \vec{R}(x) \Leftrightarrow V(x)/T$, where $V(x)$ is potential energy and T is temperature. Z_C is a classical partition function.

ME works best for problems, such as deconvolution, where $\vec{R}(x)$ is a broad function and $f(x)$ contains comparatively sharp features. However, for problems where $\vec{R}(x)$ is sharp and $f(x)$ broad, ME tends to produce spurious structure. An extreme example is the data interpolation problem, in which the goal is to infer a density function from knowledge of its values at a finite number of points. Then $\vec{R}(x)$ consists of a set of δ-functions, and the ME solution is nonsense. Figure 1 shows a test density function and three data analysis problems corresponding to interpolation (I), Gaussian deconvolution (G), and exponential deconvolution (E). The data are taken to be exact and measured at 32 equally spaced data points. The corresponding ME images are shown in Fig. 2 displayed in 128 pixels. The ME image for the G data is credible, but it exhibits overly sharp and occasionally spurious structure. The ME image for the E data exhibits sharp edges reflecting the sharp feature in the resolution function. The ME image for the I data equals the data for measured pixels and equals the default model value of 1.0 for unmeasured pixels. The I and E ME images are not credible because they clearly reflect how the data were measured.

3. Quantum Statistical Mechanics & Fisher Information

A general relation exists between the calculus of variations and eigenvalue problems, such that variational principles can be associated with most of the differential equations used in physics (see, e.g. Mathews, 1964). However, physicists are not generally aware that the variational functional appropriate to the Schrödinger equation has a significance in statistics (Frieden, 1989). The kinetic energy term is formally analogous to *Fisher information*, which was originally introduced (Fisher, 1925) as a measure of the inverse uncertainty in determining a position parameter by maximum likelihood estimation. De Bruijn's identity of information theory relates Fisher information to entropy like the surfaces to volumes of sets (see e.g., Dembo, 1991). A tantalizing proposition is that such statistics and information theory concepts may provide the foundations for quantum physics. A counterargument is that the Hilbert space structure and associated operator algebra of quantum mechanics may be more fundamental than the Schrödinger equation. Nevertheless, a principle of *minimum Fisher information* (MFI) provides a convenient derivation of a Schrödinger-like wave equation applicable to density function estimation.

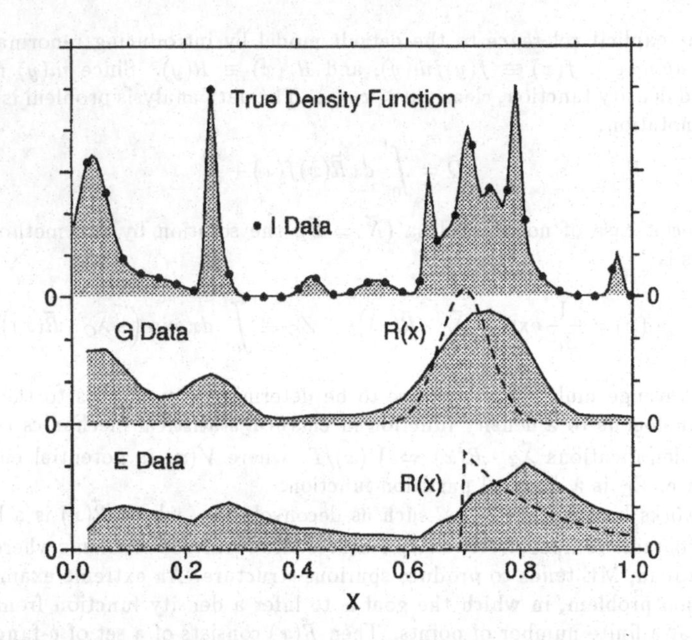

Fig. 1. Exact Data Example - Top curve is the test density function, and dots are the interpolation (I) data. Middle curves are the Gaussian (G) data (solid) and resolution function $R(x)$ (dashed $\times 0.35$). Bottom curves are the exponential (E) data and resolution function. Data are exact and measured at 32 points. The Gaussian has a standard deviation of 0.044 and the exponential has a decay constant of 0.15.

Applying MFI to the data analysis problem defined by Eq. (3), extremize

$$Q_1 \equiv \frac{1}{4}\int_0^1 dx \frac{1}{f(x)}\left(\frac{\partial f(x)}{\partial x}\right)^2 + \vec{\lambda}\cdot\int_0^1 dx \vec{R}(x) f(x) - E\int_0^1 dx f(x) \quad . \tag{5}$$

The first term is Fisher information, the second term imposes the constraints due to the data with Lagrange multipliers $\vec{\lambda}$, and the third term is the normalization constraint with Lagrange multiplier E. Defining a *wave function* by $\psi(x) \equiv \pm\sqrt{f(x)}$, the Euler-Lagrange equation is

$$\frac{\partial Q_1}{\partial \psi(x)} - \frac{d}{dx}\left(\frac{\partial Q_1}{\partial(d\psi(x)/dx)}\right) = 0 \quad . \tag{6}$$

This results in a Schrödinger-like wave equation,

$$-\frac{d^2\psi(x)}{dx^2} + \vec{\lambda}\cdot\vec{R}(x)\psi(x) = E\psi(x) \quad . \tag{7}$$

The data constraints $\vec{\lambda}\cdot\vec{R}(x)$ are analogous to potential energy $V(x)$, and E is analogous to total energy. Eq. (7) may also be written in matrix form as $\mathbf{H} \mid \psi >= E \mid \psi >$ using the Dirac *bra ket* notation $\psi(x) \equiv < x \mid \psi >$. The matrix $\mathbf{H}(x, x')$ is a *Hamiltonian*.

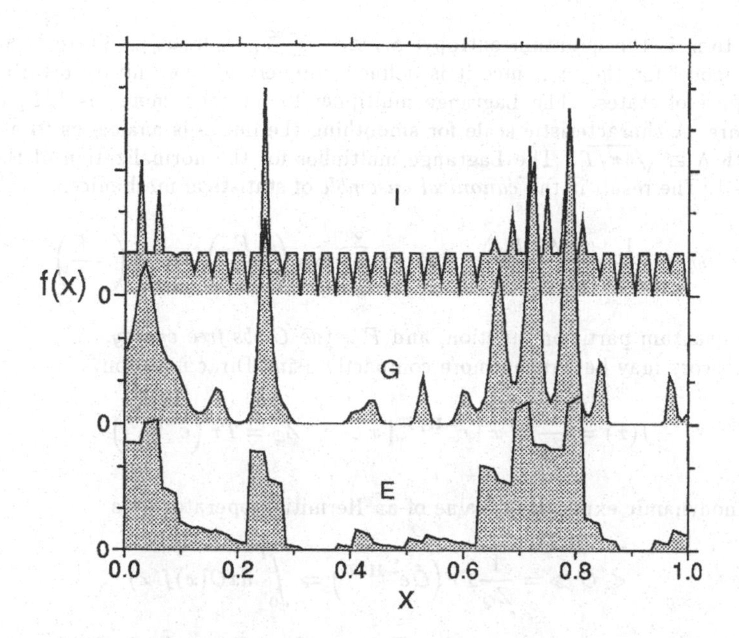

Fig. 2. Maximum Entropy Images - For exact data example in Fig. 1, displayed in 128 pixels.

Solving the Schrödinger equation is an eigenvalue problem subject to a requirement that the solutions form a Hilbert space in $0 \leq x \leq 1$. For example, for $V(x) = 0$ a complete orthonormal set of solutions is $\psi(x) = \sqrt{2}\cos(k_n x); \sqrt{2}\sin(k_n x)$ where wave vectors are $k_n = 2\pi n$ and eigenenergies are $E_n^o = k_n^2$. For $V(x) \neq 0$, the $\psi(x)$ for the ground state is nodeless and would provide an image corresponding to the MFI principle.

However, the ME principle is certainly successful for many data analysis problems. Fortunately, it is not necessary to make a choice between these two variational principles, since quantum statistical mechanics (see, e.g., McQuarrie, 1976) provides a seamless interpolation between the MFI principle in the quantum limit and the ME principle in the classical limit. The image is taken to be a weighted sum of eigensolutions,

$$f(x) = \sum_{n=0}^{\infty} w_n \psi_n^2(x) \quad . \tag{8}$$

The weights, w_n, are to be interpreted as the probabilities for the n'th eigensolutions, and they should be determined by maximizing the quantum entropy in the Hilbert space subject to any applicable constraints. One constraint is that the image should be normalized to unity. Higher energy solutions have increasing numbers of nodes enabling them to describe increasingly sharp structure in $f(x)$. A second constraint on the average energy would, therefore, act as a low-pass spatial filter resulting in smoothing of the image.

Therefore, maximize

$$Q_2 \equiv \sum_{n=0}^{\infty} \left[-w_n \ln(w_n) - \frac{1}{T} E_n w_n + (-\ln Z_Q + 1) w_n \right] \quad . \tag{9}$$

The first term is the quantum entropy, $S_Q \equiv -\sum_{n=0}^{\infty} w_n \ln(w_n)$. There is no need for a default model for the w_n, since it is defined uniquely by the unitary invariance of the Hilbert space of states. The Lagrange multiplier for average energy is $1/T$, where T is temperature. A characteristic scale for smoothing the image is analogous to a de Broglie wavelength $\Lambda \equiv \sqrt{4\pi/T}$. The Lagrange multiplier for the normalization of the image is $-\ln Z_Q + 1$. The result is the *canonical ensemble* of statistical mechanics,

$$w_n = \frac{1}{Z_Q} \exp\left(-\frac{E_n}{T}\right) \qquad Z_Q \equiv \sum_{n=0}^{\infty} \exp\left(-\frac{E_n}{T}\right) = \exp\left(-\frac{F}{T}\right) \quad . \tag{10}$$

Z_Q is the quantum partition function, and F is the *Gibbs free energy*.

This theory may be written more compactly using Dirac notation,

$$f(x) = \frac{1}{Z_Q} < x \mid e^{-\mathbf{H}/T} \mid x > \qquad Z_Q = Tr\left(e^{-\mathbf{H}/T}\right) \quad . \tag{11}$$

The thermodynamic expectation value of an Hermitian operator $\hat{O}$ is

$$\ll \hat{O} \gg = \frac{1}{Z_Q} Tr\left(\hat{O} e^{-\mathbf{H}/T}\right) \Rightarrow \int_0^1 dx O(x) f(x) \quad . \tag{12}$$

In the following, the term *observable* will be used to denote both the operator and its expectation value. Here, the rightarrow denotes the special limit where $\hat{O} \Rightarrow O(x)\delta(x-x')$. For example, setting $O(x) \Rightarrow \delta(x-x_o)$ gives the image $f(x_o)$, and setting $O(x) \Rightarrow R_k(x)$ gives the k'th fit to the data, $\ll R_k \gg$. The image is an implicit function of $\vec{\lambda}$ and Λ. In the classical limit ($\Lambda \to 0, T \to \infty$), the images reduce to the ME results given by Eq. (4) with the identifications $\vec{\lambda}_C \Leftrightarrow \vec{\lambda}/T$ and $Z_Q \Leftrightarrow Z_C/\Lambda$. In the quantum limit ($\Lambda \to \infty, T \to 0$), they reduce to the MFI images.

4. Bayesian Statistical Inference

The optimal $\vec{\lambda}$ and Λ are to be determined from the data by the application of Bayesian statistical inference (see, e.g., Loredo, 1990). Bayes theorem is

$$P[\vec{\lambda}, \vec{D}, \Lambda] = P[\vec{\lambda} \mid \vec{D}, \Lambda] \times P[\vec{D}, \Lambda] = P[\vec{D} \mid \vec{\lambda}, \Lambda] \times P[\vec{\lambda}, \Lambda] \quad . \tag{13}$$

In Bayesian terminology $P[\vec{\lambda} \mid \vec{D}, \Lambda]$ is termed the *posterior probability*, $P[\vec{D}, \Lambda]$ is the *evidence*, $P[\vec{D} \mid \vec{\lambda}, \Lambda]$ is the *likelihood function*, and $P[\vec{\lambda}, \Lambda]$ is the *prior probability*. The most probable $\vec{\lambda}$ is determined from the maximum of the posterior probability, and the most probable Λ is determined from the maximum of the evidence. The data are embodied in the likelihood function.

The prior probability is determined by an analogy between quantum statistical mechanics and statistical inference. In statistical mechanics, the system probability is proportional to the partition function, $Z_Q = \exp(-F/T)$, which counts the number of different eigensolutions the system can have within the constraints. In Bayesian inference the system probability is proportional to the product of the likelihood function and the prior probability. Since the data constraints have already been identified with the potential energy, the likelihood function should be proportional to $\exp(- \ll V \gg /T)$. That leaves a prior probability $P[\vec{\lambda}, \Lambda] \propto \exp(-I_Q - F_o/T)$. A subscript "$o$" will be used to denote a quantity evaluated at the default model value of $\vec{\lambda} = \vec{0}$. The quantum generalization of an information measure is

$$I_Q \equiv \frac{\delta F}{T} - \frac{\vec{\lambda} \cdot \ll \vec{R} \gg}{T} = \delta \left[\frac{1}{T} \ll -\frac{\partial^2}{\partial x^2} \gg - S_Q \right] \quad . \tag{14}$$

Here "δ" denotes a change in this quantity from the $\vec{\lambda} = \vec{0}$ value, so that $I_Q^o = 0$. The right hand side is a simple sum of the expectation values for Fisher information (kinetic energy) and quantum negentropy. In the classical ($\Lambda \to 0$) limit, I_Q reduces to the information measure in ME, $-S_C$.

QSI obeys *duality* in the sense of Legendre transformations. The information, I_Q, and the change in free energy, δF, are conjugate potentials. The Lagrange multipliers, $\vec{\lambda}$, and the observables , $\ll \vec{R} \gg$, are conjugate variables. First derivatives are related by

$$\frac{\partial \delta F}{\partial \vec{\lambda}} = \ll \vec{R} \gg \qquad \frac{\partial I_Q}{\partial \ll \vec{R} \gg} = -\frac{\vec{\lambda}}{T} \quad . \tag{15}$$

Second derivatives are related by

$$-T \frac{\partial^2 \delta F}{\partial \lambda_k \partial \lambda'_k} = \mathbf{K}_{k,k'} \qquad \frac{\partial^2 I_Q}{\partial \ll R_k \gg \partial \ll R_{k'} \gg} = \mathbf{K}^{-1}_{k,k'} \quad . \tag{16}$$

Here $\mathbf{K}$ is the *susceptibility* (or *Hessian*) matrix,

$$\mathbf{K} = -T \frac{\partial \ll \vec{R} \gg}{\partial \vec{\lambda}} \quad . \tag{17}$$

A susceptibility in statistical mechanics is the derivative of the expectation value of an observable (e.g. $\ll R_k \gg$) with respect to a *field* (e.g. $\lambda_{k'}$).

QSI also obeys differential geometry (Amari, 1985; Levine, 1986; Balian, 1986; Rodriguez, 1989). $\mathbf{K}^{-1}$ is a Riemann metric in the space of observables, infinitesimal changes in $\ll \vec{R} \gg$ and $\vec{\lambda}$ are covariant and contravariant expressions of the same vector, and the *divergence* between two sets of observables is

$$D^{1,2} \equiv I_Q^1 + \frac{\vec{\lambda}^2 \cdot \ll \vec{R} \gg^1}{T} - \frac{\delta F^2}{T} \quad . \tag{18}$$

Infinitesimal changes in the tangent plane define a Riemann distance,

$$ds^2 = \sum_{k,k'} \mathbf{K}^{-1}_{k,k'} d \ll R_k \gg d \ll R_{k'} \gg \quad , \tag{19}$$

and the Riemann invariant volume for integration over the observables is proportional to $\sqrt{\det(\mathbf{K}^{-1})}$. Hence, the prior probability must be

$$P[\ll \vec{R} \gg, \Lambda] = \frac{1}{\sqrt{(2\pi)^{N_d} \det(\mathbf{K})}} \exp\left(-I_Q - \frac{F^o}{T}\right) \quad . \tag{20}$$

The factor $(2\pi)^{N_d}$ ensures that the prior probability is normalized to unity in a Gaussian approximation.

The *fluctuation-dissipation theorem* (Kubo, 1959) of quantum statistical mechanics relates $\mathbf{K}$ to a correlation function, which may be written in two ways,

$$\mathbf{K}_{k,l} = \int_0^1 d\tau \ll \delta R_k e^{-\mathbf{H}\tau/T} \delta R_l e^{+\mathbf{H}\tau/T} \gg = \int_0^1 dx \int_0^1 dx' \delta R_k(x) \delta R_l(x') \Theta(x,x') \quad . \tag{21}$$

Here $\delta R_k(x) \equiv R_k(x) - \ll R_k \gg$. Eq. (21) may be derived by second-order quantum perturbation theory. The first expression for $\mathbf{K}$ is the same as a susceptibility in the Kubo theory for linear response (see, e.g., Fick, 1990). The second expression introduces a *spatial correlation function* $\Theta(x,x')$ (discussed further in Sec. 7), whose width characterizes prior correlations between points in the image. It is a strictly positive real symmetric function satisfying $\int_0^1 dx \Theta(x,x') = f(x')$. Consequently, $\mathbf{K}$ must be a positive definite matrix. Combined with Eq. (16), this implies that I_Q must be a strictly positive convex function of the observables.

The behavior of the eigenvalues of $\mathbf{K}$ will be important to determining the optimal Λ from the evidence. In the classical limit of $\Lambda \to 0$, $\Theta(x,x') \to f(x)\delta(x-x')$. The peak in $\Theta(x,x')$ broadens with increasing $\Lambda \neq 0$. In the default model limit of $\vec{\lambda} = \vec{0}$, the variance of the peak is given by $\Lambda/\sqrt{12\pi}$. In the quantum limit of $\Lambda \to \infty$, $\Theta(x,x') \to f(x)f(x')$. At fixed $\ll \vec{R} \gg$, the eigenvalues of $\mathbf{K}$ are maximal at $\Lambda = 0$, decrease monotonically with increasing Λ, and go to zero at $\Lambda = \infty$.

Consider the application of QSI to the exact data example considered in Figures 1,2. Then the effect of the likelihood function is simply to set $\ll \vec{R} \gg \to \vec{D}$ in Eq. (20). In principle, the image should be obtained by marginalizing over Λ. In practice, a single Λ_{opt} may be used instead of marginalizing if the evidence, $P[\vec{D},\Lambda]$, has a single sharp peak. Then Eq. (20) separates into a product of three factors whose behavior may be understood using the dependence of the eigenvalues of $\mathbf{K}$ on Λ. The Riemann volume factor $1/\sqrt{\det(\mathbf{K})}$, favors the simpler model of larger Λ. This means fewer quantum states contributing to the image or fewer independent components of the image. The Riemann volume factor is therefore equivalent to an *Occam factor* in the Bayesian literature (Bretthorst, 1988), since it implements the principle of Occam's razor. Eq. (16) implies that the *data factor*, $\exp(-I_Q)$, favors the more complex model of smaller Λ. This means more quantum states contributing to the image. The factor $\exp(-F^o/T)$ is a slowly varying Jeffrey's prior for Λ which is independent of the data.

Figure 3 shows the evidence, Occam factor and data factor for the interpolation problem as a function of Λ. The $\Lambda_{opt} \approx 0.6$, corresponding to a variance of the spatial correlation function given by $\Lambda_{opt}/\sqrt{12\pi} \approx 0.1$. Comparable values of Λ_{opt} are found for all three data analysis problems in Figure 1. Figure 4 shows the optimal QSI images for the data in Figure 1. They are clearly more credible than the corresponding ME images shown in Figure 2. Credibility is quantitatively expressed by the many orders of magnitude larger Bayesian evidences for the QSI images relative to the ME images.

5. Noisy Data

Consider the extension of QSI to data subject to Gaussian independent noise defined by $E(\hat{N}_k) = 0$ and $Cov(\hat{N}_k \hat{N}_{k'}) = \delta_{k,k'} \hat{\sigma}_k^2$ in Eq. (1). It is most convenient to renormalize to

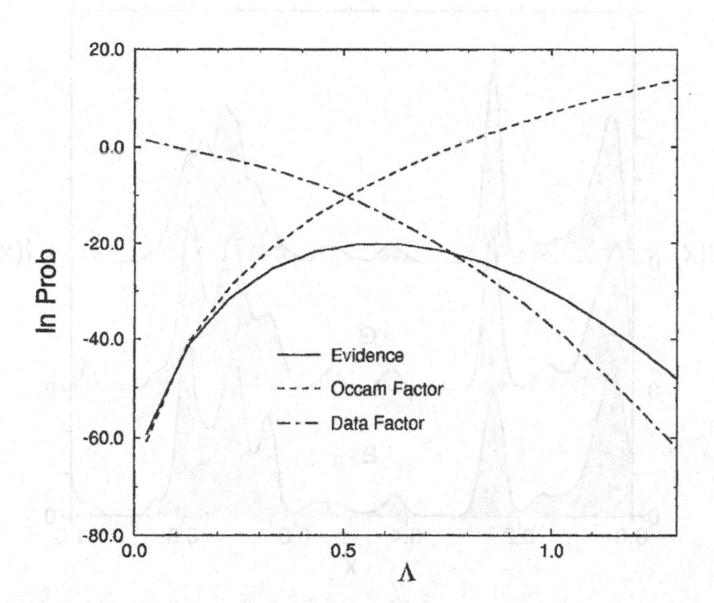

Fig. 3. Optimization of de Broglie Wavelength - The evidence, occam factor, and data factor for the interpolation problem in Fig. 1 as a function of Λ.

new variables $D_k \equiv \hat{D}_k/\hat{\sigma}_k$ and $R_k(x) \equiv \hat{R}_k(y)/\hat{\sigma}_k$. The likelihood function is then

$$P[\vec{D} \mid \vec{\lambda}, \Lambda] = \frac{1}{\sqrt{(2\pi)^{N_d}}} \exp\left(-\frac{\chi^2}{2}\right) \quad \chi^2 = (\vec{D} - \ll \vec{R} \gg)^\dagger (\vec{D} - \ll \vec{R} \gg) \quad . \tag{22}$$

Bayes theorem for noisy data becomes

$$P[\ll \vec{R} \gg \mid \vec{D}, \Lambda] \times P[\vec{D}, \Lambda] = \frac{1}{(2\pi)^{N_d}\sqrt{\det(\mathbf{K})}} \exp\left(-I_Q - \frac{\chi^2}{2} - \frac{F^o}{T}\right) \tag{23}$$

The most probable image is obtained from the minimum of

$$Q_3 \equiv I_Q + \frac{\chi^2}{2} \quad , \tag{24}$$

which is given by the solution to

$$\frac{\partial Q_3}{\partial \ll \vec{R} \gg} = -\frac{\vec{\lambda}}{T} - \vec{D} + \ll \vec{R} \gg = \vec{0} \quad . \tag{25}$$

A sub(super)script "f" denotes a quantity evaluated at the *final* solution, $\vec{\lambda}^f$, to Eq. (25). The second derivative

$$\frac{\partial^2 Q_3}{\partial \ll \vec{R} \gg \partial \ll \vec{R} \gg} = \mathbf{K}^{-1} + \mathbf{1} \quad , \tag{26}$$

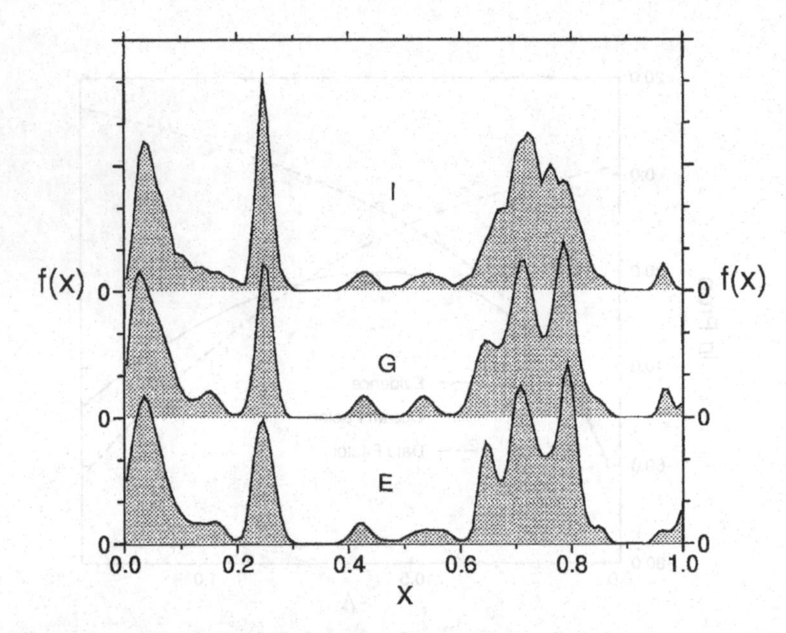

Fig. 4. Quantum Statistical Inference Images - For the exact data example in Fig. 1, displayed in 128 pixels.

is strictly positive implying that a unique solution to Eq. (25) exists.

For typical data analysis problems, the eigenvalue spectrum of $\mathbf{K}^f$ is a very steep function varying over many orders of magnitude. Independent measurements can be defined as eigenvectors of $\mathbf{K}^f$, and their quality can be ranked according to the size of their eigenvalues. *Good measurements* may be defined as eigenvectors of $\mathbf{K}^f$ whose eigenvalues are much greater than 1. The data for good measurements are dominated by signal, so that the typical values of $|\ll R_k \gg| \approx | D_k |$ are large compared to one. *Bad measurements* may be defined by eigenvectors of $\mathbf{K}^f$ whose eigenvalues are much less than 1. The data for bad measurements are dominated by noise so that the typical values of $| D_k |$ are of order 1 and the typical values for $|\ll R_k \gg|$ are much less than one. In other words, QSI fits the good measurements and ignores the bad ones. The *number of good measurements* may be defined by

$$N_g^f \equiv Tr\left[\mathbf{K}^f \cdot (\mathbf{1} + \mathbf{K}^f)^{-1}\right] \quad . \tag{27}$$

An error scaling consistency argument suggests that

$$\chi_f^2 \approx N_d - N_g^f \quad , \tag{28}$$

so that the χ^2 for the most probable image is less than the number of data. Numerical experiments suggest that N_g satisfies inequalities,

$$N_g^{\vec{\lambda}\neq\vec{0},\Lambda\neq 0} \leq N_g^{o,\Lambda\neq 0} \leq N_g^{o,\Lambda=0} \quad . \tag{29}$$

Therefore, the initial good measurements can be identified from the eigenvectors of the classical susceptibility, $\mathbf{K}^{o,\Lambda=0}$. They can often be calculated *a priori* from the resolution functions and noise, before any data are acquired. Throwing out the initial bad measurements can dramatically reduce the computational and data acquisition tasks without affecting the results.

The evidence is given by

$$P[\vec{D},\Lambda] = \frac{1}{(2\pi)^{N_d}\sqrt{\det(\mathbf{K}^f+1)}} \exp\left(-I_Q^f - \frac{\chi_f^2}{2} - \frac{F^o}{T}\right) \quad . \tag{30}$$

Only good measurements contribute to the Occam factor $1/\sqrt{\det(\mathbf{K}^f+1)}$. The data factor is now $\exp(-I_Q^f - \chi_f^2/2)$. Again, the Occam factor favors small Λ, while the data factor favors large Λ.

Figure 5 shows noisy data sets for the test density function in Fig. 1 and the same Gaussian (G) and exponential (E) resolution functions. The data are measured at 64 equally spaced points with the noise levels indicated. Figure 6 shows the corresponding ME images obtained by setting $\Lambda = 0$ in QSI, although similar results would be obtained from using the MEMSYS routines (Skilling, 1989). The ME images clearly show excessive overfitting and noise artifacts. Figure 7 shows the corresponding QSI images for the same data, which are clearly superior to the ME images. Although χ^2 is larger for the QSI images than for the ME images, the evidence for them is greater because they correspond to the simpler model of larger Λ. The data sets are placed in order of increasing information content, as measured by the value of I_Q^f. This ranking also corresponds to the visual quality of the images in Figure 7 and to the values of N_g^f. Note that the lineshape can be more important than the statistical errors in determining the information content of the data, with sharper lineshapes generally yielding more information (everything else being equal).

6. Statistical and Incompleteness Errors

The goal of a statistical inference procedure is to use the data and prior knowledge to make predictions for unmeasured observables. This should include estimates of the reliability of those predictions. For any observable, $\ll O \gg$, these may be calculated from the conditional probability, $P[\ll O \gg \mid \vec{D}]$, using Bayes theorem. In the absence of data on that observable, the most probable value for $\ll O \gg$ is obtained at $\vec{\lambda}^f$ and $\lambda_O = 0$. This section summarizes the results for the errors on these estimates.

For two observables, $\ll A \gg$ and $\ll B \gg$, define generalized susceptibilities,

$$K_{A,B}^f \equiv \int_0^1 d\tau \ll \delta A e^{-\mathbf{H}\tau/T} \delta B e^{+\mathbf{H}\tau/T} \gg^f = \int_0^1 dx \int_0^1 dx' \delta A(x) \delta B(x') \Theta_f(x,x') \quad , \tag{31}$$

with $\delta A(x) \equiv A(x) - \ll A \gg_f$. Then the composite covariance $C[\delta \ll A \gg \delta \ll B \gg]$ is given by

$$C = K_{A,B}^f - \vec{K}_{A,R}^{f\dagger} \cdot (1+\mathbf{K}^f)^{-1} \cdot \vec{K}_{R,B}^f \quad . \tag{32}$$

In the absence of measurements the covariance is maximal, given by $K_{A,B}^o$. The second term in Eq. (32) says that good measurements reduce the covariance for unmeasured quantities, as should be required of any valid statistical inference procedure.

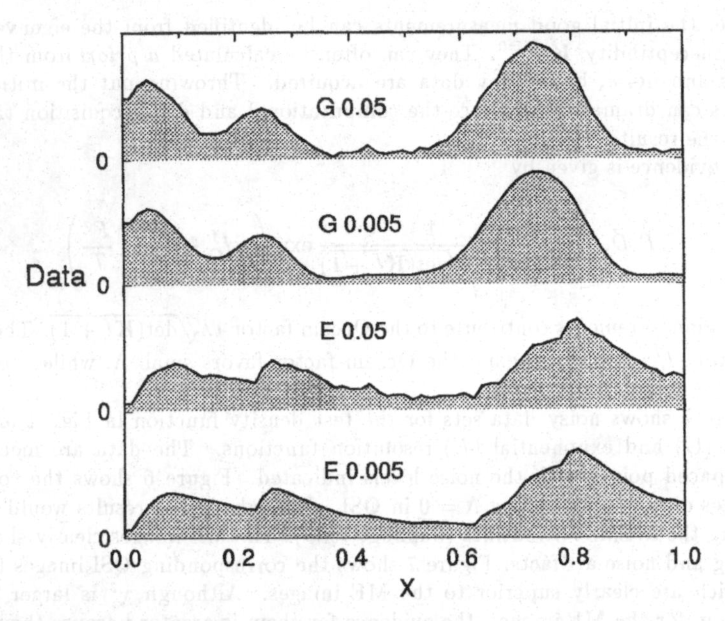

Fig. 5. Noisy Data Example - The test density function is convolved with the same resolution functions in Fig. 1 at 64 equally spaced data points, and then Gaussian independent noise is added. The notation "E 0.05" stands for the exponential resolution function with a noise standard deviation of 0.05.

Insight can be gained by separating the covariance into the sum of a *statistical covariance*,

$$C_S \equiv \vec{K}^{f\dagger}_{A,R} \cdot (\mathbf{K}^f + \mathbf{K}^f \cdot \mathbf{K}^f)^{-1} \cdot \vec{K}^f_{R,B} \quad , \tag{33}$$

and an *incompleteness covariance*,

$$C_I \equiv K_{A,B} - \vec{K}^{f\dagger}_{A,R} \cdot (\mathbf{K}^f)^{-1} \cdot \vec{K}^f_{R,B} \quad . \tag{34}$$

These definitions make sense in several respects. If the statistical errors on the data are zero, C_I remains finite while C_S goes to zero. If $\ll A \gg$ is one of the measurements, C_I goes to zero while C_S remains finite. Bad measurements are the dominant contribution to C_S, which cancel the reduction in C_I.

An example is the covariance on the fits to the data, $\delta \ll \vec{R} \gg^\dagger \cdot \delta \ll \vec{R} \gg$. Then $C_I = 0$ and $C_S = N_g^f$. Combined with Eq. (28), it implies that the average fit to the data satisfies $E(\chi^2) \approx N_d$, which corresponds to the intuitive expectation. Another example is the covariance on points in the image. Define $\vec{\Gamma}(x) \equiv \int_0^1 dx' \Theta(x,x') \delta \vec{R}(x')$. Then the covariance of two points in the image is

$$C[\delta f(x_1) \delta f(x_2)] = \Theta_f(x_1, x_2) - f(x_1) f(x_2) - \vec{\Gamma}(x_1)^\dagger \cdot (1 + \mathbf{K}^f)^{-1} \cdot \vec{\Gamma}(x_2) \quad . \tag{35}$$

The case $x_1 = x_2$ would give the errors on points. The first term on the r.h.s. is the spatial correlation function. In the absence of data it indeed characterizes prior correlations

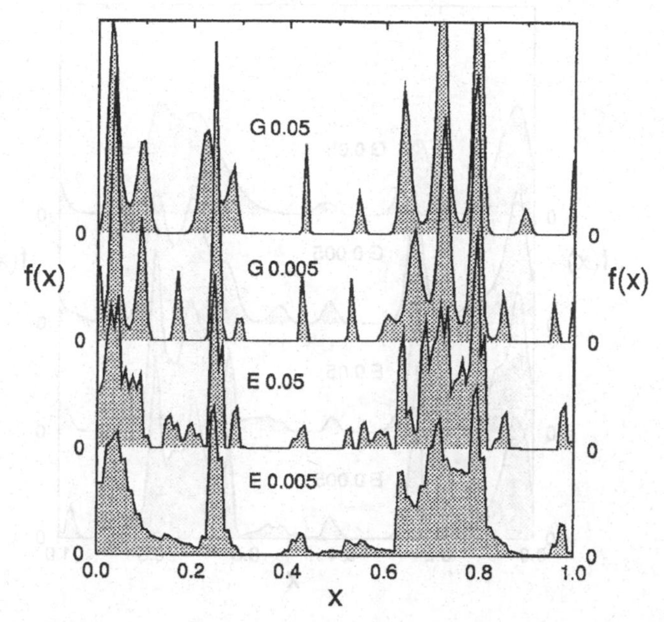

Fig. 6. Maximum Entropy Images - For the noisy data example in Fig. 5.

between points in the image. In the classical ME limit of zero Λ, it approaches a δ- function, so that the errors on individual points diverge. For $\Lambda \neq 0$, the errors on points are finite and they decrease with increasing Λ. The second term preserves the normalization of the image. The third term states how good measurements reduce the covariance of points in the image.

Generally, the optimal QSI images are smoother and have smaller errors than the ME images for the same data. If the object consists of a small number of isolated δ-functions, $\Lambda_{opt} \approx 0$ (T large) so that QSI may reduce to ME with maximal errors. If the object is smooth but has sharper structure than the resolution function, the QSI image may have $\Lambda_{opt} \neq 0$ even though it may resemble a ME image at $\Lambda_{opt} = 0$; however, the errors on the QSI image will be smaller than errors on the ME image. If the data are close to the default model predictions, Λ_{opt} will be large ($T \approx 0$) and the errors will be small.

7. Prior Probability in Image Space

The above formulation of QSI using the method of Lagrange multipliers may be derived from a prior probability in image space. This is more fundamental since it can also be applied to non-linear data analysis problems which do not satisfy Eq. (1).

Let the space $0 \leq x \leq 1$ be divided into M pixels of width $\Delta x = 1/M$, such that $f_i \equiv \Delta x f(x_i)$, and $\sum_i f_i = 1$. Denote the potential energy for the i'th pixel by $V^i \equiv V(x_i)$. Then the f_i's and V^j's are Legendre transform conjugate variables such that

$$I_Q = \frac{\delta F}{T} - \sum_{i=1}^{M} \frac{V^i f_i}{T} \qquad \frac{\partial \delta F}{\partial V^i} = f_i \qquad \frac{\partial I_Q}{\partial f_i} = -\frac{V^i}{T} \ . \tag{36}$$

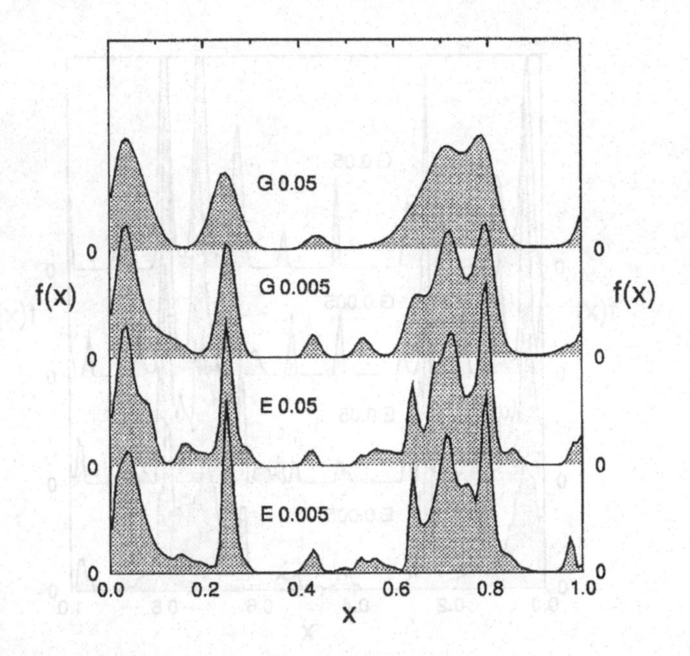

Fig. 7. Quantum Statistical Inference Images - For the noisy data example in Fig. 5.

The Riemann metric in image space is

$$\mathbf{g}_{ij} = -T\frac{\partial^2 \delta F}{\partial V^i \partial V^j} \qquad \mathbf{g}^{ij} = \frac{\partial^2 I_Q}{\partial f_i \partial f_j} \quad . \tag{37}$$

It is related to the spatial correlation function by

$$\mathbf{g}_{ij} = \Delta x^2 \Theta(x_i, x_j) - f_i f_j \qquad \sum_{i=1}^{M} \mathbf{g}_{ij} = 0 \quad , \tag{38}$$

where

$$\Theta(x, x') = \frac{1}{Z_Q}\int_0^1 d\tau < x \mid e^{-\mathbf{H}\tau/T} \mid x' >< x \mid e^{-\mathbf{H}(1-\tau)/T} \mid x' > \quad . \tag{39}$$

The susceptibility $\mathbf{K}$ given by Eq. (21) is then the Riemann *inner product* of two observables. The $\mathbf{g}_{ij}$ has a zero eigenvalue corresponding to the requirement that the image is normalized to unity. Removing this zero eigenvalue to define a $\mathbf{g}'$, leads to a final expression for the prior probability in image space,

$$P[\vec{f}, \Lambda] = \frac{1}{\sqrt{(2\pi)^{M-1}\det(\mathbf{g}')}} \exp\left(-I_Q - \frac{F^o}{T}\right)\delta\left(1 - \sum_{i=1}^{M} f_i\right) \quad . \tag{40}$$

In the classical ($\Lambda \to 0$) ME limit, $\mathbf{g}_{ij} \to \delta_{ij} f_i - f_i f_j$ and $I_Q \to -S_C$. The prior probability reduces to

$$P[\vec{f}, \Lambda] \to \frac{1}{\sqrt{(2\pi)^{M-1} \prod_{i=1}^{M} f_i}} \exp(S_C)\, \delta\left(1 - \sum_{i=1}^{M} f_i\right) , \tag{41}$$

and the image becomes

$$f_i = \frac{\exp\left(-V^i/T\right)}{Z} \qquad Z = \sum_{i=1}^{M} \exp\left(-\frac{V^i}{T}\right) . \tag{42}$$

Previous implementations of ME (e.g. Skilling, 1989) assumed a prior probability proportional to $\exp(\alpha S_C)$, where α is a statistical regularization parameter to be determined from the data. However, the classical limit of QSI yields a ME method in which $\alpha = 1$ is known *a priori*.

More generally, QSI has important differences with the traditional literature (Titterington, 1985) on regularizing ill-posed inverse problems. The QSI image is normalized to unity, whereas the normalization is a free parameter in the traditional approach. A lack of knowledge of how to normalize the data may require the introduction of a data normalization *hyperparameter* to be optimized by Bayesian means. The QSI regularizing functional, I_Q, is not an explicit function of the image, but rather it is an implicit function of the potential energy. Although QSI has no statistical regularization parameter, it introduces a smoothing parameter Λ which equals zero in the ME limit. Eqs. (27-29) say that N_g^o is maximal in the absence of data, whereas prior ME formulations start at $\alpha = \infty$ corresponding to $N_g^o = 0$. So QSI provides criteria for the *a priori* identification of good measurements in an experiment, whereas traditional regularization approaches do not. This should have especially important implications for the design of experiments.

The expression of QSI in terms of a reduced set of observables and Lagrange multipliers presented in Sections 3&4 can be derived starting from the image space formulation of Bayes theorem using Eq. (40) as the prior probability. Define a noisy data problem as a set of data $\ll R_k \gg$ with errors σ_k. The $P[\ll \vec{R} \gg, \Lambda]$ in Eq. (20) is the evidence in the zero noise limit of this image space problem, at least within Gaussian approximations for the integrals. The Lagrange multipliers are obtained from the $\sigma_k \to 0$ limit of $\lambda_k = (\ll R_k \gg - \sum_i R_k(x_i) f_i)/\sigma_k^2)$.

8. Conclusion

Did the founders of quantum mechanics, in solving the riddles of the atom, serendipitously discover how to reason from incomplete and uncertain information? QSI is the first explicit application of quantum theory to human reasoning beyond the physical domain. The compelling motivations for QSI include the inadequacies of the maximum entropy method for a broad class of density function estimation problems, the unquestioned success of quantum statistical mechanics in physics, and the Bayesian hypothesis that probabilities can represent degrees-of-belief. This paper has proposed that a subset of quantum mechanics provides a natural prescription for the statistical inference of *continuous* density functions. It has demonstrated applications to the interpolation and deconvolution of data, and the results are superior to the maximum entropy method.

The successful validation of QSI would provide a powerful new tool for data analysis. QSI can best be tested and developed by experience with diverse data analysis problems. A vast reservoir of knowledge about quantum calculations may be applicable to the development of robust QSI codes. So far QSI has used only commuting observables, real wave functions, the time-independent Schrödinger equation, etc. This is only a small fraction of the rich structure of quantum theory. There is enormous potential for the application of additional quantum concepts (such as gauge fields, path integrals, semiclassical approximations, etc.) to information theory and statistics. In addition, the process of abstracting quantum probability theory from its physical origins within the framework of differential geometry may suggest more rigorous mathematical formulations for quantum physics.

ACKNOWLEDGMENTS. This research was supported by the U. S. Dept. of Energy. I thank R. Balian, C. Campbell, C. Rodriguez, J. Skilling for helpful comments on the manuscript.

REFERENCES

Amari, S.: 1985, *Differential-geometrical Methods in Statistics*, Springer-Verlag, Berlin.

Balian, R.; Alhassid, Y.; Reinhardt, H.: 1986, 'Dissipation in Many-Body Systems: A Geometric Approach Based on Information Theory', *Physics Reports* **131**, **1**.

Bretthorst, G. L.: 1988, *Bayesian Spectrum Analysis and Parameter Estimation*, Springer-Verlag, Berlin.

Dembo, A.; Cover, T. M.; Thomas, J. A.: 1991, 'Information Theoretic Inequalities', *IEEE Trans. Info. Theory* **37**, 1501.

Fick, E.; Sauermann, G.: 1990, *The Quantum Statistics of Dynamic Processes*, Springer-Verlag, Berlin.

Fisher, R. A.: 1925,'Theory of Statistical Estimation', *Proc. Cambr. Phil. Soc.* **22**, 700.

Frieden, B. R.: 1989, 'Fisher Information as the basis for the Schrödinger wave equation', *Am. J. Phys.* **57**, 1004.

Jaynes, E.T.: 1989, 'Clearing Up Mysteries – The Original Goal', in J. Skilling (ed.), *Maximum Entropy and Bayesian Methods*, Kluwer, Dordrecht.

Jaynes, E. T.: 1983, in R. D. Rosenkrantz, (ed.), *E. T. Jaynes: Papers on Probability, Statistics and Statistical Physics*, D. Reidel Publishing Co., Dordrecht.

Kubo, R.: 1959, 'Some Aspects of the Statistical-Mechanical Theory of Irreversible Processes', in Britten & Dunham, (eds.), *Lectures in Theoretical Physics*, **1**.

Levine, R. D.: 1986, 'Geometry in Classical Statistical Thermodynamics', *J. Chem. Phys.* **84**, 910.

Loredo, T.: 1990, 'From Laplace to Supernova SN 1987A: Bayesian Inference in Astrophysics',in P. Fougere (ed.), *Maximum Entropy and Bayesian Methods*, Kluwer, Dordrecht., 81.

J. Mathews, R. L. Walker: 1964, *Mathematical Methods of Physics*, W. A. Benjamin, New York, p. 315.

McQuarrie, D. A.: 1976, *Statistical Mechanics*, Harper & Row, New York.

Rodriguez, C.: 1989, 'The Metrics Induced by the Kullback Number', in J. Skilling (ed), *Maximum Entropy and Bayesian Methods*, Kluwer, Dordrecht.

Skilling, J.; Gull, S.: 1989, 'Classic MaxEnt', in J. Skilling (ed.), *Maximum Entropy and Bayesian Methods*, Kluwer, Dordrecht, 45-71.

Tikochinsky, Y.; Tishby, N. Z.; Levine, R. D.: 1984, 'Consistent Inference of Probabilities for Reproducible Experiments', *Phys. Rev. Letts.* **52**, 1357.

Titterington, D. M.: 1985, 'Common Structure of Smoothing Techniques in Statistics', *Int. Statist. Rev.*, **53**, 141.

DUAL MEASUREMENTS AND INFORMATION IN QUANTUM OPTICS

A. Vourdas
Department of Electrical Engineering & Electronics
University of Liverpool
Liverpool, L69 3BX, United Kingdom

C. Bendjaballah
Laboratoire des Signaux et Systèmes du CNRS
École Supérieure d'Électricité
Plateau de Moulon, 91192 Gif–sur–Yvette Cedex, France

ABSTRACT. Quantum systems defined in a finite dimensional Hilbert space $\mathcal{H}_D$ spanned by complete, finite Fourier transform states $|\lambda; n)$, (resp. $|\tilde{\lambda}; n)$) with $n \in [1, D]$, (mod. D), are examined. The main properties of dual observables are briefly reviewed. The information $I(\mathbf{\Omega}, \mathbf{\Lambda})$ associated with a measurement $\mathbf{\Lambda}$ of a system described by a state $\mathbf{\Omega}$, is introduced and its properties are analyzed. The inequality

$$I(\mathbf{\Omega}, \mathbf{\Lambda}) + I(\mathbf{\Omega}, \widetilde{\mathbf{\Lambda}}) \leq \log D$$

is demonstrated for dual measurements $\left(\mathbf{\Lambda}, \widetilde{\mathbf{\Lambda}}\right)$. The results are then applied to some operators of interest in quantum optics. For instance, it is shown that under some conditions, the number and "phase" observables for a harmonic oscillator are complementary and the inequality is "equivalent" to the uncertainty relation. Quantum correlations and the case $D \to \infty$ are also considered.

1. Introduction

In a quantum system defined by the density operator $\boldsymbol{\Omega}$, the measurements of non commuting observables $\mathbf{A}$, $\mathbf{B}$ are subject to the uncertainty relation $\sigma_{\mathbf{A}}^2 \, \sigma_{\mathbf{B}}^2 \geq \frac{1}{4} \left|\mathrm{Tr}\, [\mathbf{A}, \mathbf{B}]\right|^2$ written in terms of the variances $\sigma_{\mathbf{X}}^2 = \mathrm{Tr}(\boldsymbol{\Omega}\mathbf{X}^2) - (\mathrm{Tr}\boldsymbol{\Omega}\mathbf{X})^2$ and the trace (Tr) of the commutator $[\mathbf{A}, \mathbf{B}] = \mathbf{AB} - \mathbf{BA}$. In general, this inequality depends on $\boldsymbol{\Omega}$ except in the case of canonically conjugate variables for which $[\mathbf{A}, \mathbf{B}] = c\mathbf{1}$, (c is a constant).

Therefore, as already argued by several authors [3–5, 8], instead of the above inequality, a more appropriate formulation of the uncertainty relations should be derived from information inequalities.

A. Mohammad-Djafari and G. Demoments (eds.), Maximum Entropy and Bayesian Methods, 183–188.

2. Dual Operators for quantum systems in $\mathcal{H}_D$

We consider a finite dimensional Hilbert space $\mathcal{H}_D$ spanned by complete, finite Fourier transform states

$$|\lambda; n\rangle = \frac{1}{\sqrt{D}} \sum_{m=0}^{D-1} \exp\left(i\frac{2\pi mn}{D}\right) |\tilde{\lambda}; n\rangle$$

with $n \in [0, D-1]$ (mod. D) [2].

The bases $\{|\lambda; n\rangle\}$ and $\{|\tilde{\lambda}; n\rangle\}$ define the *dual* bases in the space $\mathcal{H}_D$ and the orthonormalised projectors $\boldsymbol{\Pi}_n = |\lambda; n\rangle\langle\lambda; n|$, $\widetilde{\boldsymbol{\Pi}}_n = |\tilde{\lambda}; n\rangle\langle\tilde{\lambda}; n|$ such as $\boldsymbol{\Pi}_m \boldsymbol{\Pi}_n = \delta_{mn}\boldsymbol{\Pi}_n$, $\sum_{n=0}^{D-1} \boldsymbol{\Pi}_n = \mathbf{1}_D$. Moreover, with the unitary operator $\mathbf{U} = \sum |\tilde{\lambda}; n\rangle\langle\lambda; n|$, we obtain $\widetilde{\boldsymbol{\Pi}}_n = \mathbf{U}\boldsymbol{\Pi}_n\mathbf{U}^\dagger$. Method of constructing dual observables from group theory is analyzed in [7].

From the definitions

$$\begin{cases} \mathbf{P}_0 = \sum_{n=0}^{D-1} n\boldsymbol{\Pi}_n \\ \mathbf{Q}_0 = \sum_{n=0}^{D-1} n\widetilde{\boldsymbol{\Pi}}_n \end{cases}$$

and using the relation $\frac{1}{D}\sum_{\ell=0}^{D-1} \exp\left(i\frac{2\pi}{D}k\ell\right) = \begin{cases} 0 & (k \neq 0) \\ 1 & (k = 0) \end{cases}$, it can easily be demonstrated [14] that

$$\mathbf{P}_1|\lambda; n\rangle \triangleq \exp\left(-i\frac{2\pi}{D}\mathbf{Q}_0\right)|\lambda; n\rangle = |\lambda; n+1\rangle$$

and

$$\mathbf{Q}_1|\tilde{\lambda}; n\rangle \triangleq \exp\left(i\frac{2\pi}{D}\mathbf{P}_0\right)|\tilde{\lambda}; n\rangle = |\tilde{\lambda}; n+1\rangle$$

Notice that these operators are "similar" to the conventional operators of the harmonic oscillator $\mathbf{p}, \mathbf{x}, \exp(ia\mathbf{p}), \exp(ib\mathbf{x})$ respectively.

Furthermore, we have $\mathbf{P}_1 = \mathbf{U}\mathbf{Q}_1$, and also

$$\begin{cases} \mathbf{U}|\lambda; n\rangle = |\tilde{\lambda}; -n\rangle \\ \mathbf{U}^2|\lambda; n\rangle = |\lambda; -n\rangle \\ \mathbf{U}^4 = \mathbf{1}_D \end{cases}$$

3. Information inequality

Let $S(\Omega) = -\mathrm{Tr}\ (\Omega \log \Omega)$ be the entropy of the quantum system.

Let A be the observable $\left(A = A^\dagger\right)$ of the spectral resolution $A = \sum_{n=0}^{D-1} \lambda_n \Pi_n$ resp. $(\widetilde{A})$ where $\widetilde{A} = \mathbf{U}A\mathbf{U}^\dagger$.

The information inequality that will be used in the following is derived from the measurement entropy $S(\Omega_A) \equiv S(\Omega, A)$, where Ω_A is the density operator of the system after the measurement A is performed. A different form of the well-known inequality $S(\Omega, A) \leq S(\Omega)$, with $I(\Omega, A) \triangleq \log D - S(\Omega, A)$ [12], is given by $I(\Omega, A) \leq I(\Omega)$.

Let us now consider two measurements of the observables $A, \widetilde{A}$ on Ω. According to [4–6, 8, 9], the information inequality takes the form

$$I(\Omega, A) + I\left(\Omega, \widetilde{A}\right) \leq \log D \tag{1}$$

from which it is easily noticed that $I\left(\Omega, \widetilde{A}\right) = 0$ iif $\left[\Omega, A\right] = 0$. This is valid also $\forall\left(\overline{\Omega}, \mathbf{X}\right)$ such that

(*i*) $\mathrm{Tr}\left(\overline{\Omega}\mathbf{X}\right) = \mathrm{Tr}(\Omega\mathbf{X})$,

(*ii*) $\left[\mathbf{X}, A\right] = 0$.

We want to apply some of these results to the observables *number – phase* which are of great importance in quantum optics [1].

4. Application to harmonic oscillator

In the quantum harmonic oscillator with Hilbert space $\mathcal{K}$, the operators $\mathbf{x}$ and $\mathbf{p}$ satisfy the commutation relation $[\mathbf{x}, \mathbf{p}] = i\hbar$. Defining $\mathbf{x} = \sqrt{\frac{\hbar}{2\omega_0}}\left(\mathbf{a} + \mathbf{a}^\dagger\right), \mathbf{p} = -i\sqrt{\frac{\hbar\omega_0}{2}}\left(\mathbf{a} - \mathbf{a}^\dagger\right)$, where the *non-Hermitian* annihilation operator $\mathbf{a} = \dfrac{\omega_0\mathbf{x} + i\mathbf{p}}{\sqrt{2\hbar\omega_0}}$ is such that $\left[\mathbf{a}, \mathbf{a}^\dagger\right] = \mathbf{1}$. The uncertainty relation takes the form $\sigma_\mathbf{p}^2\sigma_\mathbf{x}^2 - (\Re e\langle\psi|\mathbf{xp}|\psi\rangle)^2 \geq \frac{\hbar^2}{4}$ and depends on the $|\psi\rangle$ state of the system. Now, let $|n\rangle$ be an eigenvector of $\mathbf{N} = \mathbf{a}^\dagger\mathbf{a}$, the resolution of identity on the Fock space spanned by $\{|n\rangle\}$ being $\sum_n |n\rangle\langle n| = \mathbf{1}_D$, the coherent states are given by

$$|\alpha\rangle = \sum_{n=0}^{\infty} \frac{\alpha^n}{\sqrt{n!}} \exp\left(-\frac{|\alpha|^2}{2}\right)|n\rangle \tag{2}$$

for any $\alpha \in \mathbb{C}$. The scalar product of such states is $\langle\alpha|\beta\rangle = \exp\left(\alpha^*\beta - \frac{|\alpha|^2}{2} - \frac{|\beta|^2}{2}\right)$.
For $|\psi\rangle \equiv |\alpha\rangle$, $\sigma_\mathbf{x}^2 = \sigma_\mathbf{p}^2 = \frac{\hbar}{2}$.

We now show that the phase and the number operators are complementary in the sense of [8]. Let D phase-modulated coherent states $|\mu_k\rangle = |\mu u^k\rangle \equiv |\mu \exp\left(-i\frac{2\pi k}{D}\right)$ where $\mu \in \mathbb{R}$,

$k \in [0, D-1]$ and $u = \exp\left(i\frac{2\pi}{D}\right)$. The Gram matrix $\mathbf{G}$ of elements $g_{ij} = \langle\mu_i|\mu_j\rangle$ is hermitian and circulant. From $\mathbf{G}\,|\beta_p\rangle = \lambda_p|\beta_p\rangle$, $(0 \le p \le D-1)$ we obtain the eigenvalues $\lambda_p = \sum_{k=0}^{D-1} g_{0k} u^{-pk}$ where $g_{0k} = \exp[-\mu^2(1-u^k)]$. Let $\mathbf{X}$ be given by $\mathbf{G} = \mathbf{X}^\dagger\mathbf{X}$, we have

$$\mathbf{G} = \sum_{p=0}^{D-1} \lambda_p|\beta_p\rangle\langle\beta_p|, \quad \mathbf{X} = \sum_{p=0}^{D-1} \xi_p|\beta_p\rangle\langle\beta_p| \equiv \sum_{p=0}^{D-1} \sqrt{\lambda_p}\exp(i\theta_p)\,|\beta_p\rangle\langle\beta_p|$$

Now, according to the quantum detection theory [3], the detection probability Q_d when applying the optimum detection operator (o.d.o.) $\mathbf{W}_m = |w_m\rangle\langle w_m|$, $m = 0, D-1$ is maximum for the matrix $\mathbf{X} = \mathbf{G}^{\frac{1}{2}}$ [11].

To calculate the λ_p, we start with $|\mu_k\rangle = \exp\left(-ik\mathrm{N}\frac{2\pi}{D}\right)|\mu\rangle = \sum_{r=0}^{\infty} \exp\left(-ikr\frac{2\pi}{D}\right)|r\rangle\langle r|\mu\rangle$

and

$$g_{0k} = \langle\mu|\mu_k\rangle = \sum_{r=0}^{\infty} |\langle r|\mu\rangle|^2 u^{kr} = \exp(-S)\sum_{r=0}^{\infty}\frac{S^r}{r!}u^{kr}$$

where $S = \mu^2$ so that

$$\lambda_p = \sum_{k=0}^{D-1}\sum_{r=0}^{\infty} |\langle r|\mu\rangle|^2 u^{kr-pk} = D\exp(-S)\sum_{n=0}^{\infty}\frac{S^{p+nD}}{(p+nD)!}$$

Then,

$$\langle n|w_m\rangle = \sum_{\ell=0}^{D-1} \xi_{\ell m}\langle n|\mu_\ell\rangle = \frac{1}{D}\exp\left(-\frac{S}{2}\right)\sqrt{\frac{S^n}{n!}}\sum_{\ell=0}^{D-1}\sum_{k=0}^{D-1}\frac{1}{\sqrt{\lambda_k}}\exp\left[\frac{2i\pi}{D}(k(\ell-m)-\ell n)\right]$$

Therefore $|\langle n|w_m\rangle| = \exp\left(-\frac{S}{2}\right)\sqrt{\frac{S^n}{n!}}\frac{1}{\sqrt{\lambda_n}} = \frac{1}{\sqrt{D}}\left[1 - \frac{S^D}{2(n+D)} + \dots\right]$.
This demonstrates that the observables $\mathbf{N}$ and $\mathbf{W}_m$ are complementary for $S^D \ll 4D$, that is verified for $S \le 1$, $\forall D \ge 1$.
Now, the quantum correlations are briefly examined.

Let the system in an arbitrary initial pure state $\Psi = |\psi\rangle\langle\psi|$.
Under the evolution operator $U(t)$, we first measure $\mathbf{N}$ and calculate $c_n = \langle n|U(t_1)|\psi\rangle$ with the measurement probability density $\mathrm{P}(n) = |c_n|^2$.
Similarly for $\mathbf{W}_m$ and $d_{m,n} = \langle w_m|U(t_2,t_1)|n\rangle$, we obtain $\mathrm{P}(n,\varphi_m) = |d_{m,n}|^2$. Since

$$\mathrm{P}(\varphi_m|\forall n) = \sum_n \left|d_{m,n}c_n\right|^2 \equiv \mathrm{P}_0(\varphi_m)$$

for the two steps measurement, we can write

$$\mathrm{P}(\varphi_m) = \left|\left[\sum_n d_{m,n}c_n\right]\right|^2 \equiv \mathrm{P}_0(\varphi_m) + f(\varphi_m)$$

where $f(\varphi_m)$ is the quantum interference term.
Specific calculations can be performed for given $\mathbf{H} = \hbar\omega_0 \mathbf{a}^\dagger \mathbf{a}$ and $\Psi = |\alpha\rangle\langle\alpha|$. In this case, we easily prove that $\mathrm{I}\Big(\mathrm{P}_0 + f(\varphi_m)\Big) \geq \mathrm{I}\Big(\mathrm{P}_0\Big)$.

The asymptotic behavior $D \to \infty$ is also interesting to consider. As $\sum_{n=0}^{\infty} \frac{S^{p+nD}}{(p+nD)!} \sim \frac{S^p}{p!}$, the limit $\lim_{D\to\infty} \frac{\lambda_p}{D} = \exp(-\mu^2)\frac{\mu^{2p}}{p!}$ yields

$$\langle\mu|\beta_r\rangle = \frac{1}{\sqrt{D}} \sum_{p=0}^{D-1} \sqrt{\frac{\lambda_p}{D}}\, u^{pr} \tag{3}$$

With $|\phi\rangle = \frac{1}{\sqrt{\frac{2\pi}{D}}}|\beta_r\rangle$, (3) becomes

$$\langle\mu|\phi\rangle = \lim_{D\to\infty} \frac{1}{\sqrt{2\pi}} \sum_{p=0}^{D-1} \exp\left(-\frac{\mu^2}{2}\right) \frac{\mu^p}{\sqrt{p!}} \exp\left(-i\varphi p\right)$$

where we set $\varphi = \frac{2\pi}{D} r$. Therefore

$$|\phi\rangle = \frac{1}{\sqrt{2\pi}} \sum_{n=0}^{\infty} \exp\left(-in\varphi\right)|n\rangle \tag{4}$$

so that the o.d.o. $\mathbf{W}$ is the phase operator.

Thus, it is proved why there is no Hermitian phase operator in $\mathcal{K}$ corresponding to an exact phase measurement, so that the usual uncertainty relation based on variances $\sigma_{\mathrm{N}}^2\, \sigma_{\mathbf{W}}^2 \geq \frac{1}{4}$ is inappropriate [10,13].

On the other hand, it can be proved that $\mathrm{I}(\omega, \boldsymbol{\lambda}) \leq \mathrm{I}(\Omega, \Lambda)$ whereas $\mathrm{Tr}(\omega) = \lim_{D\to\infty} \mathrm{Tr}(\Omega)$ and $\mathrm{Tr}(\boldsymbol{\lambda}) = \lim_{D\to\infty} \mathrm{Tr}(\Lambda)$.
However, in this case, for a system in the density Ψ, the differential entropy is simple to write

$$\mathrm{S}(\Psi, \mathbf{W}) = -\frac{1}{2\pi} \int_0^{2\pi} d\varphi\, p(\varphi) \log p(\varphi)$$

where with (4), $p(\varphi) = \left|\langle\psi|\phi\rangle\right|^2$.

5. Conclusions

It is shown that in some sense the operators number and phase can be considered as complementary. The information inequality is then applied to replace the uncertainty relation which is meaningless in this case.

REFERENCES

[1] P. Carruthers, M.M. Nieto, *Review of Modern Physics* **40** (1968), 411.

[2] J. Schwinger, **Quantum Kinematics and Dynamics** (Benjamin, New York 1970).

[3] C.W. Helstrom, **Quantum Detection and Estimation Theory** (Academic Press, New York 1976).

[4] D. Deutsch, *Physical Review Letters* **50** (1983), 631.

[5] H. Partovi, *Physical Review Letters* **50** (1983), 1883.

[6] M. D. Srinivas, *Pramāṇa* **24** (1985), 673.

[7] R. Balian, C. Itzykson, *Comptes Rendus de l'Académie des Sciences* **303** (1986), 773.

[8] K. Kraus, *Physical Review* **D–37** (1987), 3070.

[9] H. Maassen, J.M.B. Uffink, *Physical Review Letters* **60** (1988), 1103.

[10] D.T. Pegg, S.M. Barnett, *Physical Review* **A–39** (1989), 1665.

[11] M. Charbit, C. Bendjaballah, C. Helstrom, *I.E.E.E. Transactions* **IT–35** (1989), 1131.

[12] A. Vourdas, *Physical Review* **A–41** (1990), 1653.

[13] H.E. Hall, *Quantum Optics* **3** (1991), 7.

[14] A. Vourdas, C. Bendjaballah *Physical Review* (1993), (to appear).

QUANTUM ENTROPIES AND THEIR MAXIMIZATIONS

Masanori OHYA
Department of Information Sciences
Science University of Tokyo
Noda City, Chiba 278, JAPAN

ABSTRACT. The notion of entropy was introduced by Clausius in order to discuss the thermal behavior of physical systems on the basis of Carnot's work for the efficiency of thermal engine. Boltzmann then tried to give a rigorous description of the entropy from the microscopic points of view, that is, the dynamical behavior of a large number of atoms. About fifty years after these works, Shannon gave a new light on the notion of entropy and reformulated the entropy within probabilistic frameworks. About twenty years before Shannon's work, von Neumann introduced the entropy for quantum state ρ. After von Neumann, quantum entropies are developed and studied extensively [1]. In this paper, we review quantum mechanical entropies and discuss maximization of the entropy and the mutual entropy.

1. Quantum Entropy of States

Von Neumann defined the entropy for a density operator ρ (say, state in the sequel) in a Hilbert space $\mathcal{H}$ such as

$$S(\rho) = -\mathrm{Tr}\,\rho\log\rho. \tag{1}$$

That is, for any complete orthonormal system (CONS) $\{x_k\}$ in $\mathcal{H}$,

$$S(\rho) = -\sum_k < x_k, \rho\log\rho x_k > . \tag{2}$$

The value of "trace" does not depend on the choice of the CONS $\{x_k\}$. This entropy contains Shannon's entropy for a probability distribution $p = \{p_k\}$ as a special case:

$$S(p) = -\sum_k p_k \log p_k.$$

The spectral set of ρ is discrete, so that we write the spectral decomposition of ρ as

$$\rho = \sum_n \lambda_n P_n,$$

where λ_n is an eigenvalue of ρ and P_n is the projection from $\mathcal{H}$ onto the eigenspace associated with λ_n. Therefore if every eigenvalue λ_n is not degenerate, then the dimension of the range of P_n is one. If a certain eigenvalue, say λ_n, is degenerate, then P_n can be further decomposed into one dimensional projections:

$$P_n = \sum_{j=1}^{dim P_n} E_j^{(n)}, \tag{3}$$

A. Mohammad-Djafari and G. Demoments (eds.), Maximum Entropy and Bayesian Methods, 189–194.

where $E_j^{(n)}$ is one dimensional projection expressed by $E_j^{(n)} = | x_j^{(n)} >< x_j^{(n)} |$ with the eigenvector $x_j^{(n)}$ $(j = 1, 2, \cdots, dimP_n)$ for λ_n. By relabeling the indexes j, n of $\{E_j^{(n)}\}$, we write

$$\rho = \sum_n \lambda_n E_n \tag{4}$$

with

$$\lambda_1 \geq \lambda_2 \geq \cdots \geq \lambda_n \geq \cdots \tag{5}$$

$$E_n \perp E_m (n \neq m). \tag{6}$$

We call this decomposition Schatten decomposition. Now, in (4), the eigenvalue of multiplicity n is repeated precisely n times. For example, if the multiplicity of λ_1 is 2, then $\lambda_1 = \lambda_2$. Moreover, this decomposition is unique if and only if no eigenvalue is degenerate.

For two Hilbert spaces $\mathcal{H}_1$ and $\mathcal{H}_2$, let $\mathcal{H} = \mathcal{H}_1 \otimes \mathcal{H}_2$ be the tensor product Hilbert space of $\mathcal{H}_1$ and $\mathcal{H}_2$ and denote the tensor product of two operators A and B acting on $\mathcal{H}_1$ and $\mathcal{H}_2$ respectively by $A \otimes B$. The reduced states ρ_1 in $\mathcal{H}_1$ and ρ_2 in $\mathcal{H}_2$ for a state ρ in $\mathcal{H}$ are given by the partial traces, which are denoted by $\rho_k = \mathrm{Tr}_{\mathcal{H}_j} \rho (j \neq k; j, k = 1, 2)$.

Theorem 1.1 For any density operator $\rho \in \Sigma(\mathcal{H})$, we have the following.

1. Positivity: $S(\rho) \geq 0$.

2. Symmetry: Let $\rho' = U^{-1} \rho U$ for an invertible operator U. Then

$$S(\rho') = S(\rho).$$

3. Concavity: $S(\lambda \rho_1 + (1 - \lambda) \rho_2) \geq \lambda S(\rho_1) + (1 - \lambda) S(\rho_2)$ for any $\rho_1, \rho_2 \in \Sigma(\mathcal{H})$ and any $\lambda \in [0, 1]$.

4. Additivity: $S(\rho_1 \otimes \rho_2) = S(\rho_1) + S(\rho_2)$ for any $\rho_k \in \Sigma(\mathcal{H}_k)$ $(k = 1, 2)$.

5. Subadditivity: For the reduced states ρ_1, ρ_2 of $\rho \in \Sigma(\mathcal{H}_1 \otimes \mathcal{H}_2)$,

$$S(\rho) \leq S(\rho_1) + S(\rho_2).$$

6. Lower Semicontinuity: If $\|\rho_n - \rho\|_1 (= \mathrm{Tr}\ | \rho_n - \rho |) \to 0$, then

$$S(\rho) \leq \liminf_{n \to \infty} S(\rho_n).$$

7. Strong subadditivity: Let $\mathcal{H} = \mathcal{H}_1 \otimes \mathcal{H}_2 \otimes \mathcal{H}_3$ and denote the reduced states $\mathrm{Tr}_{\mathcal{H}_k} \rho$ and $\mathrm{Tr}_{\mathcal{H}_{ij}} \rho$ by ρ_{ij} and ρ_k, respectively. Then

$$\begin{array}{ll} (i) & S(\rho) + S(\rho_2) \leq S(\rho_{12}) + S(\rho_{23}), \\ (ii) & S(\rho_1) + S(\rho_2) \leq S(\rho_{13}) + S(\rho_{23}). \end{array}$$

The proof of the theorem is given in [1] (see also [3]).

For Shannon's entropy, the monotonicity is satisfied, that is, the inequality $\max\{S(p), S(q)\} \le S(r)$ holds. This monotonicity is not always satisfied in quantum systems. Indeed, take

$$x = \frac{(x_1 \otimes x_2 + y_1 \otimes y_2)}{\sqrt{2}},$$

where x_k is in $\mathcal{H}_k$ and $< x_k, y_k > = 0 (k = 1,2)$, and put

$$\rho = | x >< x | \quad \text{and} \quad \rho_k = \text{Tr}_{\mathcal{H}_j}\rho \quad (j \neq k; j,k = 1,2).$$

Then $S(\rho) = 0$ because ρ is a pure state (i.e., $\rho^2 = \rho$), and $S(\rho_k) = \log 2$ because of $\rho_k = (| x_k >< x_k | + | y_k >< y_k |)/2$.

2. Quantum Relative Entropy

The relative entropy (Kullback-Leibler information) in a classical system for two probability distributions $p = \{p_k\}$, $q = \{q_k\}$ is given by

$$S(p|q) = \begin{cases} \sum_k p_k \log \frac{p_k}{q_k} & (q_k = 0 \Longrightarrow q_k = 0) \\ +\infty & (\text{ otherwise }). \end{cases}$$

The relative entropy in a quantum system is defined for two states ρ and σ such as

$$S(\rho|\sigma) = \begin{cases} \text{Tr}\, \rho(\log\rho - \log\sigma) & \text{when } \rho << \sigma \\ +\infty & (\text{ otherwise }) \end{cases}$$

where $\rho << \sigma$ means that if $\text{Tr}\, \sigma A = 0$ for a positive bounded operator A, then $\text{Tr}\, \rho A = 0$.

The fundamental properties of the relative entropy are the following:

Theorem 2.1

1. Positivity: $S(\rho|\sigma) \ge 0, \; = 0$ iff $\rho = \sigma$.

2. Joint convexity: For any states $\rho_1, \sigma_1, \rho_2, \sigma_2$ and $\lambda \in [0,1]$

$$S(\lambda\rho_{+}(1-\lambda) \quad \rho_2 \quad |\lambda\sigma_1 + (1-\lambda)\sigma_2) \le \lambda S(\rho_1|\sigma_1) + (1-\lambda) S(\rho_2|\sigma_2).$$

3. Symmetry: For any invertible operator U

$$S(U\rho U^{-1} | U\sigma U^{-1}) = S(\rho|\sigma).$$

4. Additivity: $S(\rho_1 \otimes \rho_2 | \sigma_1 \otimes \sigma_2) = S(\rho_1|\sigma_1) + S(\rho_2|\sigma_2)$.

5. Lower semicontinuity: If $\lim_{n\to\infty} \|\rho_n - \rho\|_1 = 0$ and $\lim_{n\to\infty} \|\sigma_n - \sigma\|_1 = 0$, then

$$S(\rho|\sigma) \le \liminf_{n\to\infty} S(\rho_n|\sigma_n).$$

6. Monotonicity: For any channel Λ (defined in §3),

$$S(\Lambda\rho|\Lambda\sigma) \le S(\rho|\sigma).$$

7. Inequality:

$$\|\rho - \sigma\| \le \sqrt{2S(\rho|\sigma)}.$$

3. Quantum Mutual Entropy

When we study information transmission from one system to another system , it is quite essential for us to ask how much information can be transmitted from one system to another. The amount of this transmitted information is expressed by so - called mutual entropy (information). The mutual entropy in Shannon's theory is given as follows: Let Σ and $\overline{\Sigma}$ be the sets of probability distributions (say, states) of an input and output systems respectively and Λ be a channel from Σ to $\overline{\Sigma}$. Then for two states p, $\overline{p} = \Lambda p$, take the joint probability distribution r such that $p_j = \sum_i r_{ij}$, $\overline{p}_i = \sum_j r_{ij}$. The mutual entropy in a classical system is given by

$$\begin{aligned} I(p;\Lambda) &= S(r|p\otimes\overline{p}) \\ &= \textstyle\sum_{i,j} r_{ij}\log\frac{r_{ij}}{p_i\overline{p}_j}. \end{aligned}$$

A fundamental inequality Shannon is

$$0 \leq I(p;\Lambda) \leq S(p).$$

That is, the information transmitted from an initial state p to the final state $\overline{p} = \Lambda p$ is less than the information carried by p itself, which opens the door to discuss communication of information.

A quantum version of the mutual entropy is given in [2,3]. In order to discuss this entropy, we have to state the definition of channel first. For two quantum systems $(\mathcal{A},\Sigma)$ and $(\overline{\mathcal{A}},\overline{\Sigma})$, the the former is an input system and the later is an output system, where $\mathcal{A}$ is the set of all bounded operators on a Hilbert space $\mathcal{H}$ and Σ is the set of all states (density operators) on $\mathcal{H}$ and $\overline{\mathcal{A}},\overline{\Sigma}$ are those on another Hilbert space $\overline{\mathcal{H}}$ (sometimes $\mathcal{H} = \overline{\mathcal{H}}$). A channel is a mapping from Σ to $\overline{\Sigma}$ whose dual map Λ^* $\overline{\mathcal{A}}$ to $\mathcal{A}$ satisfies the completely positivity: $\sum_{i,j}^n A_i^*\Lambda^*(\overline{A}_i^*\overline{A}_j)A_j \geq 0$ for any $n \in N$ and any $A_i \in \mathcal{A}$, $\overline{A}_j \in \overline{\mathcal{A}}$. An input state $\rho \in \Sigma$ is sent to an output system through a channel Λ, then the mutual entropy is defined by

$$I(\rho;\Lambda) = \sup\{\sum_n \lambda_n S(\Lambda E_n|\Lambda\rho);\, E = \{E_n\}\}$$

where the supremum is taken over all Schatten decompositions $E = \{E_n\}$ of ρ. Then the fundamental inequality of Shannon's type is satisfied [2].

Theorem 3.1 $\quad 0 \leq I(\rho;\Lambda) \leq S(\rho).$

We often need some special channels such as:

Definition 3.1 Let Λ be a channel from Σ to $\overline{\Sigma}$.
(1) Λ is stationary if $\Lambda\circ\alpha_t = \bar{\alpha}_t\circ\Lambda$ for any $t \in R$, where $\alpha_t, \overline{\alpha_t}$ $(t \in R)$ are some evolutions of input and output systems, respectively.
(2) Λ is ergodic if Λ is stationary and $\Lambda(exI(\alpha)) \subset exI(\bar{\alpha})$, where $I(\alpha)$ is the set of all stationary states with respect to α and $exI(\alpha)$ is the set of all extremal elements in $I(\alpha)$.
(3) Λ is orthogonal if any two orthogonal states ρ_1, $\rho_2 \in \Sigma$(denoted by$\rho_1 \perp \rho_2$) implies $\Lambda\rho_1 \perp \Lambda\rho_2$.
(4) Λ is deterministic if Λ is orthogonal and bijective.
(5) For a subset $\mathcal{S}$ of Σ, Λ is chaotic for $\mathcal{S}$ if $\Lambda\rho_1 = \Lambda\rho_2$ for any ρ_1, $\rho_2 \in \mathcal{S}$.

(6) Λ is chaotic if Λ is chaotic for Σ.

The quantum mechanical mutual entropy $I(\rho;\Lambda)$ can be computed for the above channels.

Theorem 3.2 (1) If a channel Λ is deterministic, then $I(\rho;\Lambda) = S(\rho)$.
(2) If a channel Λ is chaotic, then $I(\rho;\Lambda) = 0$.
(3) If ρ is faithful state (i.e., if Tr $\rho A^*A = 0$, then $A = 0$) and every eigenvalue of ρ is nondegenerate, then $I(\rho;\Lambda) = S(\Lambda\rho)$.

This mutual entropy can be used to study optical communication processes, irreversible processes and others [2,3,4].

4. Maximization of Quantum Entropy $S(\rho)$

Let us maximize the entropy $S(\rho)$ of a qunatum state ρ, a density operator, under a constraint with a certain positive observable A

$$\mathrm{Tr}\ \rho A = a(>0). \tag{7}$$

Take a Schatten decomposition

$$\rho = \sum \lambda_n E_n. \tag{8}$$

First we consider the case that the set of $\{E_n\}$ of (8) is fixed and $\{\lambda_n\}$ is changed by which ρ is changed. In this case, the discussion is going in a parallel way as the classical commutative case. That is, we have

Proposition 4.1 Under the condition above, $\rho_0 = \sum \lambda_n^{(0)} E_n$ gives the maximum value of $S(\rho)$, where

$$\lambda_n^{(0)} = \frac{e^{-\beta a_n}}{\sum e^{-\beta a_n}},\ a_n = \mathrm{Tr}\ AE_n$$

and β is the constant determined by the constraint (7).
Secondly we consider the case that ρ itself is changed. Then we obtain the following result.

Theorem 4.2 If the spectrum of A is bounded below, then a state $\rho_0 = \frac{e^{-\beta A}}{\mathrm{Tr}\ e^{-\beta A}}$ gives the maximum of $S(\rho)$ where β is the constant determined by (8).
Proof : Since the spectrum of A is lower bounded, the operator $e^{-\beta A}$ is a trace class operator, and $\rho_0 = \frac{e^{-\beta A}}{\mathrm{Tr}\ e^{-\beta A}}$ is a state for any positive constant β. The equality

$$\mathrm{Tr}\ \rho \log \rho_0 = -\beta \mathrm{Tr}\ \rho A - \log \mathrm{Tr}\ e^{-\beta A}$$

is equal to

$$\mathrm{Tr}\ \rho_0 \log \rho_0 = -\beta \mathrm{Tr}\ \rho_0 A - \log \mathrm{Tr}\ e^{-\beta A}$$

due to the constraint

$$\mathrm{Tr}\ \rho A = a = \mathrm{Tr}\ \rho_0 A$$

because the constant β is determined so that Tr $\rho_0 A = a$ is satisfied. The positivity of the relative entropy $S(\rho|\rho_0)$ implies

$$\begin{aligned} S(\rho) &= -\mathrm{Tr}\ \rho\log\rho \\ &= -S(\rho|\rho_0) - \mathrm{Tr}\ \rho\log\rho_0 \\ &\le -\mathrm{Tr}\ \rho\log\rho_0 \\ &= -\mathrm{Tr}\ \rho_0\log\rho_0 = S(\rho_0) \end{aligned}$$

for any density operator ρ. Hence, $S(\rho_0)$ gives the maximum value of $S(\rho)$. □

5. Maximization of Quantum Mutual Entropy

The maximization of mutual entropy is a bit more complicated. Here we just mention a result concerning this problem. Its full proof will be given in other occasion. When we change the channel Λ freely, the maximum value of $I(\rho;\Lambda)$ is obviously $S(\rho)$ due to Theorem 3.1. Therefore we have to consider this maximization problem under a constraint. For a certain positive observable A, put

$$a(\rho) = \mathrm{Tr}\ \rho A. \tag{9}$$

Then a constraint is given by

$$a(\rho) \le \alpha \tag{10}$$

with some positive constant α. The maximization problem of the mutual entropy is given as follows: For fixed Λ and α, maximize $I(\rho;\Lambda)$ under the constraint (10).

Let $\alpha_0 \equiv \min_n \mathrm{Tr}\ AE_n$ and $\alpha_1 \equiv \max_n \mathrm{Tr}\ AE_n$ for a Schatten decomposition $\{E_n\}$ of ρ. One solution of this problem is expressed as

Theorem 5.1 Let $\{E_n\}$ be fixed and $\{\lambda_n\}$ be changed. Then

1. $C(\Lambda;\alpha)$ is nondecreasing and concave function of α for $\alpha \ge \alpha_0$;
2. there exists ρ^* such that $I(\rho^*;\Lambda) = C(\Lambda;\alpha)$ for any α with $\alpha_0 \le \alpha \le \alpha_1$ if and only if there exists a positive constraint γ such that

$$\begin{aligned} D_n(\rho^*) &= C(\Lambda;\alpha) + \gamma(\alpha - \alpha_n) \quad \text{when } \lambda_n > 0 \\ &\le C(\Lambda;\alpha) + \gamma(\alpha - \alpha_n) \quad \text{when } \lambda_n = 0 \end{aligned}$$

where

$$\begin{aligned} D_n(\rho^*) &= \textstyle\sum_m \lambda_m \mathrm{Tr}\ (\Lambda E_m) E_n \log \frac{\lambda_m \mathrm{Tr}\ (\Lambda E_m) E_n}{\mathrm{Tr}\ \rho(\Lambda E_n)}, \\ \alpha_n &= \mathrm{Tr}\ \rho E_n. \end{aligned}$$

References I here refere only my papers and see [1,3] for complete references:

[1] M. Ohya and D. Petz, Quantum Entropy and its Use, Springer-Verlag, 1993.

[2] M. Ohya, On compound State and Mutual Information in Quantum Information Theory, IEEE Trans. Information Theory, Vol.29, No.5, 770-777, 1983.

[3] M. Ohya, Some Aspects of Quantum Information Theory and their Applications to Irreversible Processes, Rep. Math. Phys., Vol.27, 19-47, 1989.

[4] M. Ohya, Fractal Dimensions of States, Quantum Probability and Related Topics. Vol.6, 359-369, 1991.

INFORMATION DYNAMICS AND ITS APPLICATION TO GAUSSIAN COMMUNICATION PROCESSES

Masanori OHYA and Noboru WATANABE
Department of Information Sciences
Science University of Tokyo
Noda City, Chiba 278, JAPAN

ABSTRACT. The "dynamics of systems" is described by the state change. One of the essential characters of a state is expressed by complexity, such as entropy, an important concept in information theory. Information dynamics introduced in [4] is for the study of the state change together with such complexities, being a synthesis of two concepts; dynamics and complexity. There are two complexities in information dynamics; (i) the first complexity is one for a state itself, (ii) the second complexity (transmitted complexity) is determined for a state and a channel.

In this paper, we apply the general frames of information dynamics to Gaussian communication processes.

Gaussian communication processes have been studied by Baker [1] et al. They used the classical mutual entropy $I(\mu;\Gamma)$ to discuss Gaussian communication processes. Their discussion has two defects [5]: (1) If we take the differential entropy as information for an input state, then this entropy often becomes less than the mutual entropy $I(\mu;\Gamma)$; (2) If we take Shannon's definition for the entropy $S(\mu)$, then it is always infinite for any Gaussian measure μ. These points are not well-matched to the thought of Shannon [7].

In order to avoid these defects and discuss Gaussian communication processes consistently, we introduced [5] two entropy functionals based on the quantum von Neumann entropy [8] and the quantum mutual entropy (information) introduced in [2]. In this paper, we show that two functionals satisfy the conditions of complexities in information dynamics.

In §1, we explain what information dynamics is.

In §2, we briefly review the conventional treatment (Baker et al.) of Gaussian communication processes.

In §3, we explain the theory of quantum entropies for density operators.

In §4, based on quantum entropies, our entropy functional for an input state and a mutual entropy functional for an input state and a Gaussian channel satisfy the conditions of complexities of information dynamics.

1. Information Dynamics

Let an input dynamical system and an output dynamical system be described by $(\mathcal{A}_1, \Sigma_1, \alpha^1)$ and $(\mathcal{A}_2, \Sigma_2, \alpha^2)$, respectively. Here $\mathcal{A}_1$ is a set of all objects to be observed and Σ is a set of all means getting the observed value for each element A in $\mathcal{A}_1$, and α describes an inner evolution of the input system. Σ_1 is called a "state space". Same for the output system $(\mathcal{A}_2, \Sigma_2, \alpha^2)$. Having a theory based on information dynamics is to give a mathematical structure to input and output triples.

In order to discuss communication systems, it is necessary to give a transformation between two systems. A map providing a bridge between two systems is called a "channel"

A. Mohammad-Djafari and G. Demoments (eds.), Maximum Entropy and Bayesian Methods, 195–203.

if it sends a state of the input system to that of the output system; $\Gamma : \Sigma_1 \to \Sigma_2$. There exist several channels, whose properties specify characters of two systems. Mathematical structure of almost all systems can be expressed by an input system, a transformation system (channel) and an output system. Let $(\mathcal{A}, \Sigma, \alpha)$ be the compound system such that $\mathcal{A} = \mathcal{A}_1 \otimes \mathcal{A}_2$, $\Sigma = \Sigma_1 \otimes \Sigma_2$ and $\alpha = \alpha_1 \otimes \alpha_2$.

Definition 1.1 For a state $\varphi \in \mathcal{S}_1$, a certain subset of Σ_1 and a channel Γ, (i) two functionals $C^{\mathcal{S}_1}(\varphi)$ from Σ_1 to R^+ and (ii) $T^{\mathcal{S}_1}(\varphi;\Gamma)$ (or $T^{\mathcal{S}_1}(\varphi;\psi)$) are complexities of systems if these functionals satisfy the following properties:

1. For any state $\varphi \in \mathcal{S}_1 \subset \Sigma_1$, $C^{\mathcal{S}_1}(\varphi) \geq 0$ and $T^{\mathcal{S}_1}(\varphi;\Gamma) \geq 0$.

2. For any bijection j from $ex\Sigma_1$ to $ex\Sigma_1$ ($ex\Sigma_1$ is the set of all extremal elements in Σ_1),

$$C^{j(\mathcal{S}_1)}(j(\varphi)) = C^{\mathcal{S}_1}(\varphi)$$
$$T^{j(\mathcal{S}_1)}(j(\varphi);\Gamma) = T^{\mathcal{S}_1}(\varphi;\Gamma).$$

3. For any $\Phi \in \mathcal{S} \subset \Sigma$, put $\varphi \equiv \Phi \uparrow \mathcal{A}_1 \in \mathcal{S}_1 \equiv \mathcal{S} \uparrow \mathcal{A}_1$ and $\psi \equiv \Phi \uparrow \mathcal{A}_2 \in \mathcal{S}_2 \equiv \mathcal{S} \uparrow \mathcal{A}_2$

$$C^{\mathcal{S}}(\Phi) \leq C^{\mathcal{S}_1}(\varphi) + C^{\mathcal{S}_2}(\psi).$$

Moreover

4. $0 \leq T^{\mathcal{S}_1}(\varphi;\Gamma) \leq C^{\mathcal{S}_1}(\varphi)$.

5. $T^{\mathcal{S}_1}(\varphi; id) = C^{\mathcal{S}_1}(\varphi)$, where "id" is an identity channel from Σ_1 to Σ_1; $id(\varphi) = \varphi$, $\forall \varphi \in \Sigma_1$.

(i) The first complexity is one for a state itself; (ii) the second complexity (transmitted complexity) is determined by both input and output states φ, ψ or an input state φ and a channel Γ. Under the above settings, information dynamics is defined as follows [4]:

Definition 1.2 Information dynamics is a dynamics described by a set

$$\{\mathcal{A}_1, \Sigma_1, \mathcal{S}_1, \alpha^1 ; \mathcal{A}_2, \Sigma_2, \mathcal{S}_2, \alpha^2 ; \Gamma ; C^{\mathcal{S}_1}(\varphi), T^{\mathcal{S}_1}(\varphi;\Gamma) ; R\},$$

where R is a certain relation among above quantities.

Therefore, for a system of interest, we have to (1) mathematically determine $\mathcal{A}_1$, Σ_1, $\mathcal{S}_1$, α^1; $\mathcal{A}_2$, Σ_2, $\mathcal{S}_2$, α^2; (2) choose Γ and R ; (3) fix $C^{\mathcal{S}_1}(\varphi)$ and $T^{\mathcal{S}_1}(\varphi;\Gamma)$.

2. Gaussian Communication Processes

We briefly review the Gaussian communication processes treated by Baker et al.

Let $\mathcal{B}$ be the Borel σ-field of a real separable Hilbert space $\mathcal{H}$ and μ be a Borel probability measure on $\mathcal{B}$ satisfying

$$\int_{\mathcal{H}} \|x\|^2 d\mu(x) < \infty.$$

Further, we denote the set of all positive self-adjoint trace class operators on $\mathcal{H}$ by $T(\mathcal{H})_+ (\equiv \{\rho \in B(\mathcal{H});\ \rho \geq 0,\ \rho = \rho^*,\ \mathrm{Tr}\,\rho < \infty\})$ and define the mean vector $m_\mu \in \mathcal{H}$ and the covariance operator $R_\mu \in T(\mathcal{H})_+$ of μ such as

$$\begin{aligned} < x_1, m_\mu > &= \int_{\mathcal{H}} < x_1, y > d\mu(y) \\ < x_1, R_\mu x_2 > &= \int_{\mathcal{H}} < x_1, y - m_\mu >< y - m_\mu, x_2 > d\mu(y) \end{aligned}$$

for any $x_1, x_2, y \in \mathcal{H}$. A Gaussian measure μ in $\mathcal{H}$ is a Borel measure such that for each $x \in \mathcal{H}$, there exist real numbers m_x and $\sigma_x (> 0)$ satisfying

$$\mu\{y \in \mathcal{H};\ < y, x > \leq a\} = \int_{-\infty}^{a} \frac{1}{\sqrt{2\pi\sigma_x}} \exp\{\frac{-(t - m_x)^2}{2\sigma_x}\} dt.$$

The notation $\mu = [m, R]$ means that μ is a Gaussian measure on $\mathcal{H}$ with a mean vector m and a covariance operator R.

Let $(\mathcal{H}_1, \mathcal{B}_1)$ be an input space, $(\mathcal{H}_2, \mathcal{B}_2)$ be an output space and we denote the set of all Gaussian probability measures on $(\mathcal{H}_k, \mathcal{B}_k)$ by $P_G^{(k)} (k = 1, 2)$. Moreover, let $\mu_1 \in P_G^{(1)} (\equiv P(\mathcal{H}_1))$ be a Gaussian measure of the input space and $\mu_0 \in P_G^{(2)} (\equiv P(\mathcal{H}_2))$ be a Gaussian measure indicating a noise of the channel. Then, a map Γ from $P_G^{(1)}$ to $P_G^{(2)}$ is defined by the Gaussian channel $\lambda : \mathcal{H}_1 \times \mathcal{B}_2 \to [0,1]$ such as

$$\begin{aligned} \Gamma(\mu_1)(Q) &\equiv \int_{\mathcal{H}_1} \lambda(x, Q) d\mu_1(x) \\ \lambda(x, Q) &\equiv \mu_0(Q^x), \\ Q^x &\equiv \{y \in \mathcal{H}_2;\ Ax + y \in Q\},\ x \in \mathcal{H}_1,\ Q \in \mathcal{B}_2, \end{aligned}$$

where A is a linear transformation from $\mathcal{H}_1$ to $\mathcal{H}_2$ and λ satisfies the following conditions: (1) $\lambda(x, \cdot) \in P_G^{(2)}$ for each fixed $x \in \mathcal{H}_1$; (2) $\lambda(\cdot, Q)$ is a measurable function on $(\mathcal{H}_1, \mathcal{B}_1)$ for each fixed $Q \in \mathcal{B}_2$. The compound measure μ_{12} derived from the input measure μ_1 and the output measure μ_2 is given by

$$\mu_{12}(Q_1 \times Q_2) = \int_{Q_1} \lambda(x, Q_2) d\mu_1(x)$$

for any $Q_1 \in \mathcal{B}_1$ and any $Q_2 \in \mathcal{B}_2$. Then, the mutual entropy (information) with respect to μ_1 and Γ is defined by the Kullback-Leibler information such as

$$\begin{aligned} \mathrm{I}(\mu_1; \Gamma) &= \mathrm{S}(\mu_{12} | \mu_1 \otimes \mu_2) \\ &= \begin{cases} \int_{\mathcal{H}_1 \times \mathcal{H}_2} \frac{d\mu_{12}}{d\mu_1 \otimes \mu_2} \log \frac{d\mu_{12}}{d\mu_1 \otimes \mu_2} d\mu_1 \otimes \mu_2 & (\mu_{12} \ll \mu_1 \otimes \mu_2) \\ \infty & (\mu_{12} \ll \mu_1 \otimes \mu_2) \end{cases} \end{aligned}$$

where $\frac{d\mu_{12}}{d\mu_1 \otimes \mu_2}$ is the Radon-Nikodym derivative of μ_{12} with respect to $\mu_1 \otimes \mu_2$.

3. Quantum Entropies

Let $B(\mathcal{H})$ be the set of all bounded linear operators on a separable Hilbert space $\mathcal{H}$, $\Sigma(\mathcal{H})$ be the set of all density operators on $\mathcal{H}$. Then a quantum dynamical system is described by $(B(\mathcal{H}), \Sigma(\mathcal{H}))$. In order to discuss communication processes, we need two dynamical systems: an input system $(B(\mathcal{H}_1), \Sigma(\mathcal{H}_1))$ and an output system $(B(\mathcal{H}_2), \Sigma(\mathcal{H}_2))$.

In communication theory, it is important to know the amount of information can be correctly transmitted from the input system to the output system. The input state generally change to the output state under the influence of noise and loss associated with a channel. The quantum mechanical channel describing this change is defined as follows [2]: A mapping Λ from $\Sigma(\mathcal{H}_1)$ to $\Sigma(\mathcal{H}_2)$ is said to be a channel.

When an input state is given by a density operator $\rho \in \Sigma(\mathcal{H}_1)$, von Neumann introduced [8] the entropy of the input state ρ such as

$$\mathrm{S}(\rho) = -\mathrm{Tr}\ \rho \log \rho.$$

When an input state ρ changes to an output state $\bar{\rho}(= \Lambda\rho)$ through a channel Λ, it is interesting for us to know how much information of ρ can be transmitted to the output system. The mutual entropy (information) represents this amount of information transmitted from ρ to $\Lambda\rho$.

We here discuss about the mutual entropy [2] in quantum dynamical systems. The mutual entropy is defined by using the compound state expressing the correlation between ρ and $\Lambda\rho$. Let us denote the Schatten decomposition [6] of ρ by

$$\rho = \sum_n \lambda_n E_n,$$

where λ_n is the eigenvalue of ρ with $\lambda_1 \geq \lambda_2 \geq \ldots$, and E_n is the projection operator to the subspace generated by the eigenvector x_n of λ_n, that is, $E_n = |x_n><x_n|$. This decomposition is unique if and only if no eigenvalue of ρ is degenerate. The compound state expressing the correlation between ρ and $\Lambda\rho$ is given by [2]

$$\sigma_E = \sum_n \lambda_n E_n \otimes \Lambda E_n$$

for a decomposition $\mathrm{E} = \{E_n\}$. This compound state does depend on the decomposition of ρ, that is, on the choice of $\mathrm{E} = \{E_n\}$.

The mutual entropy introduced in [2] with respect to an input state ρ and a communication channel Λ is defined by

$$\begin{aligned} \mathrm{I}(\rho;\Lambda) \ &\equiv\ \sup\{\mathrm{S}(\sigma_E|\sigma_0); E = \{E_n\}\}, \\ &\quad \sup\{\sum_n \lambda_n \mathrm{S}(\Lambda E_n|\Lambda\rho);\ E = \{E_n\}\}, \end{aligned}$$

where $\sigma_0 = \rho \otimes \Lambda\rho$ is a trivial compound state and $\mathrm{S}(\sigma_E|\sigma_0)$ is relative entropy of σ_E from σ_0 defined by

$$S(\sigma_E|\sigma_0) \equiv \text{Tr } \sigma_E(\log \sigma_E - \log \sigma_0).$$

Then this mutual entropy satisfies the same property of the mutual information of Shannon.

4. Application to Gaussian Communication Processes

In order to discuss communication processes based on "information dynamics", it is necessary to be able to treat two complexities in these communication systems consistently. In this section, we indicate how information dynamics can be used in Gaussian communication processes.

As explained in Introduction, conventional treatment of Gaussian communication processes reviewed in §2 has two defects: (1) If we take the differential entropy as information for an input state, then this entropy often becomes less than the mutual entropy $I(\mu;\Gamma)$; This means that the differential entropy is not compatible with the mutual entropy. (2) Shannon's type entropy $S(\mu)$ is always infinite for any Gaussian measure μ. These points are not well-matched to Shannon's communication theory.

Based on the quantum entropies, we introduced in [5] two functionals characterizing Gaussian states and Gaussian channels in order to discuss Gaussian communication processes consistently. Besides our entropy functionals resolved the above difficulties, they are shown to be the complexities in information dynamics.

Let $P_{G,1}^{(k)}$ be the set $\{\mu = [0, R] \in P^{(k)};\ \text{Tr } R = 1\} (k = 1, 2)$. We assume that $A^*A = (1 - \text{Tr } R_0)I_1$ holds for the covariance operator R_0 of μ_0. We define a transformation Γ from $P_{G,1}^{(1)}$ to $P_{G,1}^{(2)}$ by

$$(\Gamma\mu_1)(Q) \equiv \int_{\mathcal{H}_1} \lambda(x, Q) d\mu_1(x)$$

for any $\mu_1 \in P_{G,1}^{(1)}$ and any $Q \in \mathcal{B}_2$. This can be expressed as

$$\Gamma(\mu_1) \equiv [0, AR_1A^* + R_0]$$

for any $\mu_1 = [0, R_1] \in P_{G,1}^{(1)}$. There exists a bijection Ξ_k from $P_{G,1}^{(k)}$ to $\Sigma(\mathcal{H}_k)$ given by

$$\text{Tr } \Xi_k(\mu_k)A_k = \int_{\mathcal{H}_k} < \xi,\ A_k\xi > d\mu_k(\xi)$$

for any $A_k \in B(\mathcal{H}_k)$ and any $\mu_k \in P_{G,1}^{(k)}$ (k=1,2). We further define a map from $\Sigma(\mathcal{H}_1)$ to $\Sigma(\mathcal{H}_2)$ such as

$$\begin{aligned} \Lambda(\rho) &\equiv \Xi_2 \circ \Gamma \circ \Xi_1{}^{-1}(\rho) \\ &= A\rho A^* + R_0 \end{aligned}$$

for any $\rho \in \Sigma(\mathcal{H}_1)$.

$$\begin{array}{ccc} P_{G,1}^{(1)} \ni \mu_1 & \xrightarrow{\Gamma} & \Gamma_{\mu_1} \in P_{G,1}^{(2)} \\ \Xi_1 \downarrow & & \downarrow \Xi_2 \\ \Sigma(\mathcal{H}_1) & \xrightarrow{\Lambda} & \Sigma(\mathcal{H}_2) \end{array}$$

Λ^* is the dual map of Λ from $B(\mathcal{H}_2)$ to $B(\mathcal{H}_1)$ given by

$$\Lambda^*(Q) = A^*QA + (\mathrm{Tr}\ R_0 Q)I_1$$

for any $Q \in B(\mathcal{H}_2)$.

Lemma 4.1 [5]: Λ is a channel from $\Sigma(\mathcal{H}_1)$ to $\Sigma(\mathcal{H}_2)$. In [5], we defined the mutual entropy functional with respect to the input Gaussian measure μ_1 and the Gaussian channel Γ as

$$\begin{aligned} \tilde{\mathrm{I}}(\mu_1;\Gamma) &\equiv \sup\{\mathrm{S}(\sigma_E|\sigma_0);\ E = \{E_n\}\} \\ &= \sup_E \mathrm{Tr}\ \sigma_E(\log\sigma_E - \log\sigma_0), \end{aligned}$$

where

$$\begin{aligned} \sigma_E &\equiv \sum_n \lambda_n E_n \otimes \Lambda E_n, \\ \sigma_0 &\equiv \Xi_1(\mu_1) \otimes \Lambda(\Xi_1(\mu_1)). \end{aligned}$$

$\sum_n \lambda_n E_n$ is a Schatten decomposition of $\Xi_1(\mu_1)$ such as

$$\Xi_1(\mu_1) = \sum_n \lambda_n E_n.$$

Moreover, the entropy functional of the input Gaussian measure $\mu_1 = [0, \Xi_1(\mu_1)]$ expressing certain "information" of μ_1 is given in [5] as

$$\tilde{\mathrm{S}}(\mu_1) \equiv -\mathrm{Tr}\ \Xi_1(\mu_1)\log\Xi_1(\mu_1).$$

Under the above settings, we show that our functionals satisfy the conditions of complexities in information dynamics.

Theorem 4.2 For any $\mu_1 \in P_{G,1}^{(1)}$ and any Gaussian channel Γ from $P_{G,1}^{(1)}$ to $P_{G,1}^{(2)}$,

1. $0 \le \tilde{\mathrm{I}}(\mu_1;\Gamma) \le \tilde{\mathrm{S}}(\mu_1)$.

2. For any bijection j from $exP_{G,1}^{(1)}$ to $exP_{G,1}^{(1)}$ (where $exP_{G,1}^{(1)}$ is the set of all extremal elements of $P_{G,1}^{(1)}$),

$$\tilde{S}(j(\mu_1)) = \tilde{S}(\mu_1)$$
$$\tilde{I}(j(\mu_1); \Gamma) = \tilde{I}(\mu_1; \Gamma).$$

3. For any $\nu \in P_{G,1}^{(1,2)} (\equiv \{\nu = [0, R] \in P_G(\mathcal{H}_1 \otimes \mathcal{H}_2);\ \mathrm{Tr}\ R = 1\})$, put $\nu_1 \equiv \nu \uparrow \mathcal{H}_1 \in P_{G,1}^{(1)}$ and $\nu_2 \equiv \nu \uparrow \mathcal{H}_2 \in P_{G,1}^{(2)}$. Then the following inequality holds.

$$\tilde{S}(\nu) \leq \tilde{S}(\nu_1) + \tilde{S}(\nu_2)$$

4. $\tilde{I}(\mu; id) = \tilde{S}(\mu)$, where "$id$" is an identity channel from $P_{G,1}^{(1)}$ to $P_{G,1}^{(2)}$; $id(\mu) = \mu$, $\forall \mu \in P_{G,1}^{(1)}$.

Proof of (1): For any $\mu_1 \in P_{G,1}^{(1)}$ and any Gaussian channel Γ from $P_{G,1}^{(1)}$ to $P_{G,1}^{(2)}$,

$$\tilde{I}(\mu_1; \Gamma) = \sup_E \mathrm{Tr}\ \sigma_E(\log \sigma_E - \log \sigma_0).$$

$f(t) = t \log t$ is a convex function and $f'(t) = \log t + 1$ holds. By using Klein's inequality, we get

$$\begin{aligned}
0 &\leq \mathrm{Tr}\ \{f(\sigma_E) - f(\sigma_0) - (\sigma_E - \sigma_0) f'(\sigma_0)\} \\
&= \mathrm{Tr}\ \{\sigma_E \log \sigma_E - \sigma_0 \log \sigma_0 - (\sigma_E - \sigma_0)(\log \sigma_0 + I)\} \\
&= \mathrm{Tr}\ \{\sigma_E \log \sigma_E - \sigma_E \log \sigma_0 - (\sigma_E - \sigma_0)\} \\
&= \mathrm{S}(\sigma_E | \sigma_0) - \mathrm{Tr}\ \sigma_E - \mathrm{Tr}\ \sigma_0 \\
&= \mathrm{S}(\sigma_E | \sigma_0).
\end{aligned}$$

For any $\mu_1 \in P_{G,1}^{(1)}$ and any Gaussian channel Γ from $P_{G,1}^{(1)}$ to $P_{G,1}^{(2)}$, we denote the Schatten decomposition of covariance operator $\Xi_1(\mu_1)$ by

$$\Xi_1(\mu_1) = \sum_n \lambda_n E_n$$

The inequality between $\mathrm{S}(\sigma_E | \sigma_0)$ and $\tilde{S}(\mu_1)$ is given as

$$\begin{aligned}
\mathrm{S}(\sigma_E | \sigma_0) &= \mathrm{Tr}\ \sigma_E(\log \sigma_E - \log \sigma_0) \\
&= \sum_m \sum_n \lambda_n < y_m^{(n)},\ \Lambda E_n y_m^{(n)} > \log < y_m^{(n)},\ \Lambda E_n y_m^{(n)} > \\
&\quad - \sum_m \sum_n \lambda_n < y_m^{(n)},\ \Lambda E_n y_m^{(n)} > \log < y_m^{(n)},\ \Lambda \Xi(\mu_1) y_m^{(n)} > \\
&= - \sum_m \sum_n \lambda_n < y_m^{(n)},\ \Lambda E_n y_m^{(n)} > \log \frac{< y_m^{(n)},\ \Lambda \Xi(\mu_1) y_m^{(n)} >}{< y_m^{(n)},\ \Lambda E_n y_m^{(n)} >}
\end{aligned}$$

$$
\begin{aligned}
&= -\sum_m \sum_n \lambda_n < y_m^{(n)}, \Lambda E_n y_m^{(n)} > \log \frac{\sum_k \lambda_k < y_m^{(n)}, \Lambda E_k y_m^{(n)} >}{< y_m^{(n)}, \Lambda E_n y_m^{(n)} >} \\
&= -\sum_m \sum_n \lambda_n < y_m^{(n)}, \Lambda E_n y_m^{(n)} > \log \lambda_n \\
&\quad + \sum_m \sum_n \lambda_n < y_m^{(n)}, \Lambda E_n y_m^{(n)} > \log \frac{\lambda_n < y_m^{(n)}, \Lambda E_n y_m^{(n)} >}{\sum_k \lambda_k < y_m^{(n)}, \Lambda E_k y_m^{(n)} >} \\
&\leq -\sum_n \lambda_n \log \lambda_n = \tilde{S}(\mu_1).
\end{aligned}
$$

Therefore we obtain the following inequality:

$$
0 \leq \tilde{I}(\mu_1; \Gamma) \leq \tilde{S}(\mu_1).
$$

Proof of (2): For any $\mu_1 \in exP_{G,1}^{(1)}$ and any bijection j from $exP_{G,1}^{(1)}$ to $exP_{G,1}^{(1)}$, it is easily seen that there exists a bijective channel Λ_j from $ex\Sigma_1$ to $ex\Sigma_1$ such as

$$
j = \Xi_1^{-1} \circ \Lambda_j \circ \Xi_1 .
$$

Then we have

$$
\begin{aligned}
\tilde{S}(j(\mu_1)) &= S(\Xi_1(j(\mu_1))) \\
&= S(\Lambda_j \Xi_1(\mu_1)) \\
&= S(\Xi_1(\mu_1)) \\
&= \tilde{S}(\mu_1).
\end{aligned}
$$

Moreover, for any $\mu_1 \in exP_{G,1}^{(1)}$, any bijection j from $exP_{G,1}^{(1)}$ to $exP_{G,1}^{(1)}$ and any Gaussian channel Γ from $P_{G,1}^{(1)}$ to $P_{G,1}^{(2)}$, we obtain

$$
\begin{aligned}
\tilde{I}(j(\mu_1); \Gamma) &= \sup_E \sum_n \lambda_n S(\Lambda \Lambda_j E_n | \Lambda \Xi_1(j(\mu_1))) \\
&= \sup_E \sum_n \lambda_n S(\Lambda E_n | \Lambda \Xi_1(\mu_1)) \\
&= \tilde{I}(\mu_1; \Gamma),
\end{aligned}
$$

where $\sum_n \lambda_n E_n$ is the Schatten decomposition of $\Xi(\mu_1)$.

Proof of (3): For any $\nu \in P_{G,1}^{(1,2)} (\equiv \{\nu = [0, R] \in P_G(\mathcal{H}_1 \otimes \mathcal{H}_2);\ \mathrm{Tr}\ R = 1\})$, put $\nu_1 \equiv \nu \uparrow \mathcal{H}_1 \in P_{G,1}^{(1)}$ and $\nu_2 \equiv \nu \uparrow \mathcal{H}_2 \in P_{G,1}^{(2)}$. Moreover we denote a bijection from $P_G(\mathcal{H}_1 \otimes \mathcal{H}_2)$ to $\Sigma(\mathcal{H}_1 \otimes \mathcal{H}_2)$ by Ξ. Then we have

$$
\begin{aligned}
0 &\leq S(\Xi(\nu) | \Xi_1(\nu_1) \otimes \Xi_2(\nu_2)) \\
&= \mathrm{Tr}\ \Xi(\nu)(\log \Xi(\nu) - \log \Xi_1(\nu_1) \otimes \Xi_2(\nu_2)) \\
&= \mathrm{Tr}\ \Xi(\nu) \log \Xi(\nu) - \mathrm{Tr}\ \Xi(\nu)\left(\log \Xi_1(\nu_1) \otimes I_2\right) - \mathrm{Tr}\ \Xi(\nu)\left(I_1 \otimes \log \Xi_2(\nu_2)\right) \\
&= \mathrm{Tr}\ \Xi(\nu) \log \Xi(\nu) - \mathrm{Tr}\ \Xi_1(\nu_1) \log \Xi_1(\nu_1) - \mathrm{Tr}\ \Xi_2(\nu_2) \log \Xi_2(\nu_2) \\
&= \tilde{S}(\nu_1) + \tilde{S}(\nu_2) - \tilde{S}(\nu).
\end{aligned}
$$

Therefore $\tilde{S}(\nu) \leq \tilde{S}(\nu_1) + \tilde{S}(\nu_2)$ holds.

Proof of (4): Put $\mathcal{H}_1 = \mathcal{H}_2 = \mathcal{H}$. For an identity channel id, there exists an identity channel Id from $\Sigma(\mathcal{H})$ to $\Sigma(\mathcal{H})$ satisfying

$$Id(\Xi(\mu)) = \Xi(\mu).$$

For the Schatten decomposition $\Xi(\mu) = \sum_n \lambda_n E_n$, we obtain

$$S(\sigma_E|\sigma_0) = \sum_n \lambda_n S(E_n|\Xi(\mu)) = -\sum_n \lambda_n E_n \log \Xi(\mu) = \tilde{S}(\mu).$$

Hence

$$\tilde{I}(\mu; id) = \tilde{S}(\mu).$$

□

References

[1] C.R. Baker, Capacity of the Gaussian channel without feedback, Inform. and Control, Vol.37, 70-89, 1978.

[2] M. Ohya, On compound state and mutual information in quantum information theory, IEEE Trans. Inform. Theory, Vol.29, 770-774, 1983.

[3] M. Ohya, Note on quantum probability, L. Nuovo Cimento Vol.38, 402-404, 1983.

[4] M. Ohya, Information dynamics and its applications to optical communication processes, Lecture Notes in Physics, Vol.378, Springer-Verlag, 1991.

[5] M. Ohya and N. Watanabe, A new treatment of communication processes with Gaussian channels, Japan Journal of Applied Math., Vol.3, No.1, 197-206, 1986.

[6] R. Schatten, Norm Ideals of Completely Continuous Operators, Springer-Verlag, 1970.

[7] C.E. Shannon, A mathematical theory of communication, Bell System Tech. J., Vol.27, 379-423 and 623-656, 1948.

[8] J. von Neumann, Die Mathematischen Grundlagen der Quantenmechanik, Springer - Berlin, 1932.

QUANTUM MECHANICS - HOW WEIRD FOR BAYESIANS?

F.H. Fröhner
Kernforschungszentrum Karlsruhe
Institut für Neutronenphysik und Reaktortechnik
Postfach 3640, W-7500 Karlsruhe 1
Germany.

ABSTRACT. Quantum mechanics is spectacularly successful on the technical level. The meaning of its rules appears, however, shrouded in mystery even today, more than sixty years after its inception. Quantum-mechanical probabilities are often said to be "operator-valued" and therefore fundamentally different from "classical" probabilities, in disregard of the work of Cox (1946) – and of Schrödinger (1947) – on the foundations of probability theory. One central question concerns the superposition principle, i. e. the need to work with interfering wave functions, the absolute squares of which are the probabilities. Other questions concern the collapse of the wave function when new data become available. These questions are reconsidered from the Bayesian viewpoint. The superposition principle is found to be a consequence of an apparently little-known theorem for non-negative Fourier polynomials published by Fejér (1915). Combined with the classical Hamiltonian equations for point particles, it yields all basic features of the quantum-mechanical formalism. It is further shown that the correlations in the spin pair version of the Einstein-Podolsky-Rosen experiment can easily be calculated classically, in contrast to EPR lore. All this demystifies the quantum-mechanical formalism to quite some extent. Questions about the origin and the empirical value of Planck's quantum of action remain; finite particle size may be part of the answer.

1. The Riesz-Fejér Theorem and Quantum-Mechanical Probabilities

In his work on Fourier series L. Fejér (1915) published a proof given by F. Riesz of the following theorem (see Appendix): Each non-negative Fourier polynomial (truncated Fourier series) of order n (maximal wave number n) can be expressed as the absolute square of a Fourier polynomial of (at most) the same order,

$$0 \le \rho(x) \equiv \sum_{l=-n}^{n} c_l\, e^{ilx} = \Big| \sum_{k=-n}^{n} a_k\, e^{ikx} \Big|^2 \equiv |\psi(x)|^2 , \tag{1}$$

where the Fourier polynomial $\psi(x)$ is not restricted to non-negative values, in contrast to the Fourier polynomial $\rho(x)$. Our notation anticipates the obvious application to quantum-mechanical probability densities ρ and probability wave functions ψ (without excluding

A. Mohammad-Djafari and G. Demoments (eds.), Maximum Entropy and Bayesian Methods, 205–212.

application to other inherently positive quantities such as intensities of classical energy-carrying waves). Fourier techniques are most convenient whenever wave or particle propagation constrained by initial or boundary conditions is to be described. Constraints such as point sources, slits, scatterers etc. define, together with the wave equation for the Fourier components, eigenvalue problems whose eigenfunctions are all those waves which are possible under the given circumstances.

In quantum mechanics it is customary to introduce (infinite) Fourier series by the familiar device of the periodicity box. We note that

(a) infinite Fourier series can be approximated by finite Fourier polynomials to any desired accuracy if the order n is chosen high enough;

(b) the transition to Fourier integrals describing arbitrary nonperiodic processes is achieved if the box is made bigger and bigger.

In view of these uneventful generalisations we may consider the Riesz-Fejér theorem as equivalent to the wave-mechanical superposition principle: Probabilities are to be calculated as absolute squares of wave functions that can be expressed as linear superpositions of orthogonal functions. In Eq. 1 the orthogonal functions are standing waves in a (one-dimensional) periodicity box. Other possible orthogonal bases are generated by unitary transformations. Historically the superposition principle had been established first, as a rather puzzling empirical feature of the quantum world, before Born found that the absolute square of the wave function can be interpreted as a probability density.

If, on the other hand, one starts with probabilities, the superposition principle, far from puzzling, appears as a theorem, valid not only in quantum mechanics but *throughout probability theory* (cf. e. g. Feller 1966 on L^2 theory). The much discussed role of the phases of the superposed functions is also clarified: *They ensure faithful reproduction of the nonnegative probability density* $\rho(x)$ in Eq. 1. Furthermore, there is no reason to consider "operator-valued" quantum-mechanical probabilities as fundamentally different from "ordinary" ones. In fact, any such difference would contradict the results of Cox (1946), Schrödinger (1946) and Rényi (1954) who found that any formal system of logical inference must be equivalent to ordinary probability theory, with probability understood as a numerical scale of rational expectation (or incomplete knowledge) in the tradition of Bernoulli and Laplace (and Heisenberg 1930), – otherwise it is bound to contain inconsistencies.

With this understanding there is nothing strange about the "collapse of the wave function" when new data become available. Their utilisation by means of Bayes' theorem inevitably changes all prior probabilities to posterior ones. As this is not a physical but a *logical* change, questions about its sudden (superluminal) occurrence throughout physical space, or about the exact time of death of Schrödinger's cat, do not arise for Bayesians - who are quite prepared to reason even backwards in time.

2. Mechanics of Particles with Uncertain Initial Coordinates

Let us consider a classical particle. Its energy as a function of its location $\mathbf{x}$ and momentum $\mathbf{p}$ is given by the Hamilton function $H = H(\mathbf{x}, \mathbf{p})$; its motion is determined by Hamilton's canonical equations

$$\frac{d\mathbf{x}}{dt} = \frac{\partial H}{\partial \mathbf{p}}, \qquad (2) \qquad\qquad \frac{d\mathbf{p}}{dt} = -\frac{\partial H}{\partial \mathbf{x}}. \qquad (3)$$

For given initial phase space coordinates, $\{\mathbf{x}(0), \mathbf{p}(0)\}$, one obtains the trajectory in phase space, $\{\mathbf{x}(t), \mathbf{p}(t)\}$, by integration of the canonical equations, for $t < 0$ as well as for $t > 0$.

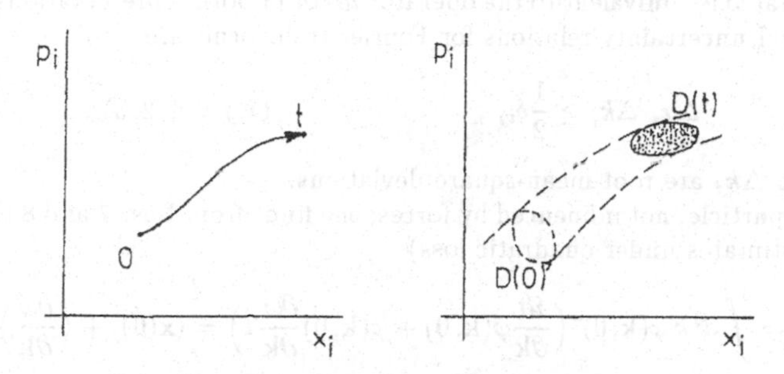

Fig. 1 – **Classical particle trajectories in phase space without and with initial uncertainty**

If the initial coordinates are uncertain, lying somewhere in a phase space domain $D(0)$, there is a multitude of possible trajectories (Fig. 1). At time t the possible values of $\mathbf{x}(t)$ and $\mathbf{p}(t)$ lie in a domain $D(t)$ that has the same size as $D(0)$: The canonical equations imply that the divergence in phase space is zero,

$$\sum_{i=1}^{3} \left(\frac{\partial}{\partial x_i} \frac{dx_i}{dt} + \frac{\partial}{\partial p_i} \frac{dp_i}{dt} \right) = 0 \tag{4}$$

(Liouville's theorem, valid already separately for each pair x_i, p_i). More generally the initial uncertainty can be described by a continuous probability density. Let us consider a time-dependent spatial probability density $\rho(\mathbf{x},t) = |\psi(\mathbf{x},t)|^2$ in a periodicity box so large that the Fourier polynomials of the Riesz-Fejér theorem can be replaced by Fourier integrals. The resulting wave function and its Fourier transform,

$$\psi(\mathbf{x},t) = (2\pi)^{-3/2} \int d^3k \, \varphi(\mathbf{k},0) \, e^{i(\mathbf{kx}-\omega t)}, \tag{5}$$

$$\varphi(\mathbf{k},t) = (2\pi)^{-3/2} \int d^3x \, \psi(\mathbf{x},0) \, e^{-i(\mathbf{kx}-\omega t)}, \tag{6}$$

both normalised to unity, are superpositions of waves propagating with phase velocities ω/k in directions $\mathbf{k}/k$. The resulting averages,

$$\langle \mathbf{x}(t) \rangle = \int d^3x \, |\psi(\mathbf{x},t)|^2 \, \mathbf{x} = \int d^3k \, \varphi(\mathbf{k},t)^* \frac{i\partial}{\partial \mathbf{k}} \varphi(\mathbf{k},t), \tag{7}$$

$$\langle \mathbf{k}(t) \rangle = \int d^3k \, |\varphi(\mathbf{k},t)|^2 \, \mathbf{k} = \int d^3x \, \psi(\mathbf{x},t)^* \frac{\partial}{i\partial \mathbf{x}} \psi(\mathbf{x},t), \tag{8}$$

show that $|\varphi|^2$ is the probability density in $\mathbf{k}$ representation corresponding to the probability density $|\psi|^2$ in $\mathbf{x}$ representation. Furthermore, the factor $\mathbf{k}$ in $\mathbf{k}$ representation is replaced by the operator $-i\partial/\partial\mathbf{x}$ in $\mathbf{x}$ representation, and the factor $\mathbf{x}$ in $\mathbf{x}$ representation is replaced by the operator $i\partial/\partial\mathbf{k}$ in $\mathbf{k}$ representation. Similarly one finds

$$\langle \omega \rangle = \int d^3x \, \psi(\mathbf{x},t)^* \frac{i\partial}{\partial t} \psi(\mathbf{x},t) = \int d^3k \, \varphi(\mathbf{k},t)^* \frac{i\partial}{\partial t} \varphi(\mathbf{k},t) \tag{9}$$

which shows that ω is equivalent to the operator $i\partial/\partial t$ in both representations. The familiar wave-mechanical uncertainty relations for Fourier transforms are

$$\Delta x_i \, \Delta k_j \geq \frac{1}{2}\delta_{ij} , \qquad (i,j = 1,2,3) , \tag{10}$$

where Δx_i and Δk_j are root-mean-square deviations.

For a free particle, not influenced by forces, one finds from Eqs. 7 and 8 the expectation values (best estimates under quadratic loss)

$$\langle \mathbf{x}(t) \rangle = \int d^3k \, \varphi(\mathbf{k},0)^* \Big(\frac{i\partial}{\partial \mathbf{k}} \varphi(\mathbf{k},0) + \varphi(\mathbf{k},0) \frac{\partial \omega}{\partial \mathbf{k}} t \Big) = \langle \mathbf{x}(0) \rangle + \Big\langle \frac{\partial \omega}{\partial \mathbf{k}} \Big\rangle_{t=0} t , \tag{11}$$

$$\langle \mathbf{k}(t) \rangle = \int d^3k \, \varphi(\mathbf{k},0)^* \mathbf{k} \varphi(\mathbf{k},0) = \langle \mathbf{k}(0) \rangle , \tag{12}$$

describing linear translation with group velocity $\langle \partial\omega/\partial\mathbf{k} \rangle$. Their time derivatives,

$$\frac{d\langle \mathbf{x} \rangle}{dt} = \Big\langle \frac{\partial \omega}{\partial \mathbf{k}} \Big\rangle , \tag{13}$$

$$\frac{d\langle \mathbf{k} \rangle}{dt} = 0 , \tag{14}$$

can be compared with the expectation values

$$\frac{d\langle \mathbf{x} \rangle}{dt} = \Big\langle \frac{\partial H}{\partial \mathbf{p}} \Big\rangle , \tag{15}$$

$$\frac{d\langle \mathbf{p} \rangle}{dt} = 0 , \tag{16}$$

that follow from Hamilton's canonical equations for a free particle. Evidently we can take $\mathbf{k} \propto \mathbf{p}$ and $\omega \propto H$ (if we disregard uninteresting additive constants). Denoting the common proportionality constant by $\hbar$ we get de Broglie's particle-wave transcription,

$$H = \hbar\omega , \tag{17}$$

$$\mathbf{p} = \hbar \mathbf{k} , \tag{18}$$

and, from Eq. 10, Heisenberg's quantum-mechanical uncertainty relations,

$$\Delta x_i \, \Delta p_j \geq \frac{\hbar}{2}\delta_{ij} , \qquad (i,j = 1,2,3) , \tag{19}$$

comparable to Liouville's theorem – see Figs. 1 and 2. The equality sign applies if $|\psi|^2$ is a (three-dimensional) Gaussian, the *maximum entropy distribution* for given $\langle \mathbf{x} \rangle$ and $\langle \mathbf{x}^2 \rangle$. Expectation values of physical quantities that depend on both $\mathbf{x}$ and $\mathbf{p}$, such as the Hamilton function $H(\mathbf{p},\mathbf{x}) = \mathbf{p}^2/(2m) + V(\mathbf{x})$ of a particle with mass m moving in a potential $V(\mathbf{x})$, can be calculated from ψ or φ with the appropriate operators. For example, the best estimate of the orbital angular momentum with respect to the origin, $\mathbf{x} = 0$, is

$$\langle \mathbf{x} \times \mathbf{p} \rangle = \int d^3x \, \psi^* \big(\mathbf{x} \times \frac{\hbar}{i} \frac{\partial}{\partial \mathbf{x}} \big) \psi = \int d^3k \, \varphi^* \Big(\frac{i\partial}{\partial \mathbf{k}} \times \hbar \mathbf{k} \Big) \varphi . \tag{20}$$

Real expectation values imply Hermitean (self-conjugate) operators. If ψ is one of the eigenfunctions of the operator, the variance vanishes. If, for instance, ψ is an eigenfunction of the operator $H = i\hbar\partial/\partial t$, with eigenvalue E, satisfying the Schrödinger equation

$$H\psi = E\psi \tag{21}$$

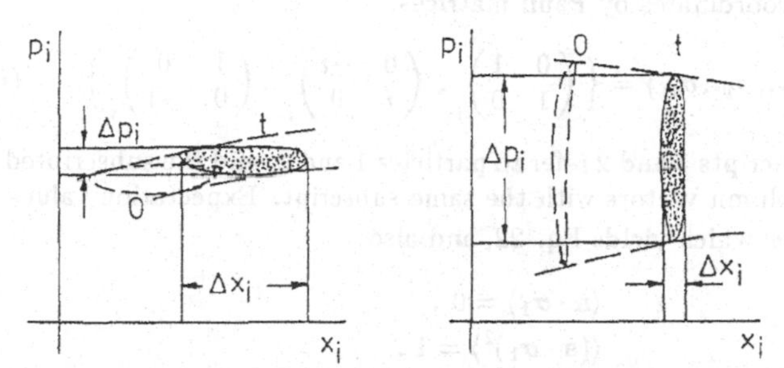

Fig. 2 – Phase space illustration of Heisenberg's uncertainty relations, Eq. 19. Left side: wave-like behaviour with well-defined momentum (wave length) but ill-defined location. Right side: particle-like behaviour with ill-defined momentum (wave length) but well-defined location.

(for given boundary conditions), one has $\langle H \rangle = E$ and var $H = \langle H^2 \rangle - \langle H \rangle^2 = 0$: The estimated energy is E without any uncertainty.

Thus we obtain, for a classical particle obeying Hamilton's canonical equations, the basic formal apparatus of quantum mechanics – complete with wave-particle duality, operator calculus including commutation rules, uncertainty relations and Schrödinger equation. All we had to do was to admit finite uncertainties of the phase space coordinates. The Riesz-Fejér theorem permits then unrestricted use of Fourier series – the proper tool for dealing with boundary conditions and similar constraints – in a way that guarantees the non-negativity of all probability densities. Planck's quantum of action appears naturally, as a "blurring" parameter, in such a probability theory of classical particles that move according to the Hamiltonian equations. Its role as a limit to attainable accuracies in phase space is clear from Heisenberg's uncertainty relations (see Fig. 2).

3. Spin Correlations

Generalisation to several indistinguishable particles and to additional attributes such as spins is straightforward. In the spin version of the famous Einstein-Podolsky-Rosen (1935) experiment one considers a particle with spin zero that decays into two particles, each with spin 1/2, flying in opposite directions. Because angular momentum is conserved, the spins of the two particles must be antiparallel, $\boldsymbol{\sigma}_1 = -\boldsymbol{\sigma}_2$. If one of the spin components of particle 1 is measured as pointing up, the same spin component of particle 2 is immediately known to be pointing down (which can be confirmed experimentally). More generally, one finds that the correlation of arbitrary spin coordinates $(\mathbf{a} \cdot \boldsymbol{\sigma}_1)$ and $(\mathbf{b} \cdot \boldsymbol{\sigma}_2)$ is given by

$$\langle (\mathbf{a} \cdot \boldsymbol{\sigma}_1)(\boldsymbol{\sigma}_2 \cdot \mathbf{b}) \rangle = -\mathbf{a} \cdot \mathbf{b} = -\cos(\mathbf{a}, \mathbf{b}) , \tag{22}$$

where **a** and **b** are unit vectors along two arbitrary analyser directions. This result is obtained quantum-mechanically if one describes the singlet state (total spin zero) by the antisymmetric fermion wave vector (wave function for the two discrete possibilities "spin up" and "spin down" of the two particles)

$$\psi = \frac{1}{\sqrt{2}} \left[\begin{pmatrix} 1 \\ 0 \end{pmatrix}_1 \begin{pmatrix} 0 \\ 1 \end{pmatrix}_2 - \begin{pmatrix} 0 \\ 1 \end{pmatrix}_1 \begin{pmatrix} 1 \\ 0 \end{pmatrix}_2 \right] , \tag{23}$$

and the spin coordinates by Pauli matrices,

$$\sigma_i = \{\sigma_{xi}, \sigma_{yi}, \sigma_{zi}\} = \left\{ \begin{pmatrix} 0 & 1 \\ 1 & 0 \end{pmatrix}_i, \begin{pmatrix} 0 & -i \\ i & 0 \end{pmatrix}_i, \begin{pmatrix} 1 & 0 \\ 0 & -1 \end{pmatrix}_i \right\} \qquad (i = 1,2) \tag{24}$$

where the subscripts 1 and 2 refer to particles 1 and 2, and the subscripted matrix operators act only on column vectors with the same subscript. Expectation values are calculated as $\langle \ldots \rangle \equiv \psi^\dagger \ldots \psi$ which yields Eq. 22, and also

$$\langle \mathbf{a} \cdot \boldsymbol{\sigma}_1 \rangle = 0 \,, \tag{25}$$

$$\langle (\mathbf{a} \cdot \boldsymbol{\sigma}_1)^2 \rangle = 1 \,, \tag{26}$$

$$\langle \sigma_{x1}^2 \rangle = \langle \sigma_{y1}^2 \rangle = \langle \sigma_{z1}^2 \rangle = \langle \boldsymbol{\sigma}_1^2 \rangle / 3 = 1 \tag{27}$$

(similarly for $\mathbf{b}$ and $\boldsymbol{\sigma}_2$). Because of $\mathrm{var}\, x \equiv \langle (x - \langle x \rangle)^2 \rangle$, $\mathrm{cov}\,(x,y) \equiv \langle (x - \langle x \rangle)(y - \langle y \rangle) \rangle$, one recognises (26) as the variance of $(\mathbf{a} \cdot \boldsymbol{\sigma}_1)$, and (22) as the covariance of the spin coordinates $(\mathbf{a} \cdot \boldsymbol{\sigma}_1)$ and $(\mathbf{b} \cdot \boldsymbol{\sigma}_2)$. The latter is numerically equal to the correlation coefficient, $r(x,y) \equiv \mathrm{cov}\,(x,y)/\sqrt{\mathrm{var}\, x \ \mathrm{var}\, y}$.

It is often stated that the correlation (22) cannot be obtained classically. Its confirmation by experiment is then taken as evidence that the spin coordinates cannot exist simultaneously before a measurement reveals one of them, in accordance with N. Bohr's (1935) epistemological interpretation of quantum mechanics but at variance with the ontological view of Einstein, Podolsky and Rosen (1935). Since, however, the quantum-mechanical result (22) does not contain Planck's constant one expects that a classical derivation is possible – as, for instance, in the case of the Rutherford scattering formula. Let us therefore consider the spin $\boldsymbol{\sigma}_1 = -\boldsymbol{\sigma}_2$ as an ordinary vector, for which all orientations are equally possible. Expectation values are then to be calculated classically as

$$\langle \ldots \rangle \equiv \int_0^\infty d\sigma_1 \, \rho(\sigma_1) \int_{-1}^{+1} \frac{d(\cos\vartheta)}{2} \int_0^{2\pi} \frac{d\varphi}{2\pi} \ldots \,, \tag{28}$$

where $\rho(\sigma_1)$ is the probability density of the length $\sigma_1 = |\boldsymbol{\sigma}_1|$ of both spin vectors and ϑ, φ are the polar angle and azimuth of $\boldsymbol{\sigma}_1$. Without any difficulty one finds

$$\langle (\mathbf{a} \cdot \boldsymbol{\sigma}_1)(\boldsymbol{\sigma}_2 \cdot \mathbf{b}) \rangle = -\langle (\mathbf{a} \cdot \boldsymbol{\sigma}_1)(\boldsymbol{\sigma}_1 \cdot \mathbf{b}) \rangle = -\frac{\langle \boldsymbol{\sigma}_1^2 \rangle}{3} (\mathbf{a} \cdot \mathbf{b}) \tag{29}$$

which, with $\langle \boldsymbol{\sigma}_1^2 \rangle / 3 = 1$ (cf. Eq. 27), is equal to the quantum-mechanical result. Hence the correlation measurements alone do not rule out the ontological viewpoint, i. e. reality of unobserved spin components. On the other hand, if one treats the quantum-mechanical spin eigenvalues, $+1$ or -1, measured along $\mathbf{a}$ and $\mathbf{b}$, as determined by hidden variables, one gets the inequalities derived by Bell (1964) that are, in fact, contradicted by experiment. That Bell's inequalities are only valid for a certain class of hidden-variable models, and hence less general than is commonly believed, was pointed out by Jaynes (1989).

4. Summary

The formalism of quantum mechanics, in the traditional axiomatic presentation, seems mysterious. It emerges naturally, however, if one handles phase space uncertainties for classical point particles wave-mechanically, by means of the Riesz-Fejér superposition theorem,

– which by the way dispels any doubts about the *linearity* of the theory. Planck's quantum of action appears automatically, as a "blurring" parameter. The *nonlocality* (instantaneous collapse of the wave function throughout physical space if new information is taken into account) follows from Born's interpretation of $|\psi|^2$ as a probability density and from the Bayesian scheme for the updating of knowledge. There is no reason to doubt that physical quantities, such as the spin coordinates in the spin version of the Einstein-Podolsky-Rosen experiment, have a *reality independent of the observer*, in obvious contrast to eigenfunction expansions and eigenvalues that reflect his choice of measurement. From this viewpoint *quantum mechanics looks much like an error propagation formalism for uncertainty-afflicted physical systems that obey the classical equations of motion.*

What remains mysterious, however, is the *irreducible uncertainty* enforced by the empirical finite and universal value of the blurring parameter $\hbar$. That this value is the same for electrons, nucleons, photons etc. is not too surprising since their mutual interactions conserve energy and momentum. Its role as a limit to the attainable information and control in microphysics has been clear ever since Heisenberg (1930) discussed his uncertainty relations: Phase space trajectories and orbits are always affected by a non-removable minimum blur. As finite particle size would produce a similar blur, one is tempted to ask if quantum mechanics can perhaps be viewed as a kind of minimum information (maximum entropy) generalisation of probabilistic Hamiltonian mechanics *from mass points to particles with finite extension* (spatial distribution) *and internal motion* (momentum distribution, spin). How this conjecture compares with others, such as zitterbewegung, granular space-time structure, or superstrings, remains to be seen.

Appendix: Proof of the Riesz-Fejér Theorem

The proof presented by L. Fejér (1915) as due to F. Riesz is given here in slightly different notation. Consider the real Fourier polynomial

$$\rho(x) = \rho(x)^* = \sum_{l=-n}^{n} c_l \, e^{ilx} \,, \qquad (c_l = c_{-l}^*) \tag{A1}$$

Defining the polynomial $g(z)$ as

$$g(z) \equiv c_n^* + \ldots + c_1^* z^{n-1} + c_0 z^n + c_1 z^{n+1} + \ldots + c_n z^{2n} \tag{A2}$$

one can write $\rho(x) = e^{-nix} g(e^{ix})$. If the Fourier polynomial is non-negative, this becomes

$$\rho(x) = |g(e^{ix})|. \tag{A3}$$

The polynomial $g(z)$ is of degree $2n$ if $c_n \neq 0$, so that $g(0) \neq 0$. If z_k is a solution of $g(z) = 0$,

$$g(z_k) = c_n^* + \ldots + c_n z_k^{2n} = 0 \,, \tag{A4}$$

then $1/z_k^*$ is another solution,

$$g(\frac{1}{z_k^*}) = [(c_n^* + \ldots + c_n z_k^{2n}) z_k^{-2n}]^* = 0 \,. \tag{A5}$$

One concludes that each root z_k is accompanied by another root $1/z_k^*$. (Recall that $z = 0$ is not a root.) On the unit circle, $z = e^{i\varphi}$, both roots coincide. If the root z_k is of multiplicity m one has $2m$ roots on the unit circle. Thus a complete set of independent roots contains, for instance, those within the unit circle and half of those on the unit circle itself, with due account of multiple roots. One gets

$$g(z) = c_n \prod_{k=1}^{n} (z - z_n)(z - \frac{1}{z_k^*}) \tag{A6}$$

and, with $z = e^{ix}$,

$$\rho(x) = |g(e^{ix})| = \left| \sqrt{c_n} \prod_{k=1}^{n} \frac{e^{ix} - z_k}{\sqrt{z_k}} \right|^2 . \tag{A7}$$

This is the absolute square of a Fourier polynomial of the same order as $\rho(x)$ so we can write

$$\rho(x) = \sum_{l=-n}^{n} c_l e^{ilx} = |\psi(x)|^2 , \qquad -\pi < x \leq \pi \tag{A8}$$

$$\psi(x) = e^{i\alpha} \sqrt{\frac{c_n}{z_1 \ldots z_n}} \prod_{k=1}^{n} (e^{ix} - z_k) , \qquad (\alpha \text{ arbitrary}) , \tag{A9}$$

which completes the (constructive) proof that each non-negative Fourier polynomial can be written as the absolute square of an unrestricted Fourier polynomial of (at most) the same order (same highest harmonic), with an arbitrary phase factor.

ACKNOWLEDGMENTS. It is a pleasure to thank R. Balian for valuable discussions.

REFERENCES

J. Aczél, *Vorlesungen über Funktionalgleichungen und ihre Anwendungen*, Birkhäuser Verlag Basel (1961); translation: *Lectures on Functional Equations and Their Applications*, Academic Press, New York (1964)

J.S. Bell, "On the Einstein-Podolsky-Rosen Paradox", *Physics* **1**(1964) 195

N. Bohr, "Can Quantum-Mechanical Description of Physical Reality Be Considered Complete?", *Phys. Rev.* **48**(1935) 696

R.T. Cox, "Probability, Frequency and Reasonable Expectation", *Am. J. Physics* **14** (1946) 1

A. Einstein, B. Podolsky, N. Rosen, "Can Quantum-Mechanical Description of Physical Reality Be Considered Complete?", *Phys. Rev.* **47** (1935) 777

L. Fejér, "Über trigonometrische Polynome", *J. reine u. angew. Math.* **146** (1915) 53; reprinted in *Gesammelte Werke*, Birkhäuser Verlag Basel (1970), vol. I, p. 842

W. Feller, *An Introduction to Probability Theory and Its Applications*, Wiley & Sons, New York etc. (1966), vol. II, p. 597

W. Heisenberg, *Die physikalischen Prinzipien der Quantentheorie*, Hirzel, Leipzig (1930)

E.T. Jaynes, "Clearing Up Mysteries – The Original Goal", in *Maximum Entropy and Bayesian Methods*, J. Skilling (Ed.), Kluwer, Dordrecht (1989)

L. Rényi, *Valószinűségszámítás* (= Probability Calculus), Budapest (1954); for a slightly generalised presentation in English see Aczél (1961), Subsection 7.1.4.

E. Schrödinger, "The Foundation of the Theory of Probability", *Proc. R. Irish Acad.* **51 A**(1947) 51, **51 A** (1947) 141; reprinted in *Gesammelte Abhandlungen*,Wien (1984) vol. 1, pp. 463, 479

Chapter 3

Time Series

SPEECH PROCESSING USING BAYESIAN INFERENCE

W.J.Fitzgerald and M.Niranjan
Department of Engineering
University of Cambridge
Cambridge UK

ABSTRACT. In this paper, we report on the application of Bayesian methods to the analysis of speech signals. Voiced speech can be modelled as a superposition of decaying sinusoids and estimates of the resonant frequencies, decay rates, phases, amplitudes as well as the number of model functions are calculated. The motivation for this model is that in speech analysis, the frequencies and decay rates correspond to formants and bandwidths which are perceptually significant parameters. Speech parameters are estimated by calculating the posterior probabilities for the model parameters, after various nuisance parameters have been marginalised, and it is shown how *model order evidence* can be calculated. Comparisons with methods such as Minimum Description Length and the Akaike Information Criteria will be made.

1. Introduction

The most widely used approaches to modelling speech signals are based on linear prediction analysis (auto-regressive modelling) of short durations of speech. Linear prediction techniques are mathematically simple and computationally efficient. They can be shown to be equivalent to a lossless acoustic tube model of the vocal tract, where the acoustic tube consists of segments of equal lengths and varying cross sectional areas. For most sound this is a good model and speech analysis algorithms based on this model can be found in many applications of speech technology. The model has limitations in modelling fricative and nasal sounds. In unvoiced fricatives, the source of excitation is not at the back of the throat (as in the sound /s/) and for nasal sounds there is extra coupling of the nasal cavity.

There are many approximations commonly used in speech analysis which arise primarily from the non-stationarity of the signal. As we move our vocal tracts to pronounce a sequence of phonemes, the spectral characteristics of the acoustic signals will change. Speech analysis algorithms deal with this non-stationarity by windowing the signal and over a typical window size of 20-25ms, assume the signal to be stationary. Most algorithms that attempt parametric modelling of such windowed signal use a fixed order model, where parameters such as window size, model order etc are fixed in an *ad hoc* manner.

In speech synthesis, it is common to factorise the auto-regressive transfer function obtained by linear prediction analysis into a number of second order terms. Expansion using partial fractions gives a sum of second order transfer functions which can model the vocal tract. The resonant frequency of each component is known as a 'formant' and is a perceptually significant parameter in speech processing. Thus the modelling process for this speech model is one of analysing a sum of decaying sinusoids, and in this paper, we apply Bayesian inference to this problem.

A. Mohammad-Djafari and G. Demoments (eds.), Maximum Entropy and Bayesian Methods, 215–223.

1.1. Bayesian Inference

A fundamental task in signal processing and science in general is to develop models for signals which are observed and to determine whether the model function that one is using to describe the data, is actually appropriate for the particular problem under investigation, and if so, to extract values for the free parameters of the model. We need a way to choose between several possible models. To attempt to solve this problem, one must enumerate the possible models and realise that in terms of real data the correct model may not be within the set choosen. All that we can do is compare various models within a set that we have defined to see which models are more plausible. This, and the problem of extracting values for model parameters, is a problem of scientific inference, and to carry out consistent reasoning and inference, one may use the Bayesian paradigm.

The framework can be broken down into essentially three levels of inference. At the first level it is assumed that one of the models within a chosen set, is the correct model with which to interpret the data and the problem of inference at this level consists of extracting values for the free parameters of the model, given the data.

The second level of inference is concerned with model selection, and here various models, within a chosen set, are ranked according to their evidence in the light of the data. (This second level of inference can be broken down into essentially two further levels, the first dealing with hyperparameters and regularisers, and the second dealing with evidence).

Let us write Bayes' theorem in the form

$$P(w,\sigma^2|D,M)=\frac{P(D|w,\sigma^2,M)P(w,\sigma^2|M)}{P(D|M)}$$

or

$$Posterior=\frac{Likelihood\times Prior}{Evidence}$$

where the posterior $P(w,\sigma^2|D,M)$ is the *a-posteriori* probability density of the model parameters, w, σ^2, given the data and the model structur e, the likelihood $P(D|w,\sigma^2,M)$ is the probability density of the data given the model and its parameters, the prior $P(w,\sigma^2|M)$ is the *a-priori* probability density of the parameters given the model structure, and the evidence $P(D|M)$ is the probability density of the data given the model structure. All of these probability densities are, of course, conditional on any prior information. The quantity σ^2 is the variance of the noise model, which unless we have information to the contrary can be assumed to come from a zero mean, possibly white, Gaussian process.

As mentioned previously, having decided on a particular model structure, one can choose the model parameters which maximise the posterior probability, the so-called MAP estimates. Alternatively, one can chose a sub-set of parameters that maximise the marginal posterior probability. These marginal estimates will in general be different from the MAP estimates, since one is basically asking a different question concerning the data.

The task of choosing between several different models, in the present case, the number of decaying sinusoids, for the model which is most plausible, or the one that ha s the greatest evidence, can be formulated in terms of maximising $P(M_i|D)$, the posterior probability density of the model

$$P(M_i|D)=\frac{P(D|M_i)P(M_i)}{P(D)}$$

The denominator is a constant for all models, and if at this stage we have no reason to assign differing priors $P(M_i)$, then the choice of model structure reduces to that model structure which has the maximum evidence $P(D|M_i)$. The evidence may be written as

$$P(D|M_i) = \int\int P(D|w,\sigma^2,M_i)P(w,\sigma^2|M_i)dwd\sigma$$

Although this integral can be carried out in some cases analytically, and in all cases numerically, it is very informative to approximate the above integral by expanding about the MAP parameters in the following form

$$P(D|M_i) \approx P(D|w_{MAP},M_i)P(w_{MAP}|M_i)\Delta w$$

for a known noise variance. The numerical details of this type of calculation wi ll be given elsewhere.

As an example of the second level of inference, consider the data shown in Fig 1, which consists of two decaying sinusoids with added white Gaussian noise $e(t)$ of unit variance, given by

$$s(t) = \exp(-0.01t)\sin(\frac{2\pi t}{200}) + \exp(-0.005t)\sin(\frac{2\pi t}{100}) + 0.01e(t)$$

with t=1, 2, ..., 1000.

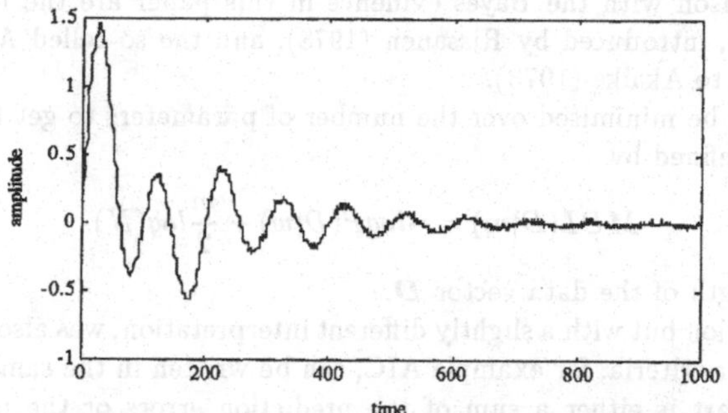

Figure 1: Two decaying sinusoids with added noise

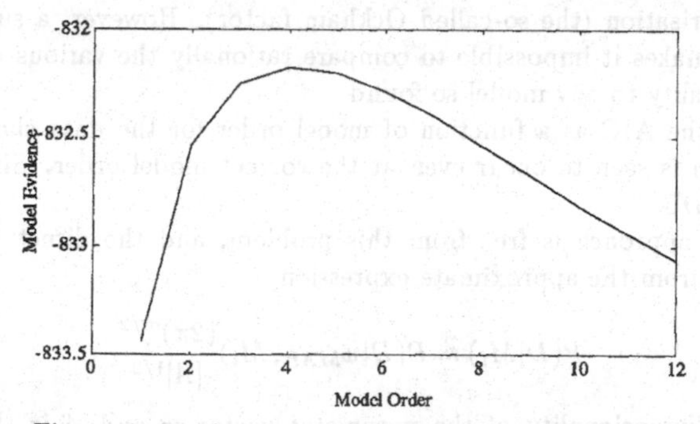

Figure 2: Evidence as a function of Model Order

Fig 2 , shows the evidence for various models (in this case varying numbers of decaying sinusoidal components) as a function of model order, and the evidence shows a maximum at model order 4, since $m/2$ decaying sinusoids may be described by an Auto-Regressive (AR) difference equation with m 'filter' coefficients $\boldsymbol{w}$

$$s_n = \sum_{j=1}^{m} w_j s_{n-j}$$

However, when noise is added, the new model should be written as an Auto-Regressive Moving-Average process as follows;

$$d_n = s_n + e_n = e_n + \sum_{j=1}^{m} w_j s_{n-j} = \sum_{j=1}^{m} w_j d_{n-j} - \sum_{j=1}^{m} w_j e_{n-j} + e_n$$

where e_n is the added noise.

It is usually assumed in speech modelling that the signal can be modelled using just the AR part of the process, but in order to analyse speech in a noisy environment, an ARMA model should be used.

In the following, the model chosen for investigation is the AR model, which in its general form is represented as a superposition of decaying sinusoids.

There exist many *ad hoc* methods for conducting model order selection, and the two chosen for comparison with the Bayes evidence in this paper are the Minimum Description Length, MDL, introduced by Rissanen (1978), and the so-called Akaike Information Criteria, AIC, due to Akaike (1973).

The MDL may be minimised over the number of parameters to get the optimal model complexity. It is defined by

$$MDL(D|m) = -logP(D|w) + \frac{m}{2}log(N)$$

where N is the length of the data vector $\boldsymbol{D}$.

The same criterion but with a slightly different interpretation, was also given by Schwartz (1978). Many other criteria, for example AIC, can be written in the same general two-part form. The first part is either a sum of the prediction errors or the negative logarithm of the likelihood, whilst the second part is designed to incorporate the necessary penalty for over-parameterisation (the so-called Ockham factor). However, a subjective choice of the second term makes it impossible to compare rationally the various criteria or claim a meaningful optimality to any model so found.

Fig 3, shows the AIC as a function of model order for the data shown in Fig 1, and no clear transition is seen to occur even at the correct model order. Similar results were obtained using MDL.

The Bayesian approach is free from this problem, and the direct calculation of the evidence is found from the approximate expression

$$P(D|M_i) \approx P(D|w_{MAP}, M_i)\frac{(2\pi)^{m/2}}{|A|^{1/2}}$$

where m is the dimensionality of the parameter vector w and A is the Hessian of the log-posterior probability.

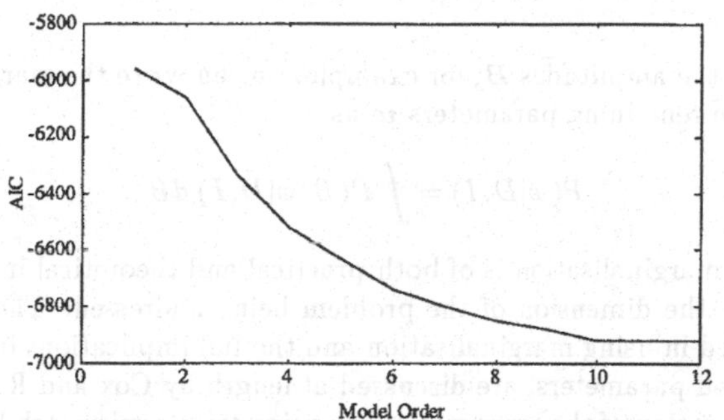

Figure 3: AIC as a function of Model Order

1.2. Parameter Estimation

In an estimation problem one assumes that the model is true for some unknown values of the model parameters, and one explores the constraints imposed on the parameters by the data, using Bayes' theorem. The hypothesis space for an estimation problem is therefore the set of possible values of the parameter vector w, and it is this vector that will form the 'hypothesis' that will be used in Bayes' theorem. The data form the sample space, and both the hypothesis space and the sample space may be either discrete or continuous.

For the purposes of this section, Bayes' theorem may be written

$$P(H|D,I) = \frac{P(H|I)P(D|H,I)}{P(D|I)}$$

Thus, $P(H|D,I)$ is the probability of H (a stated hypothesis, say) given knowledge of D and any prior information. At this level of inference $P(D|I)$ is a normalization term and we can write;

$$P(H|D,I) \propto P(H|I)P(D|H,I)$$

$P(D|H,I)$ is known as the likelihood function. Within the framework of Bayes' theorem, we are then able to integrate out any unwanted or 'nuisance' parameters, a process known as marginalisation. If such marginalisation can be done analytically then the complexity of the problem can be reduced considerably.

We may write a general model function as,

$$s(t) = \sum_{k=1}^{m} B_k G_k(t, w)$$

where B and $G(t,w)$ are amplitudes and model functions respectively, and the data we are trying to account for by this model can be written as

$$d(t_i) = s(t_i) + e(t_i) \quad \text{i} = 1, 2, 3, \ldots\ldots, \text{N}$$

where e is a noise vector. Therefore the hypothesis we are trying to test using Bayes' theorem is whether or not the model function given above accounts for the data. The posterior probability may be written $P(H|D,I)$ and this is taken to mean the probability of the model parameters given the data and prior information.

Integrating out the amplitudes $\boldsymbol{B}$, for example, we can write the marginalised posterior probabilities for the remaining parameters $\boldsymbol{w}$ as

$$P(w|D,I) = \int P(B,w|D,I)\,dB$$

This process of marginalisation is of both practical and theoretical interest, since it can significantly reduce the dimension of the problem being addressed. The conditions under which one is justified in using marginalisation and the full implications in terms of possible bias in the estimated parameters, are discussed at length by Cox and Reid (1989).

¿From a consideration of the error statistics arising from a mismatch between the model and the data, the likelihood function for a general model function given by $\sum_{j=1}^{m} B_j G_j(t_i, w)$ $= \boldsymbol{B}\ \boldsymbol{G}$ can be written

$$P(D|HI) = (2\pi)^{-N/2}\sigma^{-N} \exp\left[-\frac{1}{2\sigma^2}(D - BG)^T(D - BG)\right]$$

where $\boldsymbol{D}$ is a data vector of length N, $\boldsymbol{B}$ is a model amplitude vector of length m and $\boldsymbol{G}$ is a matrix of model functions of order $N \times m$.

To consider the marginal probabilities for the parameters independent of amplitudes, the following integrals have to be undertaken,

$$P(w|DI) \propto \int_{-\infty}^{\infty} \dots \int_{-\infty}^{\infty} dB_1 \dots dB_m P(D|HI) P(B_1, B_2, \dots, B_m, w|I)$$

Making the substitutions $\mathcal{P} = \boldsymbol{G}^T\boldsymbol{D}$ for the model functions projected onto the data, and $\mathcal{Q} = \boldsymbol{G}^T\boldsymbol{G}$ for the model functions projected onto themselves, and carrying out the integration, the required marginal posterior probability may be written

$$P(w|DI) = \frac{P(w|I)}{P(D|I)}(2\pi)^{(m-N)/2}\frac{\sigma^{m-N}}{|\mathcal{Q}|^{1/2}} \exp\left[-\frac{1}{2\sigma^2}(D^2 - \mathcal{P}^T\mathcal{Q}^{-1}\mathcal{P})\right]$$

It should be noted that the function $\mathcal{P}$ can be called a generalised periodogram, since when the model function is a sinusoid, $\mathcal{P}$ is the periodogram.

If the variance is unknown, it too can be treated as a nuisaince parameter and integrated out of the problem. In this case, a Jeffreys prior should be used, since the variance is a scale parameter. However, for only a small number of data points, the final difference between the assumption of a uniform prior and a Jeffreys prior is very small, Lasenby and Fitzgerald (1991). Integrating the noise variance between 0 and ∞ with a Jeffreys prior $1/\sigma$ gives

$$P(w|DI) = \frac{P(w|I)}{P(D|I)}\frac{\pi^{(m-N)/2}}{|\mathcal{Q}|^{1/2}}\frac{\Gamma[(N-m)/2]}{2}(D^2 - \mathcal{P}^T\mathcal{Q}^{-1}\mathcal{P})^{(m-N)/2}$$

which is of the form of the Student t-distribution, as has been noted many times in the statistical literature, Zellner (1971), West and Harrison (1989), Duijndam (1988), Broemeling (1985) and Bretthorst (1989).

The formalism can also readily deal with coloured noise by replacing the noise variance above by the coloured noise covariance matrix. This has the effect of renormalising the model functions and the data, Fitzgerald (1991), Whalen (1971).

The model function amplitudes, B, can be obtained using maximum likelihood estimation, by differentiating the likelihood, or log-likelihood, with respect to the parameters, giving;

$$B_{ML} = \frac{G^T D}{G^T G}$$

and similarly the maximum likelihood estimation of the noise variance is given by;

$$\sigma^2_{ML} = \frac{(D - GB)^T (D - GB)}{N}$$

These estimates can, in general, be different from the marginal estimates since one is asking different questions in both case. In fact, the estimates of the amplitudes are the same in both cases, but the marginal estimation of the noise variance is given by;

$$\sigma^2_{marg} = \frac{D^T D - D^T G(G^T G)^{-1} G^T D}{N\sqrt{(det(G^T G))}}$$

This is to be compared with the maximum likelihood estimate for the variance given by

$$\sigma^2_{ML} = \frac{D^T D - D^T G(G^T G)^{-1} G^T D}{N}$$

which differs from the marginal estimator by a model dependent determinant term, which can have a significant effect on the parameter estimates for some models functions.

One reason for marginalising nuisance parameters would be in the case where the *maximum a posteriori* estimates cannot be solved analytically, or if one is, for some reason, interested in solving a somewhat different problem to that usually considered. However, in most cases the integrations cannot be carried out analytically and one would be forced to integrate numerically (This is the main reason for the development of fast integration methods e.g. Bayes Four, Naylor and Shaw (1990)).

2. Results

A software speech synthesiser was used to generate the sentence 'We were away', and the time series of this data is shown in Fig 4. The frequency-time spectrogram of part of this data is shown in Fig 5, which clearly shows the complex frequency content of the speech signal. The synthetic speech consists of four decaying sinusoids, with time varying parameters with a very small amount of added white Gaussian noise.

Initially, the speech signal was broken up into segments, each segment corresponding to particular glottal pulses, and the evidence as a function of model order was then calculated for each of the segments assuming th at the data could be represented as a superposition of decaying sinusoids (which is in fact the way the synthetic speech was generated). The evidence for various model orders for all the segments of speech was calculated and clearly showed that model order 8 was the most plausible model. This, as explained earlier corresponds to a model with 4 decaying sinusoids. However, both AIC and MDL for all the segments of the data failed to show any well defined cut-off point corresponding to the correct model.

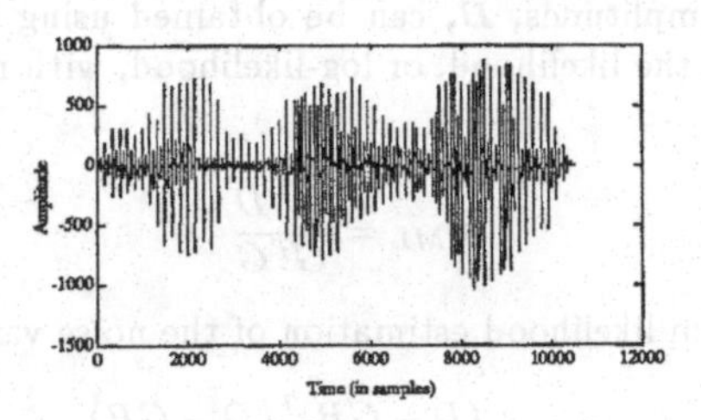

Figure 4: Speech Data

Having determined the correct model order for the data, the first level of inference was then investigated and the model parameters were found using both MAP estimation, by searching the likelihood surface using a variety of search methods ranging from that of Nelder and Mead, the approach of Hooke and Jeeves, and Genetic algorithms, and also by calculating various marginal estimators for the model parameters.

For this speech data, the MAP and marginal parameter estimates were very similar , and corresponded extremly well with the known model parameters.

The spectrogram reconstructed from the parameter estimates is shown in Fig 5. It should be pointed out that had the model order been obtained from any of the more *ad hoc* methods, then any method based solely on least squares would have reached totally different results, and the least squares residuals would have favoured higher order models thereby giving a spectrogram with more artifacts.

It must also be mentioned that choosing to display the data using the spectrogram is purely for visual purposes and does not imply that the Fourier transform is appropriate for such non-stationary data, quite the contrary.

3. Conclusions

This paper has shown that the two levels of Bayesian inference can yield extreme ly useful results in model-based signal processing in general, and in the present context, for speech processing in particular. An introduction to the evidence formulation has been given, and various paramete r estimators, e.g MAP and marginal, have been introduced, and they have been appli ed here to the model order selection and parameter estimation of models containing a superposition of decaying sinusoids. The analysis has been carried out using a block approach, and the next phase of the work will be to develop recursive methods, along the lines of Kalma n filtering, which will be useful for real-time implementation and model parameter and model order tracking. The Kalman filter can be derived within the Bayesian frame work, and it would be interesting to consider other 'filters' suited to marginal param eter tracking, for example.

References

[1] Akaike, H., Information theory and the extension of the maximum likelihood principle, 2nd Inter.Symp. on Information Theory (Petrov, B.N.and Csaki, F. eds.), Akademiai

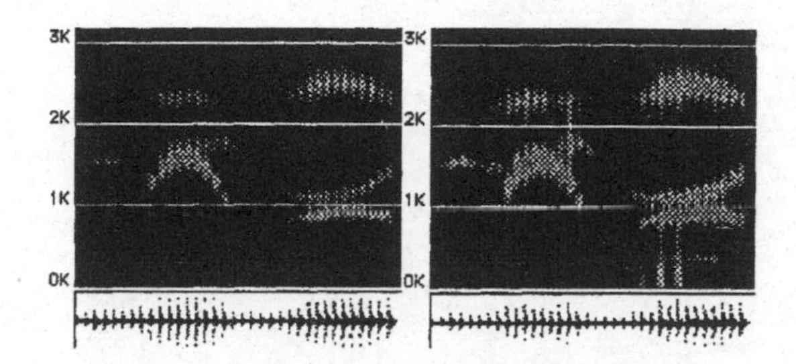

Figure 5: Original and Reconstructed Spectrogram

Kiado, Budapest, 267-281, 1973.

[2] Bretthorst, L., Bayesian Spectrum Analysis and Parameter Estimation, Lecture notes in statistics, Springer Verlag 1989.

[3] Broemeling, L.D., Bayesian Analysis of Linear Models, Marcel Dekker, Inc, 1985.

[4] Cox, D.R., and Reid, N., The Canadian Journal of Statistics, 17, 229, 1989.

[5] Duijndam, A.J.W., Geophysical Prospecting, 36,878, 1988.

[6] Naylor, J.C. and Shaw, J.E.H., Bayes Four User Manual, Nottingham Polytechnic, 1990.

[7] Rissanen, J., Automatica, 14, pp 465-471, 1978.

[8] Schwarz, G., The Annals of Statistics, 6, 461, 1978.

[9] West, M., and Harrison, J., Bayesian Forcasting and Dynamic Models, Springer Verlag, 1989.

[10] Zellner, A., An Introduction to Bayesian Inference in Econometrics, New York, John Wiley and Sons, 1971.

[11] Fitzgerald, W.J., Bayesian Data Analysis, Proc. I.O.A. Vol 13 part 9 pp 212-219, 1991.

[12] Lasenby, J., and Fitzgerald, W.J., A Bayesian Approach to High-Resolution Beamforming, IEE Proc. F, Vol 138, Number 6, pp 539-544, 1991.

A BAYESIAN METHOD FOR THE DETECTION OF UNKNOWN PERIODIC AND NONPERIODIC SIGNALS IN BINNED TIME SERIES

P.C. Gregory
Department of Physics
University of British Columbia
6224 Agricultural Road
Vancouver, British Columbia V6T 1Z1

T.J. Loredo
Department of Astronomy
Space Sciences Building
Cornell University
Ithaca, New York 14853

ABSTRACT. We show that the method we have proposed for the detection and measurement of an unknown periodic signal using arrival times of *individual* events can also be applied to time series consisting of irregularly spaced *binned* samples with gaps. As in our earlier work we again assume a Poisson sampling distribution. The same method can also be used to detect significant *nonperiodic* variation in a signal. We demonstrate this explicitly here with numerical simulations. The approach should be useful to anyone interested in the general problem of inferring the shape of an unknown function from discrete samples.

1. INTRODUCTION

At the MaxEnt 1991 meeting we presented a Bayesian method for the detection and measurement of a periodic signal in an arrival time series when we have no prior knowledge of the existence of such a signal or of its characteristics, including its shape (Gregory and Loredo, 1992a & b, henceforth referred to as GL1 and GL2). Given some data, our goal was to develop Bayesian answers to the questions, "Is there evidence for a periodic signal? If so, what are the best estimates of its period and shape?" Our initial motivation was the detection of pulsars in X-ray astronomy data consisting of a set of arrival times for individual photons, some of which are background events. The lightcurves of X-ray pulsars exhibit a very wide range of shapes, and thus demand use of a very flexible signal model.

To address the detection problem, we used Bayes's theorem to compare a constant rate model for the signal to models with periodic structure. The periodic models describe the signal plus background rate as a stepwise distribution in m bins per period, for various values of m. Such a model is capable of approximating a lightcurve of essentially arbitrary shape. Additionally, many of the calculations required for a Bayesian analysis can be performed analytically with this model, making numerical calculations more efficient than would be possible with, say, Fourier series models.

A. Mohammad-Djafari and G. Demoments (eds.), Maximum Entropy and Bayesian Methods, 225–232.

The Bayesian posterior probability for a periodic model contains a term which quantifies Ockham's razor, penalizing successively more complicated periodic models for their greater complexity even though they are assigned equal prior probabilities. The calculation thus balances model simplicity with goodness-of-fit, allowing us to determine both whether there is evidence for a periodic signal, and the optimum number of bins for describing the structure in the data. Unlike the results of traditional "frequentist" calculations, the outcome of the Bayesian calculation does not depend on the number of periods examined, but only on the range examined. Once a signal is detected, Bayes's theorem can be used to estimate the frequency and shape of the signal.

Our initial presentation of the method assumed that the data consisted of the locations or times of individual discrete events. One of the objectives of this paper, which is discussed in section 2, is to show that the method is also directly applicable to binned data. Consequently the range of applications of the method is much greater than originally conceived. Our second goal, which is presented in section 3, is to show how the same method can be used to reliably detect a nonperiodic signal. The stepwise periodic models referred to above transform to nonperiodic models when the parameter representing the trial period is set equal to the data duration. In this case a signal means any significant variation in the Poisson time series compared to a constant rate model.

2. APPLICATION TO BINNED DATA

All real detectors have finite temporal resolution, reporting the arrival time of an event to within some interval (bin size) Δt, with $\Delta t > 0$. Thus all real discrete time series are binned. But depending on how Δt compares with the timescales of interest, the effect of binning may or may not be significant, calling for different methods of analysis.

In any search for periodicity or time variation using discrete events, there are at least two signal timescales of interest: the typical time between events (*i.e.*, the reciprocal of the average event rate, $\bar{r}$); and the timescale of possible variations of the rate, τ. Clearly, if $\Delta t \gtrsim \tau$, information about interesting temporal variation is being destroyed by the limited detector resolution. Analysis of such data will be severely complicated by the need to "deconvolve." In any case, most experiments are designed to sample on timescales that are small compared with the physical timescales of interest, so we confine our analysis to cases where $\Delta t \ll \tau$.

In our earlier work, we assumed that, in addition, $\Delta t \ll 1/\bar{r}$, so that most time bins are empty, and there is never more than one event per bin. Our analysis held even in the case of infinitesimal Δt, in which case the precise arrival time of each event is known. We refer to time series with $\Delta t \ll 1/\bar{r}$ as *arrival time series.* They are common in high energy astrophysics, where fluxes are small and detectors may have limited effective area, so the expected event rate is small.

But it is also often the case, in astrophysics and elsewhere, that $\Delta t \gtrsim 1/\bar{r}$, so that there may be more than one (perhaps thousands) of events in a bin. As long as $\Delta t \ll \tau$, such *binned time series* can provide useful information about the physical timescales of interest, even though knowledge of the arrival times of individual events is lost. Indeed, if $\tau \gg 1/\bar{r}$, little may be gained by knowing the precise arrival times of individual events, and the expense of recording, storing, and analyzing individual arrival times is thus unjustified.

In this work, we seek a general method for detecting and measuring varying signals of unknown shape using *binned* data, to complement our earlier work on arrival time series.

In fact, our earlier method for analyzing arrival time series can be simply generalized for application to binned time series, as we now show.

Consider data from a counting experiment which is sampled at Z irregular time intervals with sample bin durations, $\{\Delta t_i\}$, for $i = 1$ to Z, centered on the sample times $\{t_i\}$. The number of events counted in each sample bin is represented by $\{n_i\}$. Any number of bins may contain zero counts. The sample bins need not necessarily be contiguous; there may be gaps in the observations. We let L equal the duration of the time series including gaps, and T equal L minus the total duration of all data gaps. A common special case is uniform sampling, where $L = T$, the t_i are evenly spaced, and the sample bin duration is constant.

Note that both the data and the model have bins. To distinguish between them, we always call the bins defining the data the *sample bins*, and those defining the model the *model bins* or *phase bins.* We will use the subscript i to refer to sample bins, and the subscript j to refer to model bins.

A full Bayesian analysis of the problem requires calculation of the following quantities:
1) The odds ratio in favor of the periodic model class over constant alternatives;
2) The probability distribution for the frequency of a hypothetical periodic signal;
3) The probability distribution for the shape of a hypothetical periodic signal, from which a shape estimate and "error bars" can be determined.

To derive equations for these quantities we need to know the likelihood function for each model and the prior probability distributions for the model parameters. We assume the same priors for the model parameters as those used by GL1 & 2. We will now show that for the binned time series, the likelihood functions for the the constant and periodic models are identical in form to those derived previously by GL1 & 2, so that the analysis of that earlier work applies to binned time series as well as to arrival time series.

From the Poisson distribution, the probability for seeing n_i events in a sample bin of duration Δt_i centered at time t_i, given the rate $r(t)$, is,

$$p(n_i \mid r, I) = \frac{[r(t_i)\Delta t_i]^{n_i} e^{-r(t_i)\Delta t_i}}{n_i!}. \tag{1}$$

As mentioned above, we have assumed that the rate does not vary substantially over the duration of a sample bin, so that the integral of the rate over the bin is well approximated by $r(t_i)\Delta t_i$. The likelihood function for r is just the joint probability for all Z samples in the time series, which is simply the product of equation (1) for each sample,

$$p(D \mid r, I) = K \left[\prod_{i=1}^{Z} r(t_i)^{n_i}\right] \exp\left[-\sum_{i=1}^{Z} r(t_i)\Delta t_i\right], \tag{2}$$

where

$$K = \left[\prod_{i=1}^{Z} \Delta t_i^{\,n_i}\right] \Big/ \left[\prod_{i=1}^{Z} n_i!\right]. \tag{3}$$

When the likelihood is used in Bayes's theorem to compute the quantities of interest, the factor K cancels out with the same factor in the global likelihood (the denominator of Bayes's theorem).

For the constant rate model, setting $r(t_i) = A$, where A is a constant, yields,

$$p(D \mid A, M_1) = K A^N e^{-AT}. \tag{4}$$

For the stepwise periodic models the rate is assumed to be a constant in each of the m bins ($m \geq 2$). This is not a single model, but a class of models, one for each choice of m. We denote the information specifying each such model by the symbol M_m. Model M_m has $(m+2)$ parameters: an angular frequency ω (or alternatively a period, P, with $\omega = 2\pi/P$); a phase, ϕ, specifying the location of the bin boundaries; and m values, r_j, specifying the rate in each phase bin, with $j = 1$ to m. The value of the subscript j corresponding to any particular time t is given by,

$$j(t) = \mathrm{int}[1 + m\{(\omega t + \phi) \bmod 2\pi\}/2\pi]. \tag{5}$$

The model assumes that all events in a particular sample bin can be associated with a unique model bin. Because the sample bins have a nonzero width, some fraction of them will straddle model bin boundaries. Provided that the duration of the model bins, P/m, is significantly larger than that of the longest sample bins, this fraction will be very small. This is just the requirement that $\Delta t \ll \tau$, discussed above. This should not be a strong restriction in practice. It merely requires that the duration of sample bins be small compared with the smallest anticipated timescale of variation for the signal. Of course, where possible it is better to use individual event locations because information is irretrievably lost when the event locations are binned.

To facilitate comparison with the constant model, it is convenient to re-express the r_j parameters. We write the rate as the time-averaged rate, A, times a normalized stepwise function that describes the shape of the periodic lightcurve. The average rate is,

$$A = \frac{1}{m} \sum_{j=1}^{m} r_j. \tag{6}$$

The lightcurve shape is completely described by the fraction of the total rate per period that is in each phase bin. These fractions are,

$$\begin{aligned} f_j &= \frac{r_j}{\sum_{k=1}^{m} r_k} \\ &= \frac{r_j}{mA}. \end{aligned} \tag{7}$$

Equations (6) and (7) let us write r_j in terms of A and f_j:

$$r_j = mAf_j. \tag{8}$$

In this way the m values of r_j are replaced by A and the m values of f_j. There are still only m degrees of freedom associated with the rate because by definition the f_j must satisfy the constraint,

$$\sum_{j=1}^{m} f_j = 1, \tag{9}$$

so only $(m-1)$ of them are free. We denote the full set of m values of f_j by the symbol $\vec{f}$.

With this form for the r_j parameters, the likelihood function for model M_m can be calculated using equation (2), giving

$$p(D \mid \omega, \phi, A, \vec{f}, M_m) = K\left[\prod_{i=1}^{Z} mAf_j(t_i)^{n_i}\right] \exp\left[-\sum_{i=1}^{Z} mAf_j(t_i)\Delta t_i\right]$$

$$= Km^N A^N \left[\prod_{j=1}^{m} f_j^{\,n_j}\right] \exp\left[-mA\sum_{j=1}^{m} f_j\tau_j(\omega,\phi)\right]. \tag{10}$$

In this equation $n_j = n_j(\omega,\phi)$ is the total number of events in model bin j (the sum of the n_i which fall in model bin j), and τ_j is the total sample integration time for model bin j (the sum of the durations of all sample bins that fall in bin j).

The $\tau_j(\omega,\phi)$ depend on ω and ϕ (and m), thus the exponential term depends on all of the model parameters. But in many cases this term will be essentially independent of all parameters except A, as we now show.

Unless the "off" times are concentrated in particular bins, we expect the integration time per bin to be approximately T/m. Thus we write the sum in the exponential as follows;

$$mA\sum_{j=1}^{m} f_j\tau_j = AT\sum_{j=1}^{m} f_j\frac{\tau_j}{T/m}$$

$$= AT\sum_{j=1}^{m} f_j s_j, \tag{11}$$

where we have defined the bin time factors $s_j(\omega,\phi)$ by

$$s_j(\omega,\phi) = \frac{\tau_j(\omega,\phi)}{T/m}. \tag{12}$$

Like the τ_j, the s_j are not new parameters; they are numbers that are determined by the data and the model parameters ω, ϕ, and m.

If the observing interval, T, is a contiguous interval containing an integral number of periods, then $\tau_j = T/m$, so $s_j = 1$ for all j. More generally, the observing interval will not be an integral number of periods, and may have gaps, so the s_j will differ from unity. But as long as the number of periods in the observing interval is large, and as long as the gaps are not somehow concentrated in certain bins, the s_j will be very close to unity, and to a good approximation the sum in equation (11) will be equal to $AT\sum_j f_j = AT$. Then the likelihood function is well approximated by,

$$P(D \mid \omega, \phi, A, \vec{f}, M_m) = K(mA)^N e^{-AT}\left[\prod_{j=1}^{m} f_j^{\,n_j}\right]. \tag{13}$$

We have now derived the likelihood functions for the constant and periodic models (equations (4) & (13)) for a binned time series. Comparing these with the expressions given in GL2 (see equations (4.1) and (4.11)), we find them to be identical except the factor Δt^N has been replaced by the factor K defined in equation (3). Both of these factors cancel out in all subsequent calculations and are thus of no importance. Therefore the final expressions given in GL2 for the odds ratio, the probability for the frequency, and the shape and uncertainty of the light-curve, are all valid for binned time series, provided the bin durations are significantly smaller than the anticipated timescale of variation of the signal.

Note that if the observing duration does not contain a large number of periods, or if there is significant dead time, the s_j may depart significantly from unity. Gregory and Loredo (GL2, Appendix B) considered this case, which leads to the following likelihood function,

$$P(D \mid \omega, \phi, A, \vec{f}, M_m) = \Delta t^N (mA)^N e^{-AT} S(\omega, \phi) \left[\prod_{j=1}^{m} f_j^{n_j} \right], \tag{14}$$

where the binning factor, $S(\omega, \phi)$, is given by

$$S(\omega, \phi) = \prod_{j=1}^{m} s_j^{-n_j}. \tag{15}$$

The s_j are completely determined by the data and the model parameters ω, ϕ and m.

3. NONPERIODIC ANALYSIS

In some situations it is important to determine if a significant *nonperiodic* variation has occurred in the signal producing a time series. The stepwise model, used for the periodic class of models, can just as easily represent a nonperiodic variable lightcurve. Again we assume no prior knowledge about the shape of this variation. Therefore, our calculations can also be used to compare a constant rate model to a variable model, or to compare both constant and variable nonperiodic alternatives to periodic models. All that is necessary is to set the period, $P = 2\pi/\omega$, equal to the total duration of the observations (including any gaps). A discussion of this approach was presented in GL2 (Appendix C). Here we illustrate this capability with two sets of simulated data.

For both simulations the time series consisted of a set of binned arrival times with no data gaps present. The data were simulated for a stepwise lightcurve consisting of 25 bins. In the first case the signal consisted of 26 events occupying bin number 12. To this was added a constant background of 214 events distributed amongst all 25 bins with equal probability. The odds ratio was calculated from,

$$O_{m1} = \frac{1}{2\pi\nu} \binom{N+m-1}{N}^{-1} \int_0^{2\pi} d\phi \, S(\phi) m^N \prod_{j=1}^{m} n_j(\phi)!, \tag{16}$$

as a function of the trial number of bins for the range $m = 2$ to 40. $S(\phi)$ is given by equation (15) and ν is the number of trial choices of m, which in this case equaled 39. This is shown plotted in Figure 1(a). The peak odds ratio, which occurs at $m = 12$, is only 0.18. The odds ratio, $O_{\rm var}$, in favor of the nonperiodic variable class of models over the constant model, is given by,

$$O_{\rm var} = \sum_{m=2}^{m_{\rm max}} O_{m1}. \tag{17}$$

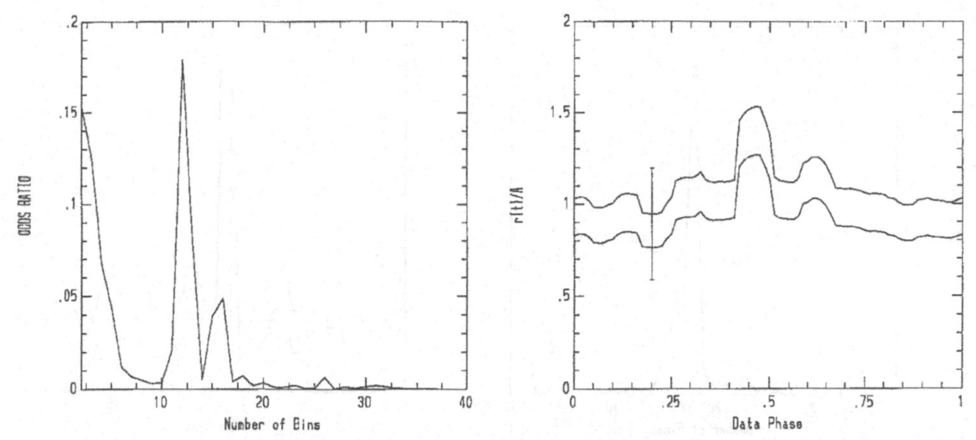

Figure 1. (*a*) Odds ratio, O_{m1}, versus m for the first simulation of a stepwise rate function with 25 bins. The ratio of signal event rate to background rate was 3.0. (*b*) Shape estimate, $r(t)/A$, for the first simulation. Solid curves show ± 1 standard deviation estimates. The error bar shows $\pm\sqrt{N}$ for a typical bin.

Note that O_{var} can exceed unity even if no single variable model is more probable than the constant model. In this particular case $O_{\text{var}} = 0.82$, indicating that the constant model is approximately 1.2 times more probable than the variable model. Thus for this level of signal to background no significant signal is indicated in the data set analyzed. Figure 1(*b*) shows a plot of the Bayesian mean lightcurve marginalized over m according to equations (7.11) and (B.7–9) in GL2. Since $O_{\text{var}} < 1$, we marginalized over the constant model, $m = 1$, as well. We calculated the mean and standard deviation for $r(t)/A$ at 80 times, and plotted the two resulting (mean$\pm$standard deviation) curves as solid lines together with a typical $\sqrt{N}$ error bar. Without prior knowledge of the shape and location of the signal it would be very difficult to persuade a seasoned experimenter that a real signal was present, which is consistent with the result from the odds ratio. In this particular simulation the ratio of the signal event rate to background rate was 3.0.

The second simulation was identical to the first, only this time the signal consisted of 30 events and the background 196 events. There is now clear evidence for a signal as the odds ratio, which is plotted in Figure 2(*a*), reaches a level of 17 for a value of $m = 22$. In this case $O_{\text{var}} = 30.8$. This conclusion is amply supported by the estimated lightcurve which is shown in Figure 2(*b*). The true underlying $r(t)/A$ is shown by the dashed curve. Note that the peak in the odds ratio occurs at $m = 22$, not at the true value of 25 (though the 25 bin model is itself also more probable than a constant model). The data, while sufficiently numerous to indicate a varying rate, are too sparse to fully resolve the signal. In this simulation the ratio of the signal event rate to the background rate was 3.8.

4. SUMMARY

In this paper we have shown that the new Bayesian method for detecting an unknown periodic signal, proposed in GL1 & GL2, is directly applicable to binned time series consisting of samples obtained at irregular intervals in some coordinate, and containing an

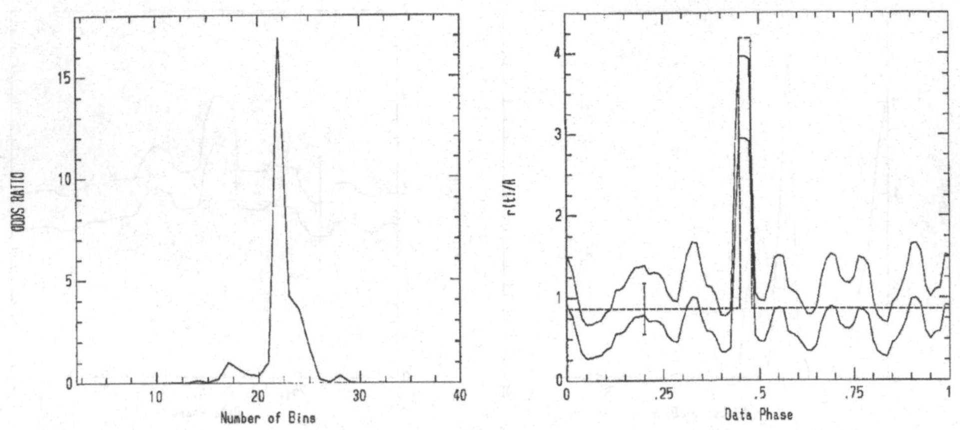

Figure 2. (*a*) Odds ratio, O_{m1}, versus m for the second simulation of a stepwise rate function with 25 bins. The ratio of signal event rate to background rate was 3.8. (*b*) Shape estimate, $r(t)/A$, for the second simulation. Solid curves show ± 1 standard deviation estimates. The error bar shows $\pm\sqrt{N}$ for a typical bin.

arbitrary number of events in each sample bin, as long as the sample bin size is small compared to the anticipated timescale of variation of the signal.

The stepwise model, used for the periodic class of models in our method, can just as easily represent a nonperiodic variable lightcurve for which we have no prior knowledge about its shape. Therefore, our method can also be used to compare a constant rate model to a variable model, or to compare both constant and variable nonperiodic alternatives to periodic models. The method is not restricted to time series, and may prove to be more generally useful for inferring the shape of an unknown function from discrete samples.

ACKNOWLEDGMENTS. This research was supported in part by a grant from the Canadian Natural Sciences and Engineering Research Council at the University of British Columbia, and by a NASA GRO Fellowship (NAG5–1758), NASA grant NAGW–666, and NSF grants AST–87–14475 and AST–89–13112 at Cornell University.

REFERENCES

Gregory, P.C., and Loredo, T. J. 1992a, 'A Bayesian Method For The Detection Of A Periodic Signal Of Unkown Shape And Period', in C. R. Smith, G. J. Erickson, & P.O. Neudorfer (ed.), *Maximum Entropy and Bayesian Methods, Seattle, 1991*, Kluwer, Dordrecht, in press.

Gregory, P.C., and T.J. Loredo 1992b, 'A New Method For The Detection Of A Periodic Signal Of Unknown Shape And Period', *Ap. J.*, in press.

NOISE IN THE UNDERDETERMINED PROBLEM

W.T. Grandy, Jr.
Department of Physics and Astronomy
University of Wyoming
Laramie, Wyoming 82071 U.S.A.

ABSTRACT. Despite considerable progress in, and development of, maximum entropy and Bayesian techniques, a complete understanding of the process combining noise with incomplete information remains elusive. The underdetermined problem with noisy data is here re-examined, and various results are discussed, compared, and criticized. Specific examples suggesting a resolution are presented.

1. Introduction

A large number of problems in data analysis can be formulated as a discrete generalized inverse problem of the form

$$d_i = \sum_{j=1}^{n} K_{ij} f_j + e_i\,, \qquad i = 1, \ldots, m\,, \tag{1}$$

where the data set $D = \{d_i\}$ is given, along with associated errors in the form of a linear noise vector $e = \{e_1, \cdots, e_m\}$. The set $\{f_j\}$ is a model, or distribution, or set of parameters, to be determined, while the matrix K is a known kernel — it can be an instrument response function, an aperture point-spread function, or even a set of function values $k_j(x_i)$. In many cases K^{-1} does not exist; but, whether it does or not, the treatment of the problem very much depends on the relation of m to n.

When the problem is overdetermined, $m > n$, the complete and correct Bayes-ian solution has been known since the work of Harold Jeffreys (1939). Namely, the posterior probability P for a set $\{f\}$, given D, e, and any other prior information I, is

$$P(f|D,e,I) \propto P(f,e|I)\; L(D|f,e,I)\,, \tag{2}$$

up to normalization. The most common scenario is that in which all that is known of the noise is its mean-square power level σ, the same for all data points. In that event, maximization of the entropy subject to this constraint leads to an independent Gaussian white-noise model,

$$P(e_1 \cdots e_m|I) = \prod_{i=1}^{m} \frac{1}{\sqrt{2\pi\sigma^2}} e^{-e_i^2/2\sigma^2}\,, \tag{3}$$

A. Mohammad-Djafari and G. Demoments (eds.), Maximum Entropy and Bayesian Methods, 233–241.

and substitution from Eq.(1) provides the likelihood function L. If this noise model is adopted, which is done from here on, then ascertaining the prior probabilities $P(f,\sigma|I)$ is the only remaining, and often contentious, formal task.

One example of this overdetermined problem is that of fitting a number of functions linearly to a set of data points, discussed by Gull (1988). Another arises in the familiar analysis of time series, and has been treated in some detail by Jaynes (1987) and Bretthorst (1988). In this latter scenario the likelihood function takes the form

$$L(R) \propto \frac{1}{\sigma^m} \exp\left\{-\frac{1}{2\sigma^2}\sum_{i=1}^{m}\left[d_i - f_i(R)\right]^2\right\}, \tag{4}$$

where $f_t(R)$ represents a model for the signal producing the data, and is generally a non-linear function of a set of parameters R. A simple example is a single-frequency signal $f(t) = A_1\cos(\omega t) + A_2\sin(\omega t)$, where the amplitudes and frequency are to be estimated. The 'art of windowing' here becomes the 'art of modelling'. One could actually attempt to fit the data exactly by employing a complete (in some sense) set of functions and the minimum mean-square error criterion, but usually there is sufficient prior information indicating a more economical choice for the signal model.

What happens in this Bayesian formulation when the noise level is very low? Clearly, as $\sigma \to 0$ in Eq.(4) the right-hand side becomes an m-dimensional δ-function, which is equivalent to the original set of equations (1) with all the $e_i = 0$. That is, the time series analysis becomes the problem of *fitting* a model to a set of precisely-known data points, according to some criterion. There is thus a well-defined limiting procedure relating the various phases of the problem over the full range of noise values.

In the underdetermined problem, $m < n$, the optimum approach is well-understood when no noise is present — it is the principle of maximum entropy (PME). The PME was designed to address the problem of incomplete information only, and undoubtedly provides the best solution possible in such cases. How, then, should one proceed in the underdetermined case in the presence of noise? As repeatedly emphasized by Jaynes, in the PME the data are presumed to be exact, and the Lagrange multipliers have no independent existence until they are created out of the data, so to speak. That is, the maxent distribution is a predictive distribution, not a sampling distribution.

Present thoughts on this question were stimulated by another remark by Jaynes (1987), in which he suggests that the time series solution described by Eq.(4) above should, as the noise goes to zero, "reduce to something like the original Burg pure maximum entropy solution" (although he notes they will not be exactly the same because we are presuming different types of data in the two cases). They cannot be related at all, of course, for one is greatly overdetermined and the other badly underdetermined. [We recall that the Burg problem refers to analysis of a *noiseless* time series when the autocovariances are known for only a relatively small number of lags (Burg, 1975).] But then we are left to wonder about what noisy scenario it *is* that has the Burg problem as its noiseless limit — and, more generally, what is the proper procedure for studying underdetermined problems in the presence of noise? Gull (1989) and Skilling (1989) have provided one answer to this general question by means of their MEMSYS algorithm, aimed primarily at image reconstruction. While this is clearly a workable procedure, we should like to seek a solution on a somewhat more transparent level.

Just as he stimulated the question, Jaynes (1982) has also recognized that the sanctity of the Lagrange multipliers may not be completely inviolate: "According to the principles of probability theory, however, this uncertainty ought to be allowed for in a quite different way; one should calculate the joint posterior distribution of the λ's ... ." Our own introspection here leads to the conclusion that errors in the data simply lead to predictable errors in the Lagrange multipliers. This can be seen both qualitatively and quantitatively. For example, in the canonical ensemble of statistical mechanics the probability for a system energy level is proportional to the Boltzmann factor $e^{-\beta E}$, where $\beta^{-1} \sim T$, the absolute temperature, and is a Lagrange multiplier. The energy is known to fluctuate, so that conservation laws alone require β to do so.

But the phenomenon can be made quantitative as well, by considering the noiseless maxent scenario

$$d_i = \sum_{i=1}^{n} K_{ij}\, f_j\,, \qquad i = 1, \ldots, m < n\,, \tag{5a}$$

with solution

$$f_j\{\lambda\} = \frac{1}{Z} \exp\Big\{\sum_i \lambda_i K_{ij}\Big\}, \qquad Z = \sum_{j=1}^{n} \exp\Big\{\sum_i \lambda_i K_{ij}\Big\}. \tag{5b}$$

Now we ask how the set $\{\lambda_i\}$ varies as the data are varied, and a short calculation yields

$$\delta d_i = \sum_k C_{ik} \frac{\partial \lambda_k}{\partial d_i} \delta d_i\,,$$
$$C_{ik}(\lambda) \equiv \sum_j K_{ij} K_{kj} f_j - \sum_q K_{kq} f_q \sum_\ell K_{i\ell} f_\ell\,. \tag{6}$$

Hence, with $\Delta_k \equiv (\partial \lambda_k / \partial d_i)\delta d_i$, variations in the Lagrange multipliers are given by the matrix inversion $\boldsymbol{\Delta\lambda} = C^{-1} \cdot \boldsymbol{\delta d}$. Perhaps it may be useful to construct a Bayesian estimation of the set $\{\lambda_i\}$ from the data, though a *caveat* is in order: we are not at all questioning the veracity of the data, only their accuracy; we are not, therefore, asking for the probability that the data are in some sense 'true'; there *is* a difference.

2. Noisy Maximum Entropy

If linear noise is added to Eq.(5a), and we consider the effective 'data' $d_i' \equiv d_i - e_i$ to be fixed, then the same formal solution described by Eq.(5b) is valid. Hence, with a common Gaussian noise level σ, the likelihood function can be written

$$L(D|\boldsymbol{\lambda}, \sigma, I) \propto \exp\left\{ -\frac{1}{2\sigma^2} \sum_{i=1}^{m} \Big[d_i - \sum_{j=1}^{n} K_{ij}\, f_j(\boldsymbol{\lambda}) \Big]^2 \right\}, \tag{7}$$

with $f_j(\boldsymbol{\lambda})$ given by Eq.(5b). Note that in arriving at Eq.(7) we have employed the usual prescription for obtaining the Lagrange multipliers:

$$d_i = \frac{\partial}{\partial \lambda_i} \ln Z(\boldsymbol{\lambda}) = \sum_{j=1}^{n} K_{ij}\, f_j(\boldsymbol{\lambda})\,. \tag{8}$$

The unnormalized posterior probability for $\boldsymbol{\lambda}$, via Bayes' theorem, is then

$$P(\boldsymbol{\lambda}|D,\sigma,I) \propto P(\boldsymbol{\lambda},\sigma|I)\, L(D|\boldsymbol{\lambda},\sigma,I). \tag{9}$$

As always, the choice of priors requires some care, though we shall adopt the Jeffreys prior $1/\sigma$ for the scale parameter σ. When no data are present, one anticipates the Lagrange multipliers to have zero expectation values. Moreover, even in the simplest problems it is possible for the set $\boldsymbol{\lambda}$ to range over the real line, and convergence difficulties are to be expected. With these points in mind, and the observation that we can always arrange things so that the λ_i are dimensionless, we adopt the prior

$$P(\boldsymbol{\lambda}|I) = \frac{e^{-\boldsymbol{\lambda}^2}}{\pi^{m/2}}\,, \tag{10}$$

where $\boldsymbol{\lambda}^2 = \lambda_1^2 + \cdots + \lambda_m^2$. Then, the joint posterior probability for the set $\boldsymbol{\lambda}$ is

$$P(\boldsymbol{\lambda}|D,\sigma,I) \propto \frac{e^{-\boldsymbol{\lambda}^2}}{\pi^{m/2}\sigma^{m+1}} \exp\left\{-\frac{1}{2\sigma^2}\sum_{i=1}^{m}\left[d_i - \sum_{j=1}^{n} K_{ij}\, f_j(\boldsymbol{\lambda})\right]^2\right\}. \tag{11}$$

The Brandeis Dice Problem

As a first, and perhaps the simplest, example we re-examine the Brandeis dice problem (Jaynes, 1963) when there is some uncertainty in the data. A die is rolled 1 000 times, say, and the number of spots 'up' recorded at each throw. In the original scenario the single (exact) datum $d = 4.5$ is obtained, rather than the value 3.5 to be expected from an honest die. The single Lagrange multiplier is found to be $\lambda_0 = 0.371\,049$ and

$$\frac{\partial}{\partial\lambda_0}\ln Z_0 = \frac{1 - 7e^{6\lambda_0} + 5e^{7\lambda_0}}{(1-e^{\lambda_0})(1-e^{6\lambda_0})}. \tag{12}$$

Note, however, that λ can range over the entire real line:

$$\lambda = \begin{cases} -\infty, & P_1 = 1;\ P_i = 0,\ i \neq 1 \\ 0, & \text{uniform distribution} \\ \infty, & P_6 = 1;\ P_i = 0,\ i \neq 6 \end{cases}. \tag{13}$$

We here consider the value of d to be a bit uncertain — perhaps because of a lack of confidence in the recorder. Let $x \equiv e^{\lambda}$, so that $x_0 \equiv e^{\lambda_0} \simeq 1.449\,25$, and

$$P(\lambda|D,\sigma,I) \propto \frac{e^{-\lambda^2}}{\sqrt{\pi\sigma^2}} e^{-\frac{1}{2\sigma^2}[F(x)(x-x_0)]^2}\,, \tag{14a}$$

with

$$F(x) \equiv 3\frac{x^4 + ax^3 + bx^2 + cx + d}{x^5 + x^4 + x^3 + x^2 + x + 1}\,,$$

$$a = x_0 - \tfrac{1}{3}\,, \quad b = ax_0 - \tfrac{1}{3}\,, \quad c = bx_0 - 1\,, \quad d = cx_0 - \tfrac{5}{3}\,. \tag{14b}$$

Neither of the polynomials defining $F(x)$ has zeros in $\mathrm{Re}\, x \geq 0$; at $x = x_0$, $F(x_0) \simeq 3.17153$, and the single maximum is $F(0.475) \simeq 5.47$.

The simplest way to study $P(\lambda|D,\sigma,I)$ is numerically, for various noise levels. Figure 1 illustrates the result of plotting the normalized probability distribution for $\sigma = 0.01$. The expected value of λ is evaluated over the range $\sigma = \{.01, 1\}$ and plotted in Figure 2. The smooth convergence of $\langle\lambda\rangle$ to λ_0 is evident.

It is also interesting to develop a sampling distribution by adding random noise a number of times to the noiseless expression (5a), and each time re-solving the noiseless equations for that datum perturbed from $d = 4.5$. Figure 3 illustrates the result of doing this for 100 samples, and the distribution tends to mimic P, though it is considerably broader.

We conclude that, in this simplest of problems, the notion of a distribution for the Lagrange multiplier leads to completely consistent results. One can also marginalize the noise, of course, but the result is a rather large error bar and not particularly illuminating. The marginalized distribution is

$$P(\lambda|D,I) \propto \frac{e^{-\lambda^2}}{(e^{\lambda} - e^{\lambda_0})F(e^{\lambda})}, \tag{15}$$

and the expectation value is found from the Cauchy principal value calculation to be $\langle\lambda\rangle \simeq 1.8412$. Thus, the probability is weighted rather strongly toward the high end of the die, so one would not go terribly wrong. But, even in this simplest of problems, the calculation tends to become tedious.

3. The Burg Problem

It is most useful for present purposes to follow the (noiseless) analysis of Jaynes (1982), in which the data pertaining to a time series $Y = (y_0, \ldots, y_T)$ of length T are given for a few $(m+1)$ values of the autocovariance,

$$a_k = \frac{1}{T+1} \sum_{t=0}^{T-k} y_\ell^* \, y_{\ell+k}\,, \qquad 0 \leq k \leq m. \tag{16}$$

Let $z \equiv e^{ix}$, and define

$$g(z) \equiv \sum_{k=-m}^{m} \lambda_k \, z^k. \tag{17}$$

Then, for $m \ll T$, the Lagrange multipliers are determined from the constraint equations:

$$\begin{aligned} a_k &= \frac{1}{2\pi} \int_0^{2\pi} \frac{e^{ikx}}{g(e^{ix})}\, dx\,, \quad -m \leq k \leq m\,, \\ &= \frac{1}{2\pi} \int_{-\pi}^{\pi} \frac{e^{-ik\omega}}{g(e^{-i\omega})}\, d\omega\,, \end{aligned} \tag{18}$$

and the power spectral density is

$$p(\omega) = \frac{1}{\sum_k \lambda_k e^{-ik\omega}}\,, \qquad |\omega| < \pi. \tag{19}$$

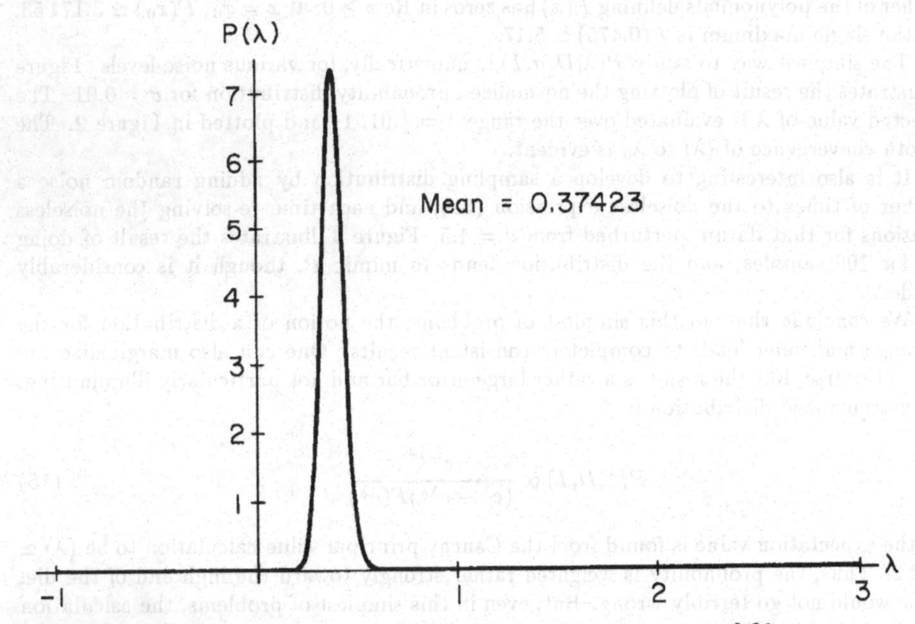

Fig. 1. The normalized posterior probability distribution of Eqs.(14) for $\sigma = 0.01$.

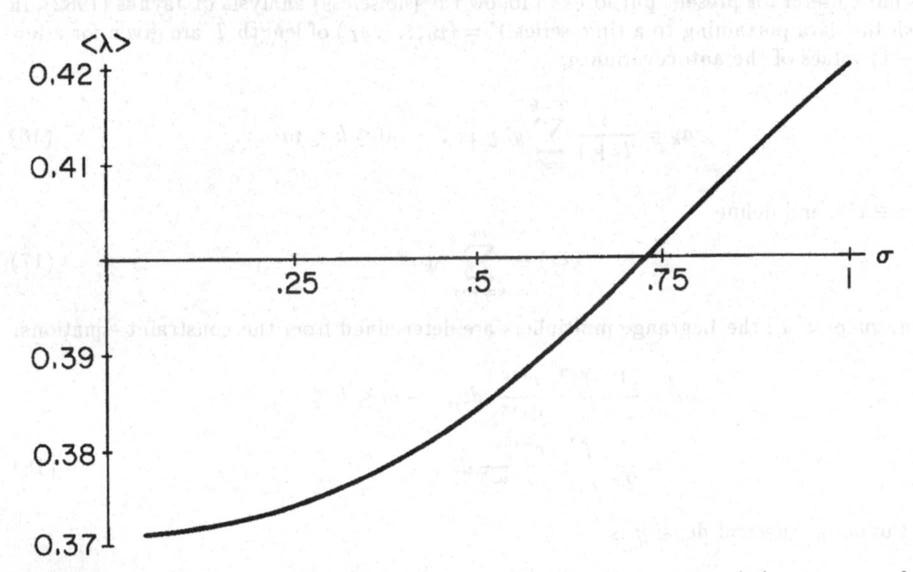

Fig. 2. Expectation values of the Lagrange multiplier in the distribution of Eqs.(14) over a range of noise levels.

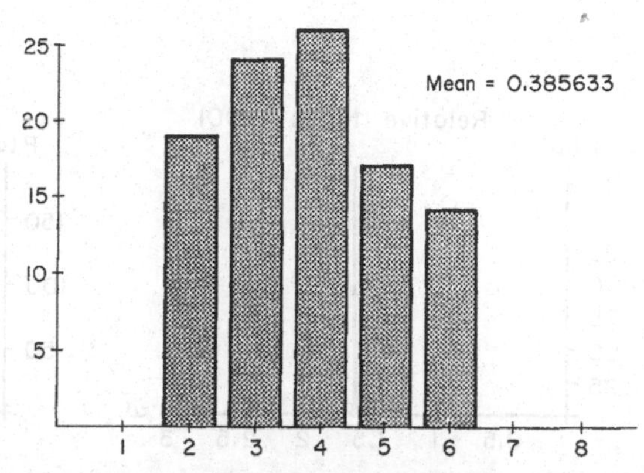

Fig. 3. A sampling approximation to the probability distribution of Eqs.(14).

In practice, the set $\boldsymbol{\lambda} \equiv \{\lambda_k\}$ is actually found from an equivalent set of linear equations, though in the true noiseless case the covariance matrix of the data can be ill-conditioned.

Proceeding in the same manner as above, the posterior probability for $\boldsymbol{\lambda}$ is

$$P(\boldsymbol{\lambda}|D,\sigma,I) \propto \frac{e^{-\boldsymbol{\lambda}^2}}{\pi^{m/2}\sigma^{m+1}} \exp\left\{-\frac{1}{2\sigma^2}\sum_{k=-m}^{m}\left[a_k + F_k(\boldsymbol{\lambda})\right]^2\right\}, \tag{20a}$$

where F_k is given by an integral around the unit circle:

$$F_k(\boldsymbol{\lambda}) = \frac{1}{2\pi i}\oint \frac{z^{-k+1}}{\sum_\ell \lambda_\ell z^{-\ell}}\,dz\,. \tag{20b}$$

In general there are m poles inside the unit circle, and therefore possibly m sharp spectral lines in $p(\omega)$. Thus, F_k is just the sum of residues of the integrand, and hence a *known* function of the Lagrange multipliers. The posterior probability $P(\boldsymbol{\lambda})$ is completely determined in principle, and Eqs.(20) provide the answer to the question raised earlier: it is this distribution for which the Burg solution provides the zero-noise limit.

By way of illustration, consider a pure sinusoidal signal $2\sin(\omega_0 t)$. If ω is not very close to zero and the time series is reasonably long, the autocovariances for lag k are $\simeq 2\cos(\omega_0 k)$. Let $\omega_0 = 1.6$ rad/sec and sample just three lags. In the event that there is absolutely no error, the problem of determining $\{\lambda_0, \lambda_1, \lambda_2\}$ is ill-conditioned, for $p(\omega)$ returns a δ-function, as we might expect. But even a small amount of noise (or computer round-off error!) allows the necessary matrix inversion, so we add a small random term to the right-hand side of Eq.(16). Indeed, a sampling approach is the most efficient procedure at this point, and Figure 4 exhibits the resulting power spectrum for four different noise levels. We see that significant noise moves the spectral peak away from the correct value of $\omega = 1.6$, as we should expect. The Lagrange multipliers tend to decrease as the noise increases. If we write the noise term as $e = A * (\text{Random}[\;] - \frac{1}{2})$, then

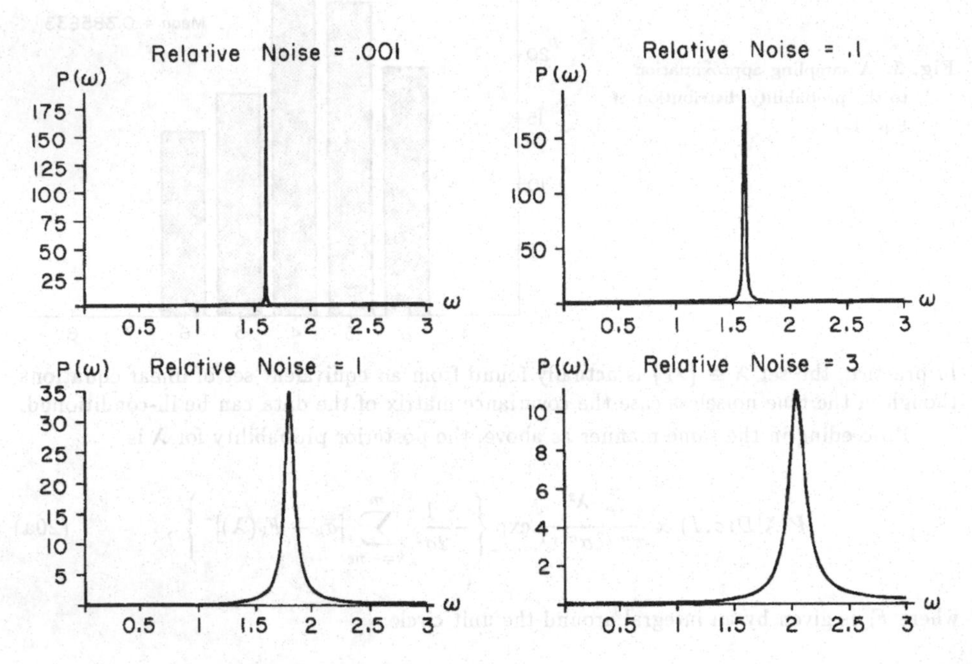

Fig. 4. Power spectra for a 3-lag example of the Burg problem for four values of the relative noise level.

A	$\{\lambda_0, \quad \lambda_1, \quad \lambda_2\}$
.001	{1039, 60.27, 518.9}
.1	{24.95, 1.787, 12.44}
1	{5.805, 2.462, 2.597}
3	{1.901, −.8904, .7931}

CONCLUSIONS

Even in this simple 3-lag problem one finds direct evaluation of the posterior probability, and subsequent computation of expectation values, to be quite tedious. This would certainly not be the recommended procedure for practical problems involving many Lagrange multipliers — at this time the method of a single Lagrange multiplier introduced years ago by Gull and Daniell (1978), and since exploited by Gull and Skilling, may still be the most efficient tool in practice. But the point to be made here is one of principle. The concept of noisy maxent appears to be consistent with accepted Bayesian precepts, and the resulting likelihood functions have perfectly well-defined limits as the noise vanishes — namely, the equations of the original PME.

REFERENCES

Bretthorst, G.L.: 1988, *Bayesian Spectrum Analysis and Parameter Estimation*, Springer-Verlag, Berlin.

Burg, J.P.: 1975, 'Maximum Entropy Spectral Analysis', *Ph.D. thesis*, Stanford University (unpublished).

Gull, S.F.: 1988, 'Bayesian Inductive Inference and Maximum Entropy', in *G.J. Erickson and C.R. Smith (eds.)*, Maximum Entropy and Bayesian Methods in Science and Engineering, Vol. 1, Kluwer, Dordrecht.

Gull, S.F.: 1989, 'Developments in Maximum Entropy and Data Analysis', in *J. Skilling, (ed.)*, Maximum Entropy and Bayesian Methods, Kluwer, Dordrecht.

Gull, S.F., and G.J. Daniell: 1978, 'Image Reconstruction from Incomplete and Noisy Data', *Nature*, **272**, 686.

Jaynes, E.T.: 1963, 'Information Theory and Statistical Mechanics', in *K. Ford (ed.)*, 1962 Brandeis Summer Institute in Theoretical Physics, Benjamin, New York.

Jaynes, E.T.: 1982, 'On the Rationale of Maximum-Entropy Methods', *Proc. IEEE* **70**, 939.

Jaynes, E.T.: 1987, 'Bayesian Spectrum and Chirp Analysis', *C.R. Smith and G.J. Erickson, (eds.)*, Maximum-Entropy and Bayesian Spectral Analysis and Estimation Problems, Reidel, Dordrecht.

Jeffreys, H.: 1939, *Theory of Probability*, Clarendon Press, Oxford.

Skilling, J.: 1989, 'Classic Maximum Entropy', in *J. Skilling, (ed.)*, Maximum Entropy and Bayesian Methods, Kluwer, Dordrecht.

Chapter 4

Inverse Problems

DECONVOLUTION WITH AN UNKNOWN POINT-SPREAD FUNCTION

M.K. Charter
Mullard Radio Astronomy Observatory
Cavendish Laboratory
Madingley Road
Cambridge CB3 0HE, U.K.

ABSTRACT. Many data analysis problems take the form of a deconvolution, and one of the common difficulties is that the point-spread function (p.s.f.) is not completely specified. Indeed, it is often obtained from measurements which are very similar to the test data themselves. The correct probabilistic approach is to form the full joint posterior for the p.s.f. and the object, and then to obtain posteriors for features of interest in the object as marginals over this joint posterior, i.e., to integrate over uncertainties in the p.s.f. If the joint posterior is sharply peaked for some particular p.s.f., one may approximate this integral by choosing the p.s.f. at its modal value. If, however, the posterior is not sharply peaked, then this procedure may give misleading error estimates. Examples of this are shown, drawn from the analysis of drug absorption in man.

1. Introduction

Many data analysis problems take the form of a deconvolution. The data consist of the object of interest, such as a line spectrum or scene, convolved with an instrumental response or point-spread function (p.s.f.), together with some measurement error or noise. Classic maximum entropy analysis (Skilling, 1989; Gull, 1989) assumes this p.s.f. to be known exactly, whereas this is often not the case. In some cases the functional form of the p.s.f., e.g., Gaussian or exponential, is known *a priori*, but it contains a parameter of unknown value, such as a dispersion or decay constant. In other cases the p.s.f. is determined from the same type of measurements as the test data, such as the spectrum of an isolated singlet line, or the blurred image of a point source. Such measurements will be, to a greater or lesser extent, noisy and incomplete so that the p.s.f. is not completely specified, but has some uncertainty associated with it.

2. Incorporating uncertainties about the p.s.f.

A full probabilistic treatment requires that uncertainties about the p.s.f. should be incorporated into the analysis. The full joint posterior $\Pr(f, \boldsymbol{\theta} \mid \mathbf{D}, H)$ for both the object of interest f and the unknown p.s.f. parameters $\boldsymbol{\theta}$ should be formed, where $\mathbf{D}$ are the data, and H is the background information about the experiment (such as the functional form of the p.s.f., the error structure in $\mathbf{D}$, etc.). The required inference $\Pr(f \mid \mathbf{D}, H)$ is then

A. Mohammad-Djafari and G. Demoments (eds.), Maximum Entropy and Bayesian Methods, 245–252.

obtained from

$$\Pr(f \mid \mathbf{D}, H) = \int \Pr(f, \boldsymbol{\theta} \mid \mathbf{D}, H)\, d\boldsymbol{\theta}, \tag{1}$$

i.e., as a marginal over this joint posterior. Expanding the integrand gives

$$\int \Pr(f, \boldsymbol{\theta} \mid \mathbf{D}, H)\, d\boldsymbol{\theta} = \int \Pr(f \mid \boldsymbol{\theta}, \mathbf{D}, H)\, \Pr(\boldsymbol{\theta} \mid \mathbf{D}, H)\, d\boldsymbol{\theta}$$

$$\propto \int \Pr(f \mid \boldsymbol{\theta}, \mathbf{D}, H)\, \Pr(\mathbf{D} \mid \boldsymbol{\theta}, H)\, \Pr(\boldsymbol{\theta} \mid H)\, d\boldsymbol{\theta}.$$

Hence if $\Pr(\mathbf{D} \mid \boldsymbol{\theta}, H)$, i.e., the evidence (Skilling, 1991), is sharply peaked around some value $\hat{\boldsymbol{\theta}}$ and the prior $\Pr(\boldsymbol{\theta} \mid H)$ is fairly flat in this region, then (1) will be well approximated by $\Pr(f \mid \hat{\boldsymbol{\theta}}, \mathbf{D}, H)$. This is, however, the only probabilistic justification for 'maximising the evidence', and if the marginalisation in (1) is not well approximated by such a procedure then misleadingly narrow posterior probability densities may result. Other, more *ad hoc* methods of choosing $\hat{\boldsymbol{\theta}}$ are also used. The actual method by which $\hat{\boldsymbol{\theta}}$ is obtained is unimportant in what follows; the essential point is that once it has been found, the p.s.f. is treated as known exactly in the rest of the analysis.

This is illustrated by the analysis of drug absorption data, which is a problem requiring the deconvolution of sparse and noisy data, often with little information about the p.s.f. Part of the development process of most drugs is to determine the proportion of an oral dose which is absorbed, i.e., reaches the bloodstream, and the rate $f(t)$ at which it does so. The aim of the analysis is therefore to reconstruct this input rate, from which one can obtain quantities such as the total amount of drug absorbed and the mean and median absorption times. This input rate can be viewed as a distribution of transit times of drug molecules, where the transit is from the point of dosing to the bloodstream, and so it corresponds to the ideal line spectrum or original scene in other deconvolution analyses. The test data are measurements of the drug's concentration in blood at various times after the oral doses were given, and correspond to the raw spectrum or blurred image. For many drugs the body is a sufficiently linear and time-invariant system that these concentrations can be calculated as the convolution of the required input rate with the p.s.f. or impulse response of the system. This impulse response can be obtained experimentally by putting an 'impulse' of drug directly into the bloodstream, i.e., giving a rapid intravenous dose, and measuring the concentration of drug in the blood at various times after dosing. Intravenous dosing is often undesirable for practical reasons, though, so in many cases such data are not available and there are no explicit measurements of the p.s.f.

In this example, the functional form of the impulse response function is assumed known, and the model described previously (Charter, 1992) is used. The approach can be extended, however, to the case where this is not so, by enlarging the hypothesis space still further to include other functions for the p.s.f., and calculating quantities of interest as marginals over the resulting expanded joint posterior density. The model used for the absorption or input kinetics has also been described previously (Charter, 1991), and the computations were performed using the `MemSys` quantified maximum entropy package (Gull and Skilling, 1991). The purpose here is to compare the inputs from two different oral formulations ('APAP' and 'NaAPAP') of the same drug, and the data were obtained after giving APAP and NaAPAP on separate occasions to the same volunteer. The experimental conditions were standardised as far as possible on each occasion, and the p.s.f. is assumed to be the

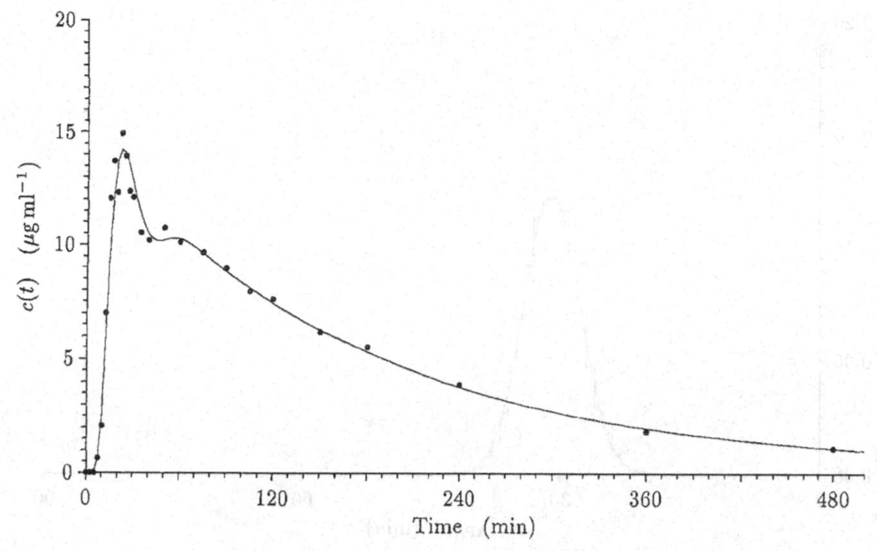

Fig. 1. Observed and predicted concentration of drug in blood after an oral dose of APAP.

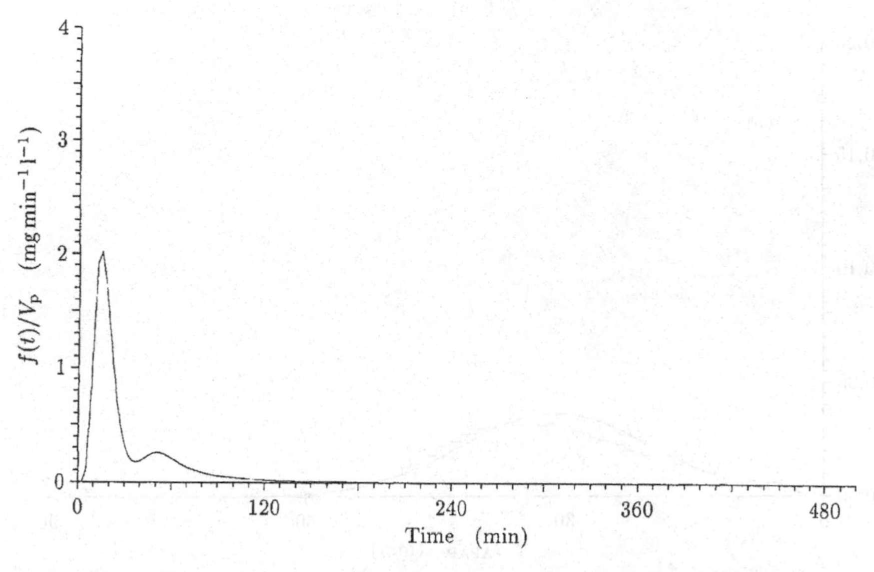

Fig. 2. Input rate $f(t)$ at the mode of the joint posterior reconstructed from the data shown in Figure 1.

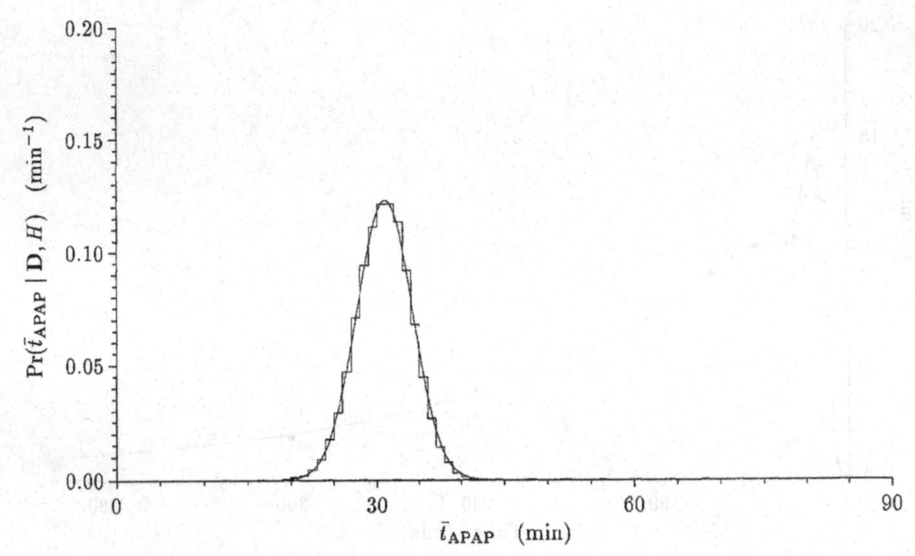

Fig. 3. Posterior probability density for the mean absorption time $\bar{t}_{\mathrm{APAP}}$ using 'best estimate' p.s.f.

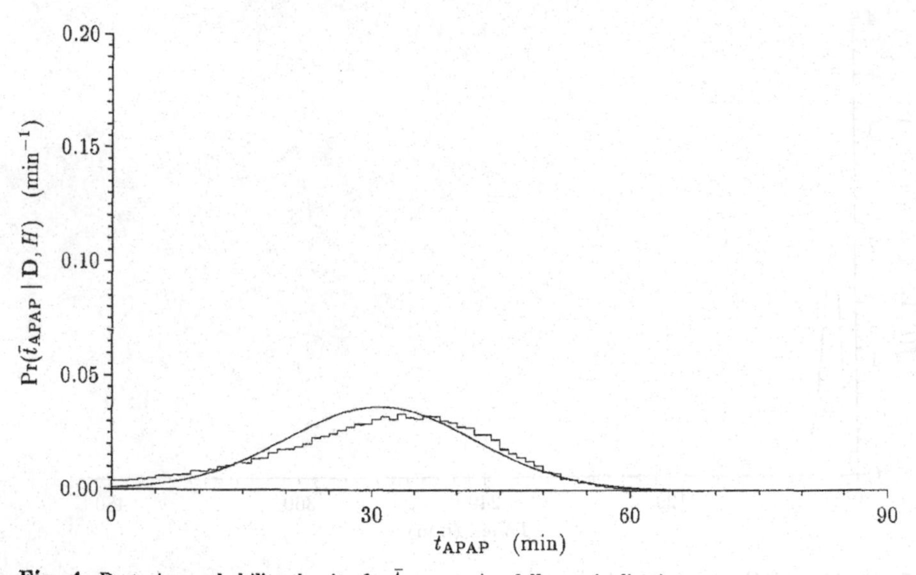

Fig. 4. Posterior probability density for $\bar{t}_{\mathrm{APAP}}$ using full marginalisation.

same for the two data sets. This is a common assumption in this type of experiment. The only information available about the p.s.f. comes from the two oral data sets; no intravenous data are used in the analysis. The data obtained after the APAP dose are shown in Figure 1, together with the concentration predicted from the input rate shown in Figure 2, which is taken from the mode of the joint posterior $\Pr(f_{\mathrm{APAP}}, f_{\mathrm{NaAPAP}}, \theta \mid \mathbf{D}, H)$. Because there are no explicit measurements of the p.s.f., $f(t)$ can be inferred only to within an unknown scaling factor. Dimensionally, this scaling factor is a volume: the constant of proportionality relating the (unknown) mass of drug reaching the bloodstream to the concentration measured. Hence the quantity plotted in Figure 2 is in fact the input rate per unit volume, or $f(t)/V_{\mathrm{p}}$, where V_{p} is the unknown volume. The data obtained after the NaAPAP dose are similar.

The posterior for the mean absorption time $\bar{t}_{\mathrm{APAP}}$ for the APAP dose, obtained by fixing the p.s.f. parameters at their modal values, is shown in Figure 3. The histograms in Figures 3 to 6 are each produced by taking about 40000 samples from the appropriate probability density, and the smooth curves are Gaussian approximations. The posterior shown in Figure 3 implies that $\bar{t}_{\mathrm{APAP}}$ is determined to within an error, i.e., the standard deviation of the approximating Gaussian, of about 3.3 minutes. By contrast, Figure 4 shows the posterior for the same quantity, obtained by marginalisation over the full joint posterior. This correctly incorporates the additional uncertainties in the p.s.f., and shows that $\bar{t}_{\mathrm{APAP}}$ is in fact much less accurately determined, with an error of about 11 minutes. The results for the NaAPAP dose are similar: with a fixed 'best estimate' p.s.f. the error for $\bar{t}_{\mathrm{NaAPAP}}$ is about 3.4 minutes, and with full marginalisation it is about 11 minutes. The posteriors for other quantities, such as the median absorption time, i.e., the time taken for half the absorbed fraction of the dose to reach the bloodstream, are similarly narrower when the p.s.f. is fixed at its 'best estimate' value.

3. Predicting well-determined quantities using the Convolution Theorem.

Although incorporation of uncertainties in the p.s.f. may widen the posteriors for features of interest in the object, some features may be well determined by the data even when those data contain very little information about the p.s.f. Such features can be predicted from the Convolution Theorem. Two examples of this are shown.

Ratio of total amounts

As mentioned earlier, the input rate is determined by the data only to within an unknown scaling factor, and hence the total amount is similarly indeterminate. Nonetheless, the ratio of total amounts for two different inputs may be well determined. This can be seen from simple application of the Convolution Theorem. If the concentration $c(t)$ of the drug in blood arising from an input $f(t)$ is

$$c(t) = R(t) * f(t)$$

where $R(t)$ is the p.s.f. and '$*$' denotes convolution, then taking Laplace transforms gives

$$\tilde{c}(s) = \tilde{R}(s)\,\tilde{f}(s) \tag{2}$$

where $\tilde{c}(s)$ is the Laplace transform of $c(t)$, and similarly for $\tilde{R}(s)$ and $\tilde{f}(s)$.

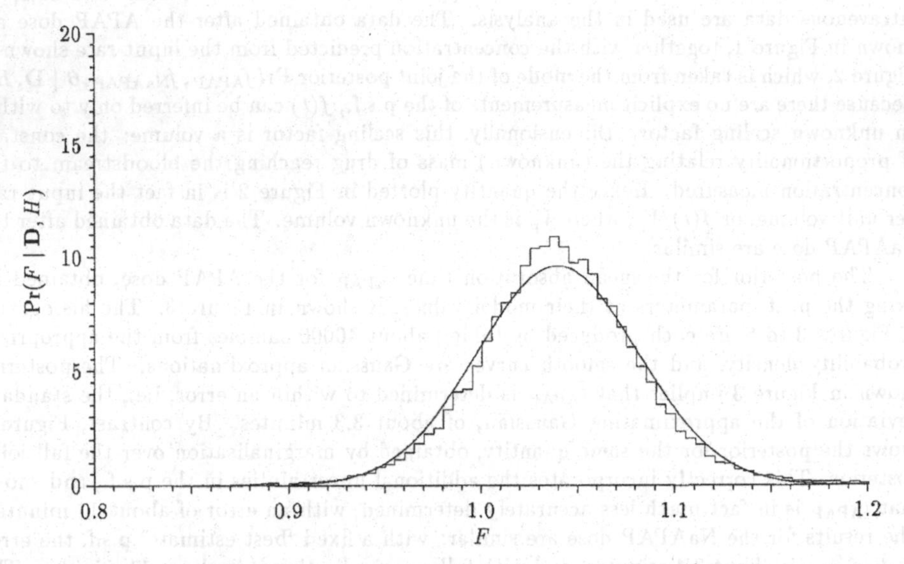

Fig. 5. Posterior probability density for the ratio F of total amounts NaAPAP/APAP using full marginalisation.

Consider two different inputs $f_1(t)$ and $f_2(t)$ of the same drug into the same individual, so that $R(t)$ can be assumed to be the same. These produce concentration functions $c_1(t)$ and $c_2(t)$ respectively, so that

$$\frac{\tilde{f}_1(s)}{\tilde{f}_2(s)} = \frac{\tilde{c}_1(s)}{\tilde{c}_2(s)}$$

and letting $s \to 0$ gives the well-known result that

$$\frac{\int_0^\infty f_1(t)\,dt}{\int_0^\infty f_2(t)\,dt} = F = \frac{\int_0^\infty c_1(t)\,dt}{\int_0^\infty c_2(t)\,dt}. \tag{3}$$

The right-hand side of (3) does not depend on $R(t)$, and so to the extent that the integrals $\int_0^\infty c_1(t)\,dt$ and $\int_0^\infty c_2(t)\,dt$ are well determined by the data [which are measurements of $c_1(t)$ and $c_2(t)$] the left hand side, i.e., the ratio F of total amounts in the inputs, is also well determined. This is shown in Figure 5, where the posterior density for F has a width of about 0.04.

Difference of Mean Absorption Times

Differentiating (2) with respect to s and dividing by $\tilde{c} = \tilde{R}\,\tilde{f}$ we obtain

$$\frac{1}{\tilde{c}}\frac{d\tilde{c}}{ds} = \frac{1}{\tilde{R}}\frac{d\tilde{R}}{ds} + \frac{1}{\tilde{f}}\frac{d\tilde{f}}{ds}.$$

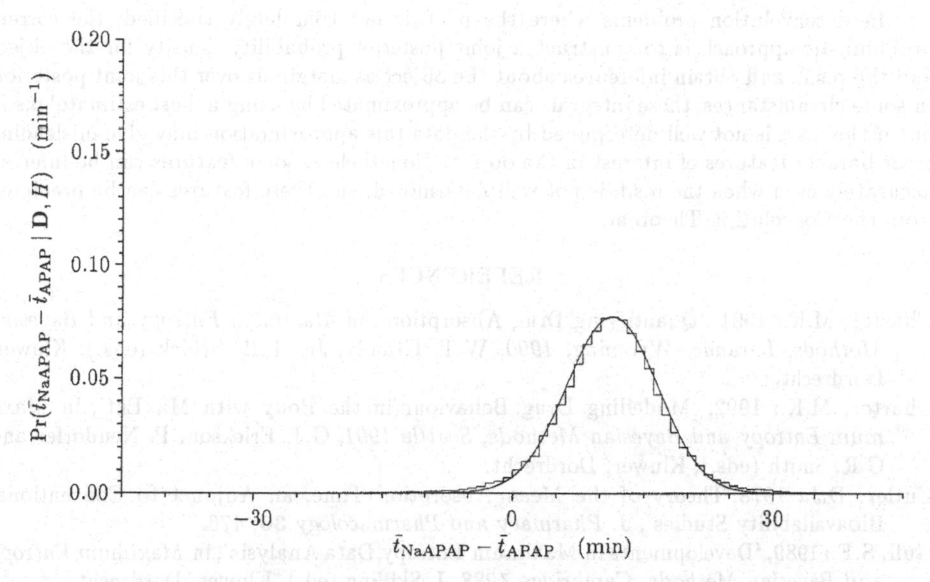

Fig. 6. Posterior probability density for the difference in mean absorption times $\bar{t}_{\mathrm{NaAPAP}} - \bar{t}_{\mathrm{APAP}}$ using full marginalisation.

Thus for two different input functions $f_1(t)$ and $f_2(t)$ as before,

$$\frac{1}{\tilde{f}_1}\frac{d\tilde{f}_1}{ds} - \frac{1}{\tilde{f}_2}\frac{d\tilde{f}_2}{ds} = \frac{1}{\tilde{c}_1}\frac{d\tilde{c}_1}{ds} - \frac{1}{\tilde{c}_2}\frac{d\tilde{c}_2}{ds}$$

and so, letting $s \to 0$,

$$\frac{\int_0^\infty t\, f_1(t)\, dt}{\int_0^\infty f_1(t)\, dt} - \frac{\int_0^\infty t\, f_2(t)\, dt}{\int_0^\infty f_2(t)\, dt} = \bar{t}_1 - \bar{t}_2 = \frac{\int_0^\infty t\, c_1(t)\, dt}{\int_0^\infty c_1(t)\, dt} - \frac{\int_0^\infty t\, c_2(t)\, dt}{\int_0^\infty c_2(t)\, dt}$$

where $\bar{t}_1$ and $\bar{t}_2$ are the mean absorption times for the two doses. Here again the right-hand side does not involve $R(t)$ and may be quite well determined by the data, as has been noted previously (Cutler, 1978). This is illustrated in Figure 6, where the posterior for $\bar{t}_{\mathrm{NaAPAP}} - \bar{t}_{\mathrm{APAP}}$ is shown, again using full marginalisation, with a width of about 5.2 minutes, although the widths of the posteriors for the individual mean absorption times $\bar{t}_{\mathrm{APAP}}$ and $\bar{t}_{\mathrm{NaAPAP}}$ were each about 11 minutes (see Figure 4). In other words, much of the uncertainty in the individual mean absorption times is correlated, because it arises from uncertainties about the p.s.f. and cancels out in the difference.

4. Conclusions

In deconvolution problems where the p.s.f. is not completely specified, the correct probabilistic approach is to construct a joint posterior probability density for the object and the p.s.f., and obtain inferences about the object as marginals over this joint posterior. In some circumstances, these integrals can be approximated by using a 'best estimate' p.s.f., but if the p.s.f. is not well determined by the data this approximation may give misleading error bars for features of interest in the object. Nonetheless, some features can be inferred accurately even when the p.s.f. is not well determined, and these features can be predicted from the Convolution Theorem.

REFERENCES

Charter, M.K.: 1991, 'Quantifying Drug Absorption', in *Maximum Entropy and Bayesian Methods, Laramie, Wyoming, 1990*, W.T. Grandy, Jr., L.H. Schick (eds.), Kluwer, Dordrecht.

Charter, M.K.: 1992, 'Modelling Drug Behaviour in the Body with MaxEnt', in *Maximum Entropy and Bayesian Methods, Seattle 1991*, G.J. Erickson, P. Neudorfer and C.R. Smith (eds.), Kluwer, Dordrecht.

Cutler, D.J.: 1978,'Theory of the Mean Absorption Time, an Adjunct to Conventional Bioavailability Studies', *J. Pharmacy and Pharmacology* **30**, 476.

Gull, S.F.: 1989, 'Developments in Maximum Entropy Data Analysis', in *Maximum Entropy and Bayesian Methods, Cambridge 1988*, J. Skilling (ed.), Kluwer, Dordrecht.

Gull, S.F. and Skilling, J.: 1991, *Quantified Maximum Entropy* `MemSys5` *Users' Manual,* Maximum Entropy Data Consultants Ltd., 33 North End, Meldreth, Royston, SG8 6NR, U.K.

Skilling, J.: 1989, 'Classic Maximum Entropy', in *Maximum Entropy and Bayesian Methods, Cambridge 1988*, J. Skilling (ed.), Kluwer, Dordrecht.

Skilling, J.: 1991, 'On Parameter Estimation and Quantified MaxEnt', in *Maximum Entropy and Bayesian Methods, Laramie, Wyoming, 1990*, W.T. Grandy, Jr., L.H. Schick (eds.), Kluwer, Dordrecht.

MAXIMUM ENTROPY AND LINEAR INVERSE PROBLEMS. A SHORT REVIEW

Ali Mohammad–Djafari
Laboratoire des Signaux et Systèmes (CNRS–ESE–UPS)
École Supérieure d'Électricité
Plateau de Moulon, 91192 Gif-sur-Yvette Cédex, France

ABSTRACT. In this paper we give a short review of the Maximum Entropy (ME) principle used to solve inverse problems. We distinguish three fundamentally different approaches for solving inverse problems when using the ME principle: a) Classical ME in which the unknown function is considered to be or to have the properties of a probability density function, b) ME in mean in which the unknown function is assumed to be a random function and the data are assumed to be the expected values of some finite number of known constraints on the unknown function, and finally, c) Bayesian approach with ME priors. In this case the ME principle is used only for assigning a probability distribution to the unknown function to translate our prior knowledge about it. In each approach, we describe the main ideas and give explicitly the hypothesis, the practical and the theoretical limitations.

1. Introduction

In many practical physical experiments, the observed data are transforms of the quantities of interest. The class of linear transforms is one of the most usual. A linear relation between the observable quantity $g(s)$ and the unobserved interesting quantity $f(r)$ is written in the form

$$g(s) = \int_D f(r)h(r,s)\,\mathrm{d}r + b(s) \tag{1}$$

where $h(r,s)$ is the instrument's function which we suppose to be known perfectly and $b(s)$ is the measurement noise. In a physical experiment we can observe $g(s)$ on a finite set of isolated points $s_m, m = 1, \cdots, M$ in the transform space:

$$g_m = g(s_m) = \int_D f(r)h(r,s_m)\,\mathrm{d}r + b(s_m) = \int_D f(r)h_m(r)\,\mathrm{d}r + b_m, \quad m = 1, \cdots, M \tag{2}$$

and to do the numerical calculation we have to discretize these equations. When discretized properly, the problem can be stated as to estimate a unknown vector $\boldsymbol{f} = [f_1, \cdots, f_N]$ representing the samples of the unknown function $f(r)$ using the data vector $\boldsymbol{g} = [g_1, \cdots, g_M]$. The relation between $\boldsymbol{f}$ and $\boldsymbol{g}$, in the case of linear inverse problems with the assumption of additive errors, is

$$g_m = \sum_{n=1}^{N} H_{mn} f_n + b_n, \quad \text{or} \quad \boldsymbol{g} = \boldsymbol{H}\boldsymbol{f} + \boldsymbol{b} \tag{3}$$

The objective of an inversion procedure is to obtain a unique and stable solution $\widehat{\boldsymbol{f}}$ from the data $\boldsymbol{g}$. This problem is an *ill–posed* one, in the sense that, in general there are infinitely

A. Mohammad-Djafari and G. Demoments (eds.), Maximum Entropy and Bayesian Methods, 253–264.

many solutions $\boldsymbol{f}$ who can provide the same data (Matrix $\boldsymbol{H}$ is, in general either singular or very ill-conditionned). So we need a prior knowledge about $\boldsymbol{f}$ to obtain a unique and stable solution. One possible prior knowledge may be "$\boldsymbol{f}$ is a positive-valued vector". The Maximum Entropy (ME) methods historically try to use this prior knowledge to solve the above inverse problems.

In this paper we will treat only the discret case. For an analyze in the continuous case see [34]. We assume also that the readers are familiar with the basic ideas of the Maximum Entropy principle. However we give here a brief description of the definitions.

Shannon (1948) defined the entropy associated to a discret random variable X with N possible outcomes $\{x_1, \cdots, x_N\}$ as

$$S(p) = -\sum_{n=1}^{N} p_n \ln p_n \tag{4}$$

where $\{p_n\} = \Pr(X = x_n)$.

If $\{p_n\}$ and $\{q_n\}$ are two distributions of a random variable X then the quantity

$$H(p,q) = \sum_{n=1}^{N} p_n \ln \frac{p_n}{q_n} \tag{5}$$

is called the relative or cross entropy of $\{p_n\}$ and $\{q_n\}$. Note that minimizing $H(p,q)$ becomes equivalent to maximizing $S(p)$ when $\{q_n\}$ is a uniform distribution.

The basic ME mathematical problem is the following: Given the prior distribution $\{q_n\}$ and a finite set of the constraints on $\{p_n\}$ in the form

$$\mathrm{E}\{\phi_m(x_n)\} = \sum_{n=1}^{N} \phi_m(x_n)\, p_n = \mu_m, \quad m = 1, \cdots, M, \tag{6}$$

find $\{p_n\}$ which satisfies these constraints and minimizes the cross entropy $H(p,q)$. The solution is given by

$$p_n = q_n \frac{1}{Z} \exp\left[-\sum_{m=1}^{M} \lambda_m \phi_m(x_n)\right], \quad n = 1, \cdots, N \tag{7}$$

where Z is the partition function

$$Z = \sum_{n=1}^{N} q_n \exp\left[\sum_{m=1}^{M} \lambda_m \phi_m(x_n)\right], \tag{8}$$

and $\lambda = [\lambda_1, \cdots, \lambda_M]$ are the Lagrange multipliers which are obtained by solving the following set of nonlinear equations:

$$\frac{1}{Z} \sum_{n=1}^{N} \phi_m(x_n)\, q_n \exp\left[\sum_{m=1}^{M} \lambda_m \phi_m(x_n)\right] = \mu_m, \quad m = 1, \cdots, M \tag{9}$$

Note that if $\{q_n\}$ is uniform, then minimizing the cross entropy $H(p,q)$ becomes equivalent to maximizing the entropy $S(p)$ and all the equations 7, 8 and 9 will hold with $q_n = 1$.

We can distinguish three fundamentally different approaches of using the ME principle when solving the ill-posed inverse problems of (2).

2. Classical Maximum Entropy

In the first one the main hypothesis is that the function $f(\boldsymbol{r})$ in equation (2) is, or has the properties of, a probability density function. In the discret case this means that, $\{f_n\}$ in equation (3) are supposed to be positive and, when normalized, one can consider them as a probability distribution, i.e.

$$f_n \geq 0, \qquad \sum_{n=1}^{N} f_n = 1 \tag{10}$$

The only problem is the fact that the noiseless data $g_m, m = 1, \ldots, M$ can not determine uniquely this distribution. So, in this approach the inverse problem (2) is solved by choosing between all the possible solutions satisfying the data constraints (2) the one who maximizes the entropy

$$S(f_n) = -\sum_{n=1}^{N} f_n \ln f_n.$$

or the one who minimizes the cross entropy

$$S(f_n, q_n) = \sum_{n=1}^{N} f_n \ln \frac{f_n}{q_n}.$$

where $\{q_n\}$ is a prior reference distribution.

Note that the hypothesis that $\{f_n\}$ is a probability distribution may correspond to a physical reality. For example $\{f_n\}$ may be the photon's energy distribution in a volume or its projection on a surface or along a line. But we are not inferring about the random variable "photon's energy", we are interested on its distribution.

To be more explicit we resume the situation:

- **Hypothesis:** $\boldsymbol{f} = \{f_n\}$ is or has the properties of a probability distribution
- **Data:** $\boldsymbol{g} = \boldsymbol{H}\boldsymbol{f}$ are exact linear constraints on $\boldsymbol{f} = \{f_n\}$
- **Mathematical problem:** Given $\boldsymbol{H}, \boldsymbol{g}$ and the prior (or template) $\boldsymbol{q} = \{q_n\}$

$$\text{maximize } -\sum_{n=1}^{N} f_n \ln \frac{f_n}{q_n} \quad \text{s.t.} \quad \boldsymbol{g} = \boldsymbol{H}\boldsymbol{f}, \quad \text{and} \quad \sum_{n=1}^{N} f_n = 1$$

- **Solution:** This is the classical Maximum Entropy problem for which the solution is:

$$f_n = \frac{1}{Z} q_n \exp\left[-\sum_{m=1}^{M} \lambda_m H_{mn}\right] = \frac{1}{Z} q_n \exp\left[-(\boldsymbol{H}^t \boldsymbol{\lambda})_n\right], \quad n = 1, \cdots, N$$

  Lagrange multipliers $\{\lambda_1, \cdots, \lambda_M\}$ are obtained by solving the following set of non-linear equations:

$$G_m(\lambda_1, \cdots, \lambda_M) = \sum_{n=1}^{N} \frac{1}{Z} q_n \exp\left[-(\boldsymbol{H}^t \boldsymbol{\lambda})_n\right] = g_m, \quad m = 1, \cdots, M$$

This system of equations can be solved by the Newton-Raphson method, which consists of developing $G_m(\boldsymbol{\lambda})$ in first order Taylor's series around an initial estimate $\boldsymbol{\lambda}^0$ and solving the resulting linear equations for $\boldsymbol{\delta} = \boldsymbol{\lambda} - \boldsymbol{\lambda}^0$, from which we can calculate a new estimate $\boldsymbol{\lambda} = \boldsymbol{\lambda}^0 + \boldsymbol{\delta}$. This new estimate will become the trial $\boldsymbol{\lambda}^0$ for the next iteration. If there exists a solution then it is unique and the iterations converge to it. For a numerical implementation and programs written in `Matlab` see the reference [33].

- **Limitations:**

  The main theoretical problem with this method is that this method does not account for the errors in the data. This fact limits its use in practical physical problems. For example the measurement noise can have as consequence that any solution can be found for the problem.

  Some practical extensions to this approach has been proposed for taking the errors into account. For more details see the references [1], [3], [4], [5], [6], [7], [8], [18], [19], [20], [22], [24], [25], [26], [27], [28], [32], [34], [37], [41], [42], [44], [43], [39], and [47].

3. Maximum Entropy in Mean

In the second one, $\{f_n\}$ are considered as outcomes of the random variables $\{f_n\}$ for which we assume to be able to define the prior distribution $\{q_{nk}\} = \{\Pr(f_n = \alpha_k)\}$ and the posterior distribution $\{p_{nk}\} = \{\Pr(f_n = \alpha_k)\}$. The data are then supposed to be the expected values $\langle g_m \rangle$ where $\langle \cdot \rangle$ means the average over $\{p_{nk}\}$.

The ME principle in this approach is used to find the probability distributions $\{p_{nk}\}$ who satisfies these data constraints and minimizes

$$H\,(p,q) = \sum_{k=1}^{K} p_{nk} \ln \frac{p_{nk}}{q_{nk}},$$

from whom we can construct an estimate $\left\{\widehat{f_n}\right\}$ for $\{f_n\}$. The solution to the inverse problem is:

$$\widehat{f_n} = \langle f_n \rangle = \sum_{k=1}^{K} f_n p_{nk}, \quad n = 1, \cdots, N \tag{11}$$

This is the reason why the methods issued from this approach are called *Maximum Entropy in mean methods*. So, explicitly, this method can be resumed as follows:

- **Hypothesis:** $\boldsymbol{f}$ is an outcome of a random vector $\boldsymbol{F}$ with the prior law $\{q_{nk}\}$
- **Data:**

$$\boldsymbol{g} = \langle \boldsymbol{H} \boldsymbol{f} \rangle = \boldsymbol{H} \langle \boldsymbol{f} \rangle, \quad \text{or} \quad g_m = \sum_{n=1}^{N} H_{mn} \sum_{k=1}^{K} f_n p_{nk}, \quad m = 1, \cdots, M$$

- **Mathematical problem:** Given $\boldsymbol{H}, \boldsymbol{g}, \boldsymbol{q} = \{q_{nk}\}$,

$$\text{maximize } H\,(p,q) = \sum_{k=1}^{K} p_{nk} \ln \frac{p_{nk}}{q_{nk}}, \quad \text{s.t.} \quad \boldsymbol{g} = \boldsymbol{H} \langle \boldsymbol{f} \rangle$$

- **Solution:** To find the solution the first step is to calculate $\{p_{nk}\}$ by

$$p_{nk} = \frac{1}{Z} q_{nk} \exp\left[\sum_{m=1}^{M} \lambda_m \sum_{n=1}^{N} H_{mn} f_n\right] = \frac{1}{Z} q_{nk} \exp\left[\boldsymbol{\lambda}^t \boldsymbol{H} \boldsymbol{f}\right]$$

and once $\{p_{nk}\}$ calculated, the estimated solution $\widehat{\boldsymbol{f}}$ is given by

$$\widehat{f}_n = \langle f_n \rangle = \sum_{k=1}^{K} f_n p_{nk}, \quad n = 1, \cdots, N$$

and we can also define the error bars:

$$\left\langle (f_n - \widehat{f}_n)^2 \right\rangle = \sum_{k=1}^{K} \left[f_n - \widehat{f}_n \right]^2 p_{nk}$$

- **Limitations:** The measured data in this approach are considered to be the mean values and the estimated solution is also a mean value. The errors on the data can not be considered directly.

- **References:** See [2], [16], [22], [45], and [23].

We may also have the data in the form $\left\langle (g_m - \langle g_m \rangle)^2 \right\rangle$ which correspond to the variances of the measurements. In this case, it is easy to show that, if $\{q_{nk}\}$ is uniform or Gaussian, then $\{p_{nk}\}$ is also Gaussian and the ME in mean estimator in this case will be equivalent to the classical Least Squares estimators.

4. Bayesian MaxEnt

In this approach,

- $\boldsymbol{f}$ is considered as an outcome of a random vector $\boldsymbol{F}$ for which we assume to be able to define a prior probability distribution $p[\boldsymbol{f} \mid I]$ which translate our uncertainty or our incomplete prior knowledge I about it.

- The data $\boldsymbol{g}$ is also considered as an outcome of a random vector $\boldsymbol{G}$ for which we are able to define the conditional probability distribution $p[\boldsymbol{g} \mid \boldsymbol{f}] = \Pr(\boldsymbol{G} = \boldsymbol{g} \mid \boldsymbol{F} = \boldsymbol{f}]$ which translate our incomplete knowledge about the errors and the noise in the measurement system.

- The next step is then to use the prior $p[\boldsymbol{f} \mid I]$, the likelihood $p[\boldsymbol{g} \mid \boldsymbol{f}]$ and the Baye's rule to determine the posterior law

$$p[\boldsymbol{f} \mid \boldsymbol{g}] \propto p[\boldsymbol{g} \mid \boldsymbol{f}] p[\boldsymbol{f} \mid I]$$

- The final step is to choose an estimator $\widehat{\boldsymbol{f}}$ and study its characteristics (error-bands, variances,$\cdots$).

Note that in this Bayesian approach the ME principle is used only in the first step which is to translate our prior knowledge I about $\boldsymbol{f}$ to a prior law $p[\boldsymbol{f} \mid I]$.

To be more explicit we are going to give three classical examples corresponding to three different state of prior knowledge I:

◇ **Example 1:**

- **Hypothesis: Prior knowledge I** We are given

$$\mathrm{E}\{\boldsymbol{f}\} = \boldsymbol{f}_0, \qquad \mathrm{E}\left\{(\boldsymbol{f}-\boldsymbol{f}_0)^2\right\} = \boldsymbol{R}_f$$

- **Data:** $\quad \boldsymbol{g} = \boldsymbol{H}\boldsymbol{f} + \boldsymbol{b}$

  We suppose that b_m's are independent and we are given the variances $\sigma_m^2, \quad m = 1, \cdots, M$. In other words, we are given

$$\mathrm{E}\{\boldsymbol{b}\} = \mathbf{o}, \qquad \mathrm{E}\left\{\boldsymbol{b}^t\boldsymbol{b}\right\} = \boldsymbol{R}_b = \operatorname{diag}\left\{\sigma_1^2, \cdots, \sigma_M^2\right\}$$

- **Mathematical Problem:** Given $\boldsymbol{f}_0, \boldsymbol{R}_f, \boldsymbol{R}_b, \boldsymbol{g}$ and $\boldsymbol{H}$, determine $\boldsymbol{f}$.
- **Solution:**

  o First, using the ME principle, we can easily obtain:

$$p[\boldsymbol{f} \mid I] = \mathcal{N}(\boldsymbol{f}_0, \boldsymbol{R}_f)$$

  o Second, knowledge of $\boldsymbol{R}_b = \operatorname{diag}\{\sigma_1^2, \cdots, \sigma_M^2\}$ will result to

$$p[\boldsymbol{g} \mid \boldsymbol{f}] = \mathcal{N}(\boldsymbol{H}\boldsymbol{f}, \boldsymbol{R}_b)$$

  o Third, using the Baye's rule we have

$$p[\boldsymbol{f} \mid \boldsymbol{g}; I] \propto \exp\left[-J(\boldsymbol{f})\right] \quad \text{with} \quad J(\boldsymbol{f}) = [\boldsymbol{H}\boldsymbol{f}-\boldsymbol{g}]^t \boldsymbol{R}_b^{-1}[\boldsymbol{H}\boldsymbol{f}-\boldsymbol{g}] + [\boldsymbol{f}-\boldsymbol{f}_0]^t \boldsymbol{R}_f^{-1}[\boldsymbol{f}-\boldsymbol{f}_0]$$

  o Finally, if we choose the maximum a posteriori (MAP) estimator we obtain

$$\widehat{\boldsymbol{f}} = \arg\max_{\boldsymbol{f}} \{p[\boldsymbol{f} \mid \boldsymbol{g}; I]\} = \arg\min_{\boldsymbol{f}} \{J(\boldsymbol{f})\} = \left[\boldsymbol{H}^t \boldsymbol{R}_b^{-1} \boldsymbol{H} + \boldsymbol{R}_f^{-1}\right]^{-1} \left[\boldsymbol{H}^t \boldsymbol{R}_b^{-1} \boldsymbol{g} + \boldsymbol{R}_f^{-1} \boldsymbol{f}_0\right]$$

  Note that, if $\boldsymbol{f}_0 = \mathbf{o}, \boldsymbol{R}_b = \sigma_b^2 \boldsymbol{I}, \boldsymbol{R}_f = \sigma_f^2 \boldsymbol{I}$, we have

$$\widehat{\boldsymbol{f}} = [\boldsymbol{H}^t\boldsymbol{H} + \alpha \boldsymbol{I}]^{-1} \boldsymbol{H}^t \boldsymbol{g}, \quad \text{with} \quad \alpha = \frac{\sigma_b^2}{\sigma_f^2}$$

  References: See [23], [17], [29], [30], and [31].

◇ **Example 2:**

- **Hypothesis:** We are given

$$\mathrm{E}\left\{S(\boldsymbol{f}, \boldsymbol{f}_0) = -\sum_{n=1}^{N} f_n \ln \frac{f_n}{f_{0n}}\right\} = s$$

- **Data:** As in Example 1.
- **Mathematical Problem:** Given $\boldsymbol{f}_0, s, \boldsymbol{g}, \boldsymbol{R}_b, \boldsymbol{H}$, determine $\boldsymbol{f}$
- **Solution:**

  ∘ First, using the MaxEnt, we have:

$$p[\boldsymbol{f} \mid I] \propto \exp\left[-\lambda S(\boldsymbol{f}, \boldsymbol{f}_0)\right]$$

  ∘ Second, as in Example 1, the knowledge of $\sigma_n^2, \quad m = 1, \cdots, M$ will result to

$$p[\boldsymbol{f} \mid I] = \mathcal{N}(\boldsymbol{f}_0, \boldsymbol{R}_f)$$

  ∘ Third, using the Baye's rule we have

$$p[\boldsymbol{f} \mid \boldsymbol{g}; I] \propto \exp\left[-J(\boldsymbol{f})\right] \quad \text{with} \quad J(\boldsymbol{f}) = [\boldsymbol{H}\boldsymbol{f} - \boldsymbol{g}]^t \boldsymbol{R}_b^{-1} [\boldsymbol{H}\boldsymbol{f} - \boldsymbol{g}] + \lambda S(f, f_0)$$

  ∘ Finally, the MAP estimate is

$$\widehat{\boldsymbol{f}} = \arg\max_{\boldsymbol{f}} \{p[\boldsymbol{f} \mid \boldsymbol{g}; I]\} = \arg\min_{\boldsymbol{f}} \left\{ J(\boldsymbol{f}) = [\boldsymbol{H}\boldsymbol{f} - \boldsymbol{g}]^t \boldsymbol{R}_b^{-1} [\boldsymbol{H}\boldsymbol{f} - \boldsymbol{g}] + \lambda S(f, f_0) \right\}$$

References: See [17], [29], [30], and [31].

◇ **Example 3:**

- **Hypothesis:** We are given

$$\mathrm{E}\left\{ S(\boldsymbol{f}) = \sum_{n=1}^{N} f_n \right\} = s, \qquad \mathrm{E}\left\{ H(\boldsymbol{f}) = -\sum_{n=1}^{N} f_n \ln f_n \right\} = h$$

- **Data:** As in the two preceding examples.
- **Mathematical Problem:** Given $\boldsymbol{f}_0, s, h, \boldsymbol{g}, \boldsymbol{R}_b$ and $\boldsymbol{H}$ determine $\boldsymbol{f}$
- **Solution:**

  ∘ First, using the ME, we have:

$$p[\boldsymbol{f} \mid I] \propto \exp\left[-\lambda_1 S(\boldsymbol{f}) - \lambda_2 H(\boldsymbol{f})\right]$$

  ∘ Second, knowledge of $\boldsymbol{R}_b$ will result to

$$p[\boldsymbol{f} \mid I] = \mathcal{N}(\boldsymbol{f}_0, \boldsymbol{R}_f)$$

  ∘ Third, using the Baye's rule we have

$$p[\boldsymbol{f} \mid \boldsymbol{g}; I] \propto \exp\left[-J(\boldsymbol{f})\right] \quad \text{with} \quad J(\boldsymbol{f}) = [\boldsymbol{H}\boldsymbol{f} - \boldsymbol{g}]^t \boldsymbol{R}_b^{-1} [\boldsymbol{H}\boldsymbol{f} - \boldsymbol{g}] + \lambda_1 S(\boldsymbol{f}) + \lambda_2 H(\boldsymbol{f})$$

  ∘ Finally, the MAP estimate is

$$\widehat{\boldsymbol{f}} = \arg\min_{\boldsymbol{f}} \left\{ J(\boldsymbol{f}) = [\boldsymbol{H}\boldsymbol{f} - \boldsymbol{g}]^t \boldsymbol{R}_b^{-1} [\boldsymbol{H}\boldsymbol{f} - \boldsymbol{g}] \lambda_1 S(\boldsymbol{f}) + \lambda_2 H(\boldsymbol{f}) \right\}$$

References: See [29], [30], [31], [32], [33], and [41].

5. Conclusions

In this paper we have tried to give a review of the ME principle usage in solving inverse problems. We have distinguished three fundamentally different approaches:

- Classical ME approach,
- ME in mean approach, and
- Bayesian approach with ME priors.

In the first one, the unknown function $f(\boldsymbol{r})$ is considered as a probability distribution function and the data g_n are considered as the exact constraints on it. The objective is to choose between all the feasible solutions the one who maximizes the entropy or minimizes its relative entropy with respect to a reference function $f_0(\boldsymbol{r})$.

In the second, the problem is the estimation of a random function $f(\boldsymbol{r})$ on which we have the data $g_n = \int_D \langle f(\boldsymbol{r})\rangle\, h_n(\boldsymbol{r})\, \mathrm{d}\boldsymbol{r}$ who can be considered either as the linear constraints on the mean $\langle f(\boldsymbol{r})\rangle$ or the mean values of linear constraints $g_n = \langle \int_D f(\boldsymbol{r}) h_n(\boldsymbol{r})\, \mathrm{d}\boldsymbol{r}\rangle$. The objective is, first determining the probability law $p[f]$ of this random function and then estimating the random function itself. This probability law is choosed to have maximum entropy or to have minimum relative entropy with respect to a reference probability law $p_0[f]$.

In the third, the problem is also the estimation of a random function on which we have expected values but corrupted by noise (errors). Maximum entropy principle is used only in the first step of the Bayesian approach, i.e. to assign the probability laws: $p[\boldsymbol{f}|I]$ representing the state of our prior knowledge I about the function, and $p[\boldsymbol{g}|\boldsymbol{f}]$ representing the state of our confidence on the data D, or the uncertainty about the errors (both measurement noise and modeling errors). The next step is then to use the Bayes rule to combine these two knowledge states and to determine the posterior law $p(\boldsymbol{f}|\boldsymbol{g}, I)$ which represents our state of knowledge about the solution. The final step is to give an estimate for the unknown function and determine its characteristics; i.e. error bars or any other statistical properties of this estimate.

In each of these three approaches we tried to give their explicit and implicit hypothesis, theoretical and practical limitations and the main references.

What is important to remind at this end is that, we must be aware of what we do in each of these methods. When we have to solve a practical inverse problem we have to start by answering to the following questions :

- What physical quantity is represented by the unknown function?
- What physical quantity is represented by the data?
- What we know about the errors or uncertainty on the data?
- What we want to know about unknown function?

When we obtained the answers to these questions we can then choose a method to solve our problem by answering the following questions :

- Does the method consider the unknown function as it must be?

- Does the method consider the data as they really are?
- Can the method give what we want to know about the unknown function?
- Can the method take into account our prior knowledge about the unknown function?
- Can the method take into account our prior knowledge about the errors on the data?

As we could see, it is possible that, starting by two fundamentally different methods, the final mathematical problem to solve or the final implementation of the optimization algorithm to solve it becomes the same. This does not mean that we can give the same interpretation to the numerical values obtained by the algorithm.

To give my own opinion on these three approaches, I can say that, in my experience, the Bayesian approach can answer all the well-posed questions which arise when we have to solve practical inverse problems and satisfies all the consistency requirements of the inference when we have to combine the prior information and the information contained in the data.

References

[1] N. Agmon, Y. Alhassid and D. Levine, "An Algorithm for Finding the Distribution of Maximal Entropy," *Journal of Computational Physics*, Vol. 30, pp: 250–258, 1979.

[2] J. M. Borwein and A. S. Lewis, "Duality relationships for entropy-like minimization problems," *SIAM J. Control and Optimization*, Vol. 29 No. 2, March 1991.

[3] R. K. Bryan, M. Bansal, W. Folkhard., C. Nave and D. A. Marvin, "Maximum-entropy calculation of the electron density at 4 A resolution of Pf1 filamentous bacteriophage," *Proc.Natl. Acad. Sci. USA*, **80**, pp. 4 728–4 731, 1983.

[4] S. F. Burch, S. F. Gull and J. Skilling, "Image Restoration by a Powerful Maximum Entropy Method," *Comput. Vis. Graph. Im. Process.*, **23**, pp. 113–128, 1983.

[5] G. J. Daniell and S. F. Gull, "Maximum Entropy algorithm applied to image enhancement," *Proc. IEE*, **127E**, pp. 170–172, 1980.

[6] T. Elfwing, "On some Methods for Entropy Maximization and Matrix Scaling," *Linear Algebra and its Applications*, **34**, pp. 321–339, 1980.

[7] J. Eriksson, "A note on solution of large sparse maximum entropy problems with linear equality constraints," *Math. Programming*, **18**, pp. 146–154, 1980.

[8] S. Erlander, "Entropy in linear programs," *Mathematical Programming*, **21**, pp. 137–151, 1981.

[9] Frieden B. R., "Band-Unlimited Reconstruction of Optical Objects and Spectra," *J. Opt. Soc. Am.*, **57**, pp. 1 013–1 019, 1967.

[10] Frieden B. R., "Restoring with Maximum Likelihood and Maximum Entropy," *J. Opt. Soc. Am.*, **62**, pp. 511–518, 1972.

[11] Frieden B. R., "Statistical Estimates of Bounded Optical Scenes by the Method of Prior Probabilities," *IEEE Trans.*, **IT–**, pp. 118–119, 1973.

[12] Frieden B. R., "Restoring with maximum entropy. III. Poisson sources and backgrounds," *J. Opt. Soc. Am.*, **68**, pp. 93–103, 1978.

[13] Frieden B. R., "Statistical Models for the Image Restoration Problem," *Comput. Graph. Im. Process.*, **12**, pp. 40–59, 1980.

[14] Frieden B. R., "Statistical Models for the Image Restoration Problem," *Comput. Graph. Im. Process.*, **12**, pp. 40–59, 1980.

[15] Frieden B. R. and Zoltani C. K., "Maximum bounded entropy: application to tomographic reconstruction," *Applied Optics*, **24**, pp. 201–207, 1985.

[16] F. Gamboa, "Méthode du maximum d'entropie sur la moyenne et applications," *Thèse*, Dépt. de Mathématiques, Univ. de Paris–Sud, 1989.

[17] S. F. Gull and J. Skilling, "Maximum entropy method in image processing," *IEE Proc.*, **131–F**, pp. 646–659, 1984.

[18] E. T. Jaynes, "On the Rationale of Maximum–Entropy Methods," *Proc. IEEE*, **70**, pp. 939–952, 1982.

[19] E. T. Jaynes, "Where do we go from here?," *Maximum–Entropy and Bayesian Methods in Inverse Problems*, C. R. Smith & T. Grandy, Jr. (eds.), pp. 21–58, 1985.

[20] R. Johnson and J. Shore, "Which is Better Entropy Expression for Speech Processing: –SlogS or logS?," *IEEE Trans*, **ASSP–32**, pp. 129–137, 1984.

[21] R. Kikuchi and B. H. Soffer, "Maximum entropy image restoration. I. The entropy expression," *J. Opt. Soc. Am.*, **67**, pp. 1 656–1 665, 1977.

[22] G. Le Besnerais, J. Navaza, G. Demoment, "Aperture synthesis using maximum entropy on the mean in radio–astronomy," *GRETSI*, 1991.

[23] V. A. Macaulay and B. Buck, "Linear inversion by the method of maximum entropy," *Maximum–Entropy and Bayesian Methods in Inverse Problems*, **C.R. Smith & T. Grandy, Jr. (eds.)**, 1985.

[24] A. Mohammad–Djafari et G. Demoment, "Maximum entropy Fourier synthesis with application to diffraction tomography," *Applied Optics*, Vol.26, No. 10, pp:1745–1754, (1987).

[25] A. Mohammad–Djafari et G. Demoment, "Tomographie de diffraction et synthèse de Fourier à maximum d'entropie," *Revue Phys. Appl.*, Vol. 22, pp:153–167, (1987).

[26] A. Mohammad–Djafari and G. Demoment, "Maximum entropy reconstruction in X ray and diffraction tomography," *IEEE Trans. on Medical Imaging*, Vol. 7, No. 4 pp:345–354, (1988).

[27] A. Mohammad–Djafari and G. Demoment, "Image restoration and reconstruction using entropy as a regularization functional," *Maximum Entropy and Bayesian Methods in Science and Engineering*, Vol. 2, pp:341–355, (1988) by Kluwer Academic Publishers.

[28] A. Mohammad–Djafari et G. Demoment, "Utilisation de l'entropie dans les problèmes de restauration et de reconstruction d'images," *Traitement du Signal*, Vol. 5, No. 4, pp:235–248, (1988).

[29] A. Mohammad–Djafari et G. Demoment, "Maximum Entropy and Bayesian Approach in Tomographic Image Reconstruction and Restoration," *Maximum Entropy and Bayesian Methods, pp:195-201*, J. Skilling (ed.) by Kluwer Academic Publishers, 1989.

[30] A. Mohammad–Djafari et G. Demoment, "Estimating Priors in Maximum Entropy Image Processing," *Proc. of ICASSP 1990*, pp: 2069–2072

[31] A. Mohammad–Djafari et J. Idier, "Maximum entropy prior laws of images and estimation of their parameters," *Proc. of The 10th Int. MaxEnt Workshop, Laramie, Wyoming*, publié dans Maximum-entropy and Bayesian methods, T.W. Grandy ed., 1990.

[32] A. Mohammad–Djafari, "Maximum Likelihood Estimation of the Lagrange Parameters of the Maximum Entropy Distributions," *Proc. of The 11th Int. MaxEnt Workshop, Seattle, USA*, 1991.

[33] A. Mohammad–Djafari, "A Matlab Program to Calculate the Maximum Entropy Distributions," *Proc. of The 11th Int. MaxEnt Workshop, Seattle, USA*, 1991.

[34] A. Mohammad–Djafari, "Maximum entropy Methods and Algorithmes for solving linear inverse problems," *Rapport interne LSS No. GPI/02/91*.

[35] D. Mukherjee and D. C. Hurst, "Maximum Entropy Revisited," *Statistica Neerlandica*, **38**, 1984.

[36] R. Narayan and R. Nityananda "Maximum entropy image restoration in astronomy," *Ann. Rev. Astron. Astrophys.*, **24**, pp. 127–170, 1986.

[37] J. Navaza, "On the maximum entropy estimate of electron density function," *Acta Cryst.*, **A–41**, pp. 232–244, 1985.

[38] J. P. Noonan, N. S. Tzannes and T. Costello, "On the Inverse Problem of Entropy Maximizations.," *IEEE Trans. Information Theory,*, pp. 120–123, 1976.

[39] J. E. Shore and R. W. Johnson, "Properties of Cross–Entropy Minimization," *IEEE Trans.*, **IT–27**, pp. 472–482, 1981.

[40] S. Sibisi, "Two–dimensional reconstructions from one–dimensional data by maximum entropy," *Nature*, **301**, pp. 134–136, 1983.

[41] J. Skilling and R. K. Bryan, "Maximum entropy image reconstruction: general algorithm," *Mon. Not. R. astr. Soc.*, **211**, pp. 111-124, 1984.

[42] J. Skilling and S. F. Gull, "Maximum entropy method in image processing," *IEE Proceedings*, **131**, pp. 646-659, 1984.

[43] J. Skilling and S. F. Gull, "Algorithms and Applications," *Maximum-Entropy and Bayesian Methods in Inverse Problems*, **C.R. Smith & T. Grandy, Jr. (eds.)**, pp. 83-132, 1985.

[44] J. Skilling and S. F. Gull, "The Entropy of an Image," *SIAM-AMS Proceedings*, **14**, 1984.

[45] C. B. Smith, "A dual method for maximum entropy restoration," *IEEE Trans. Pattern Anal. Machine Intelligence*, **PAMI-1**, pp. 411-414, 1979.

[46] H. J. Trussell, "The Relationship Between Image Restoration by the Maximum A Posteriori Method and a Maximum Entropy Method," *IEEE Trans.*, **ASSP-28**, pp. 114-117, 1980.

[47] S. J. Wernecke and L. R. D'Addario, "Maximum Entropy Image Reconstruction," *IEEE Trans.*, **C-26**, pp. 351-364, 1977.

[48] A. Wehrl, "General properties of entropy," *Reviews of Modern Physics*, Vol. 50 No. 2, April 1978.

[49] X. Zhuang, K. B. Yu and R. M. Haralick, "A Differential Equation Approach to Maximum Entropy Image Reconstruction," *IEEE Trans. on Acoustics, Speech, and Signal Processing*, Vol. ASSP-35, No. 2, pp. 208-218, 1987.

BAYESIAN MAXIMUM ENTROPY IMAGE RECONSTRUCTION FROM THE SCATTERED FIELD DATA

Maï Khuong Nguyen * † and Ali Mohammad-Djafari *
*Laboratoire des Signaux et Systèmes (CNRS–ESE–UPS)
École Supérieure d'Électricité
Plateau de Moulon, 91192 Gif–sur–Yvette Cédex, France
† Université de Cergy-Pontoise, 95014 Cergy-Pontoise Cédex

ABSTRACT. The image reconstruction employing the inverse scattering phenomena consist in reconstructing the image of an object from the scattered field measured behind the object. This reconstruction runs up against the nonuniqueness in the solution of the inverse scattering. In this communication, we propose to solve the ill-posed inverse problem by a statistical regularization method based on the Bayesian maximum *a posteriori* estimation with the maximum entropy *a priori* laws. The obtained results show the interesting possibilities of this method in image reconstructions.

1. Introduction

In an imaging system using the inverse scattering, the object is illuminated by an electromagnetic wave of known characteristics. A multidimensional inverse scattering of wavefields is considered as a mean to obtain information about the geometric structure or contour, the interior parameters or the reflectivity of the scatterer in a homogeneous environment. By definition, the scattered field is the difference between the fields in absence and in presence of the object. This field behaves as if it was created by an equivalent current source $\vec{J}$ which is dependant on the difference of dielectric properties (conductivity and permittivity) between the object and the background medium [2], the amplitude :

$$\begin{cases} J \neq 0 & \text{inside the object} \\ J = 0 & \text{outside the object.} \end{cases}$$

So the object is entirely represented by the equivalent current distribution.

The aim of our study is to determine the shape and the location of the object, hence to determine the equivalent current distribution. This distribution is related with the scattered field $\vec{E}$ by the following expression :

$$\vec{E}(\vec{r}) = -i\omega_0\mu \iiint_{(v)} \overline{\overline{G}}(\vec{r},\vec{r'})\,\vec{J}(\vec{r'})\,dv \;+\; \vec{B} \tag{1}$$

where

◇ $\vec{E}$ is a vectorial and complex representation of the scattered field,

◇ $\vec{J}$ is the vector which represents the equivalent current distribution,

A. Mohammad-Djafari and G. Demoments (eds.), Maximum Entropy and Bayesian Methods, 265–272.

◇ $\overline{\overline{G}}(\vec{r},\vec{r'})$ is Green's dyadic which is given by

$$\overline{\overline{G}}(\vec{r},\vec{r'}) = \left[\overline{\overline{I}} - \frac{1}{k^2}\overline{\overline{\vec{\text{grad}}_r \vec{\text{grad}}_{r'}}}\right]\Phi$$

◇ Φ is Green's scalar function having the following form

$$\Phi = \frac{e^{-ik|\vec{r}-\vec{r'}|}}{4\pi \mid \vec{r}-\vec{r'} \mid}$$

◇ $\overline{\overline{I}}$ is the identity dyadic,

◇ k is the wave number,

◇ $\vec{r}(x,y,z)$ is the vector which marks an observation point,

◇ $\vec{r'}(x',y',z')$ is the vector which locates a point inside the source,

◇ v is the volume of the object,

◇ ω_0 is the pulsation of the incident wave,

◇ μ is the magnetic permeability of the medium,

◇ $\vec{B}$ is a vectorial and complex representation of the noise.

The principle of the imaging system can be seen in Fig.1.

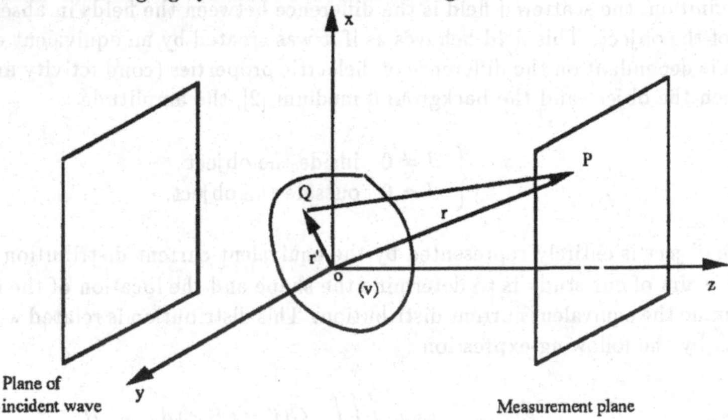

Figure 1. Principal schema of the imaging system.

Unfortunately, the reconstruction of the equivalent current distribution from the scattered field is an ill-posed inverse problem because of the nonuniqueness in the solution.
This nonuniqueness exactly reflects the ambiguity of a physical phenomenon in the problems of "Inverse Source" and of "Inverse Scattering" : indeed, there are an infinity of current densities which can create the same field or there are several objects which can produce the

same scattered field; therefore, the scattered field data are not sufficient to specify a unique reconstruction [3], [5]. To solve in a satisfactory way these inverse problems, we propose a statistical regularization method using the Bayesian approach with the maximum entropy (ME) principle to take into account the prior information in a coherent manner [10].

2. Bayesian maximum entropy image reconstruction

The integral equation (1) relating the data to the unknown parameters, when discretized, has the following form :

$$e = Gj + b \tag{2}$$

where

- $\diamond$ $e = \{e_1, e_2, ..., e_M\}$ is the data vector whose element e_k is the scattered field sample at the kth point on the measurement surface and M is the number of measurement points;

- $\diamond$ $j = \{j_1, j_2, ..., j_N\}$ is the unknown vector whose element j_k represents the unknown equivalent current distribution at the kth pixel of the object and N is the number of pixels;

- $\diamond$ $G = \{g_{m,n}, m = 1, ..., M, n = 1, ..., N\}$ is a matrix of size $M \times N$ whose elements are calculated from Green's dyadic $\overline{\overline{G}}(\vec{r}, \vec{r'})$;

- $\diamond$ $b = \{b_1, b_2, ..., b_M\}$ is the noise vector whose elements represent the noise at the different points of measurement.

The proposed method is based on the Bayesian maximum *a posteriori* estimation (MAP) with the ME *a priori* laws. The theoretical foundation of the estimation method is not the aim of this paper and can be found elsewhere [7], [8], [10]-[12]. Here, we develop the method in the case of image reconstruction from the scattered field. However, the principle can be summarized as follows :

- Assign a prior probability law $p(j)$ to the unknown object j according to our prior knowledge about it.

- Assign a conditional probability law $p(e/j)$ to the measurements according to our knowledge of the noise statistics.

- Bayes's theorem yields the *a posteriori* probability law $p(j/e)$ proportional to

$$p(j/e) \propto p(j)\, p(e/j). \tag{3}$$

- The MAP solution is obtained by maximizing the *a posteriori* likelihood :

$$\widehat{j} = \arg\max_{j} \{p(j/e)\}. \tag{4}$$

In the Bayesian approach, the main difficulty is how to deduce an *a priori* probability law $p(j)$ for the image from the prior knowledge. In general, the prior information is not directly given in a probability form and does not yield a unique prior law either. In the cases where the prior knowledge takes the form of constraints on the expectations, the application of the ME principle leads to choose, among the possible distributions, the law which maximizes the Shannon entropy

$$\hat{p}(j) = \arg\max_{p \in P} \left\{ - \int p(j) \log p(j) \, dj \right\} \tag{5}$$

where P is the set of probability distributions satisfying these constraints.
The justification for the use of ME lies in the fact that the ME distribution exactly reflects the state of prior knowledge, in other words it contains no more information than the possessed prior knowledge. Furthermore, the ME principle permits to translate by a coherent manner this information into a probability law.
Indeed, let us consider the case where the available prior information is a global constraint on the expectations such as :

$$\begin{cases} E\{H(j)\} = h \\ E\{S(j)\} = s \end{cases} \tag{6}$$

where H(j) and S(j) are two known functions.
The maximization of entropy (5) subjected to constraints (6) gives a solution in the exponential form for the *a priori* probability density function [11]:

$$p(j) = \frac{1}{Z(\lambda, \mu)} \exp\left[-\lambda H(j) - \mu S(j)\right] \tag{7}$$

where the partition function $Z(\lambda, \mu)$ is given by

$$Z(\lambda, \mu) = \int \exp\left[-\lambda H(j) - \mu S(j)\right] dj$$

and the Lagrange multipliers λ, μ are deducible from (h, s) by solving the system of equations

$$\begin{cases} \frac{\partial \ln Z(\lambda,\mu)}{\partial \lambda} = h \\ \\ \frac{\partial \ln Z(\lambda,\mu)}{\partial \mu} = s. \end{cases} \tag{8}$$

If we assume that the additive noise is white, zero mean and of the known variance σ^2, the ME principle leads to a Gaussian distribution for the probability law of measurement $p(e/j)$:

$$\mathrm{p}(e/j) \propto \exp[-\mathrm{Q}(j)] \tag{9}$$

where $\quad Q(j) = [e - Gj]^t W [e - Gj] \quad$ and $\quad W = \mathrm{diag}\left[\frac{1}{\sigma_1^2}, \ldots, \frac{1}{\sigma_M^2}\right].$

Taking into account expressions (7) and (9) for $p(j)$ and $p(e/j)$ respectively, the MAP solution (4) is equivalent to

$$\hat{j} = \arg\min_{j} \{F(j) = Q(j) + \lambda H(j) + \mu S(j)\} \tag{10}$$

in which both the noise and the prior information are present.

So, the image to be recovered is one that satisfies criterion (10). In summary, the Bayesian estimation with the ME principle gives a solution regularized thanks to the prior knowledge which is introduced in the form of the regularizing functions H and S or equivalently which is translated into the prior probability law in a coherent way.

An elaborate research of the possible forms for H and S shows that under a scale invariance axiom (i.e. when we change the scale of the image, the form of the prior probability law must remain invariant) and with the hypothesis of the independence of pixels, the admissible forms for H and S are restricted to simple combinations of power and logarithmic functions [11]. In our study, we have considered the following two typical forms of (H, S) :

$$H(j) = \sum_{k=1}^{N} \log j_k, \quad S(j) = \sum_{k=1}^{N} j_k \tag{11}$$

if the pixel values must be positive, else

$$H(j) = \sum_{k=1}^{N} j_k^2, \quad S(j) = \sum_{k=1}^{N} j_k \, . \tag{12}$$

3. Numerical simulation process

According to the above mentioned method, the reconstruction simulation process is carried out as follows :

1. Choose an original object (j).
2. Calculate the scattered field (e).
3. Reconstruct the object $\hat{j}$ using criterion (10).

The reconstruction procedure has been illustrated in a two-dimensional problem. Let us use as the incident wave an x axis polarized plane-wave propagating in the z axis, i.e., the amplitude of incident wave

$$E_i(\vec{r}) = E_{ix} \exp(-ikz) \qquad \text{and} \qquad E_{iy} = E_{iz} = 0.$$

The frequency of the incident wave is 2.5 Ghz.

The scattered field is calculated in a plane parallel to the plane of the incident wave and placed behind the object (Fig. 1), at a measurement distance of five wavelengths. The calculation of one component of relation (1), E_x and J_x for example, is as follows :

$$E_x(\vec{r}) = -i\omega_0\mu \iiint_{(v)} \left(\Phi - \frac{1}{k^2} \frac{\partial^2 \Phi}{\partial x \partial x'} \right) J_x(\vec{r'}) \, dv \, . \tag{13}$$

Each step is detailed below :

Step 1: in our simulation the object is made of a number of pixels traversed by an equivalent current distribution $\boldsymbol{j} = \{j_1, j_2, ..., j_N\}$ with j_k is different to zero inside the object and equal to zero outside the object. In the paper, we have considered $\boldsymbol{j}$ complex.

Step 2: the scattered field is calculated by (13) on a plane surface and the values are arranged in vector $\boldsymbol{e}$. To simulate the measurement noise, a Gaussian random noise is generated with respect to a given signal-to-noise ratio and added to vector $\boldsymbol{e}$.

Step 3 : minimisation of the function $\{\boldsymbol{F}(\boldsymbol{j}) = \boldsymbol{Q}(\boldsymbol{j}) + \lambda \boldsymbol{H}(\boldsymbol{j}) + \mu \boldsymbol{S}(\boldsymbol{j})\}$ provides the estimated object $\widehat{\boldsymbol{j}}$. This step uses the conjugate gradient technique that considerably economizes the occupied memory in multidimensional reconstruction.

Another difficulty is to determine the parameters λ and μ. In general, the expectations h and s in (8) are not known in reality. In our study, we have constructed an algorithm for estimating λ and μ from the estimated image $\widehat{j}$ of each iteration by the moment method [11].

4. Results

For validation of the method, a number of reconstructions were realized.

Fig.2 : Reconstructions of the inhomogeneous compound object from the simulated data for different levels of noise (the signal-to-noise ratio $\frac{S}{N}$ varying between 20dB and 5dB):

$$\left(\frac{S}{N} = 20\text{dB}, \frac{S}{N} = 10\text{dB}, \frac{S}{N} = 5\text{dB}\right).$$

In order to estimate the reconstruction quality, we have calculated the mean energy of reconstruction error for the image :

$$\mathrm{D} = \frac{\sum_{k=1}^{N} (j_k - \widehat{j}_k)^2}{N}$$

whith N the number of image pixels.

The following table shows the quality measurements of reconstructions in presence of noise :

Levels of noise	20 dB	10 dB	5 dB
D of the Fig.2	0.04892	0.04975	0.06525

Furthermore, the absolute original and reconstructed (maximal and minimal) values are visualized and prove the good similarity between them.

5. Conclusions

The numerical simulation results show the high quality of the Bayesian ME reconstruction. Even in the case of an important noise or missing data , the absolute values of pixels can be changed but the object shape is always well reconstructed. These interesting results can be explained by the fact that in the case of missing or noisy data the good reconstruction is realized as a result of the appropriate regularization introduced in a coherent way by the ME principle.

Although there exists a relation in the Fourier domain between the scattered field and the equivalent current distribution and most of image reconstruction methods often use the Fourier data [2], [9], we here propose a statistical method which directly operates in the spatial domain. So we can eliminate the errors coming from the numerical calculation of direct and inverse Fourier transformations and from the truncation of signals.

These advantages show the interesting possibilities of the statistical method with the ME *a priori* laws in image reconstructions.

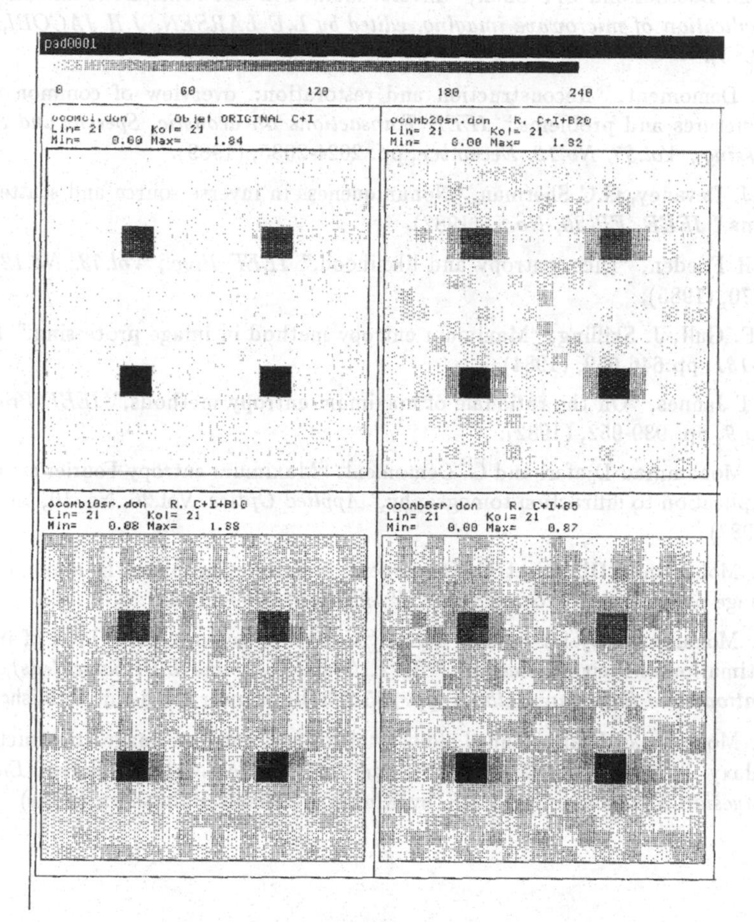

Fig.2 : Reconstructions of the inhomogeneous compound object from the simulated data for different levels of noise : (a) is the original object; (b),(c) and (d) are the reconstructed objects with $\frac{S}{N} = 20\text{dB}, \frac{S}{N} = 10\text{dB}, \frac{S}{N} = 5\text{dB}$ respectively.
The different grey levels present the range of absolute values.

References

[1] A.T. Bajkova, "The generalization of maximum entropy method for reconstruction of complex functions," *Atron. and Astrophys. Trans.1*, pp:313-320, (1992).

[2] M. Baribaud and M.K. Nguyen, " Maximum entropy image reconstruction from microwave scattered fiel distribution," *Proc. of the 18th European Microwave Conference, Sweden*, (1988).

[3] W.M Boerner and C.Y Chan, " Inverse method in electromagnetic imaging," *Medical application of microwave imaging, edited by L.E LARSEN, J.H JACOBI, M.S.E.E*, (1985).

[4] G. Demoment, "Reconstruction and restoration: overview of common estimation structures and problems," *IEEE Transactions on acoustic. Speech. and Signal processing , Vol.37, No.12. December*, pp. 2024-2035, (1989).

[5] A.J. Devaney, G.C Sherman, "Nonuniqueness in inverse source and scattering problems," *IEEE, EP-30, No.5*, (1982).

[6] B.R Frieden, "Dice, entropy and likelihood," *IEEE Proc., Vol.73, No.12*, pp:1764-1770, (1985).

[7] S.F. Gull, J. Skilling, "Maximum entropy method in image processing," *IEE Proc., F-131*, pp:646-659, (1984).

[8] E.T Jannes, "On the rationale of maximum entropy methods," *IEEE Proc., Vol.70, No.9*, pp. 939-952, (1982).

[9] A. Mohammad-Djafari and G. Demoment, "Maximum entropy Fourier synthesis with application to diffraction tomography," *Applied Optics*, Vol.26, No. 10, pp:1745-1754, (1987).

[10] A. Mohammad-Djafari et G. Demoment, " Estimating priors in maximum entropy image processing," *Proc. of ICASSP 1990*, pp. 2069–2072.

[11] A. Mohammad-Djafari and J. Idier, "Maximum entropy prior laws of images and estimation of their parameters ," *W.T. Grandy, Jr. and L.H. Schick(eds), Maximum Entropy and Bayesian Methods*, pp. 285-293, by Kluwer Academic Publishers, (1991).

[12] A. Mohammad-Djafari and J. Idier, "Maximum entropy image construction of the galaxy M51," *W.T. Grandy, Jr. and L.H. Schick(eds), Maximum Entropy and Bayesian Methods*, pp. 313-318, by Kluwer Academic Publishers, (1991).

MAXIMUM ENTROPY IMAGE RECONSTRUCTION IN EDDY CURRENT TOMOGRAPHY

Mila Nikolova and Ali Mohammad-Djafari
Laboratoire des Signaux et Systèmes (CNRS–ESE–UPS)
École Supérieure d'Électricité
Plateau de Moulon, 91192 Gif–sur–Yvette Cédex, France

ABSTRACT. In the area of non destructive control of conducting media, eddy current testing is of major importance. Recently, some authors have used eddy current data to get tomographic images of anomalies inside a conductive medium. In this paper, the same imaging technique is reformulated as an inverse problem, namely, the inversion of a functional of a Fourier-Laplace Transform (FLT). As the problem is very complex and particularly ill-conditionned, usual techniques approximate the FLT by a simple Fourier transform (FT), which amounts to partially or totally neglect attenuation.

In this paper the inverse problem is directly coped in the space domain and attenuation is fully taken into account. Prior information is introduced using maximum entropy principles, as a particular case of application of the maximum entropy technique exposed in [1].

1. Introduction

The object of Eddy Current Tomography is to provide a representation of the inner structure of a conductive medium in order to detect and characterize anomalies such as crack, notches, corrosion, stress. The basic idea is to get the variation of the conductivity or permeability of the medium. Given incident field, Maxwell's equations provide an implicit relation between the conductivity and the permeability of the medium on the one hand, and the field diffracted by the inhomogeneities, on the other hand:

$$[\Delta + k^2(r)]\vec{\mathcal{E}}(r) = 0, \tag{1}$$

$$k^2(r) = \omega^2\mu(r)\epsilon_0 + i\omega\mu(r)\sigma(r), \tag{2}$$

where $r = [x, y, z]^t$ is the position vector, ω is the pulsation of the field, ϵ_0 is the dielectric permittivity of the medium, $\mu(r)$ is the magnetic permeability, $\sigma(r)$ is the electric conductivity, $k^2(r)$ is the wave number, $\vec{E}(r,t)$ is the electric field which is supposed time-harmonic, $\vec{E}(r,t) = \Re\,[\vec{\mathcal{E}}(r)\,e^{-i\omega t}]$. From Equations (1)-(2) it is obvious that the presence of a defect, equivalent to a variation of $\sigma(r)$ in the material, leads to a change in the total field $\vec{\mathcal{E}}(r)$. In the following we are interested in testing linear, isotropic, highly conductive and non-magnetic materials, so $\mu = \mu_0$ is constant and the parameter of interest is the relative variation of the conductivity. The object function—the "image" of the interior structure—is defined as

$$f(r) = \frac{\sigma_0 - \sigma(r)}{\sigma_0},$$

where σ_0 is the value of conductivity in homogenous zones and $\sigma(r)$ is the conductivity inside the anomaly and is space-varying. $f(r)$ is a contrast function and in the homogenous zone it fulfills $f(r) = 0$.

A. Mohammad-Djafari and G. Demoments (eds.), Maximum Entropy and Bayesian Methods, 273–278.

The relation between the diffracted (anomalous) field $\vec{\mathcal{E}}_A$ and the object function is given by a Fredholm equation of the second type :

$$\vec{\mathcal{E}}_A(r) = \vec{\mathcal{E}}(r) - \vec{\mathcal{E}}_0(r) = \int_\Omega G(r, r\prime)\, f(r)\, k_2^2\, \vec{\mathcal{E}}(r\prime) dr\prime, \tag{3}$$

where Ω designate the anomaly, k_2 is the propagation constant in the homogeneous embedding and $k_2^2 \simeq i\mu_0\sigma_0\omega^2$, $\vec{\mathcal{E}}_0$ and $\vec{\mathcal{E}}$ are respectively the incident and the total field and $G(r, r\prime)$ is a Green function. Since defects are of small size or of weak diffraction, (3) is linearized thanks to the Born approximation which supposes that inside the defect $\vec{\mathcal{E}}(r) \simeq \vec{\mathcal{E}}_0(r)$.

The configuration we consider is presented on Fig. (a) and the relevant direct problem is developped in [3] and [4]. A two-dimensional case is treated—the anomaly is cylindrical along the z-axis and we reconstruct its profile in the (x, y) plane. A time-harmonic normally incident electrical field $\vec{\mathcal{E}}_0$ is emitted. The measured quantity is the voltage induced into a sensing coil along a line under the conductive material and parallel to it at height $-y_0$ and this quantity is directly related to the diffracted (anomalous) magnetical field $\mathcal{H}_A$. Since $\vec{\mathcal{E}}_0$ is of normal incidence and $-y_0$ is fixed, the relation between the electrical and the magnetical field is given by $\frac{\partial \mathcal{E}_A}{\partial x} = -i\mu_0\omega\mathcal{H}_{A\,y}$, where $\mathcal{H}_{A\,y}$ is the component of $\mathcal{H}_A$ along the y-axis. The integral formulation relating the contrast function and the magnetic field (e.g. the observation equation) is

$$\mathcal{H}_{A\,y}(x, y_0) = -\frac{T}{\delta^2\pi\mu_0\omega}\int_{\mathrm{R}} \underbrace{\frac{e^{-i\beta_1 y_0}}{\beta_1+\beta_2}}_{C(f,u)} u e^{iux} \underbrace{\iint_\Omega f(x\prime, y\prime) e^{-iux\prime} e^{i(k_2+\beta_2)y\prime} dx\prime dy\prime}_{\mathcal{F}(u)}\, du, \tag{4}$$

where $\delta = (\frac{1}{2}\omega\mu_0\sigma_0)^{-\frac{1}{2}}$ is the skin depth and decrease as the pulsation ω increase, u is the space frequence corresponding to the x-axis, $\beta_1 = \sqrt{k_1^2 - u^2}$ and $\beta_2 = \sqrt{k_2^2 - u^2}$ are the propagation factors respectively in the air and in the conductive medium with k_1—the propagation constant in the air, $k_1^2 = \mu_0\epsilon_0\omega^2$ and $T \simeq \frac{2k_1}{k_2}$ is the transmission coefficient on the air-metal interface. Note that $k_2 = \frac{1}{\delta}(1+i)$ in the conductive material. β_2 and k_2 are complex, so that $\mathcal{F}(u)$ results from a FT along the x-axis and a Laplace transform (LT) along the y-axis of the object function $f(x, y)$; the LT is calculated on the the algebraic curves defined by $p(u) = -i[k_2(\omega) + \beta_2(\omega, u)]$. Therefore $\mathcal{H}_{A\,y}$ is proportionnal to the inverse FT of a functional of $\mathcal{F}(u)$—the FLT of the image—along the algebraic curves $R \mapsto R \times C, u \mapsto (u, p)$ as defined. In order to cover the Fourier-Laplace domain better, measures are sampled for a set of electrical fields with different pulsations, so that in the Fourier-Laplace domain data are available along a set of parallel curves.

2. Maximum Entropy solution

The inverse problem is the one of resolution of (4): given the measures $\mathcal{H}_{A\,y}(x, y_0)$, one has to recover the object function $f(x, y)$. The resolution of the integral equation (4) is an extremely ill-conditionned problem [6]. Firstly, a laterally attenuating function $C(f, u) = \frac{\exp(-i\beta_1 y_0)}{\beta_1+\beta_2}$, $y_0 < 0$ multiplies the FLT of the image $\mathcal{F}(u)$, so the measures contain a relatively poor amount of information about the object function. Secondly, the higher the field pulsation, the sharper vertical resolution gets, but more important is the skin-effect, which strongly limits the exploration depth. And thirdly, our problem is a missing data one, because the Fourier-Laplace domain is covered sparsely and irregularily.

The matrix representation of the problem can be written

$$\mathcal{H} = \boldsymbol{B}\, F + n \tag{5}$$

where $\mathcal{H}$ is a complex vector containing the measures sampled on the measurement line for all the incident field pulsations, $\boldsymbol{B}$ is a complex projection matrix, issued from (4), F is a vector representation for the raster of the unknown real-valued image and n is a noise. The matrix $\boldsymbol{B}$ depends on the considered configuration, the material and the operating field pulsations.

For the resolution of our problem we have applied the maximum entropy technique exposed in detail in [1]; see also [2]. Let us briefly sum it up. A maximum a posteriori solution is seeked:

$$\hat{f} = \arg\max_F [\ \ln p(F = f/\mathcal{H} = h)]. \tag{6}$$

The noise n being supposed $\mathcal{N}(0, \sigma^2 I)$, the likelihood is

$$\ln p(\mathcal{H} = h/F = f) \propto -\frac{1}{\sigma^2}\|h - \boldsymbol{B} f\|^2 \tag{7}$$

A Maximum Entropy prior law is established in the following manner. Given the means

$$E_F[\phi_1(f)] = s_1 \quad \text{and} \quad E_F[\phi_2(f)] = s_2 \tag{8}$$

with $\phi_1(.)$, $\phi_2(.)$ some functionals, maximum entropy prior law for F is of the form

$$p(F = f) = \frac{1}{Z}\exp[-\lambda\phi_1(f) - \mu\phi_2(f)], \tag{9}$$

where λ and μ are Lagrange multipliers and Z is a normalizing constant. Under the hypothesis that the pixels of F are independent and interchangeable, we can write:

$$\phi_1(f) = \sum_i \phi_1(f_i) \quad \text{and} \quad \phi_2(f) = \sum_i \phi_2(f_i). \tag{10}$$

In order to satisfy some scale invariance requirements, $\phi_1(.)$ and $\phi_2(.)$ must appropriately combine powers and/or logarithms of f [1]. Given the nature of the problem, pixels of the image are constrained to take positive values only. The particular choice we did is

$$\phi_1(f_i) = \ln(f_i) \quad \text{and} \quad \phi_2(f_i) = f_i, \tag{11}$$

so the prior law is

$$p(F = f) = \frac{1}{Z}\exp[-\lambda\sum_i \ln(f_i) - \mu\sum_i f_i]. \tag{12}$$

There is an analytical relation between λ and μ and the mean m and the variance v of the image [1]:

$$\lambda = -\frac{m^2 - v}{v} \quad \text{and} \quad \mu = \frac{m}{v}. \tag{13}$$

The posterior distribution to maximize is

$$p(F = f/\mathcal{H} = h; \lambda, \mu) \propto \exp[-\frac{1}{\sigma^2} \|h - \boldsymbol{B} f\|^2 - \lambda \sum_i \ln(f_i) - \mu \sum_i f_i]. \quad (14)$$

The MAP solution is estimated via a generalized maximum likelihood criterion. In every iteration the mean and the variance of the currently estimated image $\hat{f}^{k-1}$ are calculated, namely

$$\hat{m}^k = m(\hat{f}^{k-1}) \text{ and } \hat{v}^k = v(\hat{f}^{k-1}), \quad (15)$$

then the Lagrange multipliers $\hat{\lambda}^k$ and $\hat{\mu}^k$ are deduced using (13), and a new estimate $\hat{f}^k$ of the image is obtained:

$$\hat{f}^k = \arg \max_F \ln p(F = f/\mathcal{H} = h; \hat{\lambda}^k, \hat{\mu}^k). \quad (16)$$

3. Numerical results

The configuration we consider is shown on Fig. (a). A time-harmonic current source is placed in the lower halfspace D_1. The anomaly Ω is assumed to be of relatively small cross section in the (x, y)-plane. Measures are taken along the probing line L, $y_0 = 1.4\,10^{-3}m$. A piece of duralumin is tested with $\sigma_0 = 1.96\,10^7\ S/m$ and $\mu_0 = 4\pi\,10^{-7}\ H/m$. We present results of numerical simulations for the particular case of anomaly shown on Fig. (c). The defect is a rectangular notch ($0.3 \times 0.5\ mm$) emerging in air. Inside the anomaly $f(x, y) = 1$. The material is illuminated as on Fig. (a). Exact synthetic data of the anomalous magnetic field diffracted by the defect were calculated by a finite element code (e.g. without the Born approximation) for 10 excitation pulses (between about 10 kHz and 200 kHz). These data are shown on Fig. (c) and were provided by the Division Ondes, LSS [5].

On Fig. (d) the result of a "usual" Fourier method of reconstruction from the exact synthetic data is shown (provided by the Division Ondes, LSS [5]). In Eq. (4), only the real part of β_2 is retained, so that for each excitation pulse, $\mathcal{F}(u)$ is approximated by a Fourier transform of the attenuated object function $e^{-\frac{1}{\delta}} f(x, y)$.

On Figs. (e) and (f) we present the reconstruction results obtained with the proposed in that paper MaxEnt algorithm from respectively the same (noisiless) data and the same data immersed in white Gaussian noise with $S/N = 5\ dB$. Note that the discussed MaxEnt algorithm is established using the Born approximation and that data are exact.

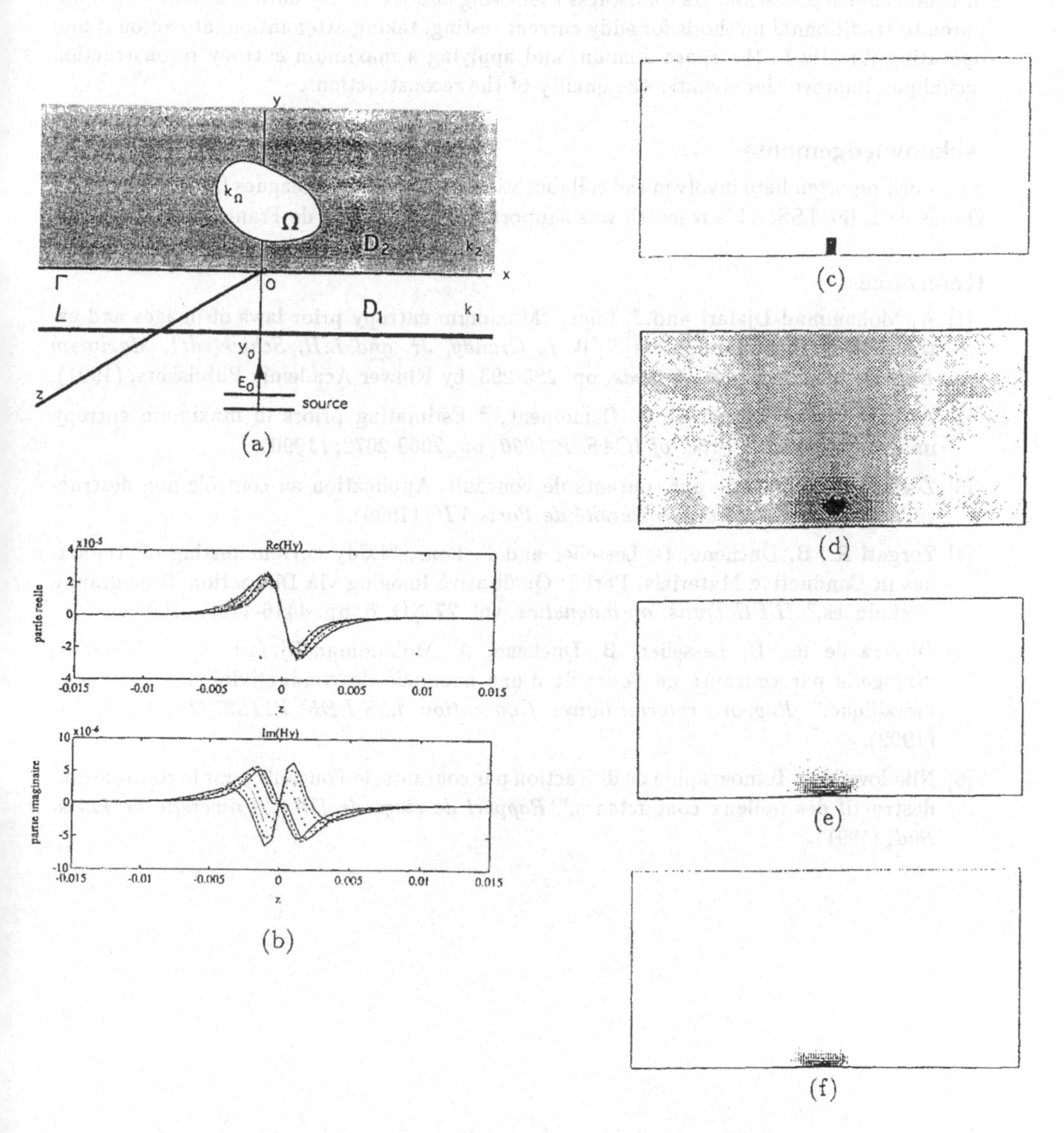

Figures: **(a)** Plane geometry configuration. **(b)** The anomalous magnetic field diffracted by the anomaly, exact synthetic data. **(c)** The original image. **(d)** A "usual" Fourier method of reconstruction from the exact synthetic data. **(e)** The MaxEnt solution from the same data. **(f)** The MaxEnt solution from very noisy synthetic data with $S/N = 5\,dB$.

4. Conclusions

The obtained numerical results confirm the capability of the algorithm to resolve extremely ill-conditionned problems. Its robustness to missing and very noisy data is significant. Compared to traditionnal methods for eddy current testing, taking attenuation into account and operating directly in the space domain, and applying a maximum entropy reconstruction technique, improve significantly the quality of the reconstruction .

Acknowledgements

The work reported here involved the collaboration with several colleagues from the Division Ondes with the LSS. This research was supported by Électricité de France [5].

References

[1] A. Mohammad-Djafari and J. Idier, "Maximum entropy prior laws of images and estimation of their parameters ," *W.T. Grandy, Jr. and L.H. Schick(eds), Maximum Entropy and Bayesian Methods*, pp. 285-293, by Kluwer Academic Publishers, (1991).

[2] A. Mohammad-Djafari et G. Demoment, " Estimating priors in maximum entropy image processing," *Proc. of ICASSP 1990*, pp. 2069-2072, (1990).

[3] Zorgati R., "Imagerie par courants de Foucault. Application au contrôle non destructif," *Thèse de Doctorat, Université de Paris VII*, (1990).

[4] Zorgati R., B. Duchene, D. Lesselier and F. Pons, "Eddy current testing of Anomalies in Conductive Materials, Part I: Qualitative Imaging via Diffraction Tomography Techniques," *IEEE Trans. on magnetics*, vol. 27 NO. 6, pp. 4416-4437, (1991).

[5] Olivera de R., D. Lesselier, B. Duchene, A. Mohammad-Djafari et M. Nikolova, "Imagerie par courants de Foucault d'une anomalie de conductivité dans un block métallique," *Rapport intermédiaire, Convention LSS-EDF P11L22/4L9034-I7205*, (1992).

[6] Nikolova M., "Tomographie de diffraction par courants de Foucault pour le contrôle non destructif des milieux conducteurs," *Rapport de stage de DEA, Université de Paris-Sud*, (1991).

Chapter 5

Applications

CONDUCTIVITY-DEPTH-IMAGING OF AIRBORNE ELECTROMAGNETIC STEP-RESPONSE DATA USING MAXIMUM ENTROPY.

P.B. Leggatt
Geophysics Department
Anglo American Corporation
Johannesburg, South Africa

N. Pendock
Department of Computational and Applied Mathematics
University of the Witwatersrand
Johannesburg, South Africa

ABSTRACT. An electromagnetic step transient waveform over an earth consisting of horizontal conductive layers, will induce a current flow in the ground, whose magnetic field can be approximately described as that of an image of the transmitter reflected at the ground surface and receding downwards as modelled by the equation.

$$t_k/\mu_0 = \int_0^{h_k} (h_k - z)\sigma(z)dz$$

h_k is half the image depth below the surface at time t_k after the step transient and $\sigma(z)$ is the conductivity of the layered earth as a function of z, the depth below the surface. The objective is to estimate $\sigma(z)$. To do so we discretize the above convolution and solve the resulting underdetermined linear system. From the many possible solutions which fit the data, the one with maximum entropy is chosen as the answer. Prior information about the nature of the layered earth can be incorporated in the inversion. We present some conductivity-depth profiles estimated from data collected by the SPECTREM airborne electromagnetic system.

1. Introduction

In the last century, the response of a thin conductive sheet to a sudden change in the field of a magnetic dipole was computed (Maxwell 1873) in terms of an image of the dipole below the sheet which recedes at a constant velocity inversely proportional to the conductance (product of the conductivity and thickness) of the sheet. Polzer (1985)

A. Mohammad-Djafari and G. Demoments (eds.), Maximum Entropy and Bayesian Methods, 281–286.

showed that an estimate of the change $\Delta\sigma(z)$ in the conductivity of a model required to reduce the step response delays Δt to zero, is given by a solution to

$$\Delta t_k = \mu_0 \int_0^\infty K_k(z)\Delta\sigma(z)dz$$

where the Frechet kernel $K_k(z)$ is the arrival time functional. It was further observed (Macnae and Lamontagne 1987) that if the penetration depth h was taken as half the image depth D, then the apparent downward velocity of the penetration depth $\partial h/\partial t$ was inversely proportional to the cumulative conductance between h and the surface. Polzer was lead to deduce the approximate result

$$t_k/\mu_0 = \int_0^{h_k} (h_k - z)\sigma(z)dz \qquad 1.1$$

Zero time is taken at the instant the transient occurs, at which moment the image dipole is at a distance below the ground surface equal to the height of the transmitter above it. Macnae and Lamontagne wrote computer software to transform step response data to apparent image depths, a process which they term "Depth Image Processing".

The SPECTREM airborne system is a development of the PROSPECT 1 EM system described by Annan (1986) and is operated by the Anglo American Geophysical Department. The transmitter function is actually not a step, but real time processing aboard the aircraft computes the unit step response from the measured response.

In a recent paper, (Macnae,Smith *et al* 1991) the depth image algorithm as applied to the SPECTREM system is described and equation 1.1 is solved by fitting a set of damped piecewise cubic polynomials to the times and penetration depths. These interpolating polynomials are then used to estimate the derivatives $\partial^2 t/\partial z^2$ and the conductivities from the relation

$$\sigma(h) = \frac{1}{\mu_0}\frac{\partial^2 t}{\partial z^2} \qquad 1.2$$

Because of noise in the data, its amplification by the differentiation process and for other less quantifiable reasons, it is necessary to smooth the results considerably. Macnae *et al* use both a first and and a second derivative smoothing procedure, the degree of smoothing being chosen by examination of the output images and adjusting the smoothing until the plotted section matches whatever preconceptions the user may have of the result. This practice, of making a computation and then manipulating the results to fit the prior preconceptions of the analysts is regrettably common in geophysical processing practice. It serves as a logically unsatisfying way of forcing a solution to be geologically plausible. A far better way to approach the problem is to pose the question "Of the many possible solutions, which is the most probable?"

2. Least Squares

We may hypothesize that the noise e_k associated with a particular observation t_k, has a zero mean Gaussian distribution with standard deviation ω_k. The result

$$t_k/\mu_0 = \int_0^{h_k} (h_k - z)\sigma(z)dz \qquad 2.1$$

can be discretized at intervals of Δz for M intervals and making specific mention of the noise as

$$t_k = \mu_0 \sum_{j=1}^{M} (h_k - j\Delta z)\sigma_j \Delta z + e_k \qquad 2.2$$

for $k = 1, \cdots, N$. We can compute a noise free approximation to t_k as

$$\hat{t}_k = \mu_0 \sum_{j=1}^{M} (h_k - j\Delta z)\sigma_j \Delta z \qquad 2.3$$

Macnae *et al* present an argument that $\omega_k \propto t_k^2$ which, since the latest time is about 100 times the earliest time, means that the noise of the latest time is 10000 times the noise of the earliest. We feel that this is certainly not supported by examination of the data and that the ratio ω_N/ω_1 is closer to 10 than 10000. Accordingly we suggest that $\omega_k^2 \propto t_k$ is a more reasonable estimate. Thus a solution for a particular t_k can be computed by a least square minimization of

$$C_k(\sigma) = \left(\frac{t_k - \hat{t}_k}{\omega_k}\right)^2 \propto \frac{(t_k - \hat{t}_k)^2}{t_k}$$

Now $e_k = t_k - \hat{t}_k$ is conditional on σ,and the general context of the problem U, so the likelihood that the data t_k, fit the model σ is

$$p(t_k \mid \sigma U) = \frac{1}{\sqrt{2\pi\omega_k^2}} e^{-\frac{1}{2}C_k(\sigma)}$$

If it is further assumed that the e_k are logically independent then the compound probability that all the data fit the model is just the product

$$P(t \mid \sigma U) = \prod_{k=1}^{N} (2\pi\omega^2)^{-1/2} e^{-1/2 \sum C_k} \qquad 2.4$$

This simple least squares procedure would be sufficient to find the model if we knew it to be unique. We have however N values of t and M values of σ with $M > N$, so there will be many plausible models.

Furthermore the likelihood that the data fit a given model is not actually what is sought. We have after all only a single set of data. What is needed is the identity of the most probable model to fit the (given) data.

3. Bayesian Conductivity Estimation

Let $P(\sigma(z) \mid t_k U)$ denote the marginal probability of occurrence of a particular conductivity distribution $\sigma(z)$ (model) given the observed delay times t_k for k=1,2 ...N. Bayes' theorem allows us to write

$$P(\sigma \mid t_k U) = \frac{P(\sigma \mid U).P(t_k \mid \sigma U)}{P(t_k \mid U)} \qquad 3.1$$

As we are not particularly concerned with the value of $P(\sigma \mid t_k U)$ but just the value of σ at this maximum, we can ignore the term $P(t_k \mid U)$ which will be constant for all sets of σ considered. The value of $P(\sigma \mid U)$ is the prior probability distribution for σ and $P(t_k \mid \sigma U)$ is the evidence for σ supplied by the data t_k which has been considered above.

In order to compute the maxima for $P(\sigma \mid t_k U)$ we need an explicit expression for the prior distribution $P(\sigma \mid U)$.

Jaynes (1968) proposed that the appropriate prior distribution should be an entropic prior, the procedure to adopt being to maximize the entropy function

$$S = -\int p(x) log \frac{p(x)}{m(x)} dx \qquad 3.2$$

subject to constraints imposed by the data. $p(x)$ and $m(x)$ are probabilities, i.e. $\sum p(x) = \sum m(x) = 1$. Skilling (1990) showed that we can relax the condition that the distributions sum to unity, in which case the correct form of the entropy is

$$S(\sigma) = \sum_{k=1}^{M} (\sigma_k - m_k - \sigma_k \log(\frac{\sigma_k}{m_k})) \qquad 3.3$$

where S is the entropy and m is the prior distribution of σ in the absence of data. In the case of positive additive σ, Skilling (*ibid*) also showed the correct prior probability to be

$$P(\sigma \mid U) = \left(\frac{\alpha}{2\pi}\right)^{\frac{M}{2}} e^{\alpha S} \qquad 3.4$$

where α can for the moment be regarded as a scaling constant. Using Bayes' theorem to combine equations 2.4 and 3.4 we see that

$$P(\sigma \mid t_k U) \propto P(\sigma \mid U).P(t_k \mid \sigma U)$$

$$P(\sigma \mid t_k U) \propto (\frac{\alpha}{2\pi})^{M/2} \prod_{k=1}^{N} (2\pi\omega^2)^{-1/2} e^{\alpha S - 1/2 \sum C_k(\sigma)}$$

This is maximized when $\alpha S - \frac{1}{2} \sum C_k(\sigma)$ is maximized. The optimal value of α being the remaining item to determine. The global maximum of the entropy is zero and occurs when $\sigma_k = m_k$ for each k.

4. Application

While studying the method of Macnae *et al* it is difficult at first to understand the need for the elaborate damping procedures used. When applied to actual measured data the reason becomes clearer and illustrates a situation common in quantitative geophysical interpretation. Usually by the time an interpretation is called for, a fair amount of ancillary information about the test site has been obtained by other earth scientists such as geologists and geochemists. This information, together with one's experience of past work on similar targets, and the objective of producing a **geologically** plausible interpretation, means that one is predisposed to reject a large class of **mathematical** solutions, which are models that "fit the data" to some desired precision but which are geologically implausible.

The Macnae solution is a admirable one, in that the program is equipped with a number of good levers to pull in order to distort the solution away from the geologically

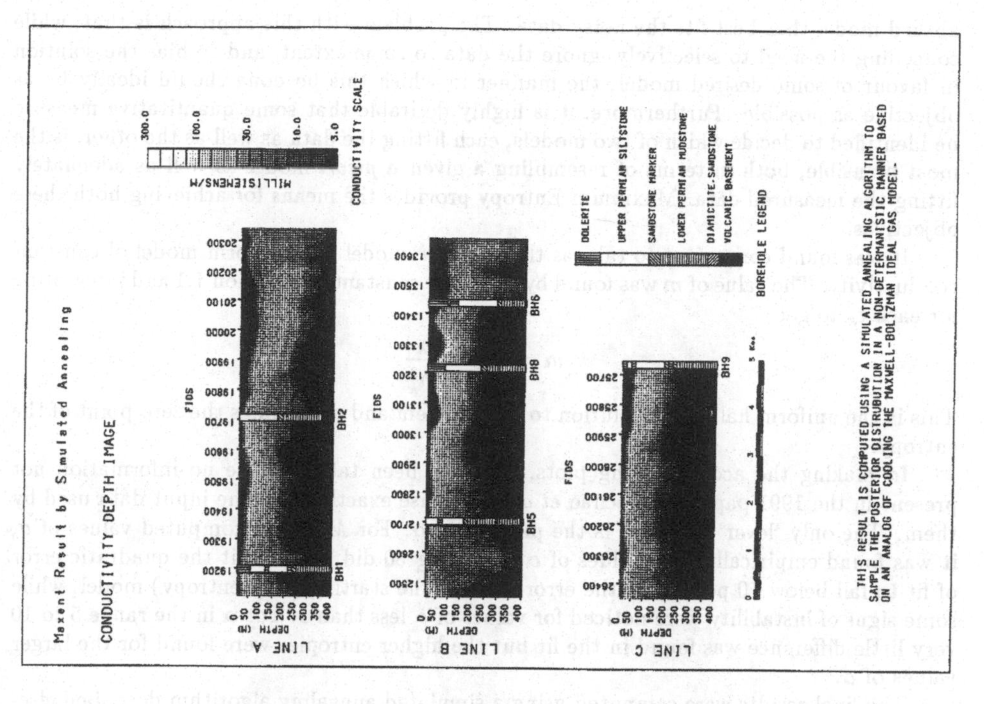
Maxent Result by Simulated Annealing
CONDUCTIVITY DEPTH IMAGE
LINE A
LINE B
LINE C
DEPTH (M)
FIDS
BH1
BH2
BH5
BH6
BH8
BH9
MILLISIEMENS/M
300.0
100.0
30.0
10.0
3.0
CONDUCTIVITY SCALE
DOLERITE
UPPER PERMIAN SILTSTONE
SANDSTONE MARKER
LOWER PERMIAN MUDSTONE
DIAMICTITE.SANDSTONE
VOLCANIC BASEMENT
BOREHOLE LEGEND
THIS RESULT IS COMPUTED USING A SIMULATED ANNEALING ALGORITHM TO
SAMPLE THE POSTERIOR DISTRUBUTION IN A NON-DETERMANISTIC MANNER BASED
ON AN ANALOG OF COOLING THE MAXWELL-BOLTZMAN IDEAL GAS MODEL.

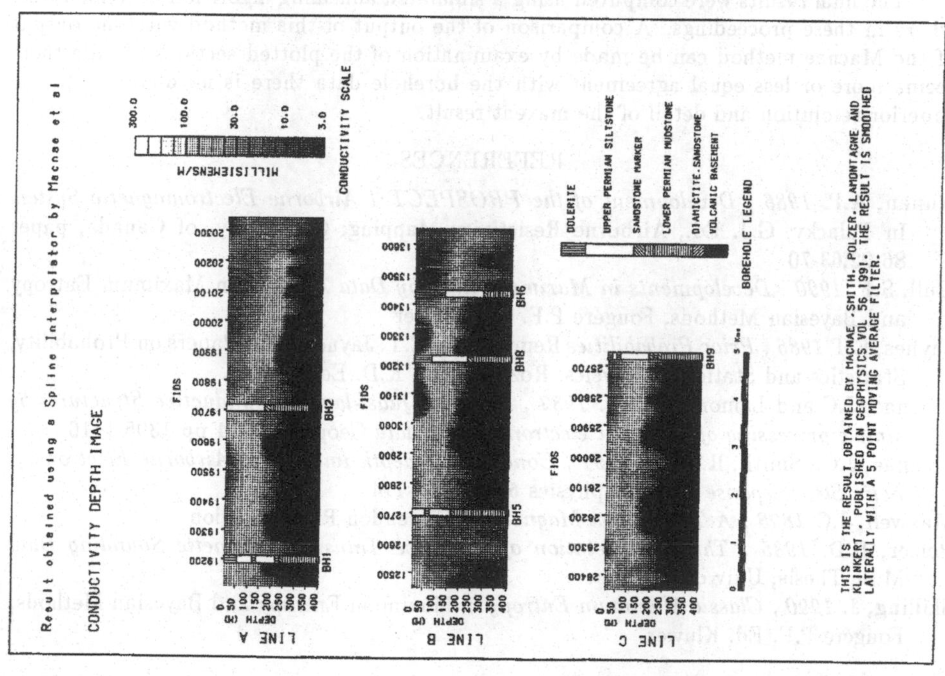
Result obtained using a Spline Interpolator by Macnae et al
CONDUCTIVITY DEPTH IMAGE
LINE A
LINE B
LINE C
DEPTH (M)
FIDS
BH1
BH2
BH5
BH6
BH8
MILLISIEMENS/M
300.0
100.0
30.0
10.0
3.0
CONDUCTIVITY SCALE
DOLERITE
UPPER PERMIAN SILTSTONE
SANDSTONE MARKER
LOWER PERMIAN MUDSTONE
DIAMICTITE.SANDSTONE
VOLCANIC BASEMENT
BOREHOLE LEGEND
THIS IS THE RESULT OBTAINED BY MACNAE, SMITH, POLZER, LAMONTAGNE AND
KLINKERT , PUBLISHED IN GEOPHYSICS VOL 56 1991. THE RESULT IS SMOOTHED
LATERALLY WITH A 5 POINT MOVING AVERAGE FILTER.

absurd model that best fits the noisy data. The problem with this approach is that, while conceding the need to selectively ignore the data to some extent, and to bias the solution in favour of some desired model, the manner in which this be done should ideally be as objective as possible. Furthermore, it is highly desirable that some quantitative measure be identified to decide which of two models, each fitting the data as well as the other, is the most plausible, both in terms of resembling a given *a priori* model as well as adequately fitting the measured data. Maximum Entropy provides the means for achieving both these objectives.

It was found convenient to take as the *a priori* model m a uniform model of constant conductivity. The value of m was found by setting σ constant in equation 1.1 and integrating for each t_k to get

$$m = \frac{2\sum_k^N t_k}{\mu_0 \sum_k^N h_k^2}$$

This is the uniform half-space solution to the problem and determines the zero point of the entropy.

In making the accompanying plots, care has been taken to use no information not present in the 1991 paper by Macnae *et al* and to use exactly the same input data used by them. The only 'lever' provided is the parameter α. For $M = 100$ computed values of σ, it was found empirically that values of α exceeding 50 did not permit the quadratic error of fit to fall below 10 percent of the error of fit of the starting (zero entropy) model, while some signs of instability were noticed for values of α less than 1. For α in the range 5 to 10 very little difference was found in the fit but the higher entropies were found for the larger values of α.

The final results were computed using a simulated annealing algorithm described elsewhere in these proceedings. A comparison of the output of this method with the output of the Macnae method can be made by examination of the plotted sections. While there seems more or less equal agreement with the borehole data there is no doubt as to the superior resolution and detail of the maxent result.

REFERENCES

Annan, A.P. *1986 , Development of the PROSPECT 1 Airborne Electromagnetic System* In Palacky, G.J. Ed., Airborne Resistivity Mapping: Geol. Surv. of Canada, paper 86-22,63-70

Gull, S.F. *1990 , Developments in Maximum Entropy Data Analysis* In Maximum Entropy and Bayesian Methods, Fougére P.F. Ed. Kluwer

Jaynes, E.T *1986 , Prior Probabilities* Reprinted in E.T. Jaynes: 1989 Papers on Probability, Statistics and Statistical Physics. Rozenkrantz, R.D. Ed. Kluwer.

Macnae, J.C and Lamontagne, Y. *1987 , Imaging Quasi-layered Conductive Structures by simple processing of transient electromagnetic data* Geophysics 44 pp 1395-1416

Macnae, J.C., Smith, R. *et al. 1991 , Conductivity-depth Imaging of Airborne Electromagnetic Step-response data* Geophysics **56** pp 102-114

Maxwell, J.C *1873 , A Treatise on Magnetism.* Clarendon Press, London

Polzer, B.D. *1985 , The Interpretation of Inductive Transient Magnetic Sounding Data* M.Sc Thesis, Univ of Toronto.

Skilling, J. *1990 , Classic Maximum Entropy* In Maximum Entropy and Bayesian Methods, Fougére P.F. Ed. Kluwer.

BAYESIAN SOURCE ESTIMATION FROM POTENTIAL FIELD DATA

Neil Pendock
Department of Computational and Applied Mathematics
University of the Witwatersrand
Johannesburg

ABSTRACT. Geophysical potential fields may be processed and interpreted by determining the magnitude and location of a configuration of equivalent sources which explain the measured field. Such a procedure is an ill-posed nonlinear inverse problem and we discuss two methods for locating the simplest configuration of sources which generate the observed field : simulated annealing, a stochastic global optimization method that minimizes a function by analogy to a statistical mechanical system and Bayesian parameter estimation and model selection.

1. Introduction

Suppose we measure some electrical, magnetic or gravitational potential field values $\{g_1, g_2, \cdots, g_n\}$ over a region. These $\{g_i\}$ comprise our data set G. We are interested in testing the hypothesis $\mathcal{H}$ that the data g_i measured at locations with spatial coordinates (x_i, y_i, z_i) were generated by a configuration of N physical sources (located at $(\alpha_j, \beta_j, \gamma_j)$) according to the Newtonian potentials

$$\phi(\mathcal{H})_i = g_i(x_i, y_i, z_i) = \sum_{j=1}^{N} \frac{c_j}{\sqrt{(x_i - \alpha_j)^2 + (y_i - \beta_j)^2 + (z_i - \gamma_j)^2}}$$

where c_j are positive numbers (mass, electrical or magnetic susceptibility). The exploration geophysicist has two reasons for wishing to estimate this source distribution :

- knowledge of the source locations and magnitudes allows the potential field to be calculated at any point and not just at the locations where the $\{g_i\}$ were measured. In particular, randomly sampled data may be gridded to form a raster which may be visualized as an image and further processed and interpreted.
- Particular source distributions may be related to geological models and thus the source estimation process is itself an interpretive exercise hopefully leading to a better geological understanding of the region over which the field was sampled.

In common with many geophysical inverse problems, the source location problem is ill-posed. Clearly, the variation of both a magnitude parameter c_j as well as the source location would make the inverse problem underdetermined. In addition, the presense of noise could make the Newtonian potential inconsistent with the measured data and finally we are attempting to estimate a discrete realization of a continuous reality which also

A. Mohammad-Djafari and G. Demoments (eds.), Maximum Entropy and Bayesian Methods, 287–293.

introduces underdeterminicity - how many sources should we have? Faced with all these problems, where should we start?

A frequentist would begin by embedding our data set G in a sample space of other hypothetical data sets $\{G_1 \ldots G_N\}$ we could have observed [but didn't] and then try to estimate the probabilities $p(G_i|\mathcal{H})$ that the data set G_i would be observed if hypothesis $\mathcal{H}$, corresponding to some distribution of sources, were true. Loredo [6] presents a critique of this frequentist approach and one of his many criticisms is the fact that the random variable is the hypothesis $\mathcal{H}$ and not the data G.

The Bayesian approach is almost exactly the opposite. Instead of the class of all possible data sets $\{G_1 \ldots G_N\}$ consistent with $\mathcal{H}$, we consider the class of all hypotheses [source distributions] $\{\mathcal{H}_1 \ldots \mathcal{H}_K\}$ consistent with the G we have observed. Our knowledge of the possible ways in which *nature* could have generated the various $\mathcal{H}_i$ (for example, based on a best guess of possible rock types occurring in the area) is used to assign *a-priori* weights to each hypothesis. In addition, the measured data will make some hypotheses more plausible than others. We then use Bayes' theorem to combine all that we know about $\mathcal{H}$ into a posterior distribution. Probability theory is now silent on which particular $\mathcal{H}$ to choose from this distribution. Historically, the distribution maximum was chosen as being the most likely but another choice could be made, for example the distribution mean. An alternate approach would be to sample the posterior distribution many times to get a "feel" for the type of distributions indicated. If we follow the MaxEnt approach and choose the hypothesis with maximum entropy relative to the prior information and consistent with the measured data we will have the most likely distribution of sources with minimal structure not infered by G. Such a source distribution is the ***most honest*** distribution we can infer from G.

2. Simulated Annealing

Simulated annealing is an optimization technique which attempts to solve combinatorial type optimization problems by analogy between optimization and crystallization. The basic idea is to identify parameter values with the states of some physical system. Initially the parameter values are assigned randomly, corresponding to a high energy material. By slowly reducing a control parameter T, corresponding to the physical property of temperature, the system settles into thermal equilibrium and reaches its ground state which corresponds to the global minimum of the optimization process.

The main appeal of simulated annealing as an optimization strategy lies in the promise of the attainment of a global minimum. In addition, Kirkpatrick *et al* [4] have shown that the performance of the scheme is essentially linear in the number of parameters being estimated.

Boltzman used a counting argument, discussed by Jaynes [3], to show that the prior probability of a physical system being in a state $\mathcal{H}$ is

$$P(\mathcal{H}) = \frac{1}{Z(T)} exp\left(\frac{-E(\mathcal{H})}{T}\right)$$

where $E(\mathcal{H})$ corresponds to the energy and T is the temperature of the physical system. $Z(T)$ is a partition function to normalize the probabilities over all possible states of the system.

If the system states and data are related by the relation

$$\phi(\mathcal{H}) = g + \eta$$

where η is a random process independent of $\mathcal{H}$ then $P(g|\mathcal{H})$ may be modelled by

$$P(g|\mathcal{H}) \propto exp\left(-\frac{1}{2}\left(\frac{||g-\phi(\mathcal{H})||_r}{\sigma}\right)^r\right)$$

where σ is the variance of η and $||\cdot||_r$ is the l_r norm [$r = 1$ if the noise is exponential, $r = 2$ if the noise is gaussian]. Applying Bayes' theorem we multiply the above two probabilities and find the state posterior distribution has the same form as the state prior distribution i.e. the Gibbs distribution

$$P(\mathcal{H}|g) \propto exp\left(\frac{-\hat{E}(\mathcal{H},\{g_i\})}{T}\right)$$

where $\hat{E}(\mathcal{H},g) = E(\mathcal{H}) + \frac{T}{2}\left(\frac{||g-\phi(\mathcal{H})||_r}{\sigma}\right)^r$

To use simulated annealing for parameter estimation, the following steps are performed :

- Choose an objective function $E(\mathcal{H})$ to be minimized. $E(\mathcal{H})$ corresponds to the energy of a physical system in state $\mathcal{H}$. If $\{g_i\}$ are a set of measured data, $\mathcal{H}$ generates mock data $\{s_i\} = \phi(\mathcal{H})$ and one possible scale invariant objective function which may be minimized is Pearson's Chi-squared statistic [3] $E(\mathcal{H}) = \sum_i \frac{(s_i-g_i)^2}{g_i}$.

- Start with a random parameter configuration $\mathcal{H}$.

- Generate random parameter perturbations and allow the initially unstable system to equilibrate by changing state. This may be achieved by the Metropolis algorithm which accepts any new state with lower energy and accept a move which increases the energy with probability a function of T, which tends to zero as T decreases. The Metropolis procedure samples from the Gibbs distribution at constant temperature, thereby simulating the average behaviour of a physical system in thermal equilibrium.

  In the potential field source location problem one possible state perturbation strategy is :

  1. Delete a source at random.
  2. Add a source at a random location.
  3. Change the depth of a source a random feasible amount.
  4. Change the magnitude parameter c of a source a random feasible amount.

2.1. Convergence considerations

Simulated annealing exhibits slow convergence. For low temperatures, equilibrium can only be attained after a large number of steps. Geman and Geman [1] show that for convergence to a stable equilibrium to be guaranteed, the decrease of temperature with time, called the cooling regime $T(t)$, must satisfy $T(t) > \frac{a}{\log t}$ for a some large positive constant. If cooling is too rapid, a metastable state may result corresponding to a local minimum of the objective function. Two remedies for this slow convergence rate have recently been proposed :

- Rothman [8] proposes replacing the Metropolis algorithm by sampling directly from the Gibbs distribution. This approach chooses a new state from a marginal distribution of other states. The method avoids exploring new states which have a high probability of rejection and produces weighted guesses that are always accepted.

- Nulton and Salamon [7] discuss simulated annealing at constant thermodynamic speed which consists of designing a cooling schedule which produces uniform changes in the value of the objective function being minimized, at each iteration.

2.2. Occam efficiency

We may identify our parameter estimation problem with many physical systems so is it possible to identify an optimal system? To do this we need a criterion for choosing between models. Occam's razor is a principle of science which holds that parameters should not proliferate unnecessarily. In the simulated annealing context this means that simpler systems are to be preferred. In the source determination problem simplicity be achieved by modifying the error term to penalize the introduction of "too many" sources :

$$E(\mathcal{H}) = \sum_i \frac{(s_i - g_i)^2}{g_i} + \alpha\theta(\mathcal{H})$$

where $\theta(\cdot)$ is some penalty function discouraging the proliferation of sources. This remedy is expedient and can of course be criticized as being *ad hoc* in the form of θ as well as introducing an arbitrary weighting parameter α.

3. Bayesian parameter estimation and model selection

An alternative to the above approach is to use Bayes' theorem to infer the simplest configuration of source locations $\mathcal{H}$ that explain g.

A fixed model $\mathcal{H}$ may be parametrized in terms of the locations and magnitudes of the sources : $\mathcal{H} = \{\xi\}$. Using Bayes' rule for inverting probabilities, the *posterior distribution* for ξ is

$$P(\xi|g,\mathcal{H}) = \frac{P(\xi|\mathcal{H})P(g|\xi,\mathcal{H})}{P(g|\mathcal{H})}$$

which may be expressed in words as

$$\textit{Posterior} = \frac{\textit{Prior} \times \textit{Likelihood}}{\textit{Evidence}}$$

Loredo [6] shows how the evidence $P(g|\mathcal{H})$ may be written as a combination of data likelihoods weighted by the model priors :

$$P(g|\mathcal{H}) = \int P(\xi|\mathcal{H})P(g|\xi,\mathcal{H})\,d\xi$$

If we have a set of hypotheses $\{\mathcal{H}_i\}$, each corresponding to a different number of sources, a second application of Bayes' theorem gives an inference rule

$$P(\mathcal{H}_i|g) \propto P(\mathcal{H}_i)P(g|\mathcal{H}_i)$$

This equation is not normalized since our hypothesis space is not completely defined. We reserve the right to add a new hypothesis after we have observed more data. In order to compare model $\mathcal{H}_1$ parameterized by $\{\xi\}$ to model $\mathcal{H}_2$ parameterized by $\{\nu\}$ we must calculate the odds ratio

$$\frac{P(\mathcal{H}_1|g)}{P(\mathcal{H}_2|g)} = \frac{P(\mathcal{H}_1)\int P(\xi|\mathcal{H}_1)P(g|\xi,\mathcal{H}_1)\,d\xi}{P(\mathcal{H}_2)\int P(\nu|\mathcal{H}_2)P(g|\nu,\mathcal{H}_2)\,d\nu}$$

If we have no preference for model $\mathcal{H}_1$ over $\mathcal{H}_2$ i.e. we assume a uniform prior for $P(\mathcal{H}_i)$, the weight of the evidence $P(g|\mathcal{H}_1)$ *vs* $P(g|\mathcal{H}_2)$ will compare $\mathcal{H}_1$ to $\mathcal{H}_2$. If the distributions are normalized and $\mathcal{H}_1$ and $\mathcal{H}_2$ predict the data equally well, the simplest model will be prefered, thus realizing Occam's razor.

With all the Bayesian machinery in place, it is now time to assign the probability distributions.

3.1. Prior distributions

The preceeding discussion introduced two prior distributions : $P(\mathcal{H})$ and $P(\xi|\mathcal{H})$. As mentioned above, the uniform distribution is appropriate for $P(\mathcal{H})$, unless specific information exits which would favour a particular hypothesis.

Skilling [9] presents a convincing argument for a general entropic prior of the form

$$\begin{aligned} P(\xi|m,\alpha,\mathcal{H}) &\propto exp\left(\alpha\sum_i \xi_i - m_i - \xi_i log\left(\frac{\xi_i}{m_i}\right)\right) \\ &\propto exp(\alpha S(\xi,m)) \end{aligned}$$

when ξ is a positive additive distribution. The prior is parametrized in terms of a model m which defines a component-wise magnitude for ξ and a parameter α which is the inverse of the expected spread of ξ about m. α has inverse dimensions to ξ and m and Skilling [10] comments that both α and m cannot be constant of nature.

The assignment of a prior to a parameter distribution is problem specific. Rather than worrying about whether the 'correct prior' has been assigned, it makes more sense to try many different priors and then use the available data to evaluate these different hypotheses.

3.2. Likelihood distributions

A distribution of sources $\mathcal{H}$ generates a positive additive distribution $\{s_i\} = \phi(\xi)$. Given two normalized positive additive distributions $\{g_i\}$ and $\{s_i\}$, Kullback [5] showed that the appropriate information theoretic measure of distance between the distributions is the "minimum discrimination information statistic"

$$K(s,g) = \sum_i s_i log\left(\frac{s_i}{g_i}\right)$$

K is not a metric since $K(s,g) \neq K(g,s)$. Jaynes [3] comments that inferences using this measure rather than the Chi-squared statistic follow from Bayes' theorem.

Since $s_i/g_i = 1 + (s_i - g_i)/s_i + (s_i - g_i)^2/s_i g_i$ a series expansion of K with $|g_i - s_i| \leq g_i$ gives

$$K = \sum_i s_i - \sum_i g_i + \sum_i \frac{(g_i - s_i)^2}{g_i} + O(\cdot)$$

thus for $s_i \approx g_i$, K behaves like χ^2 while for large deviations $|g_i - s_i|$ the two functions may be very different. For non-normalized positive additive distributions the Kullback statistic may be extended to

$$K(s,g) = \sum_i g_i - \sum_i s_i + \sum_i s_i log\left(\frac{s_i}{g_i}\right)$$

K has a global maximum of zero at $s_i = g_i$. For the forward model $\phi(\xi) = g + \eta$, l_2 minimization of $||g - \phi(\xi)||$ leads to a Gaussian maximum likelihood estimation while l_1 minimiziation leads to exponential maximum likelihood estimation. Thus minimization of $K(s,g)$ indicates a likelihood function of the form $P(g|\xi) = \pi exp(-K(s,g))$ where π is an appropriate normalizing constant. $-K(s,g)$ is simply the relative cross entropy $S(s,g)$.

3.3. Parameter estimation

Bayes' theorem expresses the posterior parameter distribution $P(\xi|g)$ as the product of the prior distribution $P(\xi)$ and the likelihood $\pi exp(S(\phi(\xi),g))$. If we create a new function

$$f(\xi,T) = \pi P(\xi) exp\left(\frac{S(\phi(\xi),g)}{T}\right)$$

then

$$\begin{aligned} f(\xi, T \to \infty) &= P(\xi) \\ f(\xi, T = 1) &= P(\xi|g) \\ f(\xi, T \to 0) &\to max\ P(\xi|g) \end{aligned}$$

thus f changes from the parameter prior distribution to the parameter posterior distribution as T is decreased from ∞ to 1. As T is further decreased to zero, f approaches the maximum of the posterior distribution.

Our algorithm to estimate the posterior parameter distribution proceeds as follows :

- start with an initial random parameter assignment ξ.
- successively change the value of a singe parameter ξ_i by sampling from the conditional distribution $f(\xi_i|\xi_{j\neq i},T)$
- Rothman [8] shows that $f(\xi_i|\xi_{j\neq i},T) \to f(\xi,T)$. When this happens, decrease T.

4. Conclusion

We have discussed how an *energy* driven simulated annealing and an *entropy* driven Bayesian parameter estimation and model selection procedure may be used to estimate optimal equivalent source distributions from potential field data. This energy / entropy parallel is discussed by Jaynes [2] who states : *thus entropy becomes the primitive concept with which we work, more fundamental even than energy.*

References

[1] S Geman & D Geman [1984], Stochastic relaxation, Gibbs distributions and the Bayesian restoration of images, IEEE PAMI **PAMI-6**.

[2] E T Jaynes [1957], Information theory and statistical mechanics I, *Phys. Rev.* **106**.

[3] E T Jaynes [1978], Where do we stand on maximum entropy?, *The Maximum Entropy Formalism*, R D Levine & M Tribus (eds.), MIT Press.

[4] S Kirkpatrick, C D Gelatt & M P Vecchi [1983], Optimization by simulated annealing, *Science* **220**.

[5] S Kullback [1959], Information theory and statistics, J. Wiley & Sons, Inc.

[6] T J Loredo [1990], From Laplace to supernova SN 1987A : Bayesian inference in astrophysics, in *Maximum Entropy and Bayesian Methods*, P F Fougère (ed.), Kluwer.

[7] J D Nulton and P Salamon [1988], Statistical mechanics of combinatorial optimization, Physical Review A **37**.

[8] D H Rothman [1986], Automatic estimation of large residual static corrections, *Geophysics* **51**.

[9] J Skilling [1989], Classic maximum entropy, in *Maximum Entropy and Bayesian Methods*, J Skilling (ed.), Kluwer.

[10] J Skilling [1990], Quantified maximum entropy, in *Maximum Entropy and Bayesian Methods*, P F Fougère (ed.), Kluwer.

CRYSTAL PHASE DYNAMICS

Andrew D. McLachlan
Medical Research Council
Laboratory of Molecular Biology
Hills Road, Cambridge CB2 2QH England

ABSTRACT. The entropy dynamics method seeks maxima for the entropy of the electron density for N atoms in a crystal cell, when the Fourier amplitudes are fixed, but their phases are unknown. By analogy with molecular dynamics, the effective potential energy is the negative entropy $V = -NS$. The kinetic energy is proportional to the squared velocities of the electron densities at grid points in the map. It reduces to a sum of Fourier mode rotor energies. Each rotor angle experiences a couple equal to the phase gradient of S, and local dynamical equilibrium yields a Boltzmann distribution of S. Trial calculations have been made of phase averages and correlations in a centrosymmetric projection of the membrane protein bacteriorhodopsin.

1. Introduction

The maximum entropy method for refining crystal electron density maps (Wilkins, Varghese & Lehmann, 1983; Gull, Livesey & Sivia, 1987) assumes that the best set of phases for given structure factor amplitudes yields the density map with the highest entropy. This criterion is closely related to the triplet and quartet probability distributions used in small-molecule direct methods (Karle & Hauptman, 1950; Giacovazzo, 1980; Bricogne, 1984). The calculation of a maximum entropy map with fixed guide phases is easy (Navaza, 1985; Prince, Sjolin & Alenljung, 1988; McLachlan, 1989; Bricogne & Gilmore, 1990) but difficult problems remain:

(1) The entropy has many maxima, sometimes separated by deep minima.
(2) Procedures for optimising phases (e.g. conjugate gradients, least squares) usually stop at the nearest maximum.
(3) There have been no systematic studies to test, whether the maximum entropy phase set is the same as, or close to the correct solution.

Most successful phasing procedures today use multiple solutions generated from random starts (Sheldrick, 1990), magic integers (Giacovazzo, 1980) or tree searches (Gilmore Bricogne & Bannister, 1990). Simulated annealing (Kirkpatrick Gelatt & Vecchi, 1983; Sheldrick, 1990; Subbiah, 1991) has been used to explore solutions, in both reciprocal space and real space.

The problem of multiple solutions also occurs in calculations of the structure of proteins, where the energy has enormous numbers of local minima. Monte Carlo methods and

A. Mohammad-Djafari and G. Demoments (eds.), Maximum Entropy and Bayesian Methods, 295–302.

molecular dynamics (McCammon & Harvey, 1987; Allen & Tildesley, 1987) are powerful tools for exploring protein conformations.

To apply a dynamical approach to the exploration of entropy maxima in the phase space we introduce a fictitious *entropic temperature* and a fictitious kinetic energy for the system of unknown phases. The kinetic energy acts as a heat reservoir which allows the phases to move freely between adjacent entropy maxima. The fictitious potential energy is the negative entropy.

2. The Rotor Model

Molecular Dynamics

Entropy dynamics, like molecular dynamics,executes a Brownian motion over the accessible neighbourhood of the current phase variables, moving continually from one maximum to another. In classical molecular dynamics N atoms, as particles with positions $\mathbf{r}_a$ and velocities $\mathbf{v}_a$, move with a known potential energy function $V(\mathbf{r})$. Typical components of V are bond stretching, van der Waals and electrostatic terms. Each atom, of mass m_a also has a kinetic energy $\frac{1}{2}m_a\mathbf{v}_a^2$, and so the total energy is the sum of V and the kinetic energy W : $E = V(\mathbf{r}) + W$. The system relaxes to a state of thermal equilibrium at a temperature T, where each kinetic degree of freedom has a mean energy of $\frac{1}{2}kT$. The temperature can be controlled over a period of time by adjusting the kinetic energy to its expected value $W = \frac{3}{2}NkT$ and the potential energy alone has the Boltzmann distribution $f(\mathbf{r}) = A\exp(-V(\mathbf{r})/kT)$, where A is a normalising factor. The system is able to jump over potential energy barriers with heights of order kT by borrowing kinetic energy from the molecular motions.

Entropy Dynamics

In the maximum entropy method (Levine & Tribus, 1979) the probability of a set of phases for N atoms distributed within a unit cell is

$$f = e^{NS} \tag{1}$$

when N is large. Here S is the entropy of the maximum entropy map generated by the prescribed Fourier amplitudes with given phases. We take the fictitious potential energy as

$$V = -NS. \tag{2}$$

Then, in equilibrium at an entropic temperature θ , Boltzmann's law gives a phase probability distribution

$$f = e^{NS/\theta}. \tag{3}$$

We specify the map as an atomic probability distribution p_j on a grid of L equally spaced lattice points in the cell, normalised to a sum of N rather than unity. Then

$$NS = -\sum_{j=1}^{L} p_j \log p_j \qquad \sum_j p_j = N. \tag{4}$$

Next we give each grid point a fictitious velocity dp_j/dt and effective mass μ, making an effective kinetic energy

$$W = \sum_j \frac{1}{2}\mu\left(\frac{dp_j}{dt}\right)^2 \tag{5}$$

This reduces to a sum over normal modes by taking Fourier components. Using a unitary transform we define coefficients

$$Q_{\mathbf{h}} = \frac{1}{\sqrt{L}} \sum_j e^{2\pi i \mathbf{h}.\mathbf{x}_j} p_j = R_{\mathbf{h}} e^{i\phi_{\mathbf{h}}} = \alpha_{\mathbf{h}} + i\beta_{\mathbf{h}}. \tag{6}$$

Here the grid has (L_x, L_y, L_z) subdivisions along the three axes, and the point j with grid indices (j_x, j_y, j_z) has cell coordinates $\mathbf{x}_j = (x, y, z) = (j_x/L_x, j_y/L_y, j_z/L_z)$. The kinetic energy reduces to

$$W = \sum_{\mathbf{h}} \frac{1}{2}\mu\left(\frac{dR_{\mathbf{h}}}{dt}\right)^2 + \sum_{\mathbf{h}} \frac{1}{2}\mu R_{\mathbf{h}}^2\left(\frac{d\phi_{\mathbf{h}}}{dt}\right)^2. \tag{7}$$

The radial term, proportional to $(dR_{\mathbf{h}}/dt)^2$ vanishes when the amplitudes are fixed. The angular term depends on the rate of change of phase, and means that each mode, **h**, behaves like a rotor with angular velocity $\omega_{\mathbf{h}} = d\phi_{\mathbf{h}}/dt$ and an effective moment of inertia

$$I_{\mathbf{h}} = \mu |Q_{\mathbf{h}}|^2. \tag{8}$$

In thermal equilibrium each rotor has an equal mean kinetic energy

$$\frac{1}{2} I_{\mathbf{h}} \omega_{\mathbf{h}}^2 = \frac{1}{2}\theta. \tag{9}$$

Thus the strong reflection phases tend to rotate slowly and the weak ones rapidly.

The fictitious entropy potential function $-NS(\phi_1, \phi_2, \ldots \phi_N)$ is a highly non-linear function of the phase invariants, involving the triplets, quartets, and higher terms. The modes are anharmonic and become mixed together during the motion because angular couples acting on each rotor depend on the phases of other rotors. Rotor modes collide and scatter, exchanging energy with the reservoir of kinetic energy.

3. Mode Dynamics

Equations of Motion

We start with given phases $\phi_{\mathbf{h}}$ and their angular velocities, $\omega_{\mathbf{h}}$. Each rotor has an angular momentum $J_{\mathbf{h}} = I_{\mathbf{h}}\omega_{\mathbf{h}}$, where the moment of inertia $I_{\mathbf{h}}$ is fixed, and it experiences a couple $C_{\mathbf{h}} = -\partial V/\partial\phi_{\mathbf{h}}$. Thus the equations of motion are

$$\frac{dJ_{\mathbf{h}}}{dt} = N\frac{\partial S}{\partial \phi_{\mathbf{h}}} = N\left(\alpha_{\mathbf{h}}\frac{\partial S}{\partial \beta_{\mathbf{h}}} - \beta_{\mathbf{h}}\frac{\partial S}{\partial \alpha_{\mathbf{h}}}\right). \tag{10}$$

These entropy gradients are the Lagrangian multipliers that belong to the standard maximum entropy solution (Wilkins Varghese & Lehmann, 1983).

DISCONTINUOUS PHASES AND STEPPED ROTORS

Although the phases of X-ray reflections often vary continuously from 0 to 2π, special reflections may be restricted to fractions of 2π, or to a sign (0 or π). Stepped phases are modelled by using a continuous rotor clock angle $\chi_{\mathbf{h}}$, with angular velocity $\omega_{\mathbf{h}}$, which carries the kinetic energy and trips the phase $\phi_{\mathbf{h}}$ when it passes certain critical markers. The entropy jumps at each marker. If the rotor has sufficient kinetic energy to jump over an upward potential energy step it continues, with reduced angular velocity. Otherwise the rotor is reflected backwards.

SIGN FLIP ENTROPIES

With a centrosymmetric object the entire motion reduces to a succession of sign flips, one mode at a time. It is also possible to control the selection of active modes. At any time a mode may be either clamped with a fixed sign, or rotating, or switched off, with an amplitude set to zero. Thus, considering a single mode $\mathbf{h}$ with all the other modes $\mathbf{k}$ set to signs $g_{\mathbf{k}}$, the variable sign $g_{\mathbf{h}}$ may take the three values $0, \pm 1$, with corresponding entropies $S_0(\mathbf{h}), S_+(\mathbf{h}), S_-(\mathbf{h})$ for the whole system. In an active mode the current actual sign may flip to the opposite value, and the subsequent motion depends on an important quantity, the *reversal entropy*, defined as

$$S_{\text{rev}}(\mathbf{h}) = S_{\text{actual}}(\mathbf{h}) - S_{\text{opposite}}(h) = \pm[S_+(\mathbf{h}) - S_-(\mathbf{h})]. \tag{11}$$

At a temperature θ the mean kinetic energy of each rotor is $\frac{1}{2}\theta$, and so modes with $|S_{\text{rev}}(\mathbf{h})| \gg \theta$ will tend to move into their maximum entropy settings, with stable fixed phases, while modes with $|S_{\text{rev}}(\mathbf{h})| \ll \theta$ will continue to flip rapidly. At a maximum of the entropy all modes will have positive values of $S_{\text{rev}}(\mathbf{h})$.

MAKING A DYNAMICS RUN

An entropy dynamics run involves several stages. The start-up, the main run, and the conclusion. In the start-up stage the active rotor modes are selected, a starting temperature is chosen and motion is initiated. The main run consists of "bursts" of phase flip events at the current temperature, in which each rotor may be either transmitted or reflected at the marker point. After a burst the temperature may be adjusted, the angular velocities reset, and the active modes reselected.

The starting temperature θ is set high (e.g. $\theta = 100 \langle |S_{\text{rev}}| \rangle$) to randomise the signs, but later an automatic control is used. We set upper and lower targets γ_{max} and γ_{min} for the fraction of rotors that flip in a burst of M events. If ρ flips are observed, and if $\rho > M\gamma_{\text{max}}$ the temperature is lowered by a certain ratio $\theta \rightarrow \theta/\epsilon_\theta$, but if $\rho < M\gamma_{\text{min}}$ the temperature is raised: $\theta \rightarrow \epsilon_\theta \theta$. Typically, $\epsilon_\theta = 1.5$.

4. Trial Calculations

ORTHORHOMBIC BACTERIORHODOPSIN

A suitable test object is the purple membrane protein bacteriorhodopsin (Henderson, 1977).The orthorhombic form, in projection at 3.9Å resolution, has been solved by electron microscope imaging (Bullough & Henderson, 1990) . The image, of two-dimensional symmetry *pgg*, yields 220 experimentally phased reflections with an average phase error of

34.7°. The cell contains four copies of the seven-helix protein subunit, with the helical rods viewed end on.

We sorted the amplitudes of the strong reflections in order and scaled each $|F|$ to a minimum of zero and a standard deviation of unity. The 21 strongest reflections, with amplitudes of 2.5 or more, give a useful trial set.

We began by setting the 21 strong signs to their correct experimental values, with all other modes off, and calculated the reversal entropies $S_{\text{rev}}(h,k)$ of these strong modes. Most of the values are positive, as expected for a maximum-entropy state, but six out of the 21 are negative. Evidently, the signs of the correct map are not the same as the maximum entropy signs, given only these 21 amplitudes. However, if we set all the 217 modes with their correct signs and recalculate S_{rev} for the top 21 modes, the values now all become positive, and the entropy, based on more complete information, does have a true local maximum.

Enumeration of Sign Combinations

The next experiment was designed to test the relationship between the correct signs and the signs that yield entropy maxima. We tried all 2097152 sign combinations of the 21 strong modes by using the Gray code (Hamming, 1986). This sequence passes once through each state by flipping just one sign at a time. The enumeration was done twice: once with all the other modes clamped at their correct signs, and once with all other modes switched off. In the clamped test the correct solution corresponds exactly to the global entropy maximum (of the restricted rotating set). In the unclamped test the correct solution lies well up in the distribution of observed entropies, but it is by no means the highest.

Another point of interest is the statistical distribution of entropies over the large number of sign sets. We define a reduced entropy $\sigma(g) = [S(g) - S_{\text{av}}]/S_{\text{dev}}$, and plot the histograms of σ. They are bell-shaped curves with a steep decline on the side of high values.

Entropic Thermal Equilibria

When entropy dynamics is run for a long time at constant temperature, the entropy and the associated phases (or signs) reach a local fluctuating equilibrium. We use reduced entropies $\sigma(g)$, and a reduced temperature $\tau = TS_{\text{dev}}/N$. An important new quantity is the statistical density of the histogram of σ, $\Omega(\sigma)$, defined so that the number of sign combinations which have entropy between σ and $\sigma + d\sigma$ is $dn = \Omega(\sigma)d\sigma$. The logarithm of Ω is a fundamental thermodynamic variable of the system, which we call the *popularity* function for the given entropy: $\Gamma(\sigma) = \log \Omega(\sigma)$. The number of times that a given value of the entropy is met during a dynamical run depends both on the shape of the popularity curve and the probability factor $\exp(\sigma/\tau)$. In large systems the sign sets are sampled in a way which depends on the popularity profile of the system. States near the entropy maximum are rare and have low popularities, so they may never be sampled at all if the temperature is too high.

Progressive Mode Selection

Once a small starting set of modes has been tested, the dynamics run has gained information about the most likely phases. To preserve and enlarge this information it is then useful to extend the active set. The most stable current modes (with large magnitudes of S_{rev}) may be clamped to preserve their good phases. We have tried several selection schemes: block extension, revision of unstable modes, alternate extension and revision. In block extension the modes are divided into groups $I, II, III \ldots$, with group I as the starting set. In the second round all group I modes are clamped and group II modes rotated: in the third the group II modes are also clamped and group III modes rotated, and so on.

We have also examined the temperature and entropy time course of typical long runs in which the active set of modes has been enlarged in blocks. The temperature passes through a series of plateaux, with a reheat whenever a new mode group is added in. The entropy decreases sharply as each new group introduces further phase and amplitude constraints, but then settles into a new range of local maxima.

5. Discussion

The entropy phase dynamics method explores the local maxima and minima of an electron density map quite efficiently without becoming trapped. In practical terms, entropy dynamics is as flexible and adaptable in use as other Monte Carlo procedures (e.g. Sheldrick, 1990). It is no better than other annealing methods at recovering from wrong decisions at the top of the cooling sequence. Multiple starting points and an intelligent tree search need to be added.

One unique feature of entropy dynamics is the analogy between the dynamical Boltzmann factor $\exp(NS/\theta)$ at an entropic temperature θ, and the joint probability $\exp(NS)$ for the same phases in the statistical theory. Varying the entropic temperature is analogous to altering N, the effective number of atoms.

In the rotor model the active amplitudes are fixed at their experimental values but the phases are completely free. This means that the R-factor (Blundell & Johnson, 1976) is zero, but the densities may be non-physical and negative. Most other methods use feasible atomic model densities with partially incorrect amplitudes, and seek to lower the R factor. Examples are molecular dynamics (Brunger, Kuriyan & Karplus, 1987; Kuriyan et al.,1989), hard sphere solvent dynamics (Subbiah, 1991), and fictitious atoms with R-factor potentials (Semenovskaya Khachaturyan & Khachaturyan, 1985).

The tests on bacteriorhodopsin show how the maximum entropy sign combinations, for a restricted set of strong reflections, but with no further amplitude information given, can be seriously wrong. On the other hand, when all the weaker reflections were clamped with their correct signs to provide extra information, the maximum entropy signs for the strong modes matched the correct solution perfectly.

Experience with other phasing methods suggests that the maximum entropy solutions, if obtainable, must converge to correct phases when the constraints are over-determined: that is, when the number of known amplitudes greatly exceeds the number of atomic degrees of freedom, $3N$ (Sheldrick, 1990).

REFERENCES

Allen, M.P. & Tildesley, D.J. (1987). *Computer Simulation of Liquids.* Oxford: University Press.

Blundell, T.L. & Johnson, L.N. (1976). *Protein Crystallography.* London: Academic Press.

Bricogne, G. (1984). *Acta Cryst.* **A40**, 410-445.
Maximum Entropy and the Foundations of Direct Methods.

Bricogne, G. & Gilmore, C.J. (1990). *Acta Cryst.* **A46**, 284-297.
A Multisolution Method of Phase Determination by Combined Maximisation of Entropy and Likelihood. I. Theory, Algorithms and Strategy.

Brunger, A.T., Kuriyan, J. & Karplus, M. (1987). *Science* **235**, 458-460.
Crystallographic R Factor Refinement by Molecular Dynamics.

Bullough, P.A. & Henderson, R. (1975). *Biophys. J.* **58**, 705-711.
Phase Accuracy in High-Resolution Electron Microscopy of Trigonal and Orthorhombic Purple Membrane.

Giacovazzo, C. (1980). *Direct Methods in Crystallography.* London: Academic Press.

Gilmore, C.J., Bricogne, G. & Bannister, C. (1990). *Acta Cryst.* **A46**, 297-308.
A Multisolution Method of Phase Determination by Combined Maximization of Entropy and Likelihood. II. Application to Small Molecules.

Gull, S.F., Livesey, A.K. & Sivia, D.S. (1987). *Acta Cryst.* **A43**, 112-117.
Maximum Entropy Solution of a Small Centrosymmetric Crystal Structure.

Hamming, R.W. (1986). *Coding and Information Theory*, p.97. Englewood Cliffs, NJ: Prentice-Hall.

Henderson, R. (1977). *Ann. Rev. Biophys. Bioeng.* **6**, 87-109.
The Purple Membrane from Halobacterium Halobium.

Karle, J. & Hauptman, H. (1950). *Acta Cryst.* **3**, 181-187.
The Phases and Magnitudes of the Structure Factors.

Kirkpatrick, S., Gelatt, C.D. & Vecchi, M.P. (1983). *Science* **220**, 671-680.
Optimization by Simulated Annealing.

Kuriyan, J. Brunger, A.T., Karplus, M. & Hendrickson, W.A. (1989). *Acta Cryst.* **A45**, 396-409.
X-ray Refinement of Protein Structures by Simulated Annealing: Test of the Method on Myohemerythrin.

Levine, R.D. & Tribus, M. (Eds.) (1979). *The Maximum Entropy Formalism.* Cambridge Mass. : MIT Press.

McCammon, J.A. & Harvey, S.C. (1987). *Dynamics of Proteins and Nucleic Acids.* Cambridge: University Press.

McLachlan, A.D. (1989).
A Statistical Potential for Modelling X-ray Electron Density Maps with Known Phases. In *Maximum Entropy and Bayesian Methods.* Edited by J. Skilling. p. 241-249. Dordrecht: Kluwer.

Navaza, J. (1985). *Acta Cryst.* **A41**, 232-244.
On the Maximum Entropy Estimate of the Electron Density Function.

Prince, E., Sjolin, L. & Alenljung, R. (1988). *Acta Cryst.* **A44**, 216-222.

Phase Extension by Combined Entropy Maximisation and Solvent Flattening.

Semenovskaya, S.V., Khachaturyan, K.A. & Kachaturyan, A.G. (1985). *Acta Cryst.* **A41**, 268-273.
Statistical Mechanics Approach to the Structure Determination of a Crystal.

Sheldrick. G.M. (1990). *Acta Cryst.* **A46**, 467-473.
Phase Annealing in SHELX-90: Direct Methods for Larger Structures.

Subbiah, S. (1991). *Science* **252**, 128-133.
Low-Resolution Real-Space Envelopes: An Approach to the Ab Initio Macromolecular Phase Problem.

Wilkins, S.W., Varghese, J.N. & Lehmann, M.S. (1983). *Acta Cryst.* **A39**, 47-60.
Statistical Geometry. I. A Self-Consistent Approach to the Crystallographic Inversion Problem Based on Information Theory.

DIFFRACTION FROM PARTICLES OF UNKNOWN SHAPE

A.L. McLean and G.J. Daniell
Department of Physics
University of Southampton
Southampton SO9 1NH, UK

ABSTRACT. We present results of an investigation of the feasibility of the determination of particle size distributions from diffraction of coherent radiation by particles of unknown shape. Contrary to our expectations it appears that some information about the particle size distribution can be extracted using Maximum Entropy data analysis. We use as a model the Fraunhoffer diffraction of coherent radiation by opaque lamellar particles on a transparent disk.

1. Introduction

The aim of this paper is to demonstrate the feasibility of solving a class of problems where we wish to determine a probability distribution from data which depend on both the distribution and a (possibly large) number of other parameters. The important point is that we are not interested in the values of those other parameters and the ideal of marginalizing them out of the problem is not computationally feasible.

We use as an example the evaluation of particle size distributions from the diffraction of coherent radiation from particles of unknown shape. This is a standard technique (Coston and George 1991, Farhadpour 1988) with applications in areas as diverse as the study of biological cells, combustion processes, atmospheric aerosols and interstellar dust. The approach usually taken is to assume that the particles are spherical and homogeneous, this allows the exact formulas of Mie (1908) to be used — or more commonly the use of approximations to the Mie formulas. It is well known that these approximations can lead to misleading results (Farhadpour 1988). We examine the Fraunhoffer diffraction regime where we can perform an analysis with non spherical particles.

2. The Model

We consider diffraction due to opaque lamellar particles on a transparent disk which are larger than the wavelength of the scattered light. Babinet's Theorem (Ditchburn 1952) states that the diffraction patterns from two complementary screens, that is screens with the positions of the opaque and transparent portions exchanged, are equal except at very small scattering angles. So opaque particles on a transparent screen are equivalent to apertures in an opaque screen. Furthermore, the lamellar particles can be thought of as the cross sections of three dimensional particles (in the large particle limit). The particles are also assumed to be sufficiently sparsely distributed, that there is no correlation between their positions, and that they do not overlap.

The diffraction pattern, $A(\underline{k})$, from a particle of shape given by the polar representation

A. Mohammad-Djafari and G. Demoments (eds.), Maximum Entropy and Bayesian Methods, 303–309.

$r = af(\theta)$ is given by its two dimensional Fourier Transform

$$A(\underline{k}) = \int_0^{2\pi} d\theta \int_0^{af(\theta)} r\, dr\, e^{i\underline{k}\cdot\underline{r}}. \quad (1)$$

Letting $\underline{k} = (k\cos\alpha, k\sin\alpha)$ we obtain

$$A(k,\alpha) = \int_0^{2\pi} d\theta \int_0^{af(\theta)} r\, dr\, e^{ikr\cos(\alpha-\theta)}. \quad (2)$$

With diffraction from a number of similar particles or apertures randomly distributed in space (with fixed orientation) the *intensity* of the resultant diffraction pattern is proportional to the *intensity* of the pattern due to a single particle (Ditchburn 1952). Furthermore, with more than one particle shape present, the *intensity* of the resulting diffraction pattern is obtained by adding the *intensities* of the diffraction patterns due to the individual shapes in proportion to the number of each shape present. If we have a single particle shape present in random orientations, with the orientations uncorrelated in space, then the diffraction pattern intensity is given by

$$I(k) = \int_0^{2\pi} d\alpha\, |A(k,\alpha)|^2. \quad (3)$$

We can perform the integration over r in equation (2) analytically, leaving two integrations to be found numerically for each k, a and $f(\theta)$. As these integrations are of periodic functions over a period the extended trapezium rule is used (Nonweiler 1983).

3. Maximum Entropy

In the 'Historic' Maximum Entropy (MaxEnt) method of solving inverse problems we have experimental data, $d_i^{(expt)}$, and a forward problem which consists of some function of a probability distribution, p_i, plus some random errors. We obtain the solution by maximizing the entropy, S, of the distribution where

$$S = -\sum_{i=1}^{M} p_i \log\left(\frac{p_i}{b_i}\right), \quad (4)$$

subject to the constraint that the χ^2 statistic,

$$\chi^2 = \sum_{i=1}^{N} \frac{1}{\sigma_i^2}\left(d_i^{(expt)} - d_i^{(calc)}\right)^2, \quad (5)$$

is equal to some target value χ^2_{targ}; p_i is also constrained to be positive. We use the criterion that χ^2_{targ} is equal to the number of data points N (the discrepancy principle).

In the cases considered, the data are a function of both the probability distribution and some other parameters, q_i, where we have no interest in the values of the q_i. Ideally we would treat this by Bayesian inference, use entropy as a prior for the probability distribution, and marginalize the other parameters out of the problem. However this is not computationally feasible for at least some of the cases we consider. We instead choose to maximize $S(\{p_i\})$

subject to $\chi^2(\{p_i\},\{q_i\}) = \chi^2_{targ}$ by varying $\{p_i\}$ and $\{q_i\}$. At the extremum, the following equations must be satisfied:

$$\frac{\partial S}{\partial p_i} - \lambda \frac{\partial \chi^2}{\partial p_i} = 0 \tag{6}$$

$$\frac{\partial \chi^2}{\partial q_i} = 0 \tag{7}$$

$$\chi^2 = \chi^2_{targ}. \tag{8}$$

The approach we adopt to solving this problem is to fix $\{q_i\}$ and maximize S subject to $\chi^2 = \chi^2_{targ}$, this ensures that equations (6) and (8) are satisfied. This is just the 'Historic' MaxEnt problem and can be solved (in the linear case) by the usual algorithm (Skilling and Bryan 1984). We then fix $\{p_i\}$ and minimize χ^2 with respect to $\{q_i\}$ (equation (7)). This process is then repeated; if both steps converge for the same values of $\{p_i\}$ and $\{q_i\}$ then a local minimum of S with respect to $\{p_i\}$ and $\{q_i\}$ subject to the constraint on χ^2 has been reached. The process is not guaranteed to converge, but in practice it frequently does.

4. Results

We can now proceed to obtain information about the particle size distribution from experimental data subject to different assumptions or knowledge about the particles in question. We test the technique by analyzing simulated experimental data. As is usual, a distribution of a size parameter with dimensions of length is found; though in certain circumstances an area weighted or volume weighted distribution may be more appropriate.

4.1. Problem type 0 — known particle shape

Assuming that the particles are all the same shape, circles of radius a say, then the problem reduces to solving a Fredholm Integral equation of the first kind, *viz*

$$I(k) = \int_0^\infty da\, I(k,a)p(a). \tag{9}$$

For the case of circles, the kernel of this equation can be found analytically. This is the assumption that experimenters usually make. They then attempt to solve the inverse problem by a variety of matrix inversion techniques or, particularly in the case of spherical particles, by using exact forms (Shifrin 1968, Acquista 1976). These methods are, of course, unstable and sensitive to the presence of experimental noise. It is a straightforward exercise to solve this problem by MaxEnt.

4.2. Problem type I — one shape, single parameter

The next level of complexity is where the particles are all the same shape, that the shape belongs to a class of shapes which can be parameterised by a single parameter, but that the parameter is unknown. As an example we assume that the particles are ellipses with an unknown ratio of minor to major axis, η, which we call the aspect ratio. The data are then given by

$$I(k) = \int_0^\infty da\, I(k,a,\eta)p(a). \tag{10}$$

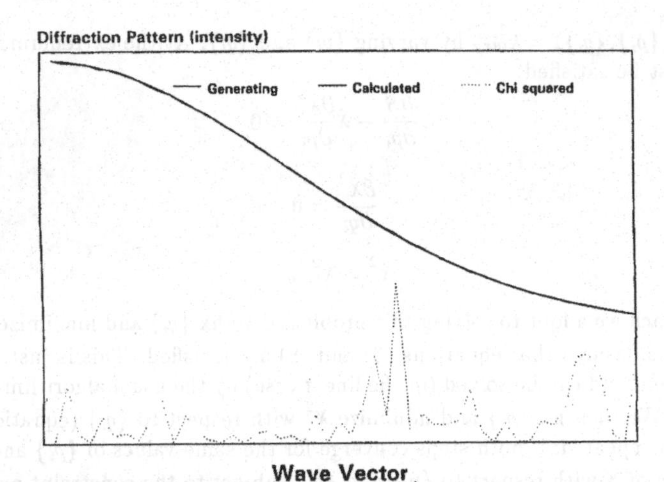

Figure 1: Data generated simulating diffraction by ellipses of aspect ratio 0.8 with a broad, flat distribution of particle sizes. The diffraction pattern is shown with the solid line, the dotted line represents the contributions to χ^2 for the MaxEnt solution.

We can maximize the entropy with respect to the size distribution and the single parameter η as described in section 3.

Note η is chosen by with the sole criterion that it maximizes the entropy of the distribution of particle sizes. We are therefore not entitled to draw any conclusions about the particle shape from the results of this analysis.

This was tested with data simulating diffraction from ellipses of aspect ratio 0.8 and a flat, broad distribution of sizes. The data and fitted data are shown in Figure 1, together with the contribution to χ^2 of each data point. The value of η which maximizes the size distribution is in fact 1.0 (circles) — emphasizing the point that we are not determining the shape.

4.3. Problem type II — distribution of shapes

Another natural extension to problem type 0, is to restrict the class of shapes considered to a small family which can be easily parameterised. For example rather than assuming the particles are all circular, we could assume that they are ellipses with a distribution of shapes. In particular, we can select a parameterization of an ellipse with unit area, $r = g(\theta, \eta)$, where η is the ratio of the length of the minor to the major axis. We then describe the particles as $r = af(\theta) = ag(\theta, \eta)$ where a is a particle size parameter. The diffraction pattern is then given by

$$I(k) = \int_0^\infty da \int_0^1 d\eta \, I(k, a, \eta) P(a, \eta), \tag{11}$$

where $I(k, a, \eta)$ is the intensity due to a single particle of size a with shape parameter η and $P(a, \eta)$ is the probability density function for the particle size and shape.

As a first step we assume that there is no correlation between the shape and size of the particles so we can write $P(a, \eta) = p(a)q(\eta)$. Now we could use MaxEnt to attempt to

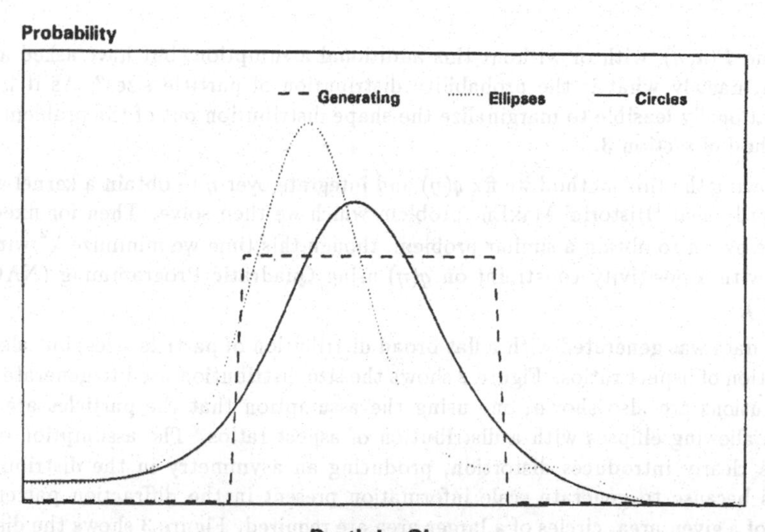

Figure 2: Particle size distributions. The data was generated by ellipses with a broad range of aspect ratios (see Figure 3). The dashed line represents the distribution used to generate data, the dotted line represents the MaxEnt solution assuming the particles to be circular and the solid line represents the MaxEnt solution assuming ellipses with a distribution of shapes.

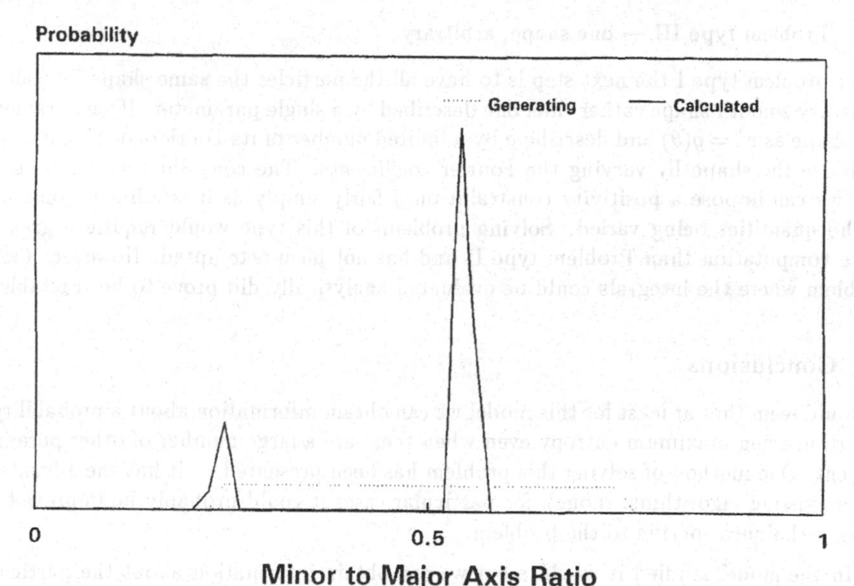

Figure 3: The shape distribution. The dashed line represents the shape distribution used to generate the data, while the solid line represents the distribution which produces a *size* distribution with maximum entropy.

determine $P(a, \eta)$, with or without this additional assumption, but have asked a different question, namely what is the probability distribution of particle sizes? As it is also not computationally feasible to marginalize the shape distribution out of the problem we apply the method of section 3.

Following the this method we fix $q(\eta)$ and integrate over η to obtain a kernel and hence a standard linear 'Historic' MaxEnt problem which we then solve. Then for fixed $p(a)$ we integrate over a to obtain a similar problem, though this time we minimize χ^2 with respect to $q(\eta)$ with a positivity constraint on $q(\eta)$ using Quadratic Programming (NAG routine E04NCF).

Test data was generated with a flat broad distribution of particle sizes, but also a broad distribution of aspect ratios. Figure 2 shows the size distribution used to generate the data. Two solutions are also shown, one using the assumption that the particles are spherical and one allowing ellipses with a distribution of aspect ratios. The assumption of circular particles clearly introduces distortion, producing an asymmetry in the distribution, this happens because to generate scale information present in the diffraction pattern due to ellipses of a given area, circles of a larger area are required. Figure 3 shows the distribution of shapes used to generate the data and the distribution required to maximize the entropy of the size distribution. Again note the sole criterion in determining the shape distribution is the entropy of the size distribution — so we are not entitled to draw any conclusions about the shapes from this analysis.

4.4. Problem type III — one shape, arbitrary

After problem type I the next step is to have all the particles the same shape, but allow an arbitrary smooth shape rather than one described by a single parameter. If we parameterize the shape as $r^2 = g(\theta)$ and describe g by a limited number of its Fourier coefficients, we can optimize the shape by varying the Fourier coefficients. The constant term fixes the area and we can impose a positivity constraint on g fairly simply as it is a linear combination of the quantities being varied. Solving problems of this type would require a great deal more computation than Problem type II and has not been attempted. However, a similar problem where the integrals could be evaluated analytically did prove to be tractable.

5. Conclusions

It would seem that at least for this model we can obtain information about a probability distribution using maximum entropy even when there are a large number of other parameters present. One method of solving this problem has been presented — it has the advantage of using existing algorithms, though for particular cases it could probably be improved upon using techniques specific to the problem.

In the model studied it would seem we can obtain information about the particle size distributions while making fewer assumptions about the particle shape. Of course in the cases examined, the simulated data are accurately described by the model used in the analysis, as this was also used to generate them. It remains to be tested on experimental data, though it would seem strange if a model allowing a distribution of ellipsoidal particles gave worse results than one assuming spheres.

References

Acquista, C.: 1976, Light scattering by tenuous particles — generalisation of Rayleigh–Gans–Rocard approach, *Appl. Opt.* **15**(11), 2932–2636.

Coston, S. D. and George, N.: 1991, Particle sizing by inversion of the optical transform pattern, *Appl. Opt.* **30**(33), 4785–4794.

Ditchburn, R.: 1952, *Light*, Blackie and Son.

Farhadpour, F.: 1988, Size analysis of micrometer sized suspensions by forward lobe light scattering in the anomalous regime, *in* P. Lloyd (ed.), *Particle Size Analysis*, John Wiley & Sons.

Mie, G.: 1908, Bieträge zur optik trüber medien, speziell kolloidaler metallösungen, *Ann. Phys.* **25**, 377–445.

Nonweiler, T.: 1983, *Computational mathematics: an introduction to numerical approximation*, Ellis Horwood, Chichester.

Shifrin, K. S.: 1968, Scattering of light in a turbid medium, NASA Technical Translation TT F-477.

Skilling, J. and Bryan, R.: 1984, Maximum entropy image reconstruction : general algorithm, *Mon. Not. R. Astr. Soc.* **211**, 111–124.

MAXENT ENHANCEMENT OF 2D PROJECTIONS FROM 3D PHASED FOURIER DATA: AN APPLICATION TO POLARIZED NEUTRON DIFFRACTION

R.J. Papoular
Laboratoire Léon Brillouin
CEN-Saclay
91191 Gif-sur-Yvette Cedex, France

E. Ressouche, J. Schweizer and A. Zheludev
Centre d'études nucléaires de Grenoble
DRFMC/SPSMS-MDN
85 X - 38041 Grenoble Cedex, France

ABSTRACT. The conventional Fourier reconstruction of a 2-dimensional projection from a given phased Fourier data set only makes use of those Fourier components related to scattering vectors perpendicular to the direction of the projection. Within this conventional framework, all other Fourier components contribute ZERO to the reconstructed projected density in direct space.
By contrast, Maximum Entropy makes use of the whole 3-dimensional available information and consequently yields much superior results even when projections onto 2-dimensional maps are considered.
Recovering 2-dimensional projections is central to understand Molecular Magnetism as studied by Polarized Neutron Diffraction. A first example, taken from real neutron data, demonstrates the power of the MaxEnt approach when applied to this emerging field of physics. As an extreme case, very reasonable reconstructions can be obtained for data sets such that not a single Fourier component is related to a scattering vector perpendicular to the direction of projection. A second example is described, pertaining to this latter case.

1. Introduction

A century-long practice of Fourier syntheses by crystallographers (see for instance Buerger, 1960), radioastronomers and others, has developed long-lived unconscious mental habits among experimentalists analyzing their scattering (Fourier) data sets.
The seminal work of Gull and Daniell (1978) stroke a first blow against those, by demonstrating how the use of Maximum Entropy considerably enhances 2-dimensional Inverse Fourier reconstructions, avoiding most of the truncation effects otherwise unavoidable. Their work was readily extended to much more general problems (Skilling & Gull, 1985, Buck & Macaulay, 1991) as the theory of image reconstruction and powerful algorithms developed (Skilling, 1989, Gull, 1989). Three years ago, we started to use part of these recent

A. Mohammad-Djafari and G. Demoments (eds.), Maximum Entropy and Bayesian Methods, 311–318.

advances, including the Memsys Cambridge Code developed by Gull and Skilling (1989), in order to improve upon our 2D/3D reconstructions of densities from 3D phased neutron scattering data, albeit spin densities (Papoular & Gillon, 1990a,b, Papoular & Schweizer, 1991) or electronic densities (Papoular *et al.*, 1992). Not surprisingly, the resulting reconstructed densities were considerably enhanced with respect to previous neutron data treatments.
One basic crystallographic concept is the 2D projection $m_P(\vec{r})$ of the 3D sought density $m(\vec{r})$ in real space, which is very valuable to locate atoms or magnetic moments as well as to evidence chemical bonding (Gillon & Schweizer, 1989) within the unit cell, when too few 3D data are available. Standard textbooks such as Buerger's explain how a 2D projection is to be obtained from **a 2D data set** using commonly-accepted practice and why the complementary 3D Fourier components are irrelevant.
In what follows, we reject this unjustified procedure and demonstrate how MaxEnt further enhances a 2D projection by making use of the allegedly irrelevant "out of plane" reflections.

2. A general problem: 2D Projections from 3D Fourier Data revisited

With obvious notations, a 3D density in real space $m(\vec{r}) = m(xyz)$ is to be retrieved from a given limited and noisy data set consisting of 3D components in Fourier space $F(\vec{K})$ related to $m(\vec{r})$ via:

$$F(\vec{K}) = \int_{-\infty}^{\infty}\int_{-\infty}^{\infty}\int_{-\infty}^{\infty} d\vec{r}\ \exp\{i\vec{K}\cdot\vec{r}\}\ m(\vec{r}) \tag{1}$$

within a noise term $\sigma(\vec{K})$, where $\vec{K}$ is the scattering vector.
In some instances, such as Polarized Neutron Diffraction (PND) applied to the study of Magnetism, the only experimentally accessible Fourier components lie very close to one plane Π and lesser physical information is then sought: the *projected density* $m_P(\vec{r})$ along the axis $\vec{z}$ normal to that plane Π.
The fact that the latter quantity $m_P(\vec{r})$, defined as:

$$m_P(\vec{r}) = m_P(xy) = \int_{-\infty}^{\infty} dz\ m(\vec{r}) \tag{2}$$

is equal to:

$$m_P(xy) = \frac{1}{(2\pi)^2}\int_{\infty}^{\infty}\int_{\infty}^{\infty} dK_x dK_y\ \exp\{-i(xK_x + yK_y)\}\ F(K_x K_y 0) \tag{3}$$

when **all** 2D Fourier components are known **without noise**, has induced experimentalists to look only for the experimentally accessible subset of those in-plane ($K_z = 0$) components.
By contrast, thanks to its intrinsically non-linear character, MaxEnt makes use of the left-over out-of-plane ($K_z \neq 0$) Fourier components to ascertain those missing or ill-defined in-plane ones, as demonstrated by the crystallographic examples discussed below. Our suggested procedure is carried out in two steps:

- First reconstruct using MaxEnt the full 3D $m(\vec{r})$ density.

- Then project $m(\vec{r})$ along the $\vec{z}$ axis to obtain an improved $m_P(\vec{r})$.

A further point to be stressed is that **the positivity requirement on m_P is not mandatory** for this procedure to be effective, only **the nonlinearity of MaxEnt is.**

Historical MaxEnt has been used in the present work. Its consists in maximizing the posterior probability:

$$\text{Prob}(m_P|\text{Data}) \propto \exp\left\{-\frac{\chi^2(\text{Data}|m_P)}{2}\right\} \cdot \exp\left\{\alpha S(m_P)\right\} \tag{4}$$

where:

$$\chi^2(\text{Data}|m_P) = \sum_{\vec{K}} \left| \frac{F(\vec{K}) - \int\int dx dy \exp\{i(xK_x + yK_y)\}\, m_P(xy)}{\sigma(\vec{K})} \right|^2 \tag{5}$$

and $S(m_P)$ is the entropy associated with m_P and the Lagrange multiplier α is determined by the restraint $\chi^2 = N$, if N is the number of independent Fourier components. If $m_P(\vec{r})$ is not known a priori to be positive definite, it is reconstructed as the difference of two strictly positive maps $m_{P\uparrow}(\vec{r})$ and $m_{P\downarrow}(\vec{r})$.

3. Application to Polarized Neutron Diffraction

The Polarized Neutron Diffraction Framework: A Bird's-Eye View.

This section explains why Discrete Phased Fourier Components are measured in this case.

These components pertain to some magnetization density in real space $m(\vec{r})$, the physical origin of which is usually connected with magnetic atoms or chemical bonding.
In a PND experiment, monochromatic neutrons are shed on a single-crystal sample and those neutrons which are elastically scattered by the latter are detected at the so-called Bragg peaks positions. Indeed, the detector measures neutron intensities, which are the squares of the moduli of the scattered amplitudes. These amplitudes are the sums of two terms, one of which is the sought Fourier component: the Magnetic Structure Factor $F_M(\vec{K})$. The second component, the Nuclear Structure Factor $F_N(\vec{K})$, is determined by another experiment, involving a so-called 4-circle diffractometer. We shall assume this $F_N(\vec{K})$ part to be known in what follows.

The single-crystal nature of the target ensures its spatial translational symmetry, which in turn yields a discrete set of scattering vectors $\vec{K}$. A given vector $\vec{K}$ is perfectly defined by the incident velocity of the neutron and the angle between its incoming and outgoing direction. The orientation of the single-crystal determines the Fourier basis, onto which the scattering vector can be decomposed into its three components K_x, K_y and K_z, which can be labelled by the three integers h, k and ℓ. It is convenient to work with centrosymmetric crystals, for which both the Nuclear and Magnetic Fourier components are real for all $\vec{K}$'s. Beside the discreteness of the non-vanishing Fourier components (Bragg peaks), it results from the spatial symmetry operations of a crystal, which are represented by a space-group (see Buerger, 1960), that minor modifications must be brought to formulae (1), (2) and (3). In particular, it is convenient to replace the Fourier exponential factor by an average over the symmetry operations of the space group, as explained in (Papoular, 1991).

The separation between the Nuclear and Magnetic structure factors for a given $\vec{K}$ is made possible through the neutron spin-sensitive scattering cross-section, which depends upon the relative orientation of the incident neutron spin $\frac{1}{2}$ and a saturating magnetic field $\vec{\mathcal{H}}$ exerted vertically on the sample. For each Bragg peak, two intensities are measured, corresponding to the two possible spin states of the incoming neutron. The ratio of these two intensities yields the so-called flipping ratio:

$$R(\vec{K}) = \left| \frac{F_{\mathrm{N}}(\vec{K}) + F_{\mathrm{M}}(\vec{K})}{F_{\mathrm{N}}(\vec{K}) - F_{\mathrm{M}}(\vec{K})} \right|^2 \tag{6}$$

from where $F_{\mathrm{M}}(\vec{K})$ is extracted unambiguously. This stems from the fact that one of the two solutions $F_{\mathrm{M}}(\vec{K})$ as obtained from (6) can most often be safely rejected as non physical.

There are three main limitations to the PND technique: 1) the limited portion of reciprocal (Fourier) space available for measurements, 2) the impossibility to extract some of the $F_{\mathrm{M}}(\vec{K})$'s because the related $F_{\mathrm{N}}(\vec{K})$'s are vanishingly small and 3) the unequal relative accuracies with which the $F_{\mathrm{M}}(\vec{K})$'s can be experimentally gathered.

TWO EXAMPLES: A LOW-TEMPERATURE MAGNETIC STUDY OF THE $[TBA]^+[TCNE]^-$ CRYSTAL

The motivation beyond the study of **TetraButylAmmonium TetraCyaNoEthylene** is the understanding of molecular magnetism at low temperature. The underlying chemistry is the capture of one electron by one acceptor TCNE molecule (C_6N_4) from a neighboring TBA donor molecule. This electron is unpaired, corresponding to a spin $\frac{1}{2}$ localized around the nuclei of the TCNE molecule. An external applied magnetic field polarizes the spin density within the molecule. How is it distributed about the nuclei? More details can be found in (Zheludev *et al*).

The crystal structure is characterized by the space group $P2_1/n$ and the unit cell contains 4 asymmetric units. The data sets discussed below were obtained at the temperature T = 1.6 K and with a vertical magnetic field $\vec{\mathcal{H}}$ = 4.65 Teslas. Neutron spectrometers belonging to the Siloé Neutron Facility (CEN-Grenoble) were used for the experimental part of this work. The $(a, b, c, \alpha, \beta, \gamma)$ lattice constants at low temperature (10 K) were found equal to (14.55 Å, 8.26 Å, 19.69 Å, 90°, 106.14°, 90°), respectively.

The $\vec{b}$ axis was set vertical and 211 unique reflections $F(\vec{K})$ were collected. Only 42 out of these were lying in the horizontal scattering plane (k = 0). Moreover, the $F(000)$ reflection, equal to the averaged magnetization per unit cell, was determined separately via a macroscopic magnetization measurement.

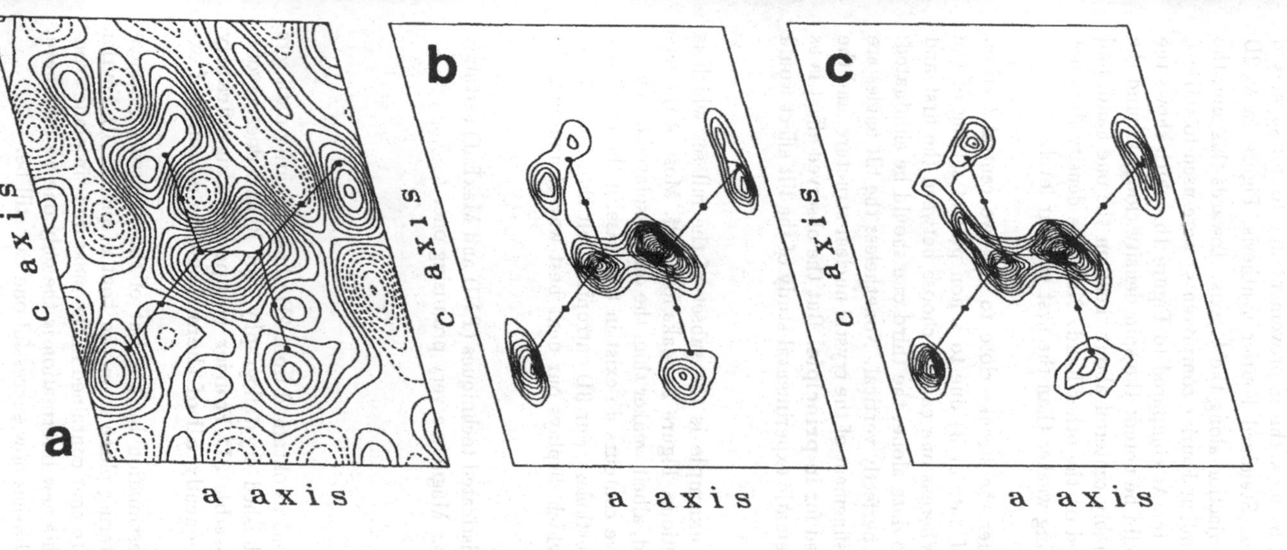

Fig. 1. Reconstruction of the 2-dimensional electronic spin density of the TCNE molecule projected along the $\vec{b}$ axis using: a) the standard Fourier synthesis formula, b) the former 2D Historical Maxent reconstruction using only the experimentally available $F(h0\ell)$ structure factors and c) the now proposed Historical Maxent reconstruction using all the experimentally measured structure factors to compute the intermediate (hidden) 3D spin density, which is subsequently projected along the $\vec{b}$ axis, yielding the map displayed in (c).

The already high-resolution 2D Maxent map in b) is now enhanced in c) due to 3D Fourier components $F(hk\ell)$ up to $k = 2$. The four Nitrogen atoms, located on the outskirts of the TCNE molecule, are now very close to the reconstructed density maxima.

The $\vec{a}$ and $\vec{c}$ axes extend from $-\frac{1}{4}$ to $\frac{1}{4}$ and from 0 to $\frac{1}{2}$. The identical gridding of a), b) and c) is 30*30 pixels. The related equidistant contour steps are $0.01, 0.025$ and $0.025\ \mu_B/\text{Å}^2$ respectively.

The latter 43 unique reflections were used to reconstruct both the 2D standard Fourier and the 2D MaxEnt projected densities.

• The purpose of this **first example** is to show the improvement due to our suggested 3D MaxEnt procedure over previous ones (Standard Fourier synthesis, Figure 1a & 2D MaxEnt, Figure 1b) to retrieve the 2D projection along the $\vec{b}$ axis. Towards this aim, the full dataset mentioned above (212 independent Fourier components) were used to retrieve the full projection as displayed in Figure 1c. As compared to Figure 1b, which shows the older 2D MaxEnt result, two features should be noted: the spin density clouds around the four Nitrogen atoms become more reasonably centered about these on the one hand, and the top left spurious peak has disappeared on the other hand. Negative density does not show up in either of these two results, being weaker than the first contour level.

The question can be asked, as to whether the streaks close to the two central Carbon atoms are: 1) real, 2) due to the lack of data or 3) due to a non perfect setting of the sample, albeit by a mere few degrees. Whereas one cannot choose between the first and the second hypotheses on the basis of the data alone, the third one should be eliminated: the axis $\vec{b}$ of the crystal can never be set perfectly vertical. Nonetheless the tilt angles are very accurately determined through the refinement of the crystal nuclear structure, and the small resulting effect can be fully corrected for **in principle**. But the observed effect is as new as our entropic tool; therefore a systematic experimental study of the tilt effect would be worthwhile.

• The data pertaining to our **second example** is the subset of the full set which is restricted to those 169 out-of-plane reflections. Figure 2 speaks for itself. Most of the features of the projected density are retrieved, albeit weaker than the ones displayed on Figure 1c. A few spurious non-negligible negative contours do exist in this case, if the positivity constraint is not enforced on $m_P(\vec{r})$. Nevertheless, our 3D entropic result shown on Figure 2b contrasts markedly with Figure 2a, which displays our next best way of handling the same data set.

The conjunction of these two highly sophisticated techniques (PND and MaxEnt) certainly makes the new emerging field of Molecular Magnets a very promising one.

4. Conclusion

Not only does MaxEnt enhance projections as obtained from 2D phased Fourier data sets within the framework of currently established crystallography, but it also does revolutionize the experimental process: **all possible 3D Fourier components significantly contribute to 2D projections.** Consequently, **all of them** should be measured if experimentally possible.
The underlying reason is the non-zero estimation by MaxEnt of those missing (unmeasured) reflections in the horizontal scattering plane from the "out of plane" 3D Fourier components. This new information adds to that contained in the measured "in plane" conventional 2D components. Sometimes, this new information is the only one we may have, and our MaxEnt-based work clearly establishes how successful one can still be.

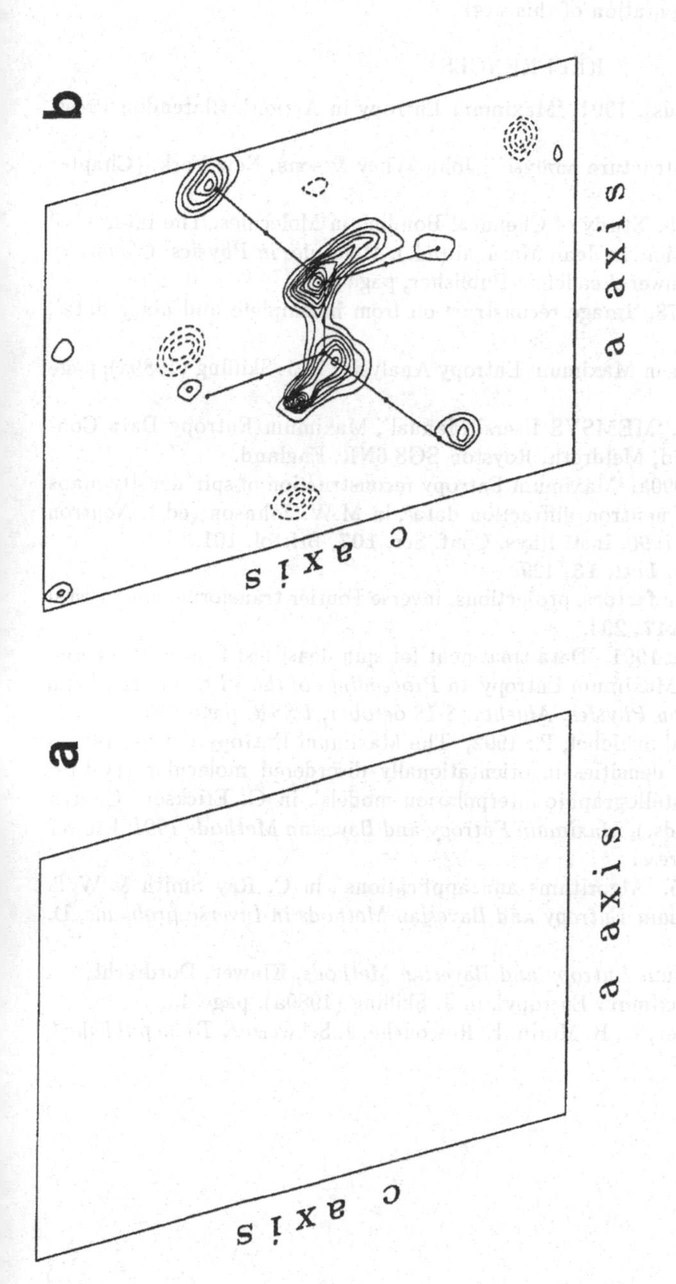

Fig. 2. Reconstruction of the 2D projected electronic spin density of TCNE using NOT A SINGLE structure factor belonging to the horizontal scattering plane. The standard Fourier synthesis yields a ZERO map (a) whereas our proposed procedure (b) retrieves most of the features obtained by using all the experimentally available information (Compare with figure 1, (c)). The gridding of the map is 30*30 pixels and the equidistant contour step is $0.025\mu_B/\text{Å}^2$. The dashed contours correspond to a negative projected density.

ACKNOWLEDGMENTS. The authors are much indebted to Dr. J. Miller (Du Pont de Nemours) and Prof. A. Epstein (Ohio University) for the TBA-TCNE single crystal used in this work. They gratefully acknowledge very relevant points raised by Drs. R. Bontekoe and J. Skilling during the presentation of this work.

REFERENCES

Buck, B. & Macaulay, V.A. (eds): 1991, 'Maximum Entropy in Action', Clarendon Press, Oxford.

Buerger, M.J.: 1960, 'Crystal-structure analysis', John Wiley & sons, New York. (Chapter 14, pp. 380-384.)

Gillon, B. & Schweizer, J.: 1989, 'Study of Chemical Bonding in Molecules: The Interest of Polarised Neutron Diffraction' in Jean Maruani (ed.), *Molecules in Physics, Chemistry and Biology*, **Vol. III**, Kluwer Academic Publisher, page 111.

Gull, S.F. & Daniell, G.J.: 1978, 'Image reconstruction from incomplete and noisy data', *Nature* **272**, 686.

Gull, S.F.: 1989, 'Developments in Maximum Entropy Analysis', in J. Skilling (1989a), page 53.

Gull, S.F. & Skilling, J.: 1989, 'MEMSYS Users' Manual', Maximum Entropy Data Consultants Ltd, 33 North End, Meldreth, Royston SG8 6NR, England.

Papoular, R.J. & Gillon, B.: 1990a, 'Maximum Entropy reconstruction of spin density maps in crystals from polarized neutron diffraction data', in M.W. Johnson (ed.), *Neutron Scattering Data Analysis 1990*, Inst. Phys. Conf. Ser. **107**, Bristol, 101.
See also: 1990b, *Europhys. Lett.* **13**, 429.

Papoular, R.J.: 1991, 'Structure factors, projections, inverse Fourier transforms and crystal symmetry', *Acta Cryst.* **A47**, 293.

Papoular, R.J. & Schweizer, J.: 1991, 'Data treatment for spin densities: Fourier inversion, multipolar expansion and Maximum Entropy' in *Proceedings of the VI-th International Neutron School on Neutron Physics, Alushta, 8-18 october, USSR*, page 170.

Papoular, R.J., Prandl, W. and Schiebel, P.: 1992, 'The Maximum Entropy reconstruction of Patterson and Fourier densities in orientationally disordered molecular crystals: a systematic test for crystallographic interpolation models', in G. Erickson, C. Ray Smith and P. Neudorfer (eds.), *Maximum Entropy and Bayesian Methods 1991*, Kluwer Academic Publisher. *In press.*

Skilling, J. & Gull, S.F.: 1985, 'Algorithms and applications', in C. Ray Smith & W.T. Grandy, Jr. (eds.), *Maximum Entropy and Bayesian Methods in Inverse problems*, D. Reidel, Dordrecht.

Skilling, J., ed.: 1989a, *Maximum Entropy and Bayesian Methods*, Kluwer, Dordrecht.

Skilling, J.: 1989b, 'Classic Maximum Entropy', in J. Skilling (1989a), page 45.

Zheludev, A., Epstein, A., Miller, J., B. Morin, E. Ressouche, J. Schweizer, *To be published.*

RELATIVE ENTROPY AND THE DEMPSTER-LAIRD-RUBIN EM ALGORITHM: APPLICATION TO THE AUTOFOCUS PROBLEM IN ISAR IMAGING

Glenn R. Heidbreder
11509 Hemingway Drive
Reston, Virginia 22094, USA

ABSTRACT. The popular Dempster-Laird-Rubin expectation-maximization (EM) algorithm obtains maximum likelihood estimates by postulating a refinement of the data (the complete data) which contains information about the unknown parameters in excess of that contained in the actual (incomplete) data. Each iteration of the algorithm involves (1) an expectation step in which the complete data log-likelihood function is averaged over the complete data distribution conditioned on the incomplete data and a prior estimate of the unknowns, and (2) a maximization step in which the new estimate of the unknown parameter vector is taken as the value which maximizes the expectation in (1). It is shown that an iteration may also be described as the minimization of the Kullback-Leibler "distance" between the complete data distribution conditioned on the incomplete data and the prior parameter vector estimate and the complete data distribution conditioned on the current unknown parameter vector estimate, ie., the entropy of the former distribution relative to the latter one. The insights which this alternative interpretation provides are applied in adapting the EM algorithm to the problem of estimating the focusing parameters in automatic focusing of inverse synthetic aperture radar (ISAR) images.

1. Introduction

The expectation-maximization (EM) algorithm of Dempster, Laird, and Rubin [1977] is a popular tool for the iterative computation of maximum likelihood estimates. In the development of the algorithm one views the available data $\boldsymbol{y}$ as incomplete and contemplates a refinement of the data (ie. a one to many mapping) $\boldsymbol{x}$, referred to as the complete data. A family of complete data likelihood functions $f(\boldsymbol{x}/\boldsymbol{\phi})$, depending on the parameters $\boldsymbol{\phi}$ to be estimated, is postulated. A complete data likelihood function is related to the incomplete (actual) data likelihood function $g(\boldsymbol{y}/\boldsymbol{\phi})$ by

$$g(\boldsymbol{y}/\boldsymbol{\phi}) = \int_{\chi(\boldsymbol{y})} f(\boldsymbol{x}/\boldsymbol{\phi}) \mathrm{d}\boldsymbol{x} \tag{1}$$

where the integration is over the space $\chi(\boldsymbol{y})$ containing all $\boldsymbol{x}$ which map into the given $\boldsymbol{y}$. The algorithm is directed at finding the value of $\boldsymbol{\phi}$ which maximizes $g(\boldsymbol{y}/\boldsymbol{\phi})$ but does so by making essential use of the associated family $f(\boldsymbol{x}/\boldsymbol{\phi})$. The space χ of the complete data

A. Mohammad-Djafari and G. Demoments (eds.), Maximum Entropy and Bayesian Methods, 319–332.

vector $\boldsymbol{x}$ provides a basis for comparison of competing hypotheses concerning the value of the parameter vector $\boldsymbol{\phi}$.

Dempster, et al. [1977] characterize an iteration of the algorithm as consisting of two steps. The first of these takes the expectation $Q(\boldsymbol{\phi}', \boldsymbol{\phi})$ of the log-likelihood $\log f(\boldsymbol{x}/\boldsymbol{\phi}')$, conditioned on the incomplete (actual) data and the prior estimate of the parameter vector. Thus for the $(p+1)$th iteration

$$Q(\boldsymbol{\phi}', \boldsymbol{\phi}^{(p)}) = \int_{\chi(\boldsymbol{y})} k(\boldsymbol{x}/\boldsymbol{y}, \boldsymbol{\phi}^{(p)}) \log f(\boldsymbol{x}/\boldsymbol{\phi}') \mathrm{d}\boldsymbol{x} \tag{2}$$

where $k(\boldsymbol{x}/\boldsymbol{y}, \boldsymbol{\phi}^{(p)})$ indicates the conditional density of $\boldsymbol{x}$. It is unnecessary to indicate restriction of the integration to $\chi(y)$ since, for all $\boldsymbol{x}$ which do not map into the given $\boldsymbol{y}, k(\boldsymbol{x}/\boldsymbol{y}, \boldsymbol{\phi}^{(p)}) = 0$. In the second step the new estimate $\boldsymbol{\phi}^{(p+1)}$ is obtained by maximizing $Q(\boldsymbol{\phi}', \boldsymbol{\phi}^{(p)})$ over $\boldsymbol{\phi}'$, ie.

$$\boldsymbol{\phi}^{(p+1)} = \arg\max_{\boldsymbol{\phi}'} Q(\boldsymbol{\phi}', \boldsymbol{\phi}^{(p)}). \tag{3}$$

It is a remarkable circumstance that one has wide latitude in choosing the refinement $\boldsymbol{x}$. It may be no more than a convenient fiction, though its mapping to $\boldsymbol{y}$ must be carefully specified as in (1). But all possible mappings $\boldsymbol{y}(\boldsymbol{x})$ cannot be equally effective. We seek some guidance in the choice of $\boldsymbol{x}$ and its associated family of likelihood functions. We first observe that the refinement must be a non-trivial one [Blahut,1987a], ie. there must be some $\boldsymbol{x}$ such that

$$\frac{f(\boldsymbol{x}/\boldsymbol{\phi})}{f(\boldsymbol{x}/\boldsymbol{\phi}')} \neq \frac{g(\boldsymbol{y}/\boldsymbol{\phi})}{g(\boldsymbol{y}/\boldsymbol{\phi}')}.$$

The alternative, namely equality of complete data and incomplete data likelihood ratios for all $\boldsymbol{x}$, would indicate that $\boldsymbol{x}$ is no more useful for the purpose of discriminating between parameter vectors $\boldsymbol{\phi}$ and $\boldsymbol{\phi}'$ than is the actual data vector $\boldsymbol{y}$. Evidently it is desirable that $\boldsymbol{x}$ be more suitable for this discrimination than is $\boldsymbol{y}$. We will return to the issue of choice of refinement but first we express the algorithm in more information theoretic terms involving likelihood ratios.

2. Expectation Maximization is Discrimination Minimization

Since the maximization of $Q(\boldsymbol{\phi}', \boldsymbol{\phi}^{(p)})$ in (3) is over $\boldsymbol{\phi}'$, it is unaffected if Q is augmented by adding any quantity not dependent on $\boldsymbol{\phi}'$. We add the entropy of the distribution of the complete data conditioned on the incomplete data and prior parameter estimate to obtain

$$\begin{aligned} Q'(\boldsymbol{\phi}', \boldsymbol{\phi}^{(p)}) &= Q(\boldsymbol{\phi}', \boldsymbol{\phi}^{(p)}) - \int k(\boldsymbol{x}/\boldsymbol{y}, \boldsymbol{\phi}^{(p)}) \log k(\boldsymbol{x}/\boldsymbol{y}, \boldsymbol{\phi}^{(p)}) \mathrm{d}\boldsymbol{x} \\ &= -\int k(\boldsymbol{x}/\boldsymbol{y}, \boldsymbol{\phi}^{(p)}) \log \frac{k(\boldsymbol{x}/\boldsymbol{y}, \boldsymbol{\phi}', \boldsymbol{\phi}^{(p)})}{f(\boldsymbol{x}/\boldsymbol{\phi}')} \mathrm{d}\boldsymbol{x} \end{aligned}$$

which we recognize as the entropy of the aforementioned conditional distribution relative to the distribution $f(\boldsymbol{x}/\boldsymbol{\phi}')$. Thus

$$\phi^{(p+1)} = \arg\max_{\phi'} Q'(\phi', \phi^{(p)}).$$

Each iteration of the algorithm is an entropy maximization step. We choose however to emphasize the ratio of likelihoods of different parameter vector values and note that $Q'(\phi', \phi^{(p)})$ is the negative of the Kullback–Leibler measure of the "distance" between $f(x/\phi')$ and $k(x/y, \phi^{(p)})$, also referred to as the "discrimination" [Blahut,1987a]. Let $D_{kf}(\phi^{(p)}, \phi')$ denote the discrimination. We have

$$D_{kf}(\phi^{(p)}, \phi') = -Q'(\phi', \phi^{(p)})$$

and (4)

$$\phi^{(p+1)} = \arg\min_{\phi'} D_{kf}(\phi^{(p)}, \phi').$$

The double subscript notation is used to remind that the discrimination is more properly a function of the probability densities $k(x/y, \phi^{(p)})$ and $f(x/\phi')$. In treating the algorithm we are most interested in the dependencies on ϕ and ϕ'. An iteration of the algorithm may be described as the selection of the current estimate of the parameter vector so as to minimize the discrimination of the x-distributions conditioned respectively on the incomplete data and the prior parameter vector estimate and on the current estimate alone. We choose $\phi^{(p+1)}$ such that $f(x/\phi^{(p+1)})$ is "closer" to $k(x/y, \phi^{(p)}$ than any other $f\ (x/\phi')$. The log-likelihood ratio

$$\log \frac{k(x/y, \phi^{(p)})}{f(x/\phi')}$$

is a sufficient statistic for deciding between the hypotheses $(y, \phi^{(p)})$ and (ϕ'). The larger its magnitude the more reliable is the decision. In setting $\phi' = \phi^{(p+1)}$ we have, on average, minimized our capability to distinguish between the hypotheses, ie. with any other choice of ϕ' the hypotheses are more distinguishable and ϕ' is farther from $\phi^{(p)}$ The idea is that when the algorithm approaches convergence at large p the hypotheses will be minimally distinguishable and $\phi^{(p+1)}$will be essentially equal to $\phi^{(p)}$. When we have adjusted ϕ' appropriately, y will contribute minimally to our knowledge of x.

3. Properties of the Algorithm

Dempster, et al. [1977] have presented several theorems describing properties of a generalized EM (GEM) algorithm. The generalized algorithm is defined as an iterative algorithm with mapping $\phi \to M(\phi)$ such that

$$D_{kf}(\phi, M(\phi)) \leq D_{kf}(\phi, \phi)$$

for every ϕ in the parameter space Ω. (Note that the definition of the EM algorithm above requires

$$D_{kf}(\phi, M(\phi)) \leq D_{kf}(\phi, \phi')$$

for every pair (ϕ, ϕ'). In the following equivalent theorems are presented in the context of discrimination minimization.

Theorem 1. The log-likelihood $L(\phi) = \log g(y/\phi)$ is non-decreasing in each iteration of a generalized algorithm, ie. $L(M(\phi)) \geq L(\phi)$ for all ϕ in the parameter vector space Ω. Equality holds if and only if both $D_{kf}(\phi, M(\phi)) = D_{kf}(\phi,\phi)$ and $k(x/y, M(\phi)) = k(x/y,\phi)$ almost everywhere.

Proof:

$$\begin{aligned} D_{kf}(\phi,\phi') &= \int k(x/y,\phi)\log\frac{k(x/y,\phi)}{f(x/\phi')}dx \\ &= \int k(x/y,\phi)\log\frac{k(x/y,\phi)}{k(x/y,\phi')}dx - \int k(x/y,\phi)\log\frac{f(x/\phi')}{k(x/y,\phi')}dx \\ &= D_{kk}(\phi,\phi') - L(\phi'). \end{aligned} \tag{5}$$

The reduction of the second integral follows from Bayes theorem, since

$$k(x/y,\phi') = \frac{f(x/\phi')}{g(y/\phi')}$$

and

$$\log\frac{f(x/\phi')}{k(x/y,\phi')} = \log g(y/\phi') = L(\phi')$$

is not a function of x. Using (5) we have

$$\begin{aligned} L(M(\phi)) - L(\phi) &= D_{kk}(\phi, M(\phi)) - D_{kf}(\phi, M(\phi)) - D_{kk}(\phi,\phi) + D_{kf}(\phi,\phi) \\ &= D_{kf}(\phi,\phi) - D_{kf}(\phi, M(\phi)) + D_{kk}(\phi, M(\phi)) \end{aligned} \tag{6}$$

since $D_{kk}(\phi,\phi) = 0$. $D_{kk}(\phi, M(\phi)) \geq 0$ since any discrimination function is non-negative [Blahut,1987a]. The quantity $D_{kf}(\phi,\phi) - D_{kf}(\phi, M(\phi))$ is non-negative by definition of a GEM. Hence

$$L(M(\phi)) \geq L(\phi).$$

Theorem 2. Suppose that $\phi^{(p)}$ for p = 0,1,2,... is a sequence of a generalized algorithm such that

(1) the sequence $L(\phi^{(p)})$ is bounded, and
(2) $D_{kf}(\phi^{(p)},\phi^{(p)} - D_{kf}(\phi^{(p)},\phi^{(p+1)}) \geq \lambda(\phi^{(p+1)} - \phi^{(p)})(\phi^{(p+1)} - \phi^{(p)})^T$ for some scalar λ and all p. (We use a row vector representation of ϕ).

Then the sequence $\phi^{(p)}$ converges to some ϕ^* in the closure of Ω.

Proof: From assumption (1) and Theorem 1, the sequence $L(\phi^{(p)})$ converges to some $L^* < \infty$. Hence, for any $\varepsilon > 0$, there exists a p(ε) such that for all $p \geq p(\varepsilon)$ and all $r \geq 1$,

$$\sum_{j=1}^{r}\left\{L(\phi^{(p+j)}) - L\,(\phi^{(p+j-1)})\right\} = L(\phi^{(p+r)}) - L(\phi^{(p)}) < \varepsilon. \tag{7}$$

From (6) and the non-negativity of its component terms we have

$$0 \leq D_{kf}(\phi^{(p+j-1)}, \phi^{(p+j-1)}) - D_{kf}(\phi^{(p+j-1)}, \phi^{(p+j)}) \leq L(\phi^{(p+j)}) - L(\phi^{(p+j-1)})$$

for $j \geq 1$, and hence from (7),

$$\sum_{j=1}^{r} \left\{ D_{kf}(\phi^{(p+j-1)}, \phi^{(p+j-1)}) - D_{kf}(\phi^{(p+j-1)}, \phi^{(p+j)}) \right\} < \varepsilon$$

for all $p \geq p(\varepsilon)$ and all $r \geq 1$, where each term in the sum is non-negative. On applying assumption (2) for $p, p+1, ..., p+r-1$ and summing we obtain

$$\lambda \sum_{j=1}^{r} (\phi^{(p+j)} - \phi^{(p+j-1)})(\phi^{(p+j)} - \phi^{(p+j-1)})^T < \varepsilon.$$

The sum is that of squares of incremental distances between members of the $\phi^{(p)}$ sequence and equals or exceeds the square of the distance between $\phi^{(p+r)}$ and $\phi^{(p)}$. Hence

$$\lambda(\phi^{(p+r)} - \phi^{(r)})(\phi^{(p+r)} - \phi^{(r)})^T < \varepsilon,$$

thus proving convergence.

Theorem 3. Suppose that $\phi^{(p)}, p = 0, 1, 2, ...$ is a sequence generated by a generalized algorithm such that the gradient of D_{kf} with respect to ϕ',

$$\nabla'_{\phi} D_{kf}(\phi^{(p)}, \phi')$$

evaluated at $\phi' = \phi^{(p+1)}$, is zero. Then for all p, there exists a $\phi_0^{(p+1)}$ on the line segment joining $\phi^{(p)}$ to $\phi^{(p+1)}$ such that

$$\begin{aligned} & D_{kf}(\phi^{(p)}, \phi^{(p)}) - D_{kf}(\phi^{(p)}, \phi^{(p+1)}) \\ & = \frac{1}{2}(\phi^{(p+1)} - \phi^{(p)}) \nabla'^2_{\phi} D_{kf}(\phi^{(p)}, \phi_0^{(p+1)})(\phi^{(p+1)} - \phi^{(p)})^T \end{aligned} \tag{8}$$

where $\nabla^2_{\phi'} D_{kf}(\phi^{(p)}, \phi_0^{(p+1)})$ is the Hessian for $D_{kf}(\phi^{(p)}, \phi')$ at $\phi' = \phi_0^{(p+1)}$ Furthermore, if the Hessian sequence is positive definite with eigenvalues bounded away from zero and $L(\phi^{(p)})$ is bounded, then the sequence $\phi^{(p)}$ converges to some ϕ^* in the closure of Ω.

Proof: Expand $D_{kf}(\phi^{(p)}, \phi')$ about $\phi^{(p+1)}$ to obtain

$$\begin{aligned} D_{kf}(\phi^{(p)}, \phi') = & D_{kf}(\phi^{(p)}, \phi^{(p+1)}) + (\phi' - \phi^{(p+1)}) \nabla'_{\phi} D_{kf}(\phi^{(p)}, \phi^{(p+1)}) \\ & + \frac{1}{2}(\phi' - \phi^{(p+1)}) \nabla^2_{\phi'} D_{kf}(\phi^{(p)}, \phi_0^{(p+1)})(\phi' - \phi^{(p+1)})^T. \end{aligned}$$

On letting $\phi' = \phi^{(p)}$ and applying the assumption of the theorem, we obtain (8). If the $\nabla'^2_{\phi} D_{kf}(\phi^{(p)}, \phi_0^{(p+1)})$ are positive definite with eigenvalues bounded away from zero, then condition (2) of Theorem 2 is satisfied and the sequence $\phi^{(p)}$ converges to some ϕ^* in the closure of Ω.

Theorem 4. Suppose that $\phi^{(p)}$, p = 0,1,2,... is a sequence generated by a generalized algorithm such that

(1) $\phi^{(p)}$ converges to ϕ^* in the closure of Ω,

(2) $\nabla'_{\phi} D_{kf}(\phi^{(p)}, \phi^{(p+1)}) = 0$ and

(3) $\nabla'^2_{\phi} D_{kf}(\phi^{(p)}, \phi^{(p+1)})$ is positive definite with eigenvalues bounded away from zero.

Then

$$\nabla_{\phi'} L(\phi^*) = 0$$
$$\nabla^2_{\phi'} D_{kf}(\phi^*, \phi^*) \text{is positive definite}$$

and

$$(\phi^{(p+1)} - \phi^*) = (\phi^{(p)} - \phi^*)[\nabla'^2_{\phi} D_{kk}(\phi^*, \phi^*)][\nabla'^2_{\phi} D_{kf}(\phi^*, \phi^*)]^{-1}. \tag{9}$$

Proof: From (5) we have

$$\nabla'_{\phi} L(\phi') = \nabla'_{\phi} D_{kk}(\phi^{(p)}, \phi') - \nabla'_{\phi} D_{kf}(\phi^{(p)}, \phi')$$

At $\phi' = \phi^{(p+1)}$ the second term on the right is zero by assumption (2) and the first term is zero in the limit as $p \to \infty$. Since $\phi^{(p+1)} \to \phi^*$ and $\phi^{(p)} \to \phi^*$ it follows that $\nabla_{\phi'} L(\phi^*) = 0$. $\nabla'^2_{\phi} D_{kf}(\phi^*, \phi^*)$is positive definite, since it is the limit of the positive definite sequence $\nabla^2_{\phi'} D_{kf}(\phi^{(p)}, \phi^{(p+1)})$. Finally a first order Taylor series expansion of $\nabla'_{\phi} D_{kf}(\phi, \phi')$ about (ϕ^*, ϕ^*) yields

$$\begin{aligned}[\nabla'_{\phi} D_{kf}(\phi, \phi')]^T &= [\nabla'_{\phi} D_{kf}(\phi^*, \phi^*)]^T \\ &\quad + (\phi' - \phi^*)\nabla'^2_{\phi} D_{kf}(\phi^*, \phi^*) + (\phi - \phi^*)\nabla_{\phi}\nabla'_{\phi} D_{kf}(\phi^*, \phi^*).\end{aligned}$$

On substituting $\phi^{(p)} = \phi$ and $\phi^{(p+1)} = \phi'$ and using assumption (2) and the equivalence (See Appendix 1)

$$\nabla_{\phi}\nabla'_{\phi} D_{kf}(\phi^*, \phi^*) = -\nabla^2_{\phi'} D_{kk}(\phi^*, \phi^*)$$

we obtain

$$(\phi^{(p+1)} - \phi^*)\nabla^2_{\phi'} D_{kf}(\phi^*, \phi^*) = (\phi^{(p)} - \phi^*)\nabla'^2_{\phi} D_{kk}(\phi^*, \phi^*)$$

from which (9) follows directly.

Theorem 1 implies that $L(\phi)$ is non-decreasing on each iteration of a generalized algorithm. It also implies that a maximum likelihood estimate is a stationary point of a generalized algorithm. Theorem 2 provides conditions under which a generalized algorithm converges. But these results fall short of implying convergence to a maximum likelihood estimator. Theorem 3 establishes conditions which guarantee convergence to a local maximum of the likelihood function and Theorem 4 addresses the issue of speed of convergence which is dependent on the eigenvalues of the matrix

$$[\nabla'^2_{\phi} D_{kk}(\phi^*, \phi^*)][\nabla'^2_{\phi} D_{kf}(\phi^*, \phi^*)]^{-1}.$$

4. The Complete Data Choice: Rate of Convergence and Algorithmic Simplicity

We have already alluded to the necessity of x being a non-trivial refinement of y. When the refinement is trivial we have

$$\begin{aligned} D_{kf}(\phi, \phi') &= \int k(x/y, \phi) \log \frac{k(x/y, \phi)}{f(x/\phi')} dx \\ &= \int k(x/y, \phi) \log \frac{f(x/\phi)}{g(y/\phi) f(x/\phi')} dx \\ &= -\int k(x/y, \phi) \log g(y/\phi') dx \\ &= -\log g(y/\phi') = -L(\phi') \end{aligned}$$

since

$$\frac{f(x/\phi)}{f(x/\phi')} = \frac{g(y/\phi)}{g(y/\phi')}.$$

There is no basis for an EM algorithm when the refinement is trivial, the discrimination function reducing to the negative of the log-likelihood function. For an arbitrary refinement

$$\begin{aligned} D_{kf}(\phi, \phi') &= \int k(x/y, \phi) \log \frac{k(x/y, \phi)}{k(x/y, \phi') g(y/\phi')} dx \\ &= -L(\phi') + D_{kk}(\phi, \phi'). \end{aligned}$$

We illustrate an iteration graphically in Fig. 1 for the case of a one dimensional parameter vector. An EM algorithm iteration moves the estimate from $\phi^{(p)}$ to $\phi^{(p+1)}$ in the direction of the maximum likelihood estimate ϕ_{ML}. A GEM algorithm moves the estimate in the same direction but possibly by a lesser amount. Fig. 1 and Theorem 4 show that convergence is most rapid when the curvature of $D_{kk}(\phi^{(p)}, \phi')$ with ϕ' is small. Indeed the zero curvature case results from a trivial refinement and a single iteration produces the maximum likelihood estimate. Why then should we choose x so that $D_{kk}(\phi, \phi')$ has significant curvature? It is because we seek an algorithm that is simpler than is direct minimization of $-L(\phi')$. To have a significant curvature we require an x whose distribution, given y, is sensitive to ϕ' (Note that for the trivial refinement case $k(x/y, \phi) = k(x/y, \phi')$ for all ϕ'). For the multiple dimensional parameter vector of interest curvature of $D_{kk}(\phi, \phi')$ is expressed by the Hessian matrix

$$\nabla'^2_{\phi} D_{kk}(\phi, \phi).$$

Our requirement is that eigenvalues of this Hessian matrix be bounded away from zero. (We note that the Hessian of D_{kk} is more familiar as the Fisher information matrix associated with $k(x/y, \phi)$ [Blahut,1987b]). If $k(x/y, \phi)$ expresses a high degree of information about ϕ (ie. $D_{kk}(\phi, \phi')$ very sensitive to ϕ'), the algorithm may require many iterations in order to converge. We are however less interested in the number of required iterations than in

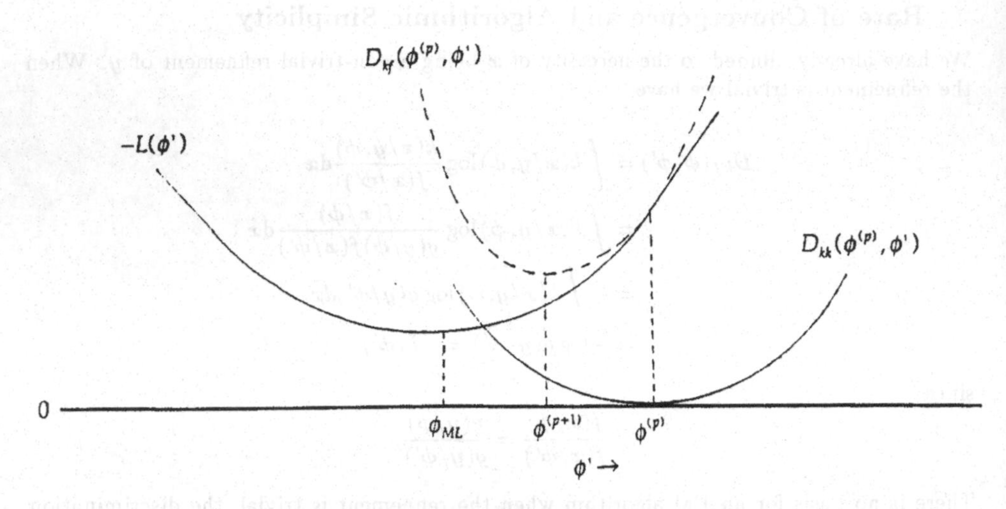

Figure 1

assurance of convergence and computational simplicity of the algorithm. We therefore seek a refinement x leading to a simple GEM algorithm.

5. Refinement of ISAR Data for Autofocusing

Dempster, et al. note that a particular application suggests a "natural" refinement. This is indeed the case with the ISAR autofocus problem. It seems intuitively apparent that one needs to determine some objective function indicative of focal quality and which one optimizes by varying focusing parameters. We are accustomed to adjusting focusing parameters so as to bring an image into focus while viewing it. This is suggestive of image parameters as missing data. To automate the focusing process one may look for an appropriate objective function to be optimized. Minimum configurational entropy and maximum contrast have been suggested. Bayesians will cite maximization of the probability of focusing parameters, given the available data. But in the latter case how should we treat the image parameters? Simply treating them as nuisance parameters is unsatisfactory because we are ultimately interested in the image parameters. The problem is really one of joint estimation of focusing and image parameters. Luttrell [1990] has outlined an elegant Bayesian (albeit computationally intensive) approach to this joint estimation. Our goal is a relatively simple algorithm by which we may estimate focusing parameters using only a minimal amount of image information. Hence our interest in the EM algorithm with its freedom of choice of missing information in the refinement. Our motivation stems from several considerations: (1) The radar data (because of target motion) are samples of the Fourier transform of target reflectivity on a non-planar surface in the transform space. Hence image reconstruction may require backprojection in three dimensions, even though only a two-dimensional image is sought. Thus iterative reconstruction of an entire image may involve excessive computation. (2) The number of focusing parameters is small. Although focusing requires highly accurate information about changes in radar-target geometry during an image frame time and the number of data points is typically the same as the number of image pixels, the changes in geometry, because of the short frame times and target inertia, can be described with a small number of parameters. (3) Minimal image information, identified with prominent scatterers, has been used successfully in focusing two dimensional radar images [Steinberg, 1981, Werness, et al., 1990].

6. A GEM Algorithm for ISAR Autofocus

We make the natural choice of complete data $\boldsymbol{x}$ by augmenting the incomplete data $\boldsymbol{y}$ with a set $\boldsymbol{z}$ of as yet unspecified image parameters. Thus

$$\boldsymbol{x} = (\boldsymbol{y}, \boldsymbol{z})$$

and

$$f(\boldsymbol{x}/\phi) = f(y, z/\phi) = p(z/\phi)l(y/\phi, z)$$

where $p(z/\phi) = p(z)$ is the density function for z and is independent of ϕ. $l(y/\phi, z)$ is a multidimensional Gaussian density whose covariances are determined by the measurement noise and whose means are functions of ϕ and z. The density for $\boldsymbol{x}$ conditioned on y and ϕ is a function of z, namely

$$k(\boldsymbol{x}/y, \phi) = h(z/y, \phi) = \frac{p(z)l(y/\phi, z)}{g(y/\phi)}.$$

For the discrimination we have

$$\begin{aligned} D_{kf}(\phi, \phi') &= \int_{\chi(y)} k(\boldsymbol{x}/y, \phi) \log \frac{k(\boldsymbol{x}/y, \phi)}{f(\boldsymbol{x}/\phi')} d\boldsymbol{x} \\ &= \int h(z/y, \phi) \log \frac{p(z)l(y/\phi, z)}{g(y/\phi)p(z)l(y/\phi', z)} dz \\ &= -\log g(y/\phi) + \int h(z/y, \phi) \log \frac{l(y/\phi, z)}{l(y/\phi', z)} dz \end{aligned}$$

Minimizing $D_{kf}(\phi, \phi')$ over ϕ' corresponds to minimizing

$$\int h(z/y, \phi) \log \frac{l(y/\phi, z)}{l(y/\phi', z)} dz.$$

We need not actually perform the minimization in a given iteration. Convergence to a local minimum of $-L(\phi')$ is assured if a reduction in discrimination is achieved at each iteration, ie. if

$$D_{kf}(\phi, \phi') \leq D_{kf}(\phi, \phi)$$

for every ϕ in the region Ω. Thus we have the requirement for a GEM algorithm mapping $\phi \rightarrow \phi'$ in an iteration that

$$D_{kf}(\phi, \phi') - D_{kf}(\phi, \phi) = \int h(z/y, \phi) \log \frac{l(y/\phi, z)}{l(y/\phi', z)} dz \leq 0.$$

We know the form of $l(y/\phi, z)$ to be

$$l(y/\phi, z) \propto \exp\left\{-\frac{1}{2\sigma^2}[y - y_0(\phi, z)]^h[y - y_0(\phi, z)]\right\}$$

where σ^2 is measurement noise variance and the superscript h indicates Hermitian transpose. Let

$$y_0 = y_0(\phi, z) \quad , \quad y_0' = y_0(\phi', z).$$

Then

$$\log\frac{l(y/\phi, z)}{l(y/\phi', z)} = -\frac{1}{2\sigma^2}\left[y_0^h y_0 - 2\mathrm{Re}\{y^h(y_0 - y_0')\} - y_0'^h y_0'\right].$$

The model data y_0 depend on ϕ and z, the latter of which we are free to choose. In the interest of simplicity we choose z to be the amplitude A and phase θ of a single scatterer at a known location z_0. Then the kth component of y_0 is

$$y_{0k} = A\exp(j\theta)\exp(-j2\pi f_k \cdot z_0)$$

where f_k, which is ϕ dependent, is the spatial frequency sampled by the kth measurement. It follows that

$$y_0^h y_0 = y_0'^h y_0' = NA^2 \tag{10}$$

where N is the total number of spatial frequencies sampled in an image frame. The log-likelihood ratio reduces to

$$\log\frac{l(y/\phi, z)}{l(y/\phi', z)} = \frac{1}{\sigma^2}\mathrm{Re}\left[y^h(y_0 - y_0')\right].$$

Let D_k represent the kth measured datum, ie. the kth component of y. Then

$$y^h y_0 = \sum_k D_k^* A\exp(j\theta)\exp(-j2\pi f_k \cdot z_0).$$

We recognize

$$\sum_k D_k \exp(j2\pi f_k \cdot z_0) = d(\phi, z_0)\exp(j\gamma(\phi, z_0))$$

as the reconstructed image value at point z_0 using focusing parameters ϕ. Thus

$$y^h y_0 = Ad(\phi, z_0)\exp\{j[\theta - \gamma(\phi, z_0)]\} \tag{11}$$

and

$$\begin{aligned}\mathrm{Re}\left[y^h(y_0 - y_0')\right] &= \mathrm{Re}\{A\exp(j\theta)[d\exp(-j\gamma) - d'\exp(-j\gamma')]\} \\ &= Ad\cos(\theta - \gamma) - Ad'\cos(\theta - \gamma')\end{aligned} \tag{12}$$

where

$$d = d(\phi, z_0) \quad , \quad d' = d(\phi', z_0)$$

$$\gamma = \gamma(\phi, z_0) \quad , \quad \gamma' = \gamma(\phi', z_0)$$

The requirement that

$$\int h(z/y,\phi) \log \frac{l(y/\phi,z)}{l(y/\phi',z)} dz = \frac{1}{g(y/\phi)} \int p(z) l(y/\phi,z) \log \frac{l(y/\phi,z)}{l(y/\phi,z)} dz \leq 0$$

is now reduced to

$$\int dz p(z) \exp\left\{-\frac{1}{2\sigma^2}\left[y^h y - 2\text{Re} y^h y_0 + y_0^h y_0\right]\right\} \text{Re}\left[y^h(y_0 - y_0')\right] \leq 0$$

Noting that $\exp(-y^h y/2\sigma^2)$is a positive constant and assuming a uniform distribution for θ, independent of A, ie. $p(z) = q(A)/2\pi$ we obtain

$$\int dA q(A) \int \frac{d\theta}{2\pi} \exp\left\{-\frac{1}{2\sigma^2}\left[y_0^h y_0 - 2\text{Re} y^h y_0\right]\right\} \text{Re}\left[y^h(y_0 - y_0')\right] \leq 0.$$

A sufficient condition for a GEM algorithm is that the inner integral be negative semidefinite. Using (10), (11), and (12) the result is

$$0 \geq \int_0^{2\pi} \frac{d\theta}{2\pi} \exp\left\{-\frac{1}{2\sigma^2}\left[NA^2 - 2Ad\cos(\theta-\gamma)\right]\right\}\left\{d\cos(\theta-\gamma) - d'\cos(\theta-\gamma')\right\}$$

$$= \exp\left\{-\frac{NA^2}{2\sigma^2}\right\} \int_0^{2\pi} \frac{d\theta}{2\pi} \exp\left[\frac{Ad}{\sigma^2}\cos(\theta-\gamma)\right]\left\{d\cos(\theta-\gamma) - d'\cos(\theta-\gamma')\right\}. \qquad (13)$$

But

$$\int_0^{2\pi} \frac{d\theta}{2\pi} \exp\left[\frac{Ad}{\sigma^2}\cos(\theta-\gamma)\right]\cos(\theta-\gamma) = I_1\left(\frac{Ad}{\sigma^2}\right) \qquad (14)$$

and

$$\int_0^{2\pi} \frac{d\theta}{2\pi} \exp\left[\frac{Ad}{\sigma^2}\cos(\theta-\gamma)\right]\cos(\theta-\gamma') = I_1\left(\frac{Ad}{\sigma^2}\right)\cos(\gamma-\gamma') \qquad (15)$$

where $I_1(\cdot)$ is the modified Bessel function of first order. Combining (13), (14), and (15) yields

$$0 \geq \exp\left[-\frac{NA^2}{\sigma^2}\right] I_1\left(\frac{Ad}{\sigma^2}\right)\left[d - d'\cos(\gamma-\gamma')\right]$$

and since the first two factors are non-negative we have that

$$d'\cos(\gamma-\gamma') \geq d$$

is a sufficient condition for a GEM algorithm. $d'\cos(\gamma-\gamma')$ is the projection of the new image phasor at point z_0 on the prior image phasor at that point. Thus any mapping $\phi \rightarrow \phi'$ placing $d'\exp(j\gamma')$ in the half plane defined by $d'\cos(\gamma-\gamma') > d$ for any ϕ in Ω suffices as a GEM algorithm iteration. On computing the gradient of $d\exp(j\gamma)$ at ϕ it is relatively straightforward to design a mapping satisfying the GEM condition.

It is significant that we have used a priori knowledge concerning the position of our single scatterer. The result of the algorithm is a maximum likelihood estimate, not of the focusing parameters alone, but of the focusing parameters and the amplitude and phase of the single scatterer. (The final values of d and γ are estimates of A and θ respectively.) It is the joint estimation using a priori information which distinguishes the process from maximum likelihood estimation of the focusing parameters treating image parameters as nuisance parameters. The use of a priori information on scatterer location, though not a fundamental requirement, is the factor most responsible for a simple algorithm. Such information may be made available if the target is cooperative. In the case of an uncooperative target, one may compute an entire unfocused image (or a subimage containing a prominent scatterer) and determine the position of the strongest response. For successful application the response must be due to a prominent scatterer. In any case it is only a local maximum of the likelihood function which is assured on using a GEM algorithm. Crude adjustment of the focusing parameters to the vicinity of the global maximum ($\phi \in \Omega$) is required before application of the algorithm.

7. Summary

It is illuminating to interpret the EM algorithm as involving relative entropy maximization or discrimination minimization at each iteration. In terms of entropy maximization, one adjusts the parameters ϕ so that the incomplete data y and a prior estimate of ϕ are maximally non-committal in providing information about the complete data x not provided by the chosen value for ϕ. On convergence ϕ is adjusted so that y is maximally non-committal in specifying x. Alternatively, in terms of discrimination minimization, ϕ is adjusted until y is of minimum additional value in discriminating between the estimate of ϕ and a competing value.

We have outlined an application of the EM algorithm to the problem of automatic focusing of ISAR images. The freedom of choice of complete data and a priori information about a prominent scatterer are used to suggest a relatively simple generalized EM algorithm.

Appendix 1: Equivalence of $\nabla_\phi \nabla_\phi D_{kf}(\phi^*, \phi^*)$ and $-\nabla^2_\phi D_{kk}(\phi^*, \phi^*)$

We show that ijth elements

$$\frac{\partial^2 D_{kf}(\phi, \phi')}{\partial \phi'_i \partial \phi'_j} \quad \text{and} \quad \frac{\partial^2 D_{kf}(\phi, \phi')}{\partial \phi'_i \partial \phi'_j}$$

are identical when evaluated at $\phi' = \phi$.

$$D_{kk}(\phi, \phi') = \int k(x/y, \phi) \log \frac{k(x/y, \phi)}{k(x/y, \phi')} dx$$

$$\frac{\partial D_{kk}(\phi, \phi')}{\partial \phi'_j} = -\int k(x/y, \phi) \frac{1}{k(x/y, \phi')} \frac{\partial k(x/y, \phi')}{\partial \phi'_j} dx$$

$$\frac{\partial^2 D_{kk}(\phi, \phi')}{\partial \phi'_i \partial \phi'_j} = -\int \frac{k^2(x/y, \phi)}{k^2(x/y, \phi')} \frac{\partial^2 k(x/y, \phi')}{\partial \phi'_i \partial \phi'_j} dx + \frac{k(x/y, \phi)}{k^2(x/y, \phi')} \frac{\partial k(x/y, \phi')}{\partial \phi'_j} \frac{\partial k(x/y, \phi')}{\partial \phi'_i} dx$$

On evaluating at $\phi' = \phi$ we obtain

$$\frac{\partial^2 D_{kk}(\phi, \phi')}{\partial \phi'_i \partial \phi'_j} = \int \frac{1}{k(x/y, \phi)} \frac{\partial k(x/y, \phi)}{\partial \phi_i} \frac{\partial k(x/y, \phi)}{\partial \phi_j} dx - \int \frac{\partial^2 k(x/y, \phi)}{\partial \phi_i \partial \phi_j} dx$$

But

$$\int \frac{\partial^2 k(x/y, \phi)}{\partial \phi_i \partial \phi_j} dx = \frac{\partial^2}{\partial \phi_i \partial \phi_j} \left[\int k(x/y, \phi) dx \right] = 0.$$

Hence, at $\phi = \phi'$,

$$\frac{\partial^2 D_{kk}(\phi, \phi')}{\partial \phi'_i \partial \phi'_j} = \int \frac{\partial k(x/y, \phi)}{\partial \phi_i} \frac{\partial \log k(x/y, \phi)}{\partial \phi_j} dx$$

$$= \int k(x/y, \phi) \frac{\partial \log k(x/y, \phi)}{\partial \phi_i} \frac{\partial \log k(x/y, \phi)}{\partial \phi_j} dx,$$

ie. the ijth element of the Fisher information matrix.

$$D_{kf}(\phi, \phi') = \int k(x/y, \phi) \log \frac{k(x/y, \phi)}{f(x/\phi')} dx$$

$$\frac{\partial D_{kf}(\phi, \phi')}{\partial \phi'_j} = -\int k(x/y, \phi) \frac{1}{f(x/\phi')} \frac{\partial f(x/\phi')}{\partial \phi'_j} dx$$

$$\frac{\partial^2 D_{kf}(\phi, \phi')}{\partial \phi_i \partial \phi'_j} = -\int \frac{\partial k(x/y, \phi)}{\partial \phi_i} \frac{\partial \log f(x/\phi')}{\partial \phi'_j} dx$$

Again, evaluating at $\phi' = \phi$ we obtain

$$\frac{\partial^2 D_{kf}(\phi, \phi')}{\partial \phi_i \partial \phi'_j} = -\int \frac{\partial k(x/y, \phi)}{\partial \phi_i} \frac{\partial \log f(x/\phi)}{\partial \phi_j} dx$$

Taking the sum of ijth elements we obtain, for $\phi' = \phi$,

$$\begin{aligned} \frac{\partial^2 D_{kf}(\phi, \phi')}{\partial \phi_i \partial \phi'_j} + \frac{\partial^2 D_{kk}(\phi, \phi')}{\partial \phi'_i \partial \phi'_j} &= \int \frac{\partial k(x/y, \phi)}{\partial \phi_i} \frac{\partial}{\partial \phi_j} \log \left[\frac{k(x/y, \phi)}{f(x/\phi)} \right] dx \\ &= -\frac{\partial}{\partial \phi_j} \log g(y/\phi) \int \frac{\partial k(x/y, \phi)}{\partial \phi_i} dx \\ &= 0 \end{aligned}$$

since

$$f(x/\phi) = k(x/y, \phi) g(y/\phi)$$

and

$$\int \frac{\partial k(x/y, \phi)}{\partial \phi_i} dx = \frac{\partial}{\partial \phi_i} \int k(x/y, \phi) dx.$$

REFERENCES

Blahut, R.E., 1987a, *Principles and Practice of Information Theory*, Addison-Wesley, Reading, MA. 107-111

Blahut, R.E., 1987b, op.cit. 304-306 (but note the factor of 2 error in the statement of Theorem 8.2.2., p 305)

Dempster, A.P., N.M.Laird and D.B.Rubin, 1977, 'Maximum Likelihood from Incomplete Data via the EM Algorithm', *Journal of the Royal Statistical Society*, Series B. vol. 29, 1-37

Luttrell, S.P.,1990, 'A Bayesian Derivation of an Iterative Autofocus/ Super-resolution Algorithm', *Inverse Problems* 6, 975-996

Steinberg, B.D., 1981, 'Radar Imaging from a Distorted Array: The Radio Camera and Experiments', *IEEE Trans. Antennas and Propagation*, AP-29 (5), 740-748

Werness, S.,W. Carrara, L. Joyce and D. Franczak, 1990, 'Moving Target Imaging Algorithm for SAR Data', *IEEE Trans. Aerospace and Electronic Systems*, vol.26 (1), 57-67.

BAYESIAN ANALYSIS FOR FAULT LOCATION BASED ON FINITE ELEMENT MODELS

Nathan Ida*, Minyan Li, and Louis E. Roemer

*Electrical Engineering Department
The University of Akron, Akron, OH 44325 USA

Electrical Engineering Department
Louisiana Tech University, Ruston, LA 71272 USA

ABSTRACT. The availability of suitable models or templates is often restricted due to computational efforts (for the Finite Element Method [FEM] template) or the limited number of experimental models which can be constructed as templates. In trying to identify the location or presence of a physical structure (or fault), high speed analysis of measured data must often be accomplished while the items are under test. This is particularly true in a manufacturing situation. This situation entails interpolation between a limited number of templates while trying to minimize computational demands. Such a situation arises when data are measured at different positions than provided by the original template.

The example demonstrated is the identification of the position of a fault (manufactured) in a steel bar, based on Hall type magnetic field sensor measurements of the magnetic field near the bar. Finite element models are used as the templates. The large volume over which the effects of the fault are sensed (compared to the volume of the fault) lends itself to Bayesian methods, given a suitable model to which to compare the experimental data (Roemer, 1991).

1. Introduction

Many measurements made of physical structures result in sensing a large volume of space. Quite often, measurements extend to regions far removed from the source of the signal. For example, the electric currents which flow around a crack produce magnetic fields which can be sensed in regions far removed from the crack. Further, these regions provide additional information on the possible configuration which gave rise to the magnetic field. Most instruments integrate some physical field quantity over a volume which is larger than we would choose. Even the Hall effect probe used in this study is larger than features which we would like to detect in future tests.

2. The Method

The object tested was a steel bar (mild steel, type 1020), 3.18 cm (1.25 inch) by 3.18 cm by 91 cm (36 inch). A slot was machined into the middle of one face of the bar, 0.64 cm (0.25 inches) transverse to the bar by 0.64 cm deep by 1.27 cm (0.5 inches) along the axis of the bar, as shown in Figure 1. A direct current of 125 Amperes was passed through the bar, and the magnetic flux density normal to the bar was sensed with a Hall effect probe. A graph of the magnetic flux density measured is shown in Figure 3.

A. Mohammad-Djafari and G. Demoments (eds.), Maximum Entropy and Bayesian Methods, 333–339.

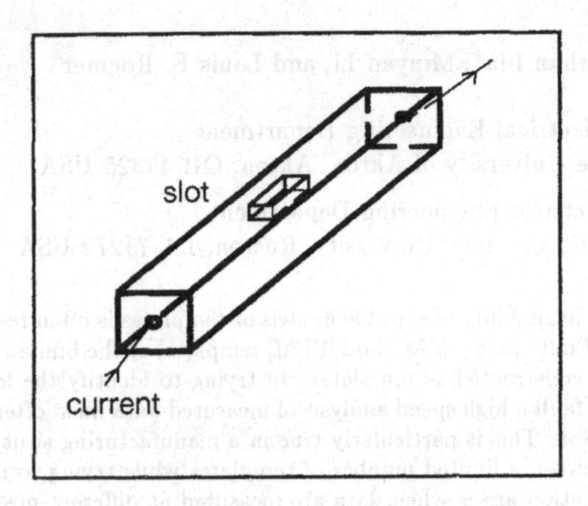

Figure 1: *Steel Bar Under Test*

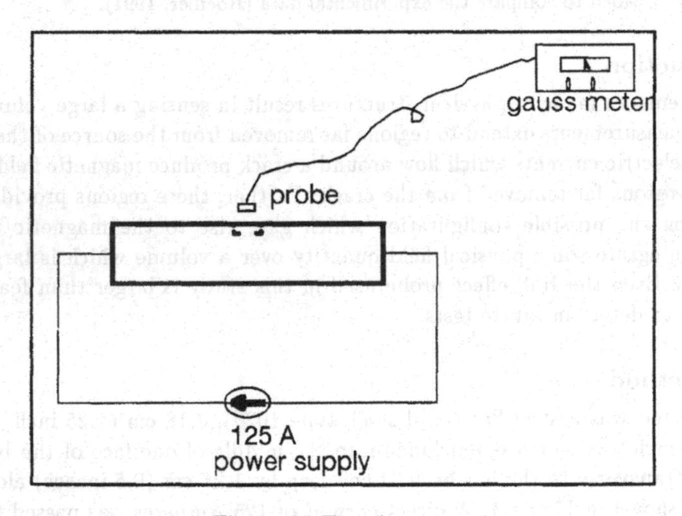

Figure 2: *Test Apparatus*

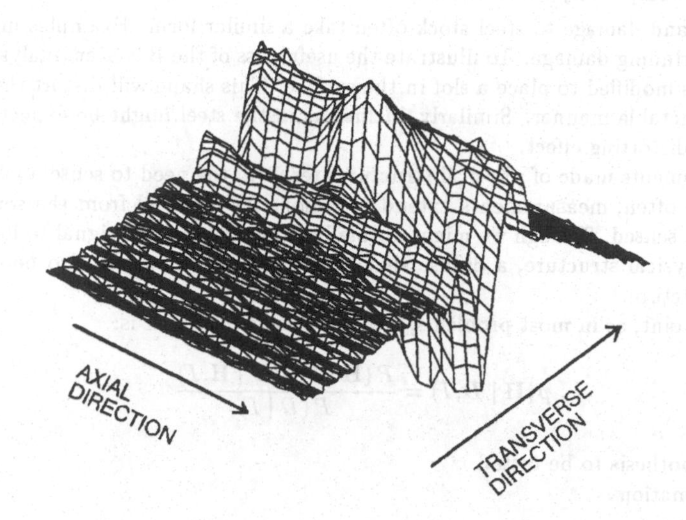

Figure 3: *Magnitude of Magnetic Flux Density Normal to Bar Surface*

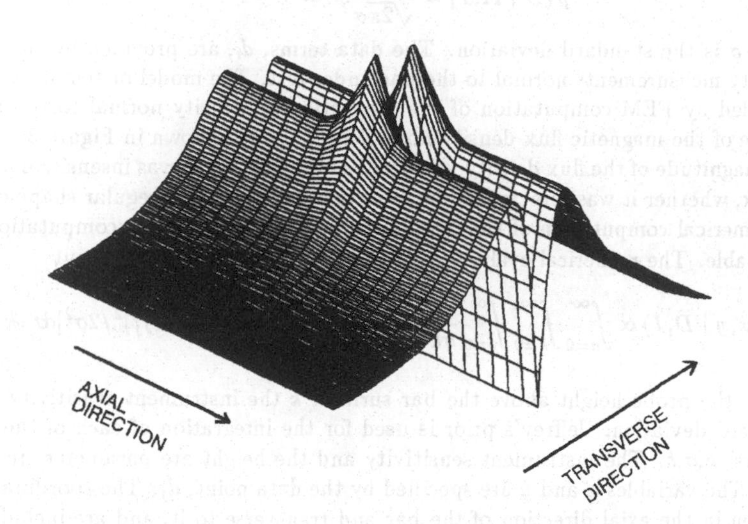

Figure 4: *Finite Element Model of Magnitude of Magnetic Flux Density Normal to Bar*

3. Test Description, fault in a steel bar

Faults, voids, and damage to steel stock often take a similar form. Examples might be inclusions and grinding damage. To illustrate the usefulness of the Bayesian analysis method, a steel bar was modified to place a slot in the surface. This shape will distort the magnetic field in a predictable manner. Similarly, inclusions in the steel might be expected to have a predictably distorting effect.
Many measurements made of physical structures result in the need to sense a large volume of space.Quite often, measurements extend to regions far removed from the source of the signal which is sensed. Though we might normally expect a localized signal to be necessary to locate a physical structure, a diffuse signal of known variation can also be effective in locating a structure.
The starting point, as in most problems, is Bayes' theorem, which is:

$$p(\mathbf{H} \mid D,I) = \frac{P(\mathbf{H} \mid I)\,p(D \mid \mathbf{H},I)}{P(D \mid I)} \tag{1}$$

where **H**=Hypothesis to be tested
I= prior information
D=data

The terms and their significance are well described in Bretthorst's book (Bretthorst, 1988). Lower case p denotes probability density and upper case P denotes probability. The likelihood function is

$$p(D \mid \mathbf{H},I) = \frac{1}{\sqrt{2\pi}\sigma} e^{-\sum(d_i - f_i)^2/2\sigma^2} \tag{2}$$

where σ is the standard deviation. The data terms, d_i, are provided by the magnetic flux density measurements normal to the bar under test. The model or template values, f_i, are provided by FEM computation of the magnetic flux density normal to the bar. The magnitude of the magnetic flux density normal to the bar is shown in Figure 3.

The magnitude of the flux density was used, as the flux meter was insensitive to the sign of the flux, whether it was directed upwards or downwards. The irregular shape of the bar makes numerical computation of the template a necessity, as analytic computations would be intractable. The numerical evaluation of the integral below is carried out

$$p(x,y \mid D,I) \propto \int_{\sigma=0}^{\infty}\int_{k=0}^{\infty}\int_{\delta=0}^{\infty} \frac{1}{\sigma k \delta}\left(\frac{1}{\sqrt{2\pi}\sigma}\right)^N exp[-\sum_j (d_j - f_j)^2/2\sigma^2]\,d\sigma\, dk\, d\delta \tag{3}$$

where δ is the probe height above the bar surface, k the instrument sensitivity, and σ is the standard deviation. Jeffrey's prior is used for the integration of each of the nuisance parameters, k,σ,δ. The instrument sensitivity and the height are parameters in the template, f_j. The variables x and y are specified by the data point, d_j. The coordinates x and y are taken in the axial direction of the bar and transverse to it, and are included in the template coordinates of f_j. Figure 5 shows the resultant probability density. The peak fell at the center of the manufactured slot, as expected. Integration of the probability density to evaluate $p(x \mid D,I)$,using the Jeffrey's prior of $1/y$, yields Figure 6. Similarly, integration of the probability density to evaluate $p(y \mid D,I)$,using the prior of $1/x$, yields Figure 7.

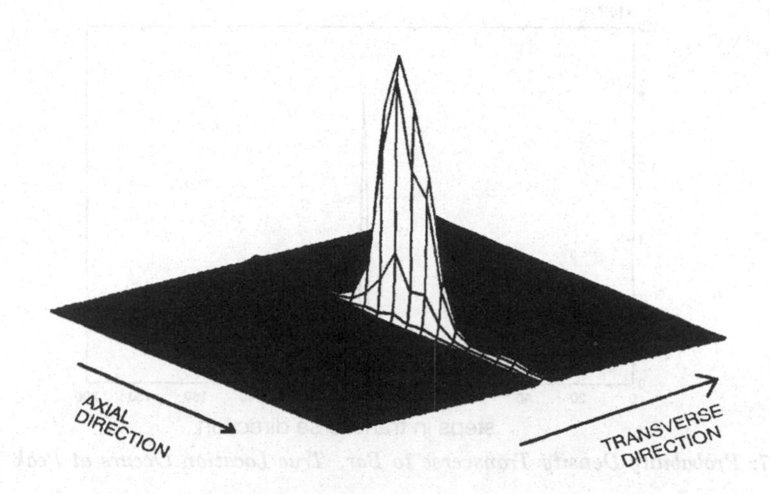

Figure 5: *Probability Density versus Axial and Lateral Position*

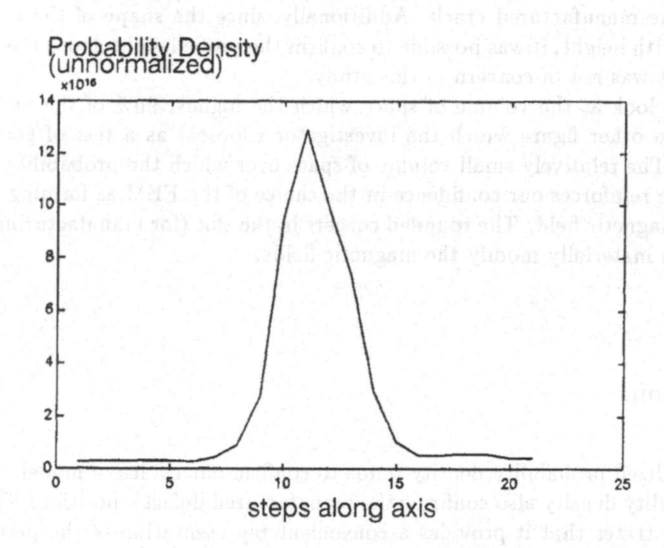

Figure 6: *Axial Probability Density. True Location Occurs at Peak Value*

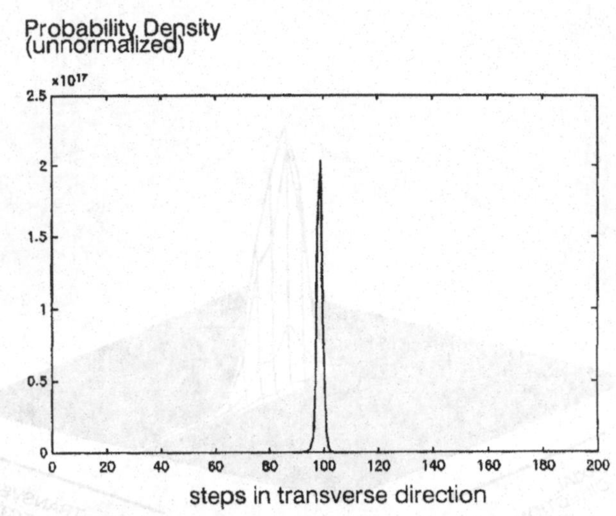

Figure 7: *Probability Density Transverse to Bar. True Location Occurs at Peak Value*

Both curves have the location of the highest probability density falling at the location of the center of the manufactured crack. Additionally, since the shape of the magnetic flux density varies with height, it was possible to confirm the probe height above the bar surface. However, height was not of concern in this study.
One might also look at the volume of space which the highest 90% of the probability occupied (or some other figure which the investigator chooses) as a test of goodness of fit for the model. The relatively small volume of space over which the probability density has significant value reinforces our confidence in the choice of the FEM as forming an accurate model for the magnetic field. The rounded corners in the slot (for manufacturing ease) were not expected to materially modify the magnetic fields.

4. Conclusions

The highly localized probability density tends to confirm our choice of model. The highest value of probability density also confirms the manufactured defect's position. The Bayesian analysis demonstrates that it provides a convenient representation of the pertinent question, "Given the data obtained, and the model which we believe to be true, what is the likelihood that we find such a structure (similar to the model) at this point?" It would seem to be the only question of interest.

References

[1] Roemer, L. and N. Ida: 1991, 'Location of Wire Position in Tyre Belting Using Bayesian Analysis', *NDT & E International,* Vol. 24, No. 2, 95-97.

[2] Bretthorst, G.Larry: 1988, *Bayesian Spectrum Analysis and Parameter Estimation,Lecture Notes in Statistics,* Springer-Verlag, New York.

[3] Silvester, P.P. and R.L. Ferrari: 1990, *Finite Elements for Electrical Engineers,* Cambridge University Press, New York.

[4] Zatsepin, N.N.. and V.E. Shcherbinnin: 1966, 'Calculation of the magnetostatic field of surface defects I: Field topography of defect models', *Defectoskopiya,* Vol. 5, 50-59.

[5] Zatsepin, N.N.. and V.E. Shcherbinnin: 1966, 'Calculation of the magnetostatic field of surface defects II, Experimental verification of the principal theoretical relationships',*Defectoskopiya,* Vol. 5, 59-65.

PYRAMID IMAGES

Tj. Romke Bontekoe
Astrophysics Division, Space Science Department of ESA
European Space Research and Technology Centre
Postbus 299, 2200 AG Noordwijk
The Netherlands.

ABSTRACT. We introduce the concept of pyramid images in problems of image reconstruction, and demonstrate that they provide a natural way to account for spatial correlation, especially when the images have many different length scales. Virtually all length scales can be accounted for because images are formed from a weighted sum of a 1×1 pixel channel, a 2×2 channel, a 4×4 channel, etc. All channels cover the entire image, and each channel can be tuned optimally for its own scale length. Diagnostics for fine-tuning may be introduced in a straightforward manner. Image reconstructions through pyramid images are only a factor 4/3 larger in size than conventional single channel methods.

Although in maximum entropy data analysis no distinction is necessary whether the scene to reconstruct is of one, two, or higher dimensionality, the very nature of the scene may demand different approaches. In spectroscopy the goal often is to resolve all spectral lines to essentially delta-functions and determine their position and amplitude. Essentially, this is an over-deconvolution since it ignores the natural line width, but for many applications this may be of no concern. The two-dimensional analogue would be a distribution of bright sources in a map.

The reconstruction of scenes of slowly varying intensity on top of which a few sharp features are present, is known to be troublesome (see e.g. Sivia 1990, Charter 1990). Although maximum entropy is supposed to give the "smoothest solution consistent with data and within the noise", a strong "ringing" may result. The results show a tendency towards a series of delta-functions instead of a smooth distribution. Sivia (1990) argues that formally such a maximum entropy solution is nevertheless valid, because there was insufficient prior information provided. This can be improved by separating broad and sharp features in different channels, to be solved for simultaneously. One channel has a few large pixels and will absorb the background, and the other has small pixels. Both channels cover the same area. The entropy is defined in the usual fashion

$$S = \sum_{l=1}^{L} h - m - h \log(h/m) \qquad (1)$$

where h is the scene of interest and m the default model, often taken a constant (Gull 1989, Skilling 1989). The summation is over all L pixel values, both channels taken together.

A. Mohammad-Djafari and G. Demoments (eds.), Maximum Entropy and Bayesian Methods, 341–344.

Broad features will now appear in the channel with large pixels because this is entropically cheaper, i.e. it requires fewer pixels to change from their default value. The adjustable parameter is the size of the large pixels. In Sivia's case the background was problematic and consequently discarded.

A different approach was taken by Charter (1990) who attempted to reconstruct all features, broad and sharp. Essentially he exploits the over-deconvolution by re-blurring the result, thereby enforcing spatial correlations. This is done via an Intrinsic Correlation Function (ICF). The reconstruction is now performed on two images. First, a *hidden image* which is *a priori* uncorrelated and controlled by the entropy. Second, a *visible image* which is a blurred version of the hidden image and which, by construction, represents a spatially correlated result. Often the ICF is chosen to be a gaussian with a width large enough to smooth out unrealistically small-scale structure, but small enough to improve the spatial resolution of the image. Theoretically, the optimum width scan be found by maximizing the 'Evidence' (see Skilling 1991) from a few reconstructions with different widths.

In its simplest form the deconvolution problem is modelled by a linear set of equations

$$D = Rf + \sigma \tag{2}$$

where D represents all measured and calibrated data (a data vector of length N), f is the unknown visible image (an image represented as vector of length M), σ represents the uncertainty of the data (a noise vector of length N). R is the $N \times M$ response matrix which describes all other features in the data aquisition process, e.g. the instrumental blur, the pixel positions etc. Note that data and image are represented as vectors of possibly different lengths. For the reconstruction of images from InfraRed Astronomical Satellite (IRAS) data there are advantages to be gained in choosing the pixel size "too small" such that the problem of Eqn. (2) becomes underdetermined, i.e. $N < M$ (Bontekoe *et al.* 1991). This, however, does not prevent reliable image reconstructions.

The intrinsic correlation function can be represented as

$$f = Ch \tag{3}$$

where h forms the hidden space image (a vector of length L). C is the $M \times L$ matrix representing the ICF operator. In single channel reconstructions $L = M$, and the matrix C is a square symmetric matrix. C has significant substructure since it correlates neighbouring pixels by rows as well as by columns. Alternatively the ICF convolution can be performed by FFTs.

Application of of this method, through the MemSys5 package, to reconstruct images from the scan data of IRAS showed, however, that more development was necessary. It was found that a single ICF width gives rise to ringing effects at scale lengths comparable to the applied ICF width, and that this ringing does not disappear by increasing the width (Bontekoe and Koper 1992). Clearly, many scale lengths are present in IRAS images.

Next we used a multi-channel method in which all channels have their individual ICF widths. In multi-channel reconstructions the hidden space dimension $L = KM$ where K is the number of channels. All channels cover the entire image. In practice the reconstructions are performed by matrix multiplications of R and C and their transposes R^T and C^T. For multi-channel reconstructions this must be done with care since the individual channels must be multiplied separately by their channel ICF and its transpose. The response matrix

R is, of course, unaffected by the details of the transformations between visible and hidden space.

The representation is conceptually simpler when each visible space channel is regarded as a separate image f_k of size M. f is the sum of all channel images

$$f = \sum_{k=1}^{K} w_k f_k \tag{4}$$

where w_k are the weights for the individual channels. The hidden space vector h can also be regarded as consisting of K separate channels h_k since, by construction, the channels do not mix. In the forward operation, from image to data, one has for each channel

$$f_k = C_k h_k \tag{5}$$

where the C_k are single channel ICF operators operating only on channel k. In the transpose operation one has the algebraic transpose of Eqns. (4) and (5), for all k:

$$f_k = w_k f \tag{6}$$

and

$$h_k = C_k^T f_k. \tag{7}$$

Note that the weights w_k are not inverted.

A disadvantage of this is that the broad structure must be described by full images in visible and hidden space, while in principle it can be represented by relatively few parameters, as was the case with the large pixels. In addition, it is difficult to find the smallest value of K, the individual ICF widths in C_k, and the channel weights w_k. Often an optimum parameter set could not be found.

In the pyramid images the channels are built up from coarse to fine pixels, and we restrict ourselves to factors of two. Channels are numbered by $k = 0, \cdots, K$ and have $2^k \times 2^k$ pixels. Therefore channel $k = 0$ has 1×1 pixel, the next 2×2, then 4×4, up to $2^K \times 2^K = M$ pixels for the highest channel. The total size of L is the sum of all these pixels and is about $4/3\ M$. The forward transform from hidden space to visible space is now in two steps. First the ICF on channel k:

$$g_k = C_k h_k \tag{8}$$

where g_k is the blurred h_k. The C_k are "narrow width" ICFs, which nevertheless, depending on the pixel size, can still give a substantial blur. For channel $k = 0$ the ICF is replaced by the identity. Next the g_k must be expanded to the size M of a visible image channel

$$f_k = E_k g_k \tag{9}$$

in which the $M \times 4^k$ matrix E_k contains either 0's or 1's, such that each pixel value of g_k is duplicated into the pixels of f_k, insofar as they overlay each other in the image. The visible space image is again the weighted sum of the f_k.

In the transpose operation

$$g_k = w_k E_k^T f \tag{10}$$

whereby the pixel values in f are added into the corresponding coarse pixels of g_k. The other transpose operation is

$$h_k = C_k^T g_k. \tag{11}$$

The advantage of pyramid images is that artifacts in the visible space image can be easily traced back to the channel where they originated. Assume a given setting of ICF widths represented in the C_k. If in f an obvious block structure is present, one can determine the size of these blocks in pixels and adjust the corresponding ICF width. Skew backgrounds are usually absorbed in one particular channel, and thus can be separated from the image. However, occasionally a background is reconstructed as an area filled with narrow spikes in the highest channel. Obviously, the "wrong" channel has been used. In such a case we decrease the weight of this channel to encourage the background in another channel. Otherwise we always set all weights equal to 1.0.

In conclusion, many different attempts of incorporating prior knowledge about what reconstructed images should look, have been made. The pyramid images have proven to be the best so far, and we expect that there will be a much wider use for this approach. More discussion on pyramid images, and examples, in relation to IRAS image reconstruction will be given in Bontekoe and Koper (in preparation).

ACKNOWLEDGMENTS. I wish to acknowledge many discussions with my long friend and colleague Do Kester (Space Research Groningen), with Enrico Koper (Sterrewacht Leiden), with Mike Perryman (ESTEC), and with John Skilling and Mark Charter (MEDC Cambridge).

REFERENCES

Bontekoe, Tj.R., Kester, D.J.M., Price, S.D., de Jonge, A.R.W., and Wesselius, P.R.: 1991, 'Image Construction from the IRAS Survey', *Astron. Astrophys.* **248**, 328–336.

Bontekoe, Tj.R. and Koper, E.: 1992, 'Pitfalls in image reconstruction' in D.M. Worall, C. Biemesderfer, and J. Barnes (eds.) *Astron. Soc. of the Pacific Conf. Series* Vol. 25, 191–195.

Charter, M.K.: 1990, 'Drug absorption in man, ...', in P.F. Fougere (ed.), *Maximum Entropy and Bayesian Methods, Dartmouth 1989*, 325–339, Kluwer, Dordrecht.

Gull, S.F.: 1989, 'Developments in maximum entropy data analysis', in J. Skilling (ed.), *Maximum Entropy and Bayesian Methods, Cambridge 1988*, 53–71, Kluwer, Dordrecht.

Sivia, D.S.: 1990, 'Applications of maximum entropy ...', in P.F. Fougere (ed.), *Maximum Entropy and Bayesian Methods, Dartmouth 1989*, 195–209, Kluwer, Dordrecht.

Skilling, J.: 1989, 'Classic maximum entropy', in J. Skilling (ed.), *Maximum Entropy and Bayesian Methods, Cambridge 1988*, 45–72, Kluwer, Dordrecht.

Skilling, J.: 1991, 'Fundamentals of MaxEnt in data analysis' in B. Buck and V. Macaulay (eds.), *Maximum Entropy in Action*, 19–40, Clarendon Press, Oxford.

FINE STRUCTURE IN THE SUNSPOT RECORD

Brian Buck & Vincent A. Macaulay
Theoretical Physics
University of Oxford
1 Keble Road
Oxford, OX1 3NP, England

ABSTRACT. A new method for the analysis of time-series, using the ideas of Maximum Entropy, has been applied to the yearly mean records of sunspot numbers. These numbers are regarded as reflecting an underlying magnetic cycle in the sun. The actual time-series considered consists of the square roots of the Wolf numbers, alternated every half cycle, so as to give a roughly sinusoidal shape with zero mean and a clear period of about 22 years. This device yields a much simpler spectrum than does an application of the method to the raw data.

The most probable spectrum is very clean, with the following features. Firstly, we see a group of four strong spectral lines, equally spaced, and with the dominant peak at 0.045 cycles per year (c/y). This corresponds to a period of 22.1 years, the well-known magnetic cycle. The next marked feature is a cluster of lines at approximately three times the frequency of the first group and therefore a prime candidate for a third harmonic. The last significant components of the spectrum are two incompletely resolved sets of lines on either side on the main group.

All this spectral information, which includes phases as well as amplitudes, will provide clues to the mechanism of the solar cycle. Here we consider chiefly the third harmonic. By taking cuts in the spectrum we can reconstruct time-series implied by selected frequency ranges and compare them. Doing this for the main group around 22 years, and for the possible third harmonic, we find that there is a component in the series having an envelope proportional to the cube of the size of the dominant signal, a local frequency three times that of the large component and a constant time delay of one and a half years. We have found that the spectral details of this third harmonic can be accounted for using a cubic function of the main signal and its time derivative.

The sunspot record shows an anomaly at the end of the 18th century. When this is removed, the spectrum becomes even cleaner. The confused group between 0.06 and 0.09 c/y now resolves into a set of sharp, regularly spaced peaks whose frequencies have a surprising ratio to those of the main group. This talk develops ideas first reported at MaxEnt91.

1. Introduction

The Wolf sunspot index is defined by $R_z = k(10g + f)$, where f is the observed number

A. Mohammad-Djafari and G. Demoments (eds.), Maximum Entropy and Bayesian Methods, 345–356.

of single spots, g is the number of groups and k is a calibration factor used to reconcile the results from different observatories. By comparison with modern data it is known that this index is closely proportional to the magnetic flux through the spots. The complete record of yearly means is shown in Figure 1. It has an obvious period of about eleven years between neighbouring maxima, apart from a few hiccups, and also a clear suggestion of longer term modulations. It is therefore of some interest to attempt to analyse this record to determine all significant Fourier components.

We have used the Maximum Entropy Method as an exploratory tool in this investigation. The philosophy is to generate possible models which can be compared using the techniques advocated by Jaynes (1987) and Bretthorst (1988). We found it convenient to represent $R_z(t)$, given at unit intervals of one year, by means of the Hartley spectrum $H(f)$ (Bracewell 1986) defined for frequencies f by

$$R_z(t) = T \int_{-\frac{1}{2}}^{\frac{1}{2}} H(f) \operatorname{cas}(2\pi f t) \, \mathrm{d}f,$$

where $\operatorname{cas}(\theta) = \cos(\theta) + \sin(\theta)$, T is the total time span in years and the integral is taken between the appropriate Nyquist limits. Knowledge of the $H(f > 0)$ and $H(f < 0)$ components is equivalent to a determination of the usual Fourier spectrum.

We assigned an entropic prior to the desired spectral image by first defining a discretized amplitude function

$$F(f_i) = \sqrt{H^2(f_i) + H^2(-f_i)} \qquad \text{for} \qquad f_i > 0,$$

and then taking its prior probability proportional to

$$\exp\left(\alpha \sum_i \left(\sqrt{F_i^2 + m_i^2} - m_i - F_i \sinh^{-1} \frac{F_i}{m_i}\right)\right)$$

(Buck and Macaulay 1992). The relative sizes of $H(f > 0)$ and $H(f < 0)$ were determined from a Gaussian likelihood function, this being equivalent to assuming a uniform prior on the phases. All spectra shown here were calculated by maximizing the posterior probability and the results are already quite good enough for some interesting conclusions to be drawn from the data.

In Figure 2, we show the spectrum obtained directly from the numbers given in Figure 1. While some gross features are readily apparent, such as the large peaks near the eleven year period, it should be noted that there are at least thirty other appreciable lines. Thus an analysis à la Bretthorst would certainly be tedious and even then it would be very difficult to disentangle interesting relationships.

We now know that the sunspot cycle is closely connected with the magnetic or Hale cycle, in which the fields associated with the spots alternate with a period of around 22 years. If we thus assume, as a tentative hypothesis, that the spot index is a manifestation of the energy in the magnetic field, then we are led to replotting the data as in Figure 3. This is a graph of the derectified square roots of the Wolf index numbers and is taken to represent the underlying magnetic field. The essential ingredient in what follows is the alternation of the data, the results being quite insensitive to the exact function of the numbers which

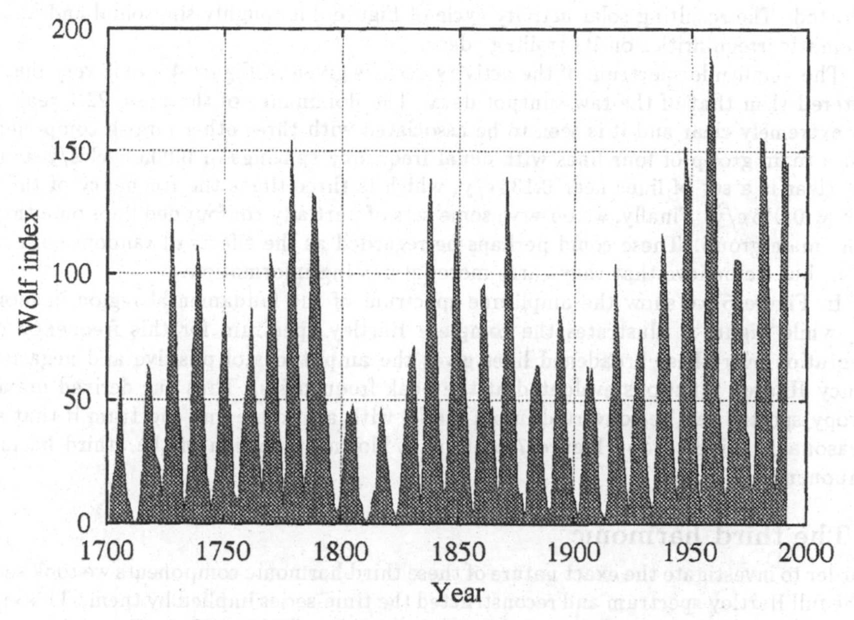

Fig. 1. The mean yearly Wolf sunspot index.

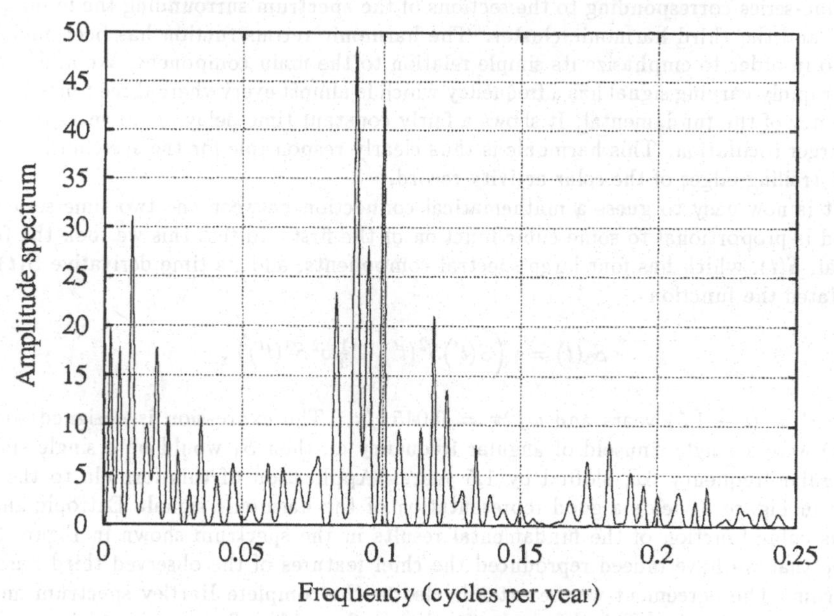

Fig. 2. Spectrum of sunspots.

is plotted. The resulting solar activity cycle of Figure 3 is roughly sinusoidal and has some systematic irregularities on its trailing edges.

The amplitude spectrum of the activity cycle is given in Figure 4 and is very much less cluttered than that of the raw sunspot data. The dominance of the basic 22.1 year line is now extremely clear and it is seen to be associated with three other largish components to form a main group of four lines with equal frequency spacings of 0.005 cycles/year (c/y). Also clear is a set of lines near 0.135 c/y, which is three times the frequency of the great peak at 0.045 c/y. Finally, we observe some sets of partially confounded lines on either side of the main group. These could perhaps be regarded as the effects of random noise in the signal, but we believe that they carry more interesting information.

In Figure 5 we show the amplitude spectrum of the fundamental region in more detail, while Figure 6 illustrates the complete Hartley spectrum for this frequency range. Integration over these broadened lines gives the amplitudes of positive and negative frequency Hartley functions evaluated at the peak frequencies. Thus our derived maximum entropy spectra can be converted into a model with a discrete line spectrum if that seems a reasonable thing to do. Figure 7 contains a blown-up version of the "third harmonic" components.

2. The third harmonic

In order to investigate the exact nature of these third harmonic components we took sections of the full Hartley spectrum and reconstructed the time-series implied by them. This enables a direct comparison of various contributions to the observed signal. In Figure 8 we present the time-series corresponding to the sections of the spectrum surrounding the fundamental group and the third harmonic cluster. The harmonic reconstruction has been multiplied by two in order to emphasize its simple relation to the main component. We find that the more rapidly varying signal has a frequency which is almost everywhere three times the local frequency of the fundamental. It shows a fairly constant time delay of 1.5 years relative to the larger oscillation. This harmonic is thus clearly responsible for the systematic shaping of the trailing edges of the solar activity record.

It is now easy to guess a mathematical connection between the two time-series: the second is proportional to some cubic function of the first. To test this we took the fundamental, $S(t)$, which has four large spectral components, and its time derivative $\dot{S}(t)$, and calculated the function

$$S_3(t) = \gamma \left(S(t')\dot{S}^2(t') - \tfrac{1}{3}\omega^2 S^3(t') \right),$$

where $t' = (t - 1.5)$ years and $\omega/2\pi = 0.045$ c/y. The expression is designed so that, if $S(t)$ were a single sinusoid of angular frequency ω, then S_3 would be a single sinusoid of angular frequency 3ω, shifted by 1.5 years. Application of this formula to the large signal in Figure 8 yields a good representation of the harmonic signal. Entropic analysis of this cubic function of the fundamental results in the spectrum shown in Figure 9 and we see that we have indeed reproduced the chief features of the observed third harmonic spectrum. The agreement we see extends also to the complete Hartley spectrum and not just to the amplitudes. This demonstrates that we have identified the time delay correctly. The location and size of the harmonic lines can now be understood in terms of interference, under cubing, between the main lines of the fundamental group.

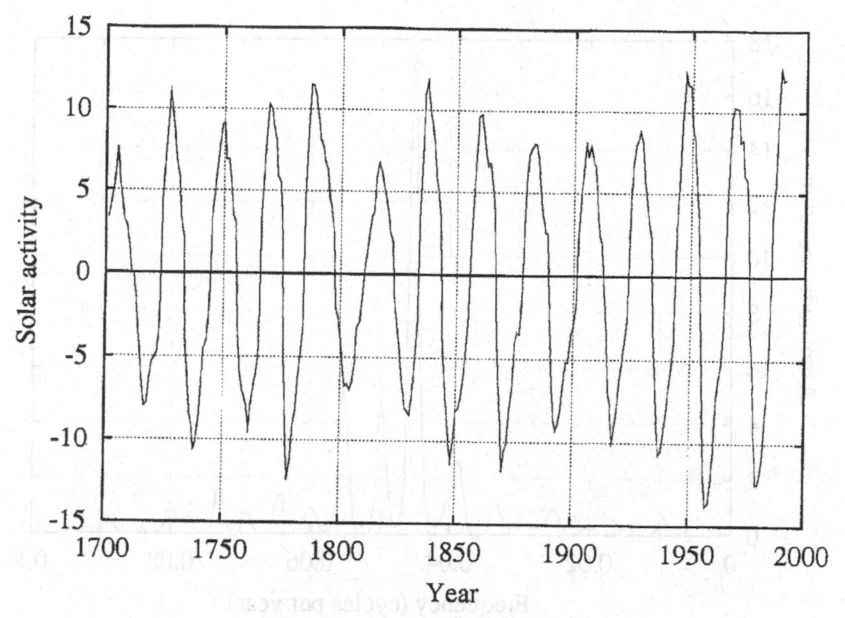

Fig. 3. Solar activity cycle.

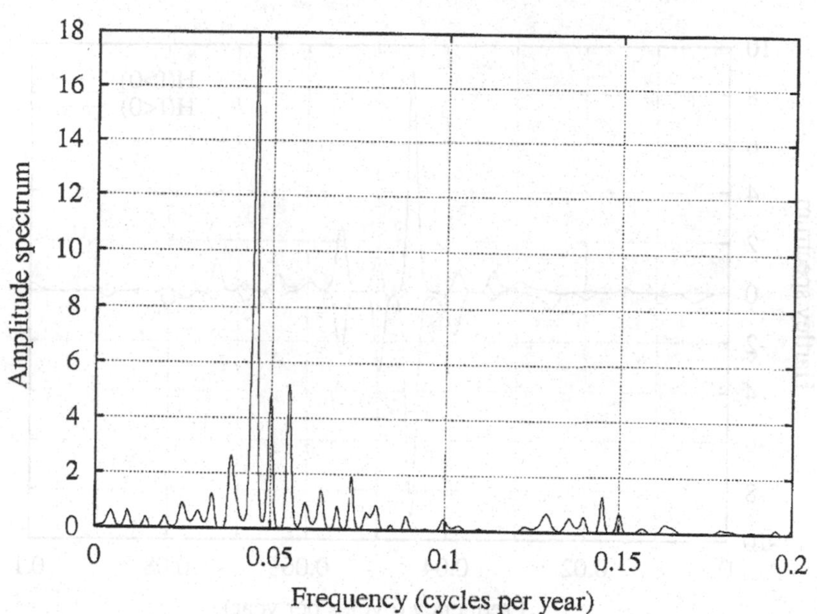

Fig. 4. Spectrum of solar activity.

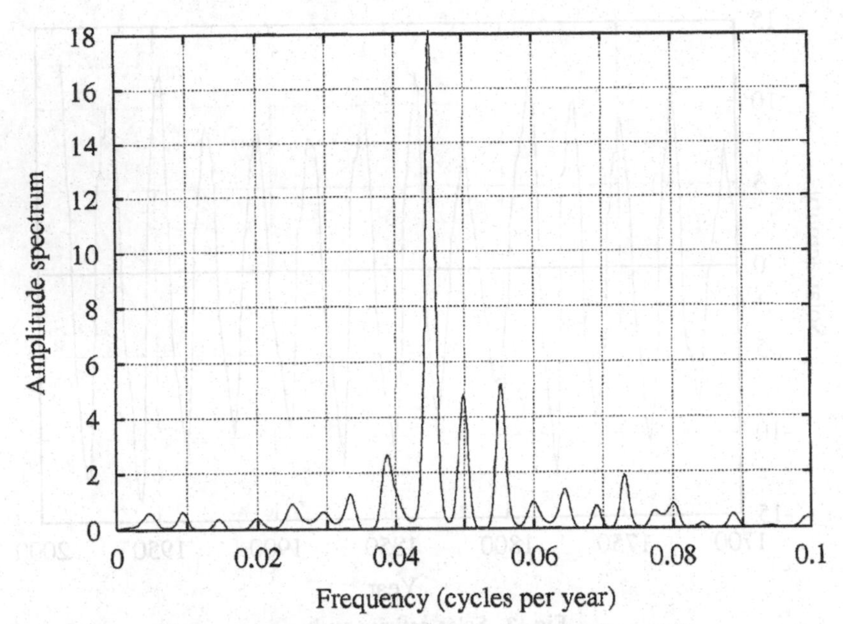

Fig. 5. Detail of solar activity spectrum: fundamentals.

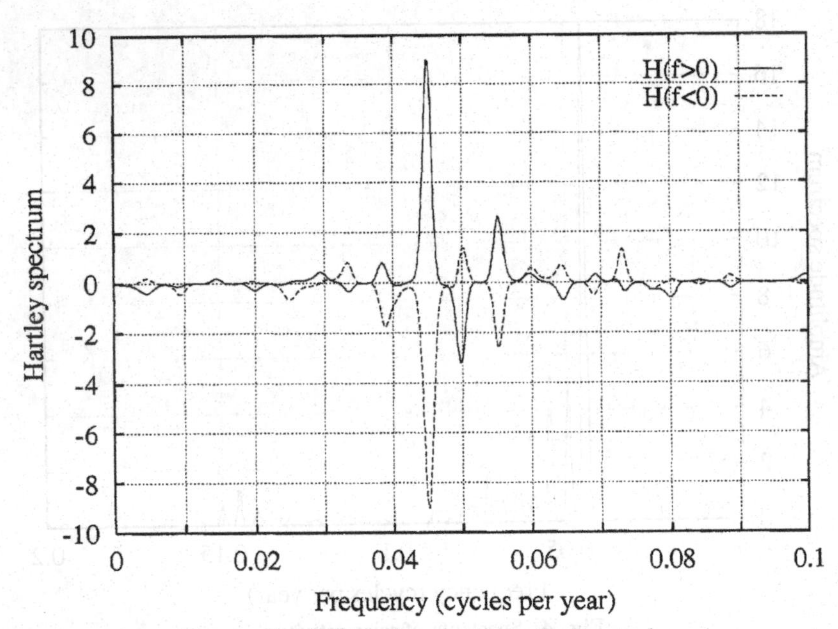

Fig. 6. Detail of Hartley spectrum: fundamentals.

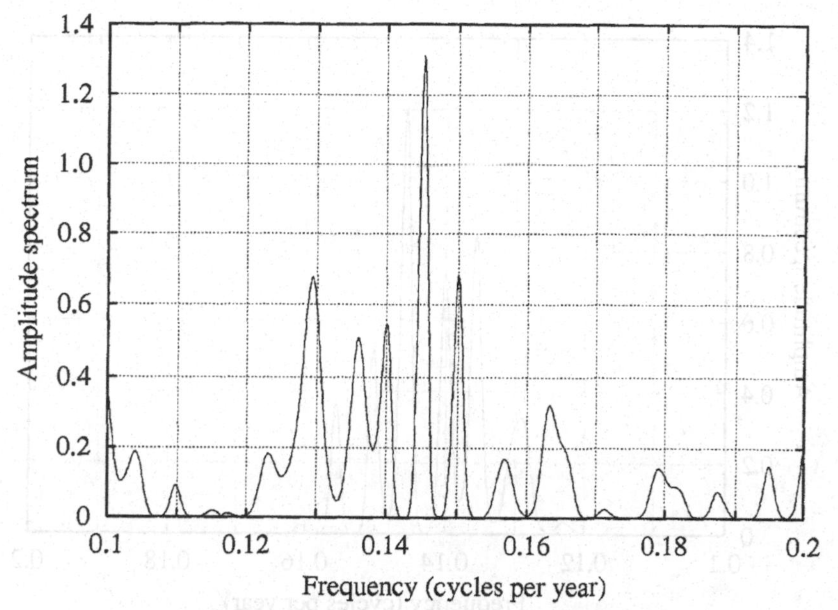

Fig. 7. Detail of solar activity spectrum: third harmonic.

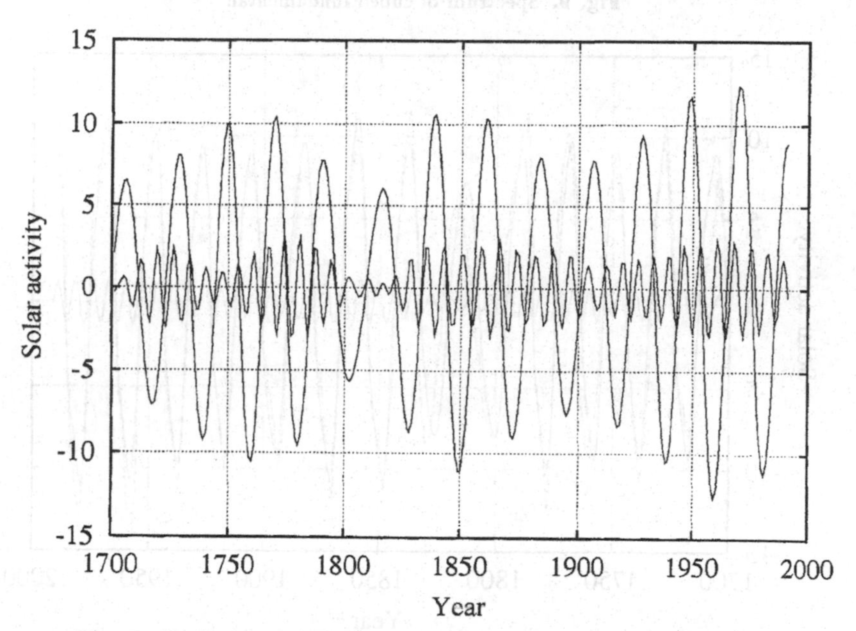

Fig. 8. Fundamental and third harmonic, the latter scaled up by two.

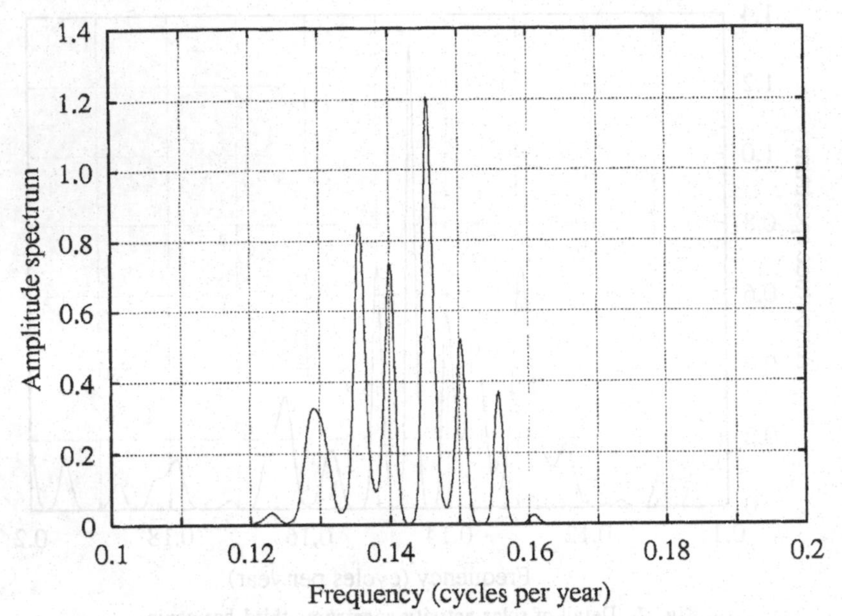

Fig. 9. Spectrum of cubed fundamental.

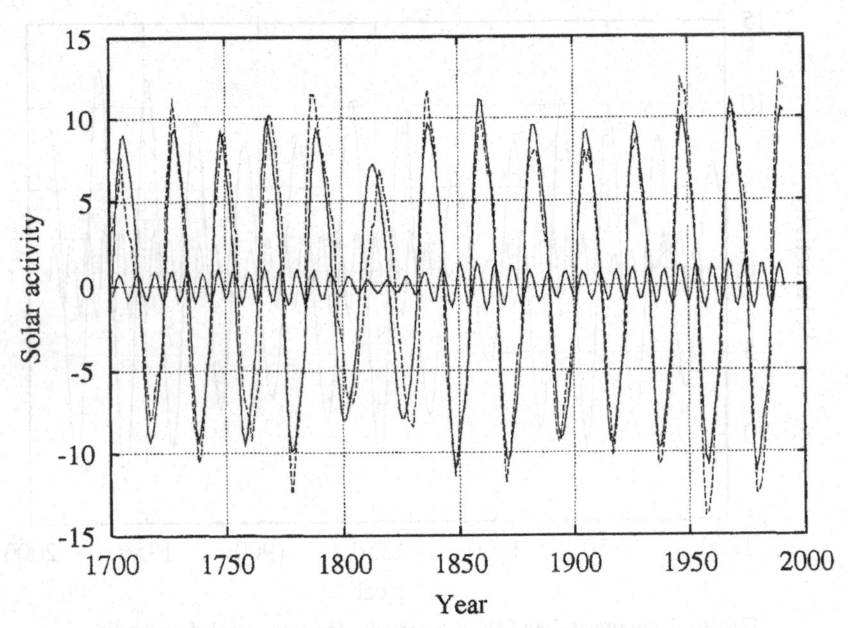

Fig. 10. Activity inferred from third harmonic (solid); data (broken).

It is even more remarkable that we can use the above cubic function to infer the structure of the fundamental time-series from the observed harmonic signal. This requires a slight change of method. We had earlier found, by a least squares analysis, that a fair account of the activity cycle can be given with just seven sinusoidal components having equally spaced frequencies from 0.040–0.055 c/y, i.e., just covering the fundamental group. If now, instead of fitting their amplitudes directly to the data, we cube the resulting signal, as above, and adjust the amplitudes to fit just the third harmonic time-series, then we obtain nearly the same amplitudes as before. The result of thus inferring the main signal, with only the third harmonic as input, is shown as the large solid-line oscillation in Figure 10. Also included, as a broken line, is the observed solar activity cycle itself. We see that the fit, both in magnitude and phase, is surprisingly good except for a certain mismatch in the years just before and just after 1800.

3. The Great Solar Anomaly

It is apparent that this mismatch is caused by a marked irregularity in the activity record as compared to the much better behaved reconstruction of the fundamental. Evidence for such an excursion in phase has been remarked on by others (Sonett 1983; Dicke 1988); in fact, even a casual inspection shows that the period from 1770 to 1810 contains both the longest and two of the shortest intervals between sunspot maxima. The phenomenon has been called the Great Solar Anomaly (GSA), and ascribed either to a disturbance on the sun, of unknown origin, or to errors in compiling the early record from archival sources.

The set of peaks below 0.035 c/y is a sequence of harmonics describing a pulse train with period 200 years. One such pulse is localized in the GSA, and is shown in Figure 11, along with the data and the time-series derived from the rest of the spectrum. A large part of the anomaly is thus accounted for. The 200 year period of the pulse train implies that the sun will be entering another anomalous phase soon.

To investigate further the nature of the GSA we did another entropic spectral analysis of the data with the Great Solar Anomaly removed. This was achieved by assigning inflated errors to the data points from 1770 to 1810, the effect being to give negligible weight to those entries. The new spectrum is exhibited in Figure 12. As compared with Figure 4, we now see three cleanly separated clusters of lines together with a series of equally spaced small peaks at the lowest frequencies. The fundamental group around 0.045 c/y now shows some evidence of an additional line. More detailed analysis (not described here) does suggest that there are in fact more lines in the main group interleaving the four obvious ones.

The third harmonic cluster still shows up strongly in the new spectrum and indeed seems little changed from the patterns evident in Figures 4 and 7. This gives added point to our reconstruction of the main signal from the third harmonic and supports the idea that the anomaly thus revealed is a real effect. Of course, the physical origin of the dominant signal itself, with its four to seven possible components, is still largely unknown, though many attempts (Foukal 1990) have been made to model the magnetic field as arising from a dynamo, powered by differential rotations in the convective outer layers of the sun. Alternatively, the sun may contain a deep, internal chronometer (Dicke 1978).

What remains for consideration here is the mysterious group of lines lying between the fundamental and third harmonic clusters. The said group consists of a well-marked, regular quintet of lines with spacings in the ratios 2 : 1 : 1 : 2. If we neglect the central peak of this group (at 0.077 c/y), for the moment, then we have four equally spaced lines

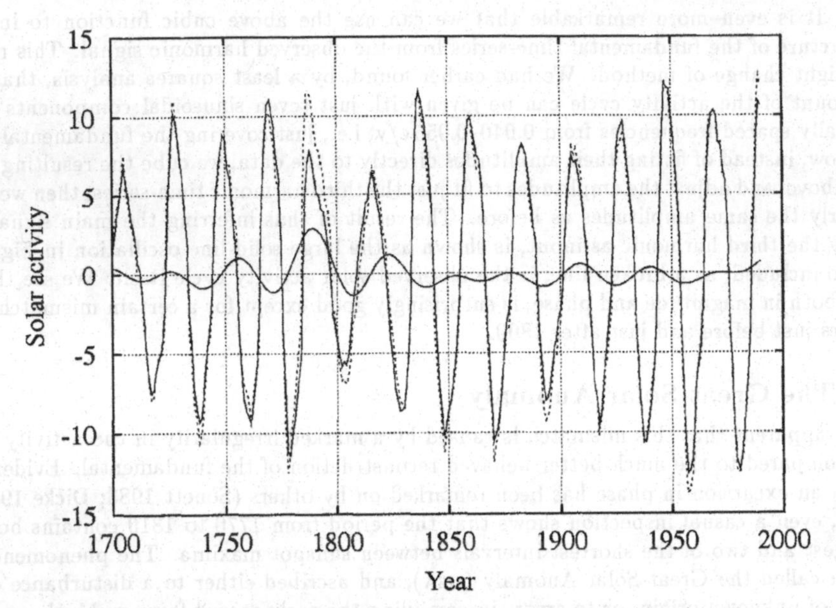

Fig. 11. The two time-series following from the spectrum when f is either less than, or greater than, 0.035 c/y (solid lines). Also shown are the data (broken line).

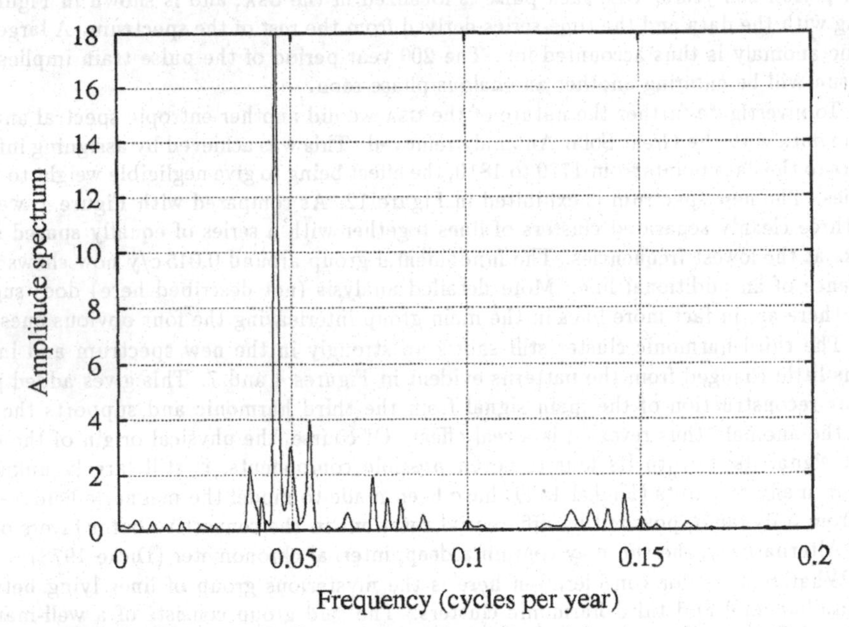

Fig. 12. Spectrum of solar activity with the Great Solar Anomaly removed.

which form a shadow or echo of the quartet of lines in the fundamental group. Each peak frequency is close to 1.62 times the frequency of a corresponding strong main line. In particular, the biggest peak of the group is at 0.0728 c/y, which is 1.618 times the frequency of the dominant signal peak at 0.045 c/y. It is thus tempting to suppose that this set of lines mirrors the fundamental lines with frequencies multiplied by the golden ratio $\phi \equiv (1+\sqrt{5})/2 \simeq 1.618034$. Taking this hypothesis seriously would lead us to predict that the central line at 0.077 c/y should have a partner in the main group at 0.0475 c/y. Fitting such a line, in addition to the other four lines suggested by the previous analyses, by the method of least squares, does give a much better account of the data. The extra line is not resolved by an entropic analysis of this discrete sum of sinusoids, but shows up as a broadening of the large component, exactly in accord with the results given earlier.

In seeking possible mechanisms that could produce this "golden harmonic", we are forced to look away from linear relations and dip into the world of non-linear systems and chaos. The third harmonic already provides a clue that non-linear processes are at work. A process which produces spectra peaked at frequencies proportional to Fibonacci numbers is the well-known non-linear circle map. The golden mean is the limit of the ratios of adjacent terms of the Fibonacci series, but even the early terms in the sequence give a good approximation to ϕ. We envisage that our harmonic is only roughly golden, corresponding to some low-lying Fibonacci numbers, but there is as yet no compelling rationale for using the circle map in a solar model. We believe nevertheless that the golden harmonic is evidence for the existence of chaotic evolution in solar activity.

4. Conclusion

In this paper, we have used Bayesian free-form spectral estimation to investigate the solar activity cycle. Most power is found in a multiplet around 0.045 c/y. A simple cubic function acts to shape this basic signal, its spectral signature being a group of peaks near 0.135 c/y. The final group shadows the main peaks at approximately 1.62 times their frequencies. It is unlikely that any linear theory could explain a factor like this, which is close to the golden ratio. The ability to resolve fine structure has resulted from using a flexible model with an entropic prior; this is a singularly effective method for generating more economical discrete models.

ACKNOWLEDGMENTS. One of us (VAM) wishes to thank the Science and Engineering Research Council of Great Britain and Merton College, Oxford for financial support.

REFERENCES

Bracewell, R. N. (1986). *The Hartley transform.* Oxford University Press, New York.

Bretthorst, G. L. (1988). *Bayesian spectral analysis and parameter estimation* (Lecture notes in statistics, Vol. 48). Springer-Verlag, New York.

Buck, B. and Macaulay, V. A. (1992). 'Entropy and sunspots: their bearing on time-series.' In *Maximum entropy and Bayesian methods, Seattle University, 1991* (ed. G. Erickson, P. Neudorfer and C. Ray Smith), pp. ?–?. Kluwer, Dordrecht (in press).

Dicke, R. H. (1978). 'Is there a chronometer hidden deep in the sun?' *Nature*, **276**, 676–80.

Dicke, R. H. (1988). 'The phase variations of the solar cycle.' *Solar Physics*, **115**, 171–81.

Foukal, P. V. (1990). *Solar astrophysics.* Wiley, New York.

Jaynes, E. T. (1987). 'Bayesian spectrum and chirp analysis.' In *Maximum-entropy and Bayesian spectral analysis and estimation problems: Proceedings of the third workshop on maximum entropy and Bayesian methods in applied statistics, Wyoming, U.S.A., August 1–4, 1983* (ed. C. R. Smith and G. J. Erickson), pp. 1–37. Reidel, Dordrecht.

Sonett, C. P. (1983). 'The Great Solar Anomaly ca. 1780–1800: an error in compiling the record?' *Journal of Geophysical Research*, **88A**, 3225–8.

TOPOLOGICAL CORRELATIONS IN 2D RANDOM CELLULAR STRUCTURES

R. Delannay
Laboratoire d'Énergétique et de Mécanique Théorique et Appliquée,
associé au C.N.R.S., U.R.A. 875, École des Mines, F-54042 Nancy-Cedex, France

G. Le Caër
Laboratoire de Science et Génie des Matériaux Métalliques,
associé au C.N.R.S., U.R.A. 159, École des Mines, F-54042 Nancy-Cedex, France

A. Sfeir
École des Mines,
F-54042 Nancy-Cedex, France.

ABSTRACT. Topological correlations in some natural and simulated cellular structures are compared with the predictions of a maximum entropy model and with the correlations calculated exactly in topological models of 2D random cellular structures.

1. Introduction

Cellular structures abound in Nature and are of interest in many scientific fields (Weaire and Rivier, 1984) such as fluid mechanics (foams), physics (magnetic domains), metallurgy (polycrystals), geology (crack networks in basalt flows), biology (biological tissues). They are also widely used as models at the atomic scale or to account for the large scale structure of the universe. The characteristics which are generally investigated in 2D random space- filling cellular structures include cell perimeter, cell area, angles at vertices as well as topological parameters such as the number n of edges of cells (n-cell) and its distribution $P(n)$ with a second moment $<(n-<n>)^2>=\mu_2$, the mean number $m(n)$ of sides of the first neighbour cells of n-cells. A well-known semi-empirical law, the Aboav-Weaire law expresses that $nm(n)$ varies linearly with n. More recently the two-cell correlation $M_k(n) = P(k)A_{kn}$ which is the average number of k-sided neighbours of an n-cell (Peshkin et al., 1991, Delannay et al., 1992) has also been investigated. In the latter cellular structures, every vertex belongs to $z_c = 3$ cells as vertices which belong to more than z_c cells are structurally unstable as their properties change by small deformations (Rivier, 1985). The constraints imposed on the $P(n)$ are:

$$<1>=\sum_n P(n)=1, \qquad <n>=\sum_n nP(n)=6, \qquad <A_{kn}>=\sum_n P(n)A_{kn}=k \quad (1)$$

Moreover, $A_{kn} \geq 0$ and $A_{kn} = A_{nk}$, while A_{kn} and $nm(n)$ are related by:

$$nm(n)=\sum_k kP(k)A_{kn}=<kA_{kn}> \qquad (2)$$

A. Mohammad-Djafari and G. Demoments (eds.), Maximum Entropy and Bayesian Methods, 357–362.

with $< nm(n) >= \mu_2 + 36$ (Weaire and Rivier, 1984). Extra constraints may also exist: $A_{33} = 0$, for 2D tessellations with convex polygons such as Voronoi or Laguerre (Rivier, 1991) tessellations or for structures in which two cells share at most one side.

Considering that all cellular structures are undistinguishable in spite of the different driving forces, apart from a specific length scale, and that they evolve to a steady state characterized by stationary distributions of cell sizes, shapes and of cell correlations, Rivier (1985) and Peshkin et al. (1991) have explored maximum entropy as an explanation of the previous similarities once the scaling state is reached. Rivier (1985, 1991) has assumed that froths, tissues are the least biased partitions of space subject to a few inevitable mathematical constraints (space filling) and possibly to some energy constraints. The purpose of the present paper is to compare the predictions of the previous maximum entropy model with results calculated from topological models of 2D cellular structures (Le Caër, 1991 a and b) and with results experimentally measured in some naturally occuring structures.

2. Maximum entropy predictions

Peshkin et al. (1991) have applied the maximum entropy principle with the previous constraints (1) imposed on the $P(n)$. If the A_{kn} are linear in n, that is if they are linear combinations of the two first constraints in (1), the number of independent constraints is reduced and the entropy still increases. The constraint matrix, which includes all constraints given by equation 1, has therefore a rank of 2. We notice that the Maxent distribution $P_m(n)$ with uniform priors, subject to the two constraints $< 1 >= 1$ and $< n >= 6$, is $P_m(n) = 0.75^{n-3}/4 (n \geq 3)$ with $\mu_2 = 12$ and is very different from the distributions which are observed in general in simulated or in actual structures. More realistic distributions may be derived with appropriate prior probabilities which are unfortunately difficult to guess. A linear dependence of $nm(n)$ with n is accounted for by the model of Peshkin et al. (1991) as shown by equation (2). The knowledge of a free parameter 'a' is still needed in order to define completely the Aboav- Weaire law:

$$nm(n) = (6 - a)n + (6a + \mu_2) \tag{3}$$

If the A_{kn} are linear in k and in n, they are expressed in a unique way (independently of the Maxent problem) as a function of a ratio of the previous parameter a to μ_2 (Delannay et al. 1992):

$$A_{kn} = n + k - 6 - (a/\mu_2)(n - 6)(k - 6) \tag{4}$$

with in general, n, $k \geq 3$. The point $(n = 6 + \mu_2/a,\ k = 6 + \mu_2/a)$ belongs to all the lines $A_{kn} = g_k(n) \quad (k \geq 3)$. However, equation (4) does not guarantee that $M_k(n)$ and $A_{kn} \geq 0$ for all allowed values of k and n. The previous model predicts in fact negative A_{kn} values for some values of k and n and for $a > 0$, in particular for A_{33}. In most natural and simulated 2D structures, 'a' is positive and of the order of 1. If the upper limit of variation of k and n is infinite, it is only for $-1/3 \leq a/\mu_2 \leq 0$ that A_{kn} (4) will satisfy to the positivity constraint whatever k and n. In other cases, we expect to observe deviations from the linear behaviour in order to prevent the A_{kn} from becoming negative. These deviations, which will occur for small and large values of k, will also propagate to other A_{kn} lines as required

by equation (1) and $nm(n) = f(n)$ may also deviate from the Aboav-Weaire line. Linear A_{kn} have been computed by Peshkin et al. (1991) in simulations with $a \leq 0$. They also report that $\mu_2 = 12.69$ (close to the value quoted above) in a case where $a = -1.33$.
We have furthermore investigated the form of the A_{kn} correlations as a function of the rank r of the constraint matrix $n_1 \leq k$, $n \leq n_2$. These r constraints are the first two in equation 1 and $r-2$ independent constraints (among the $n_2 - n_1 + 1$ possible ones): $< A_{ki,n} >= k_i$ (eq. 1) with $3 \leq i \leq r$. The correlations can be calculated from:

$$A_{jn} = \mathbf{L}\mathbf{B}^{-1}\mathbf{C} \tag{5}$$

$\mathbf{L}$ and $\mathbf{C}$ are $(1 \times r)$ row and $(r \times 1)$ column matrices with elements $(1, j, A_{k3,j}, \ldots, A_{kr,j})$ where $j = k$ and n for $\mathbf{L}$ and $\mathbf{C}$ respectively. The matrix $\mathbf{B}$ is a $(r \times r)$ symmetric matrix with $B(1,i) = 1$, except $B(1,1) = \epsilon$ (a parameter), $B(2,2) = 6$, $B(2,i) = k_i$ for $i \geq 3$ and $B(i,j) = A_{ki,kj}$ for $i,j \geq 3$. The Aboav-Weaire law, which does not require that the A_{kn} are linear in k and in n, will be valid if and only if eq. 3 is valid for $n = k_i$ $(3 \leq i \leq r)$ while ϵ and a are related by: $a = \mu_2\epsilon/(1-6\epsilon)$. Equation 5 can also be written as (see also Delannay et al., 1992):

$$A_{jn} = n + j - 6 + \sum_{i,k} C_{ik}\{f_i(j)f_k(n) + f_i(n)f_k(j)\} \qquad (1 \leq i,\ k \leq r) \tag{6}$$

where $C_{ik} = B^{-1}(i,k)/2, f_1(n) = 0, f_2(n) = n-6, f_i(n) = A_{ki,n} - k_i$ for $3 \leq i \leq r$. All the coefficients A_{kn} can therefore be completely determined as soon as all the independent constraints and one extra information (another A_{ij}, some value related to eq. 3) are known. Equations 5 and 6 may be used to determine an effective rank from a set of experimental correlations A_{ij} (Le Caër and Delannay, to be published).

3. Comparison with some available topological correlations

We have calculated exactly the A_{kn} for a set of topological models of 2D cellular structures that we have recently introduced (Le Caër, 1991a,b, Delannay et al. 1992). If we consider a lattice in which every site is characterized by its valence z which is the number of edges merging in that vertex, every vertex which belongs to more than three cells is structurally unstable (see introduction). The construction method is based on rules which allow removal of this unstability. The stable configuration is obtained by adding $z-3$ sides at every vertex. Every added side is connected at least to one added side for $z > 4$. The latter rule produces a set of $Q(z)$ (= a Catalan number C_{z-2}) possible stable configurations, called states, which have been enumerated as a function of z (Le Caër, 1991 b). The last ingredient of the method is a criterion for distributing the various states on the lattice sites. A cell of the topological cellular model is finally associated with every polygon of the mother lattice. For triangular tessellations in which the vertices of every triangle belong to z_1, z_2, z_3 triangles respectively, the number of cell sides of the associated structure may vary between 3 and $z_1 + z_2 + z_3 - 6$. An equivalent method of construction of the cellular structure uses the Euler's diagonal triangulation of the dual lattice (Le Caër, 1991b). We will only consider the case of a distribution of independent and equiprobable states (DIES) on the various lattice sites for which exact calculations of $P(n)$ and of the A_{kn} can be simultaneously performed (Le Caër and Delannay, to be published). The previous topological models have

P(3)	P(4)	P(5)	P(6)	P(7)
0.045554	0.136662	0.218659	0.236880	0.185860
0.045538	0.136770	0.218686	0.236267	0.186454
P(8)	P(9)	P(10)	P(11)	P(12)
0.109329	0.048105	0.015306	0.003280	0.000364
0.109631	0.047561	0.015369	0.003379	0.000344

Table 1: Distribution of the number of cell sides $P(n)$: second and fifth rows $P_{\text{DIES}}(n)$ for a triangular lattice (Le Caër, 1991 b), third and sixth rows Maxent distribution $P_m(n)$

always $A_{33} = 0$ and have $A_{34} = 0$ only if all values of z are larger than 3. For various reasons (two cells share at most one side, convexity, energy), the constraint $A_{33} = 0$ holds in almost all natural structures.

$P(n)$ as well as A_{kn} and $nm(n)$ are easily calculated for a cellular structure associated with a DIES on a triangular lattice for which $z = z_1 = z_2 = z_3 = 6$ ($3 \leq n \leq 12$, $\mu_2 = 18/7$, table 1). Deviations from the Aboav- Weaire law occur mainly for $n = 3$ and for $n > 9$. It is therefore not possible to define a unique 'a' value from equation 3. In all cases where such deviations occur, we have chosen to define a_w as $(216 + 12\mu_2 - < n^2 m(n) >)/\mu_2$ which is calculated from a weighted least-squares fit of $nm(n)$ to equation 3 with weights $w_n = P(n)$. For $z = 6$, the previous relation yields $a_w = 1$. Figure 1a shows the $A_{kn} = g_k(n)$ curves which almost all intersect at the point (8.4, 8.4) close to the point expected for linear A_{kn} ($n = A_{kn} \approx 8.57$, section 2). We have thus calculated the Maxent distribution $P_m(n)$ subject to constraints 1 ($r = 7$) and the maximum entropy S_m using Mathematica (Wolfram, 1991). The two distributions differ little (Table 1) although significantly while $S_{\text{DIES}} = 1.872603$ and $S_m = 1.872610$.

We have also compared the correlations A_{kn} in the previous topological models to the correlations measured in some natural and simulated structures. There is a fair overall quantitative agreement between the A_{kn} of an epidermal epithelium of a cucumber, the A_{kn} of a topological model associated with a ferromagnetic Ising model on a square lattice, the A_{kn} of a Voronoi tessellation generated from a 2D hard-disk fluid (Delannay et al., 1992). All have similar values of a_w and $\mu_2(< 1)$. Figure 1b compares, without any adjustable parameter, the correlations measured on a planar cut of an alumina polycristal (a_w = 1.19, $\mu_2 = 2.58$, $a_w/\mu_2 = 0.46$, Righetti, Liebling, Le Caër, Mocellin, to be published) with the correlations calculated exactly from a DIES on a lattice derived from a square lattice by adding diagonals in such a way that all vertices have $z = 5(a_w = 1.2126, \mu_2 = 2.5867, a_w/\mu_2 = 0.4688)$. Finally, the correlations A_{kn} (figure 1a) for a DIES on a triangular lattice ($a_w = 1$, $\mu_2 = 2.5714$, $a_w/\mu_2 = 0.3888$) are in very good agreement with the correlations computed in a 2D Voronoi tessellation associated with a Poisson point process ($a_w = 0.676$, $\mu_2 = 1.781$, $a_w/\mu_2 = 0.379$, Le Caër and Delannay, to be published). This agreement is not inconsistent with a trend towards the constrained maximization of entropy as the ratios a_w/μ_2 are similar while both a_w and μ_2 differ.

4. Conclusion

Our results are not inconsistent with the idea that the topological properties of some of the investigated 2D cellular structures are determined by a constrained maximisation of entropy. However, these properties disagree with the Maxent model in its linear version although equation 4 provides a useful and easily performed comparison with experimental or simulated data. A striking similarity exists between the properties of topological models and of some simulated and natural structures. The $P(n)$ distribution slightly deviates from the Maxent distribution. We don't know at the present time if the evolution of the considered structures is hindered so that some mechanisms prevent them to reach the Maxent state or if some constraints have not yet been identified. Further theoretical work is necessary in order to take full profit of the heuristic power of the Maxent method. Unfortunately, no theory is presently able to predict both $P(n)$ and A_{kn} beforehand.

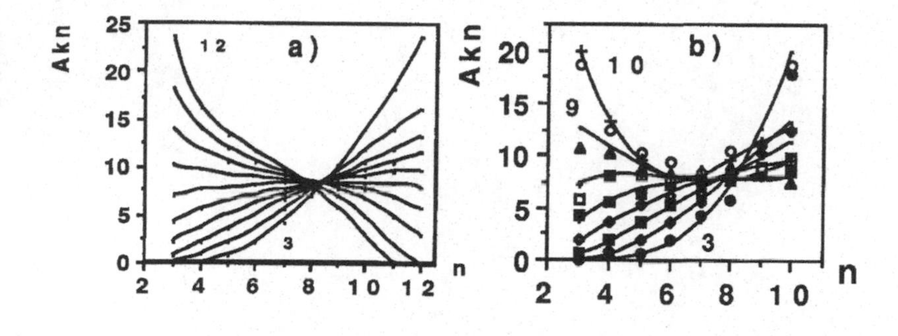

Figure 1

Figure 1: Correlations A_{kn} as a function of the number of cell sides n for : a) a DIES distribution on a triangular lattice ($z = 6$) b) a DIES distribution (solid lines) on a lattice with $z = 5$ (Le Caër, 1991b) and for an alumina cut with 4310 cells. In both cases, solid lines have been drawn for the sake of clarity; k increases from 3 to k_1 (=12, a, =10, b) as indicated in the left region by bold figures while the order is reversed in the right region.

References

[1] Delannay, R., Le Caër, G. and Khatun, M.: 1992, "Random cellular structures generated from a 2D Ising ferromagnet", J. Phys. A: Math. Gen., **25**, 6193.

[2] Le Caër, G.: 1991 a, "Topological models of cellular structures", J. Phys. A: Math. Gen., **24**, 1307 and 2677.

[3] Le Caër, G.: 1991 b, "Topological models of 2D cellular structures: II $z \geq 5$", J. Phys. A: Math. Gen., **24**, 4655.

[4] Peshkin, M.A., Strandburg K.J. and Rivier, N.: 1991, "Entropic predictions for cellular networks", Phys. Rev. Lett. **67**, 1803.

[5] Rivier, N.: 1985, "Statistical crystallography: Structure of random cellular networks", Phil. Mag. B **52**, 795.

[6] Rivier, N.: 1991, "Geometry of random packings and froths", in D. Bideau and J.A. Dodds (eds.), *Physics of Granular Media*, Nova Science Publishers, New-York.

[7] Weaire, D. and Rivier, N.: 1984, "Soap, Cells and Statistics-Random Patterns in Two Dimensions", Contemp. Phys. **25**, 59.

[8] Wolfram, S.: 1991, *Mathematica*, 2nd ed., Addison-Wesley Publishing Co, Redwood City.

ANALYSIS OF DNA OLIGONUCLEOTIDE HYBRIDIZATION DATA BY MAXIMUM ENTROPY

J K Elder
Department of Biochemistry
University of Oxford
Oxford OX1 3QU UK

ABSTRACT. Oligonucleotide array hybridization allows one to read a DNA sequence by interpreting the pattern generated when the sequence hybridizes to an array of known oligonucleotides. Because of noise and cross-hybridization, there is uncertainty about the frequency of occurrence of each oligonucleotide in the sequence, and it is necessary to set up a model relating sequences to the data sets they are expected to generate. This model has parameters specifying the noise level and the pattern of cross-hybridization; we use the maximum entropy algorithm MemSys5 to estimate cross-hybridization parameters from data sets corresponding to known sequences, and to choose between different models. To determine an unknown sequence from a data set, a maximum entropy reconstruction provides estimates of the noise level and the frequency of occurrence of each oligonucleotide. This information allows candidate sequences to be evaluated against the data and probabilities to be assigned to them.

1. Introduction

The need for DNA sequencing methods capable of generating the large amount of information required by efforts to map and sequence the human genome has promoted interest in the development of new sequencing techniques. Among these is oligonucleotide array hybridization, in which a set of known oligonucleotides is immobilized on an array and many copies of an unknown DNA sequence are allowed to hybridize to it. Each copy hybridizes to an oligonucleotide which is complementary to, or related to the complement of, a subsequence in the unknown sequence; the aim is to determine this sequence from the pattern of hybridization on the array. The method can also be used to compare related sequences and to detect mutation.

The accuracy of DNA sequence information is of critical importance, and a study of errors in the current sequence databases has drawn attention to the need for quantitative estimates of sequence quality (Kristensen *et al.*, 1992). Proposals for obtaining such estimates from consensus sequence data have recently been made (Churchill and Waterman, 1992).

In conventional DNA sequencing methods, the reliability of the sequence data is usually treated in a rudimentary way. Ambiguity codes are used to flag uncertain base assignments,

A. Mohammad-Djafari and G. Demoments (eds.), Maximum Entropy and Bayesian Methods, 363–371.

but a quantitative estimate of a sequence's reliability is rarely given. The nature of the data generated by such methods allows one to read a sequence without the use of a statistical model. For example, in a sequencing gel autoradiograph, the sequence is represented by an ordered pattern of bands, and it is thus possible to read the sequence (or at least an approximation to it) directly from the data. The application of a statistical model to such data would allow quantitative estimates of the reliability of sequences to be made, but the complicated relationship which exists between sequence and data makes it difficult to set up a model which is realistic.

When analyzing oligonucleotide hybridization data, we are faced with the complementary situation. Because of the form the data take, it is not possible to read a sequence directly, and we are forced to consider in detail the relationship between sequence and data. This relationship is, however, more straightforward than that which exists in conventional methods of sequencing, since the physical form of the data consists of a regular array of cells, and is therefore better suited to image processing and statistical analysis techniques than is, say, a sequencing gel.

Some studies have assumed that it is possible to determine the frequency of occurrence of sequence oligonucleotides with certainty, and have accordingly developed reconstruction algorithms (Bains and Smith, 1988; Bains, 1991; Pevzner, 1989). Under this assumption every possible sequence is either consistent or inconsistent with the data: a consistent sequence is one which contains precisely the complements of the array oligonucleotides where hybridization has taken place, with the same frequencies as indicated by the array data. Thus, every sequence oligonucleotide hybridizing to the array is accounted for in the data, and *vice versa.* An inconsistent sequence is one for which this is not true. There may be more than one sequence consistent with the data; for example the sequences **agag** and **gaga** each comprise the trinucleotides **aga** and **gag**.

Such an approach is satisfactory when applied to an ideal data set: one which contains no cross-hybridization or noise. Since real data sets do contain these features, we must acknowledge and deal with their presence. Cross-hybridization takes place when the target sequence hybridizes to an oligonucleotide containing bases not complementary to the corresponding target bases, and occurs because discrimination between complementary and non-complementary bases is imperfect. The presence of cross-hybridization is analogous to that of a point spread function in image processing, and the result is a blurred set of hybridization data. Superimposed on this data is random noise, which further hinders interpretation. In order to read sequences from such data sets and provide quantitative estimates of their reliability, we need to set up a model which takes account of cross-hybridization and noise.

2. The Model and Its Parameters

Suppose the array consists of N oligonucleotides of length s, and we view the sequence which will hybridize to it as consisting of overlapping oligonucleotides, also of length s (Figure 1). Since there are four choices for each base, there are $4^s = M$ possible sequence oligonucleotides. A typical array would consist of an exhaustive set of oligonucleotides, so that $M = N$. In general, however, M and N may be different. For example, the array may include oligonucleotides containing degenerate bases (those which can hybridize to more than one normal base), or we may choose to view the sequence as consisting of subsequences of length $s + 2$, in order to take account of the effect of neighbouring bases

```
tcgaaacgtactggc

tcgaaa
 cgaaac
  gaaacg
   aaacgt
    aacgta
     acgtac
      cgtact
       gtactg
        tactgg
         actggc
```

Fig. 1. A DNA sequence and its representation as a set of overlapping hexamers ($s = 6$).

on hybridization.

The model we will adopt assumes that the set of N observed array oligonucleotide hybridization intensities $\mathbf{D}$ is related to the set of M sequence oligonucleotide frequencies $\mathbf{f}$ by an $N \times M$ response matrix $\mathbf{R}$, and is subject to Gaussian noise of standard deviation σ. Thus

$$\mathbf{D} = \mathbf{R}\mathbf{f} + \sigma\mathbf{e} \tag{1}$$

or

$$D_k = \sum_{j=1}^{M} R_{kj} f_j + \sigma e_k, \qquad e_k \in \mathcal{N}(0,1), \qquad k = 1, \ldots, N.$$

Given this model, we need to estimate $\mathbf{R}$ and σ before reconstructing the sequence.

The response matrix $\mathbf{R}$ has NM elements, but the number of parameters needed to characterize it is likely to be much smaller, since many of the elements may be zero, or identical to each other. In any case, the presence of noise will limit the measurement of weak elements of $\mathbf{R}$. We can explore different structures for $\mathbf{R}$ by choice of model. Here we will allow at most p mismatches to occur between any array oligonucleotide and any sequence oligonucleotide, where p determines the model ($1 \le p \le s$). If we allow a separate parameter for each possible set of mismatch positions, there are $\sum_{k=0}^{p} \binom{s}{k} = M'$ response coefficients r_i to determine. Thus, as the number of permitted mismatches p increases, so does the number of model parameters, and the model achieves ever better fits. There comes a point, however, when the model starts to fit to noise as well as to the true signal in the data. This tradeoff between simplicity of model and goodness of fit is an example of Ockham's Razor, which states that one should choose the simplest model that fits noisy data adequately (Garrett, 1991). Bayesian probability theory is able to resolve this tradeoff quantitatively.

We estimate $\mathbf{R}$ using a data set generated by a known sequence, and recast the model $\mathbf{R}\mathbf{f} = \mathbf{D}$ in the form $\mathbf{R}'\mathbf{r} = \mathbf{D}$, where the distribution to be inferred is now $\mathbf{r}$, and the $N \times M'$ response matrix $\mathbf{R}'$ is constructed from the known sequence oligonucleotide frequencies $\mathbf{f}$.

For example,

$$\begin{pmatrix} r_1 & r_2 & 0 \\ 0 & r_1 & r_2 \\ r_2 & 0 & r_1 \end{pmatrix} \begin{pmatrix} f_1 \\ f_2 \\ f_3 \end{pmatrix} = \begin{pmatrix} D_1 \\ D_2 \\ D_3 \end{pmatrix}$$

would be recast as

$$\begin{pmatrix} f_1 & f_2 \\ f_2 & f_3 \\ f_3 & f_1 \end{pmatrix} \begin{pmatrix} r_1 \\ r_2 \end{pmatrix} = \begin{pmatrix} D_1 \\ D_2 \\ D_3 \end{pmatrix}.$$

To estimate $\mathbf{r}$, we use Bayes' theorem

$$\Pr(\mathbf{r} \mid \mathbf{D}, p) = \frac{\Pr(\mathbf{r} \mid p) \Pr(\mathbf{D} \mid \mathbf{r}, p)}{\Pr(\mathbf{D} \mid p)}$$

and the maximum entropy method (MaxEnt), as implemented by the quantified maximum entropy package `MemSys5` (Gull and Skilling, 1991). MaxEnt assigns an entropic prior

$$\Pr(\mathbf{r} \mid p) \propto \exp(\alpha S(\mathbf{r})),$$

where α is a regularization parameter whose value is determined by the noise level in the data (Gull, 1989), and $S(\mathbf{r})$, the entropy of $\mathbf{r}$ relative to a default distribution $\mathbf{m}$, is defined as

$$S(\mathbf{r}) = -\sum_{i=1}^{M'} (r_i - m_i - r_i \log(r_i/m_i)). \tag{2}$$

Such a prior is appropriate when the distribution is positive and additive (Skilling, 1989), as is the vector of response coefficients $\mathbf{r}$. When estimating $\mathbf{r}$, we impose the constraint

$$\sum_{i=1}^{M'} r_i = 1.$$

For each value of p, `MemSys5` calculates the global likelihood $\Pr(\mathbf{D} \mid p)$, also called the evidence. To compare models with different values of p, we need to evaluate $\Pr(p \mid \mathbf{D})$. Applying Bayes' theorem,

$$\Pr(p \mid \mathbf{D}) = \frac{\Pr(p) \Pr(\mathbf{D} \mid p)}{\Pr(\mathbf{D})}.$$

Since we have no preference for a particular value of p we assign a uniform prior to $\Pr(p)$, and so $\Pr(p \mid \mathbf{D}) \propto \Pr(\mathbf{D} \mid p)$.

The response matrix is characteristic of the oligonucleotide array and the conditions under which it is used. For a given set of conditions it need be estimated only once, and can then be used repeatedly on data sets generated by unknown sequences. Given such a data set, we revert to the original model (1) and use MaxEnt to estimate the sequence oligonucleotide intensities and the noise level in the data.

The sequence oligonucleotide intensities $\mathbf{f}$ are positive and additive, so that an entropic prior is again appropriate. MaxEnt takes no account of correlations which exist between elements of $\mathbf{f}$, but since $\mathbf{f}$ represents a set of overlapping oligonucleotides, it certainly contains correlations. We therefore apply MaxEnt to a hidden version $\mathbf{h}$ of $\mathbf{f}$, whose components are uncorrelated (Gull, 1989). The visible distribution $\mathbf{f}$ is related to $\mathbf{h}$ by an intrinsic correlation matrix $\mathbf{C}$:

$$\mathbf{f} = \mathbf{Ch}.$$

The elements of $\mathbf{C}$ represent the degree of correlation between pairs of oligonucleotides; by making $\mathbf{C}$ the identity matrix, no correlations are assumed. With sequence oligonucleotides of length s, the jth oligonucleotide consists of bases $b_j, \ldots, b_{j+s-1}$. There are first order correlations with oligonucleotides $j-1$ and $j+1$, which consist of bases $b_{j-1}, \ldots, b_{j+s-2}$ and $b_{j+1}, \ldots, b_{j+s}$ respectively, since both have $s-1$ bases in common with oligonucleotide i. Similarly there are second order correlations with oligonucleotides $j-2$ and $j+2$, and so on up to order $s-1$. We can represent the degree of correlation by coefficients $c_0, c_1, \ldots, c_{s-1}$, construct the intrinsic correlation matrix $\mathbf{C}$ accordingly, and explore different choices of $\mathbf{c}$ by examining the evidence $\Pr(\mathbf{D} \mid \mathbf{c})$, as calculated by `MemSys5`. The presence of oligonucleotide j implies the presence (ignoring end effects) of one of 4^i possible oligonucleotides $j+i$, and one of 4^i possible oligonucleotides $j-i$, where $0 \leq i < s$, so in order to preserve flux between hidden and visible space, we impose the constraint

$$c_0 + 2\sum_{i=1}^{s-1} 4^i c_i = 1. \tag{3}$$

The estimated oligonucleotide intensities are not constrained to integer values, and therefore cannot be used directly to reconstruct sequences. They can, however, be used to focus attention on oligonucleotides which are likely to be present in the sequence. This information becomes particularly useful when reconstructing sequences of realistic length, when it is feasible to evaluate only a small subset of all candidate sequences against the data. In choosing the most probable value of the regularization parameter α in the entropic prior (2), MaxEnt also provides an estimate of the noise level σ^2 (Gull, 1989), and this allows us to calculate the relative likelihoods of sequences (§3).

3. Evaluation of Sequences

Given a data set and a model relating sequences to data, we now wish to assign probabilities to sequences which could have given rise to the data. Suppose we evaluate sequences of length n, and H_i is the hypothesis that the true sequence is sequence i, where $i = 1, \ldots, 4^n$. We wish to find the posterior probability $\Pr(H_i \mid \mathbf{D})$ of each hypothesis H_i, by modulating the prior probability according to Bayes' theorem:

$$\Pr(H_i \mid \mathbf{D}) \propto \Pr(H_i) \Pr(\mathbf{D} \mid H_i).$$

Unless we have further information, each sequence is equally probable before seeing the data, and a uniform prior is appropriate, so we assign a prior probability

$$\Pr(H_i) = 4^{-n}, \qquad i = 1, \ldots, 4^n$$

to each hypothesis. We therefore only need to calculate $\Pr(\mathbf{D} \mid H_i)$, the likelihood of the data $\mathbf{D}$ given the hypothesis H_i. According to our model,

$$\Pr(\mathbf{D} \mid H_i) = \left(2\pi\sigma^2\right)^{-N/2} \exp\left(-\frac{1}{2\sigma^2}\sum_{k=1}^{N}(D_k - F_k)^2\right),$$

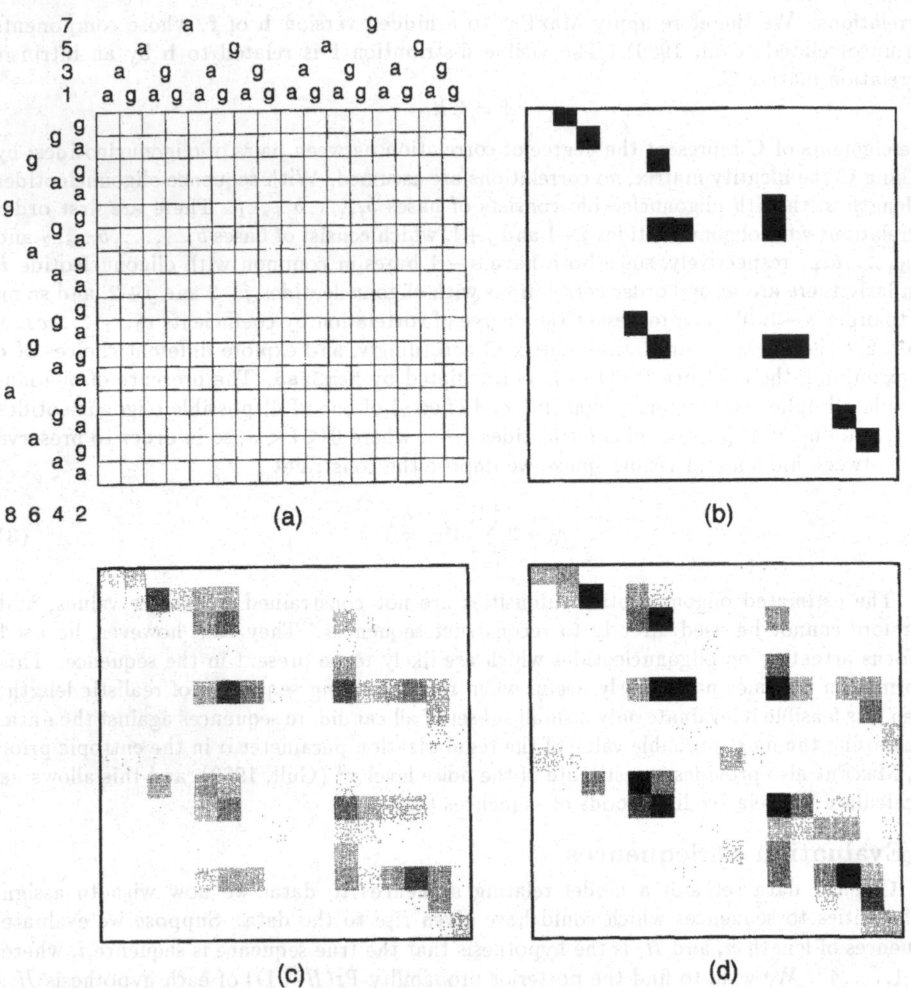

Fig. 2. The oligonucleotide array: (a) layout of the oligonucleotides; (b) ideal data; (c) data; (d) MaxEnt reconstruction. The oligonucleotide in any cell can be determined from the layout: for example, that in the top left cell is agagagag and that in the top right cell is gggggggg.

where

$$F_k = \sum_{j=1}^{M} R_{kj}\, f_j, \qquad k = 1, \ldots, N.$$

The noise level σ^2 is provided by the MaxEnt reconstruction of $\mathbf{f}$ (§2).

4. An Example

We apply the above methods to a pilot set of data which was previously analyzed using a simple response matrix (Southern *et al.*, 1992).

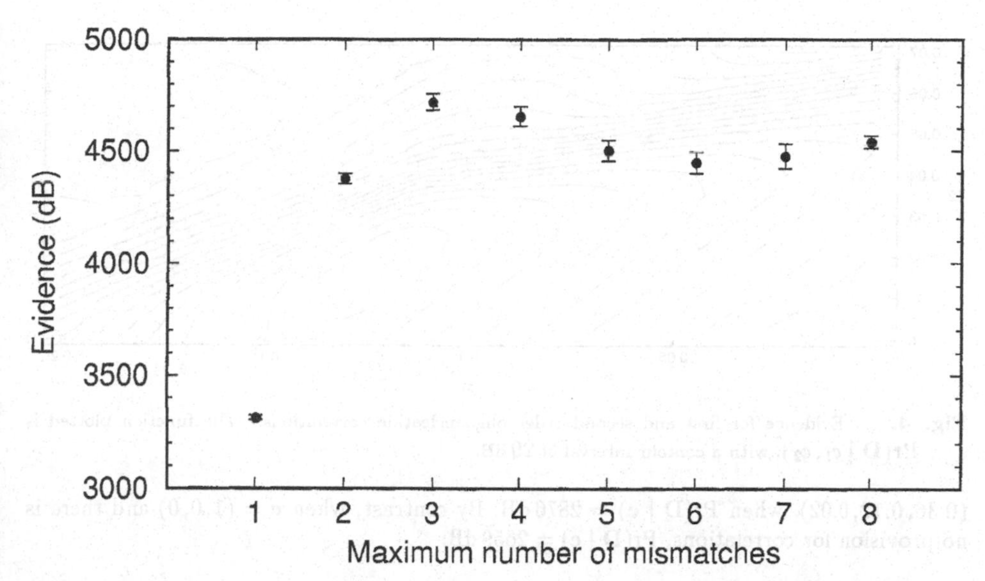

Fig. 3. Evidence for different numbers of base mismatches in the response matrix. The function plotted is $\Pr(\mathbf{D} \mid p)$.

The array consists of all octapurines (layout shown in Figure 2a), to which a sequence of 20 pyrimidines has been allowed to hybridize. Since the array contains only purines (bases a and g) and the sequence contains only pyrimidines (bases t and c), $M = N = 2^8 = 256$.

In the ideal data set for this sequence (Figure 2b), hybridization occurs in only 13 cells: those whose octapyrimidines are complementary to the 13 overlapping octapurines which make up the sequence. Figure 2c shows the set of integrated intensities for the array of experimental data. Because of cross-hybridization and noise, the image is much harder to interpret than the ideal one.

The response matrix was estimated from a data set generated by a different known sequence also of length 20, although the length need not have been the same as that of the unknown sequence. The evidence for numbers of mismatches ($1 \leq p \leq 8$) is shown in Figure 3. A maximum of three mismatches is the most probable, and we use the maximum entropy reconstruction of the response coefficients obtained with this model to form the response matrix.

The oligonucleotides in the sequence overlap by seven bases, so there are correlations of order zero to seven. However, since the strongest correlations are likely to occur between oligonucleotides with a large overlap, we restrict correlations to at most second order. The sequence consists only of pyrimidine bases, so the constraint on correlation coefficients (3) becomes

$$c_0 + 4c_1 + 8c_2 = 1.$$

The coefficient c_0 is thus defined by c_1 and c_2, and we explore values for $\mathbf{c}$ by using MaxEnt to estimate the evidence $\Pr(\mathbf{D} \mid c_1, c_2)$ (Figure 4). The best set of values occurs at $\mathbf{c} =$

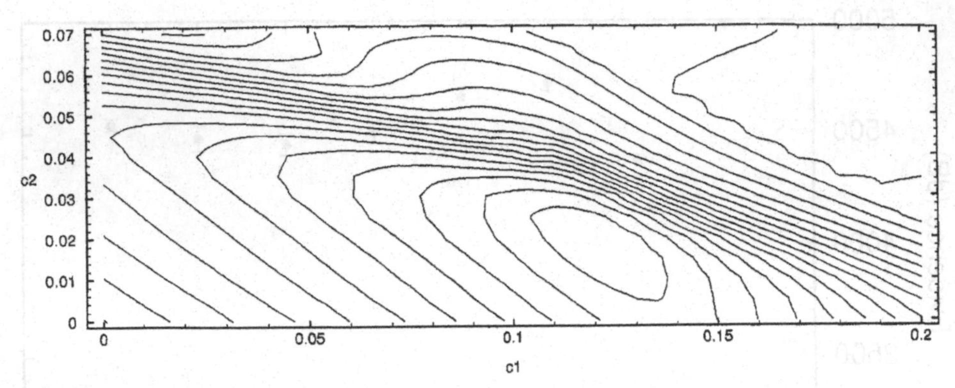

Fig. 4. Evidence for first and second order oligonucleotide correlations. The function plotted is $\Pr(\mathbf{D} \mid c_1, c_2)$, with a contour interval of 20 dB.

(0.36, 0.12, 0.02), when $\Pr(\mathbf{D} \mid \mathbf{c}) = 2876\,\mathrm{dB}$. By contrast, when $\mathbf{c} = (1,0,0)$ and there is no provision for correlations, $\Pr(\mathbf{D} \mid \mathbf{c}) = 2659\,\mathrm{dB}$.

Rank	Sequence	Likelihood Ratio		Probability
1	tttcccttccttcctctctc	1 : 2	1.055×10^5	0.9999905
2	tttcccttccttcctctctt	2 : 3	2.070×10^{15}	0.0000095
3	cttcccttccttcctctctc	3 : 4	1.810	
4	tttcccttccttcctctcct	4 : 5	3.433	
5	tttcccttccttcctctccc	5 : 6	3.902×10^4	
6	cttcccttccttcctctctt	6 : 7	6.809×10^{10}	
7	cttccttcccttcctctctc	7 : 8	3.541×10^2	
8	cttccttcccttcctctctt	8 : 9	1.214×10^1	
9	cttcccttccttcctctcct	9 : 10	8.207	
10	cttcccttccttcctctccc			

Table. 1. The ten most favoured sequences and their likelihood ratios. The first sequence is the correct one.

The MaxEnt reconstruction of **f** using the best set of correlations **c** is shown in Figure 2d. In the original analysis (Southern *et al.*, 1992), candidate sequences were ranked, but the lack of an estimate of the noise level did not allow them to be assigned probabilities. We can now use the noise level estimated by `MemSys5` to calculate the likelihood of each sequence, and hence its probability. We also use the reconstruction **f** to consider only a subset of possible sequences, by restricting attention to the 100 strongest oligonucleotide intensities in **f** (from a total of 256), and evaluating sequences containing only those oligonucleotides. This cutoff, while conservative, reduces the number of candidate sequences from $2^{20}(= 1048576)$ to 6472. The ten most favoured sequences are shown in Table 1.

These results are, of course, conditional on the choice of model and the values of its parameters. The use of Bayes' theorem and maximum entropy allow us to evaluate these choices, and their effect on the results, in a quantitative manner.

ACKNOWLEDGMENTS. I thank U Maskos and E M Southern for use of the oligonucleotide hybridization data. This work was supported by the Medical Research Council Human Genome Mapping Project.

REFERENCES

Kristensen, T., Lopez, R. and Prydz, H: 1992, 'An estimate of the sequencing error frequency in the DNA sequence databases', *DNA Sequence* **2**: 343–346.

Churchill, G.A. and Waterman, M.S.: 1992, 'The accuracy of DNA sequences: estimating sequence quality', *Genomics* **14**(1): 89–98.

Bains, W. and Smith, G.C.: 1988, 'A novel method for nucleic acid sequence determination', *Journal of Theoretical Biology* **135**(3): 303–307.

Bains, W.:1991, 'Hybridization methods for DNA sequencing', *Genomics* **11**(2): 294–301.

Pevzner, P.A.: 1989, 'l-tuple DNA sequencing: computer analysis', *Journal of Biomolecular Structure and Dynamics* **7**(1): 63–73.

Garrett, A.J.M.: 1991, 'Ockham's razor', in *Maximum Entropy and Bayesian Methods, Laramie, 1990*, W.T. Grandy and L.J. Schicke (eds.), pp. 357–364. Dordrecht: Kluwer.

Gull, S.F. and Skilling, J.:1991, *Quantified Maximum Entropy MemSys5 Users' Manual.* Royston: Maximum Entropy Data Consultants. Version 1.01.

Gull, S.F.: 1989, 'Developments in maximum entropy data analysis', in *Maximum Entropy and Bayesian Methods, Cambridge, 1988*, J. Skilling (ed.), pp. 53–71. Dordrecht: Kluwer.

Skilling, J.:1989, 'Classic maximum entropy', in *Maximum Entropy and Bayesian Methods, Cambridge, 1988*, J. Skilling (ed.), pages 45–52. Dordrecht: Kluwer.

Southern, E.M., Maskos, U. and Elder, J.K.: 1992, 'Analyzing and comparing nucleic acid sequences by hybridization to arrays of oligonucleotides: evaluation using experimental models', *Genomics* **13**(4): 1008–1017.

Chapter 6

Image Restoration and Reconstruction

BAYESIAN IMAGE CLASSIFICATION USING MARKOV RANDOM FIELDS

Z. KATO, J. ZERUBIA, M. BERTHOD
INRIA,
2004 Route des Lucioles, BP 93
06902 Sophia Antipolis Cedex, FRANCE

ABSTRACT. In this paper, we present two relaxation techniques: Deterministic Pseudo-Annealing (DPA) and Modified Metropolis Dynamics (MMD) in order to do image classification using a Markov Random Field modelization. For the first algorithm (DPA), the basic idea is to introduce weighted labelings, which assign a weighted combination of labels to any object, or site, to be labeled, and then to build a merit function of all the weighted labels in such a way as this merit function takes the values of the probability of a global assignment of labels (up to a monotonic transform) for any weighted labeling which assigns the value 1 to one label and 0 to the others at any site. As for the second method (MMD), it is a modified version of the Metropolis algorithm: at each iteration the new state is chosen randomly but the decision to accept it is purely deterministic. This is of course also a suboptimal technique which gives faster results than stochastic relaxation. These two methods have been implemented on a Connection Machine CM2 and simulation results are shown with a SPOT image and a synthetic noisy image. These results are compared to those obtained with the Metropolis algorithm, the Gibbs sampler and ICM (Iterated Conditional Mode).

1. Introduction

Markov Random Fields (MRF) have become more and more popular during the last few years in image processing. A good reason for that is that such a modelization is the one which requires the less a priori information on the world model. Many standard computer vision problems, specifically early vision ones such as image classification, can thus be expressed quite naturally as combinatorial optimization problems. Many heuristics have been proposed to solve them : Iterated Conditional Modes (ICM) [5], Graduated Non-Convexity [7], Mean Field Annealing [21], Simulated Annealing [12, 18] ...

We propose here two different annealing techniques. The first approach, which we propose to call Deterministic Pseudo Annealing (DPA) is related to Relaxation Labeling, a quite popular framework for a variety of computer vision problems [10]. The second approach (MMD) is a modified version of the Metropolis algorithm [18].

2. Probabilistic Modelization

Herein, we are interested with the following general problem : we are given a set of units (or sites) $S = S_i$, $1 \leq i \leq N$, each of which may belong to any one of M classes, or equivalently take any label from 1 to M. We are also given a MRF on these units, defined as usual by

A. Mohammad-Djafari and G. Demoments (eds.), Maximum Entropy and Bayesian Methods, 375–382.

a graph, and the so-called clique potentials. Let c denote a clique of this graph, and $\mathcal{C}$ the set of all cliques. Also $C_i = \{c : S_i \in c\}$. The number of sites in the clique is its degree : $\deg(c)$.

A global discrete labeling L assigns one label L_i ($1 \leq L_i \leq$M) to each site S_i in $\mathcal{S}$. The restriction of L to the sites of a given clique c is denoted by L_c. The definition of the MRF is completed by the knowledge of the clique potentials V_{cL} (shorthand for V_{cL_c}) for every c in $\mathcal{C}$ and every L in $\mathcal{L}$, where $\mathcal{L}$ is the set of the M^N discrete labelings.

2.1. The Model

Given $Y = \{y_i\}$ a set of image data where y_i stands for the greylevel at pixel i. A very general problem is to find the labeling $\hat{L}$ which maximizes $P(L \mid Y)$. Bayes theorem tells us that $P(L \mid Y) = \frac{1}{P(Y)} P(Y \mid L)P(L)$. Actually $P(Y)$ does not depend on the labeling L and we have the assumption that $P(Y \mid L) = \prod_i P(y_i \mid L_i)$. It is then easy to see that the global labeling which we are trying to find is given by:

$$\hat{L} = \max_{L \in \mathcal{L}} \prod_i P(y_i \mid L_i) \prod_{C \in \mathcal{C}} \exp(-V_{cL}) . \tag{1}$$

It is obvious from this expression that the a posteriori probability also derives from a MRF. The energies of cliques of order 1 directly reflect the probabilistic modeling of labels without context, which would be used for labeling the pixels independently. But it is not very difficult to prove that it is always possible, by suitable shifts on the clique potentials to keep only the potentials of maximal cliques. The procedure to do so directly derives from the proof of the Hammersley-Clifford theorem given by [6]. The problem at hand is thus strictly equivalent to maximizing $f(L) = \sum_{c \in \mathcal{C}} W_{cL}$ where $W_{cL} = -V_{cL}$. Let us assume that $P(y_i \mid L_s)$ is Gaussian, the class L_i is represented by its mean value μ_{L_i} and its deviation σ_{L_i}. We get:

$$\hat{L} = \min_{L \in \mathcal{L}} \left(\sum_i \left(\log(\sqrt{2\pi}\sigma_{L_i}) + \frac{(y_i - \mu_{L_i})^2}{2\sigma_{L_i}^2} \right) + \sum_{C \in \mathcal{C}} V_{cL} \right) \tag{2}$$

Using the above equation, it is easy to define the local energy of any labeling L at site i:

$$\mathcal{E}_i(L) = \log(\sqrt{2\pi}\sigma_{L_i}) + \frac{(y_i - \mu_{L_i})^2}{2\sigma_{L_i}^2} + \sum_{C_i} V_{C_i L} \tag{3}$$

2.2. Changing the Problem to Approximately Solve it

The key point here is to cast this discrete, combinatorial, optimization problem into a more comfortable maximization problem on a compact subset of $\mathcal{R}^N$. Let us define a real function $f(X)$ ($X \in \mathcal{R}^{NM}$) as follows :

$$f(X) = \sum_{c \in \mathcal{C}} \sum_{l_c \in L_c} W_{cl_c} \prod_{j=1}^{\deg(c)} x_{c_j, l_{c_j}} \tag{4}$$

where c_j denotes the jth site of clique c, and l_{c_j} the label assigned to this site by l_c. It is clear from Equation (4) that f is a polynomial in the $x_{i,k}$'s; the maximum degree of f is the maximum degree of the cliques. Moreover, it is clear that f is linear with any $x_{i,k}$. Let us now restrict X to $\mathcal{P}_{NM}$, a specific compact subset of $\mathcal{R}_{NM}$ defined by $\forall i,k \ : x_{i,k} \geq 0$ and $\forall i \ : \sum_{k=1}^{M} x_{i,k} = 1$. It admits many maxima on $\mathcal{P}_{NM}$, but the absolute maxima X^* is on the border, which directly yields a solution to our problem. The difficulty is of course that f is not concave but convex with a tremendous number of such maxima.

The basic idea in DPA is to change temporarily the subset on which f is maximized so that f becomes concave, to maximize f, and to track this maximum while slowly changing the constraints until the original ones are restored so that a discrete labeling can be deduced.

First of all, when $deg(c) = 2$ (cliques of order 2), f is a quadratic form, and can always be written as $f = X^tAX$, where A is an $NM \star NM$ symmetric matrix. Besides, after suitable shift, A has non-negative entries. After Perron-Frobenius [2], A has a unique real non-negative eigenvector, which maximizes f under constraints different from the preceding ones : $\forall i,k \ : x_{i,k} \geq 0$ and $\forall i \ : \sum_{k=1}^{M} x_{i,k}^2 = 1$. We call $Q^{NM,d}$ the compact subset of $\mathcal{R}^{NM}$ so defined.

Moreover, the vector X can be obtained very efficiently by using the iterative power method : start from any X^0, and apply

$$X^{n+1} = \frac{AX^n}{\|AX^n\|_{L^2}} \simeq AX^n \qquad (5)$$

A fundamental point is the following : if f is a polynomial with non-negative coefficients and maximum degree d, then f has a unique maximum on $Q^{NM,d}$

The iterative power method can be readily extended, becoming : select $X = X^0$, and apply

$$X^{n+1} \simeq (\nabla f(X^n))^{\frac{1}{d-1}} \qquad (6)$$

This simply means that, at each iteration, we select on the pseudo-sphere of degree d the point where the normal is parallel to the gradient of f. Obviously, the only stable point is singular, and thus is the maximum we are looking for. We have only proved experimentally that the algorithm does converge very fast to this maximum.

The second key point is now to decrease d down to 1. More precisely, we define an iterative procedure, as follows :

1. set $\beta := d$, select some X
2. *while* $(\beta > 1)$ *do*
   - find $X^\star$ which maximizes f on $Q^{NM,\beta}$, starting from X
   - decrease β by some quantity
   - project $X^\star$ on the new $Q^{NM,\beta}$ giving X
3. for each S_i, select the label with value 1

This iterative decrease of β can be compared, up to a point to a cooling schedule, or better to a Graduated Non-Convexity strategy [7]. The last step (projection) is necessary, as changing β changes $Q^{NM,\beta}$.

3. Modified Metropolis Dynamics

The proposed algorithm is a modified version of the Metropolis dynamics [18]. The choice of the new label state is done randomly using a uniform distribution but the rule to accept a new state is deterministic. The parallel algorithm is the following:

1. Pick up randomly an initial configuration L^0, with $k = 1$ and $T = T(1)$.
2. Using a uniform distribution, pick up a global state L^t so that $\forall i (1 \leq i \leq N) : 1 \leq L_i^t \leq M$ and $L_i^t \neq L_i^k$.
3. In order to get the convergence of the parallel algorithm, we can partition the entire image into disjoint regions $\mathcal{R}_n$ such that pixels which belong to the same region are conditionally independent given the data of all the other regions:
$$\mathcal{S} = \bigcup_n \mathcal{R}_n \quad \text{and} \quad \mathcal{R}_n \cap \mathcal{R}_m = \emptyset \ (n \neq m) \tag{7}$$
4. For each site S_i, the local energy $\mathcal{E}_i(L^{t'})$ is computed in parallel using Equation (3) with $L^{t'} = [L_1^k, L_2^k, \ldots, L_{i-1}^k, L_i^t, L_{i+1}^k, \ldots, L_N^k]$ Calling $\Delta\mathcal{E}_i = \mathcal{E}_i(L^{t'}) - \mathcal{E}_i(L^k)$, a new label state at site S_i is accepted according to the following rule:
$$L_i^{k+1} = \begin{cases} L_i^t & \text{if } \Delta\mathcal{E}_i \leq 0 \text{ or } \left[\Delta\mathcal{E}_i > 0 \text{ and } \alpha \leq \exp\left(-\frac{\Delta\mathcal{E}_i}{T}\right)\right] \\ L_i^k & \text{otherwise} \end{cases}$$
where α is a constant threshold ($\alpha \in (0,1)$), chosen at the beginning of the algorithm.
5. Decrease of the temperature $T = T(k+1)$ (k = number of iterations) and goto 2 if the number of modified sites > *threshold*.

There is no explicit formula to get the threshold α. For image classification, α is chosen nearly equal to zero if the image is very noisy, otherwise α is equal to 0.5 (for more details about the relationship between α and $\Delta\mathcal{E}_{min}$ see [16]).

4. Experimental Results

All the algorithms used for the simulation have been implemented on a Connection Machine CM2 [13] with 8K processors.

4.1. Comparison with Other Methods

Now, we present the results obtained on a noisy checkerboard image (128 × 128) with a SNR= −5dB with 2 classes (see Fig. 1, Table 1) and on a SPOT image (256 × 256) with 4 classes (see Fig. 2, Table 1). For the simulations, the following parameters have been used:

- *Temperature*: For ICM the temperature is constant. For the other algorithms, we have used an exponential schedule: $T_{k+1} = 0.95T_k$. The initial temperature is $T_0 = 10$ for the synthetic image and $T_0 = 4$ for the SPOT image. For DPA, β decrease from 2 to nearly 1.

A	VPR	Nb. of Iter	Total time	Time per It.	error
ICM	2	9	0.249 sec.	0.027 sec.	2622
Metropolis	2	90	26.32 sec.	0.293 sec.	208
Gibbs	2	86	48.51 sec.	0.564 sec.	269
MMD	2	79	2.37 sec.	0.030 sec.	77
DPA	2	40	17.86 sec.	0.446 sec.	175
B					
ICM	8	6	0.875 sec.	0.146 sec.	—
Metropolis	8	52	61.39 sec.	1.181 sec.	—
Gibbs	8	57	235.68 sec.	4.135 sec.	—
MMD	8	30	5.75 sec.	0.192 sec.	—
DPA	8	22	75.55 sec.	3.434 sec.	—

Table 1: Results A: on the checkerboard image with 2 classes, B: on the SPOT image with 4 classes

- *Mean and deviation of each class*: They are computed using a supervised learning method.

- *Choice of the clique-potentials*: We have used second order clique-potentials. If two neighboring sites have different labels, the energy is 2.0 for the synthetic image and 0.5 for the other one. If they have the same label, the energy is 0.2 (resp. -0.5).

- *Initialization of the labels*: Random values are assigned to the labels for the initialization of Gibbs, Metropolis and MMD. As for ICM and DPA, the initial labels are obtained using only the Gaussian term in Eq. (2).

- *Initialization of α for MMD*: $\alpha = 0.1$ for the checkerboard and $\alpha = 0.8$ for the SPOT image.

5. Conclusion

In this paper, we have presented two suboptimal methods: Deterministic Pseudo Annealing and Modified Metropolis Dynamics applied to image classification. These techniques compare favorably with the classical relaxation algorithms. They represent a good trade off for image processing problems.

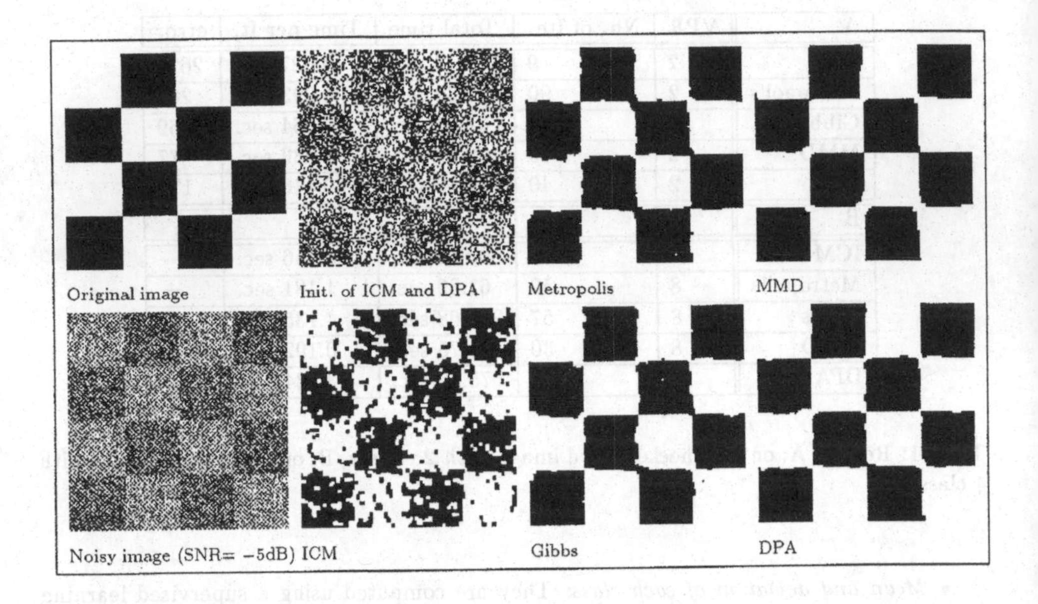

Figure 1: Results with the checkerboard image with 2 classes

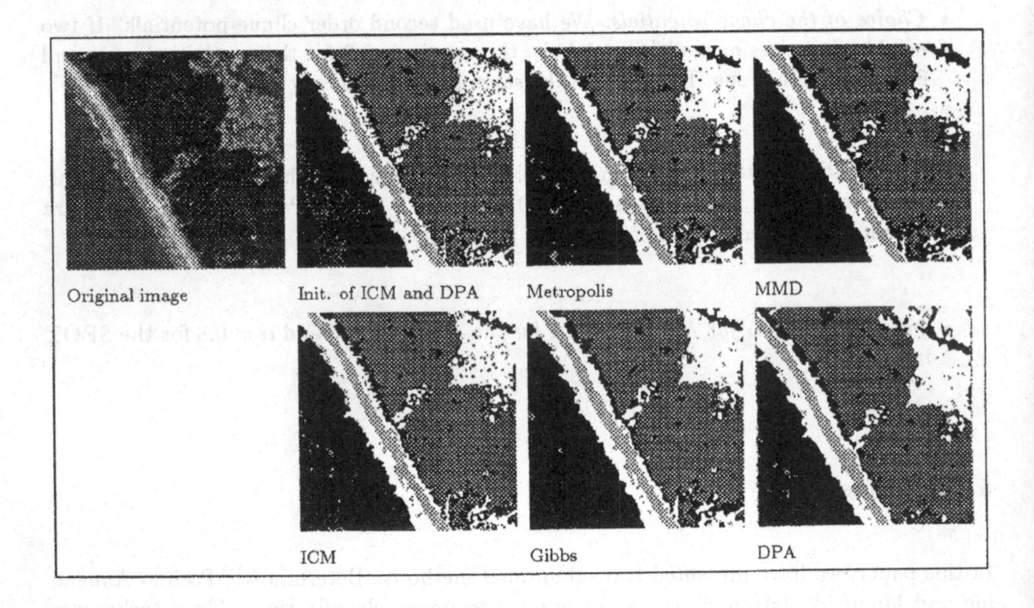

Figure 2: Results with the SPOT image with 4 classes

References

[1] R. Azencott. Markov fields and image analysis. *Proc. AFCET, Antibes*, 1987.

[2] A. Berman and R. Plemmons. *Non-negative Matrices in the Mathematical Sciences.* New York, 1979.

[3] M. Berthod. *L'amélioration d'étiquetage : une approche pour l'utilisation du contexte en reconnaissance des formes.* Thèse d'état, Université de Paris VI, December 1980.

[4] M. Berthod. Definition of a consistent labeling as a global extremum. In *Proc. ICPR6*, pages 339–341, Munich, 1982.

[5] J. Besag. On the statistical analysis of dirty pictures. *Jl. Roy. Statis. Soc. B.*, 1986.

[6] J. E. Besag. Spatial interaction and the statistical analysis of lattice systems (with discussion). *Jl. Roy. Statis. Soc. B. 36.*, pages 192–236, 1974.

[7] A. Blake and A. Zisserman. Visual reconstruction. *MIT Press, Cambridge - MA*, 1987.

[8] P. Chou and C. Brown. The theory and practice of bayesian image labeling. *International Journal of Computer Vision*, 4:185–210, 1990.

[9] H. Derin, H. Elliott, R. Cristi, and D. Geman. Bayes smoothing algorithms for segmentation of binary images modeled by markov random fields. *IEEE trans. on Pattern analysis and mach. intel., Vol 6*, 1984.

[10] O. Faugeras and M. Berthod. Improving Consistency and Reducing Ambiguity in Stochastic Labeling: an Optimization Approach. *IEEE Transactions on PAMI*, 4:412–423, 1981.

[11] D. Geiger and F. Girosi. Parallel and deterministic algorithms for MRFs : surface reconstruction and integration. In *Proc. ECCV90*, Antibes, France, 1990.

[12] S. Geman and D. Geman. Stochastic relaxation, Gibbs distributions and the Bayesian restoration of images. *IEEE Trans. on Pattern Analysis and Machine Intelligence*, 6:721–741, 1984.

[13] W. D. Hillis. *The Connection Machine.* Cambridge, MA, 1985.

[14] R. Hummel and S. Zucker. On the foundations of relaxation labeling processes. *IEEE trans. on Pattern Analysis and Machine Intelligence*, 5(3), 1983.

[15] F. C. Jeng and J. M. Woods. Compound Gauss - Markov Random Fields for Image Estimation. *IEEE Trans. Acoust., Speech and Signal Proc.*, ASSP-39:638–697, 1991.

[16] Z. Kato. Modélisation Markovienne en classification d'image. mise en oeuvre d'algorithmes de relaxation. *Rapport de stage de DEA, Université de Nice*, 1991.

[17] P. V. Laarhoven and E. Aarts. Simulated annealing : Theory and applications. *Reidel Pub., Dordrecht, Holland*, 1987.

[18] N. Metropolis, A. Rosenbluth, M. Rosenbluth, A. Teller, and E. Teller. Equation of state calculations by fast computing machines. *J. of Chem. Physics, Vol. 21, pp 1087-1092*, 1953.

[19] A. Rangarajan and R. Chellappa. Generalised Graduated Non-Convexity algorithm for Maximum A Posteriori image estimation. *Proc. ICPR*, pages 127–133, Jun. 1990.

[20] R. H. A. Rosenfeld and S. W. Zucker. Scene labeling by relaxation operation. *IEEE Trans. on Systems Mans and Cybernetics*, 1982.

[21] J. Zerubia and R. Chellappa. Mean field approximation using compound Gauss-Markov Random field for edge detection and image restoration. *Proc. ICASSP, Albuquerque, USA*, 1990.

GLOBAL BAYESIAN ESTIMATION, CONTRAINED MULTISCALE MARKOV RANDOM FIELDS AND THE ANALYSIS OF VISUAL MOTION

Patrick Pérez †, Fabrice Heitz ‡ and Patrick Bouthémy ‡
†IRISA/CNRS, ‡IRISA/INRIA,
Campus de Beaulieu, Rennes, France
E-mail : perez@irisa.fr, heitz@irisa.fr

ABSTRACT. In this paper we investigate a new approach to multigrid bayesian image analysis based on Markov Random Fields (MRF) models. The multigrid algorithms under concern are based on constrained optimization schemes. The global optimization problem associated to MRF modeling is solved sequentially over particular subsets of the original configuration space. Those subsets consist of constrained configurations describing the desired resulting field at different scales. The constrained optimization can be implemented via a coarse to fine multigrid algorithm defined on a sequence of consistent multiscale MRF models. The proposed multiscale paradigm yields fast convergence towards high quality estimates when compared to standard monoresolution or multigrid relaxation schemes.

1. Introduction

Markov Random Field (MRF) models have been successfully introduced in several important low-level image analysis problems. When they are associated to the Maximum A Posteriori (MAP) criterion, they lead to the minimization of a global energy function which may exhibit local minima, [3]. This minimization is generally performed using deterministic, [1] or stochastic, [3] relaxation algorithms which can be significantly sped up by using multigrid techniques, [5]. Yet in multigrid implementation of statistical models such as MRF, the key problem remains the derivation of the model parameters at different scales. When global mathematical consistency is not guaranteed, the parameters and the neighborhood structure associated to the model can only be adjusted over scale in an ad-hoc way. The main contribution of this paper is the presentation of a new class of globally consistent multiscale models. Other approaches of this problem may be found in [6] and [2]. The model we consider here is related to a multiscale constrained exploration of the original configuration space Ω on nested subspaces $\Omega^n \subset \Omega^{n-1} \subset \cdots \subset \Omega^i \subset \cdots \subset \Omega^1 \subset \Omega^0 = \Omega$ corresponding to finer and finer configurations. Two different versions of this model will be considered in our paper. They are associated to two different sequences of constrained subspaces. The first model corresponds to piece-wise constant constrained configurations and the second one is related to a bilinear interpolation constraint on the label field. The new multiscale algorithms have been compared to standard relaxation schemes in visual motion analysis. It shows fast convergence towards quasi optimal estimates (close to those obtained using time consuming stochastic relaxation).

A. Mohammad-Djafari and G. Demoments (eds.), Maximum Entropy and Bayesian Methods, 383–388.

2. Multiscale markov random fields models

Let $o = \{o_s, s \in S\}$ designate an observation field defined on a rectangular lattice S, and $e = \{e_s, s \in S\}$ denote an unobserved discrete-valued label field, defined on the same lattice S. Let Ω be the space of all possible configurations e. If (e, o) is assumed to be a MRF with respect to neighborhood system $\mathcal{G} = \{\mathcal{G}_s, s \in S\}$, the best estimate of e according to the MAP criterion is given by, [3] :

$$\hat{e} = \arg\min_{e \in \Omega} U(o,e) \quad \text{with} \quad U(o,e) = \sum_{c \in \mathcal{C}} V_c(o,e) \tag{1}$$

where $\mathcal{C}$ denotes the set of cliques associated to neighborhood system $\mathcal{G}$ and the potential function V_c is locally defined on the clique c. In the following we will focus on the 8-neighborhood system.

Let us assume that the size of the rectangular lattice S is $(2^m + 1) \times (2^{m'} + 1)$:

$$S = \{s = (p,q),\ p = 0 \cdots 2^m,\ q = 0 \cdots 2^{m'}\}$$

and let us define coarse grids S^i $(i = 0 \cdots \inf(m, m'))$ by :

$$S^i = \{s = (2^i k, 2^i k'),\ k = 0 \cdots 2^{m-i},\ k' = 0 \cdots 2^{m'-i}\}.$$

Γ^i will denote the space of configurations defined on S^i. Let $s = (p,q) \in S$. Its nearest neighbors belonging to the coarse grid S^i are $s_1 = (p', q')$, $s_2 = (p', q' + 2^i)$, $s_3 = (p' + 2^i, q')$, and $s_4 = (p' + 2^i, q' + 2^i)$ with $p' = 2^i.\lfloor p/2^i \rfloor$ and $q' = 2^i.\lfloor q/2^i \rfloor$ (see Fig. 1).

The constrained subspace Ω^i is defined as the space of configurations $e \in \Omega$ for which the label $e_{(p,q)}$ at the site $s = (p,q)$ of the original fine grid S is *constrained* to be a function of the labels of the coarse grid sites s_1, s_2, s_3, s_4, and of the location (p,q). More precisely (Fig. 1) :

$$e_s = g^i(e_{s_1},\ e_{s_2},\ e_{s_3},\ e_{s_4},\ p,\ q) \tag{2}$$

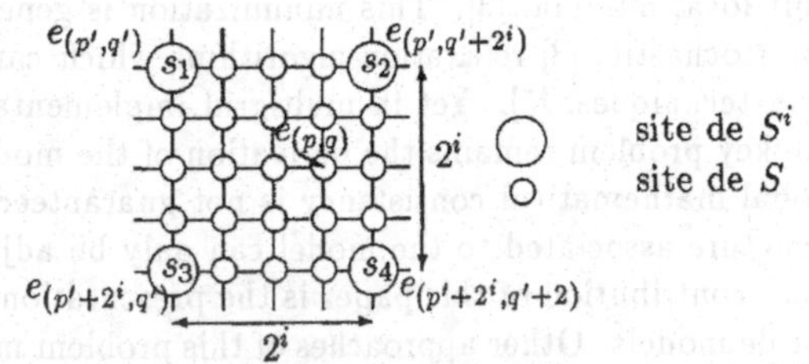

Figure 1: Location of a site $s = (p,q)$ of S with respect to grid S^i $(i = 2$ here)

It turns out that $e \in \Omega^i$ is completely defined by the value of its labels on S^i : there is a bijective link between configurations $e \in \Omega^i$ and their restrictions $e^i \in \Gamma^i$. Let Φ^i denote the bijection from Γ^i in Ω^i thus defined (Φ^i depends on function g^i) :

$$\begin{array}{rccl} \Phi^i : & \Gamma^i & \longrightarrow & \Omega^i \\ & e^i & \longmapsto & e = \Phi^i(e^i) \end{array} \tag{3}$$

Φ^i corresponds to an interpolation process defined by function g^i (equ. 2).

The energy of $e \in \Omega^i$ may be rewritten only as a function of its labels on S^i. Hence, we can derive from the energy function $U(e,o)$ a coarse energy function at scale i which is defined on coarse configurations of Γ^i :

$$U^i(e^i,o) \triangleq U(e,o) \quad \text{with} \quad e = \Phi^i(e^i)\,. \tag{4}$$

It is easy to verify that, if $\mathcal{G}$ corresponds to a 8-neighborhood system, the energy function U^i at scale i can be expressed as a sum of potential functions associated to *the same 8-neighborhood system on the coarse grid* S^i. More precisely these potential functions depend on 4 site cliques associated to a 8-neighborhood system $\mathcal{G}^i$:

$$U^i(e^i,o) = \sum_{\begin{array}{|c|c|}\hline s_1 & s_2 \\ \hline s_3 & s_4 \\ \hline\end{array} \in C^i} V^i_{\{s_1,s_2,s_3,s_4\}}(e^i_{s_1},e^i_{s_2},e^i_{s_3},e^i_{s_4},o) \tag{5}$$

where C^i is the set of cliques defined by $\mathcal{G}^i$ on S^i.

To take benefit from the sequence of multiscale models previously defined, instead of handling the original optimization problem (equ. 1) defined over Ω we consider the following sequence of problems :

$$\hat{e}(i) = \arg\min_{e\in\Omega^i} U(e,o), \quad i = n,\cdots,0 \tag{6}$$

These minimizations are equivalent to a minimization on the coarse MRF models, i.e. :

$$\hat{e}^i = \arg\min_{e^i\in\Gamma^i} U^i(e^i,o), \quad i = n,\cdots,0 \tag{7}$$

A standard coarse to fine technique is used to this end : starting from a coarse scale n, the optimization problem (equ. 1) is first solved in subset Ω^n (equ. 6) by solving the equivalent problem (7). An estimate of $\hat{e}^n$ is obtained by a deterministic relaxation algorithm known as ICM, [1]. At level i if $\tilde{e}^i$ designates the estimate of $\hat{e}^i$ (obtained after convergence of the deterministic relaxation at that level), the algorithm at resolution level $i-1$ is initialized by $[\Phi^{i-1}]^{-1} \circ \Phi^i(\tilde{e}^i)$, which is just an interpolation of $\tilde{e}^i$ on the finer grid S^{i-1} according to function g^i.

Two different multiscale models corresponding to two different sequences of constrained subspaces have been investigated. The first one corresponds to piece-wise constant constrained configurations in Ω^i : $\forall e \in \Omega^i$, $\forall s = (p,q) \in S$, $e_s = e_{s_1}$ (see Fig. 1). The second model corresponds to bilinear interpolation constraints of the form :

$$\forall e \in \Omega^i,\ \forall s = (p,q) \in S,\ e_s = \alpha(p-p') + \beta(q-q') + \gamma(p-p')(q-q') + \delta \tag{8}$$

with $\alpha = 2^{-i}.(e_{s_3} - e_{s_1})$ $\beta = 2^{-i}.(e_{s_2} - e_{s_1})$ $\gamma = 4^{-i}.(e_{s_1} + e_{s_4} - e_{s_2} - e_{s_3})$ $\delta = e_{s_4}$

The multiscale relaxation algorithms associated to the piece-wise constant constraint (**MSR1**) and to the bilinear constraint (**MSR2**) have been compared to standard relaxation algorithms for different MRF models used in visual motion analysis. Five algorithms are concerned : a standard monoresolution deterministic relaxation (**DR**) algorithm known as ICM, [1]; a monoresolution stochastic relaxation algorithm (**SR**) based on the Gibbs sampler, [3]; a standard coarse to fine multigrid relaxation (**MGR**) algorithm in which the same model is considered at each resolution (same parameters, same neighborhood system and same potential functions) and a pyramid of observations is constructed, [5]; the two propose multiscale relaxations (**MSR1** and **MSR2**).

3. Application to visual motion analysis

The multiscale approach has been applied to different MRF models in motion detection, [4], motion based segmentation, [4] and optical flow measurement. In this paper, the application of the method to a discrete model for optical flow measurement is detailed. The MRF model is associated to a 8-neighborhood and specified by the following energy function :

$$U(\vec{\omega}_t, f_t, f_{t+dt}) = \sum_{s\in S}\{f_t(s) - f_{t+dt}(s+\vec{\omega}_t(s).dt)\}^2 + A^2 \sum_{\{s_1,s_2\}\in C} \| \vec{\omega}_t(s_1) - \vec{\omega}_t(s_2) \|^2 \quad (9)$$

where $f_t(s)$ denotes the observed intensity function at time t at the pixel site s and $\vec{\omega}_t(s)$ the velocity vector of the site s at time t.

The first term in the energy (known as the "displaced frame difference") expresses the constant brightness assumption for a physical point over time. The second term balances the first one through the weighting parameter A; it can be interpreted as a regularization term which favors smooth solutions.

For the first multiscale model (piece-wise constant constraint) the coarse energy function at level i is :

$$U^i(\vec{\omega}^i_t, f_t, f_{t+dt}) = \sum_{s\in S^i}\sum_{r\in B^i_s}\{f_t(r) - f_{t+dt}(r+\vec{\omega}^i_t(s).dt)\}^2 + A^2 \sum_{\{s_1,s_2\}\in C^i} q^i_{\{s_1,s_2\}} \| \vec{\omega}_t(s_1) - \vec{\omega}_t(s_2) \|^2 \quad (10)$$

where $q^i_{\{s_1,s_2\}} = 1$ if $\{s_1, s_2\}$ is diagonal and $q^i_{\{s_1,s_2\}} = 2^i + 2(2^i - 1)$ if $\{s_1, s_2\}$ is either horizontal or vertical. B^i_s designates the $2^i \times 2^i$ square block of S which upper-left corner is $s \in S^i$.

For the second multiscale model (bilinear interpolation constraint) the coarse energy function at level i is :

$$U^i(\vec{\omega}^i_t, f_t, f_{t+dt}) = \sum_{\begin{array}{|c|c|}\hline s_1 & s_2 \\ \hline s_3 & s_4 \\ \hline\end{array}\in C^i} \sum_{s\in B^i_{s_1}} \{f_t(s) - f_{t+dt}(s + \vec{\omega}_t(s).dt)\}^2$$

$$+A^i \sum_{\begin{array}{|c|c|}\hline s_1 & s_2 \\ \hline s_3 & s_4 \\ \hline\end{array}\in C^i} [3(\alpha_1^2 + \alpha_2^2 + \beta_1^2 + \beta_2^2) + B^i(\gamma_1^2 + \gamma_2^2) + C^i(\gamma_1(\alpha_1 + \beta_1) + \gamma_2(\alpha_2 + \beta_2))] \quad (11)$$

where $\vec{\omega}_t(s)$ is deduced from $\vec{\omega}^i_t(s_1)$, $\vec{\omega}^i_t(s_2)$, $\vec{\omega}^i_t(s_3)$ and $\vec{\omega}^i_t(s_4)$ according to equ. 8 (interpolation coefficients indexed by 1 correspond to the first component of velocity vectors, and those indexed by 2 correspond to second component) and :

$$A^i = 4^i.A^2 \qquad B^i = [(2^i + 1)(2^{i+1} + 1) + (2^i - 1)(5 \times 2^{i+1} + 1)]/6 \qquad C^i = [3(2^i - 1) + 2].$$

The results on a benchmark of 41 short sequences appear in Fig. 3. The sequences were composed of two frames of 64×64 real world images with a synthetic motion field (Fig. 2).

The multiscale algorithms require the least number of iterations to converge. On the average, the MSR methods find configurations close to the best estimates obtained by SR (even sometimes better for MSR2), with an improvement of *up to two orders of magnitude* on the number of iterations.

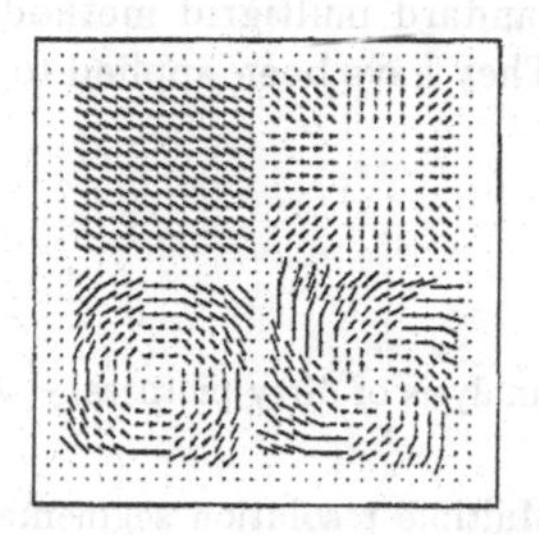

Figure 2: Synthetic motion field we try to estimate

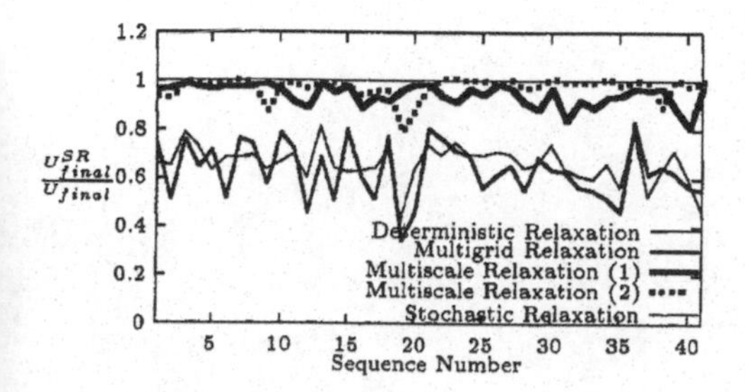

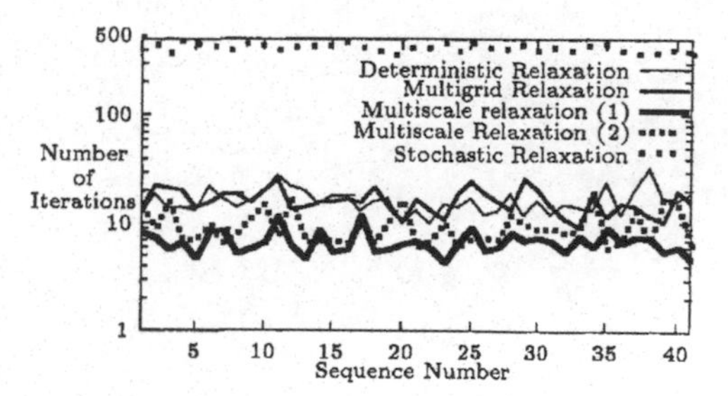

Figure 3: Ratio of the final energy value reached with SR to the one obtained with the other methods. Number of iterations at conv

4. Conclusion

In this paper we have developped a new and general approach to multiscale MRF modeling. Unlike previous approaches, the proposed paradigm enables to derive the parameters of coarse scale models in a consistent way. This approach can easily be interpreted as a constrained optimization. Two particular versions of the general model have been described. They compare favorably with standard multigrid methods : they provide quasi-optimal estimates and fast convergence. They have been applied to different issues in visual motion analysis.

References

[1] J. BESAG. – On the statistical analysis of dirty pictures. – *J. Royal Statist. Soc.*, Vol. 48, Serie B, No 3: pages 259–302, 1986.

[2] C. BOUMAN and B. LIU. – Multiple resolution segmentation of textured images. – *IEEE Trans. Pattern Anal. Machine Intel.*, Vol. 13, No 2: pages 99–113, Feb. 1991.

[3] S. GEMAN and D. GEMAN. – Stochastic relaxation, Gibbs distributions and the bayesian restoration of images. – *IEEE Trans. Pattern Anal. Machine Intel.*, Vol. 6, No 6: pages 721–741, November 1984.

[4] F. HEITZ, P. PEREZ, and P. BOUTHEMY. – Constrained multiscale markov random fields and the analysis of visual motion. – Technical Report Num. 1615, INRIA-Rennes, February 1992.

[5] J. KONRAD and E. DUBOIS. – Multigrid Bayesian estimation of image motion fields using stochastic relaxation. – In *Proc. 2nd Int. Conf. Computer Vision*, pages 354–362, Tarpon Springs, Florida, Dec. 1988.

[6] B.C. LEVY. – Multiscale models and estimation of discrete Gauss-Markov random fields. – In *2nd SIAM Conf. Linear Algebra in Systems Control and Signal Processing*, San Francisco, CA, Nov. 1990.

CRITICAL PROPERTIES OF THE 2D ISING MODEL: A MARKOVIAN APPROACH

Marc SIGELLE and Alain MARUANI
TELECOM Paris Département IMAGES
46 rue Barrault 75634 Paris Cedex 13 - France

ABSTRACT. We develop here an application of Markov Random Fields to the understanding and estimation of 2D Ising model properties with nearest-neighbor square-lattice and zero magnetic field. Using gauge invariance yields easily to the well-known phase transition properties of the model and to an analytical estimation of the spontaneous magnetization below phase transition. The estimated critical temperature derived from this model is more accurate than that of Mean Field Theory.

1. Introduction

The Ising model, a key model to Ferromagnetism Phase Transition understanding, has been recently studied for its Markov Random Field properties in Image Processing [3]. Its critical properties are related to the "Choice of Hyperparameters" problem in Regularization Theory [6]. Exact critical temperature calculation, performed by Onsager [5] and then by Landau [4], is especially complex. We present here a bayesian approach enabling to describe its critical properties.

2. A markovian analysis of the Ising model

Let us recall that the 2D Ising model concerns a square lattice of interacting spins, such that the total energy of the system may be written as a sum on all couples of neighbor sites (r,s) : $U = \sum_{(rs)} U_{rs}$, where the U_{rs} interaction energy terms write : $U_{rs} = -J\sigma_r\sigma_s$ $(\sigma_i = \pm 1)$ and J is the Ising-Heisenberg coupling parameter. One can also write such terms in the Potts model frame [7, 6] : $U_{rs} = -K\delta(x_r, x_s)$ $(x_i = 0$ or $1)$ where δ is the Kronecker symbol: $\delta(x_r, x_s) = 1$ if $x_r = x_s$ [0 otherwise] and where $K = 2J$ is the Potts coupling constant. The Ising model is the prototype of Markov Random Fields [3], whose fundamental property is : the probability that the spin located at particular site s has value x_s conditionnally to the rest of the lattice, S_s, is equal to its conditional probability knowing neighborhood configuration N_s. According to Boltzmann law (with Boltzmann constant k) and Hammersley-Clifford theorem [1], this probability at temperature T is

$$P(x_s|S_s) = P(x_s|N_s) = \frac{1}{Z_{N_s}} \exp \frac{-U(x_s|N_s)}{kT} \tag{1}$$

where $U(x_s|N_s)$ is the sum of energy terms involving site s given the neighbour configuration N_s, i.e. the conditional energy of site s, and where Z_{N_s} is the local partition function of this configuration: $Z_{N_s} = \sum_{\{x_s\}} \exp \frac{-U(x_s|N_s)}{kT}$

A. Mohammad-Djafari and G. Demoments (eds.), Maximum Entropy and Bayesian Methods, 389–392.

(a summation running on all possible values of spin x_s). The probability $P(x_s|N_s)$ resembles thus a "local" version of the global Boltzmann distribution law. Let p be the probability that this particular spin s points upwards. Then, Bayes rule yields :

$$p = P(x_s = 1) = \sum_{\{N_s\}} P(x_s = 1|N_s)P(N_s) = \sum_{\{N_s\}} \frac{\exp \frac{-U(x_s|N_s)}{kT}}{Z_{N_s}} P(N_s) \qquad (2)$$

where $P(N_s)$ is the probability of configuration N_s and where the sum runs on all local neighborhood configurations N_s. In order to simplify notations and to standardize notations with respect to previous work, we let further on $\frac{K}{kT} \rightarrow K$, where, again, K is the Potts constant . Noting for example 3(1) a local configuration with 3 neighbour spins having value 1 and investigating all their occurences yields, with no other assumption than that of a Markov Random Field:

$$\begin{aligned} p \;=\; & \frac{e^{4K}}{1+e^{4K}} P(4(1)) + C_4^1 \frac{e^{3K}}{e^{3K}+e^K} P(3(1)) + C_4^2 \frac{e^{2K}}{2e^{2K}} P(2(1)) \\ & + C_4^3 \frac{e^K}{e^{3K}+e^K} P(3(0)) + \frac{1}{1+e^{4K}} P(4(0)) \end{aligned}$$

We assume now further that spins belonging to N_s are uncorrelated (physical, strong hypothesis) and that the probability p is translation-invariant on the spin lattice (weak, simplifying stationarity hypothesis). The probabilities $P(N_s)$ follow then a Bernoulli law, and the previous relation writes:

$$\begin{aligned} p \;=\; & \frac{e^{4K}}{1+e^{4K}} p^4 + C_4^1 \frac{e^{3K}}{e^{3K}+e^K} p^3(1-p) + C_4^2 \frac{e^{2K}}{2e^{2K}} p^2(1-p)^2 \\ & + C_4^3 \frac{e^K}{e^{3K}+e^K} p(1-p)^3 + \frac{1}{1+e^{4K}} (1-p)^4 \end{aligned} \qquad (3)$$

which defines in an auto-coherent way the probability law of a particular spin as a function of temperature (recall that T is implicitely contained in K): $p = \varphi(p)$, where $\varphi(p)$ is the right hand side of (3).

2.1. Gauge invariance properties of equation (3) and resolution

Inspection of eq.(3) shows at once that :
- (a) It is of third degree instead of fourth as could seem to be the case.
- (b) $p = \frac{1}{2}$ is a solution for all K, i.e. at every temperature.
- (c) The graph of function φ is symmetric around point $(\frac{1}{2}, \frac{1}{2})$.

These properties derive basically from the gauge invariance property of Ising model with zero magnetic field. The related gauge transformation exchanges spin values 0 and 1, i.e. is associated to the transformation $x_s \rightarrow 1 - x_s$. One has correspondingly : $p \rightarrow 1 - p$ and $U_{rs} \rightarrow U_{rs}$, which imply :

$$p = \varphi(p) \quad (a) \qquad \text{and} \qquad 1 - p = \varphi(1-p) \quad (b)$$

Adding (a) and (b) gives $\varphi(p) + \varphi(1-p) = 1$, from which $\varphi(\frac{1}{2}) = \frac{1}{2}$ is trivial. It also proves that φ is an odd degree polynom, as can be seen by considering φ on $\mathbf{R}$ and examining its behaviour at infinity. Now, substracting (a) and (b) (which consists in computing the nonzero spontaneous magnetization $\mu = 2p-1$ occurring below critical temperature) yields :

$$2p - 1 = \frac{e^{4K}-1}{e^{4K}+1}[p^4 - (1-p)^4] + C_4^1 \frac{e^{3K}-e^K}{e^{3K}+e^K}[p^3(1-p) - p(1-p)^3] \qquad (4)$$

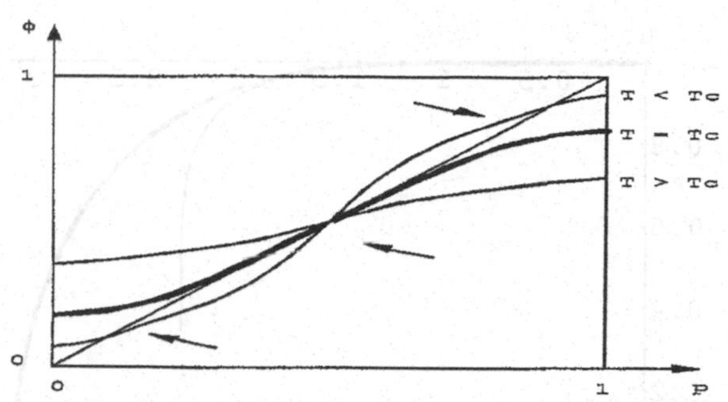

Fig. 1 Graphical interpretation of equation (3). Solutions are indicated by arrows

Noting that $p^2 - (1-p)^2 = 2p - 1$ gives immediately, that apart solution $p = \frac{1}{2}$, non-trivial solutions verify :

$$(p^2 + (1-p)^2) \tanh 2K + 4p(1-p) \tanh K = 1 \tag{5}$$

This is a second degree equation for p whose solutions are symmetric around $p = \frac{1}{2}$. The critical constant value K_c is such that $p = \frac{1}{2}$ is (double) solution of (5) i.e.

$$\frac{1}{2} \tanh 2K_c + \tanh K_c = 1 \tag{6}$$

Numerical computation gives : $e^{K_c} \approx 1.921$, $K_c \approx 0.653$ and $T_c/J \approx 3.063$, which compares favourably with Onsager's exact result $K_c = log(1+\sqrt{2}) \approx 0.882$ i.e. $T_c/J \approx 2.269$. Studying the stability of solution $p = \frac{1}{2}$ is the basis of phase transition properties of the Ising model.

2.2. Interpretation of equation (3)- The spontaneous magnetization

Equation (3) has one or three real roots according to the derivative value $\varphi'(p)$ at $\frac{1}{2}$ (See Fig. 1). Moreover, the behaviour of spontaneous magnetization μ can be found at any temperature. Equation (5) writes, in terms of μ :

$$(\frac{\mu^2+1}{2}) \tanh 2K + (1-\mu^2) \tanh K = 1 \tag{7}$$

whose real solutions are $\mu = \pm\, [1 - \frac{(1-\tanh K)^2}{\tanh^3 K}]^{\frac{1}{2}}$.

Putting $x = e^{-K}$ [8], one finds then for the positive spontaneous magnetization below critical temperature

$$\mu = [1 - \frac{4x^4(1+x^2)}{(1-x^2)^3}]^{\frac{1}{2}} \tag{8}$$

whose low temperature expansion $\mu = 1 - 2x^4 - 8x^6 - 20x^8 + O(x^{10})$ agrees up to 6th order with the exact result [8, 2]

$$\mu = [\frac{(1+x^2)}{(1-x^2)^2}(1 - 6x^2 + x^4)^{\frac{1}{2}}]^{\frac{1}{4}} \tag{9}$$

whose expansion is $\mu = 1 - 2x^4 - 8x^6 - 34x^8 + O(x^{10})$. Comparison of both results is shown on Fig. 2.

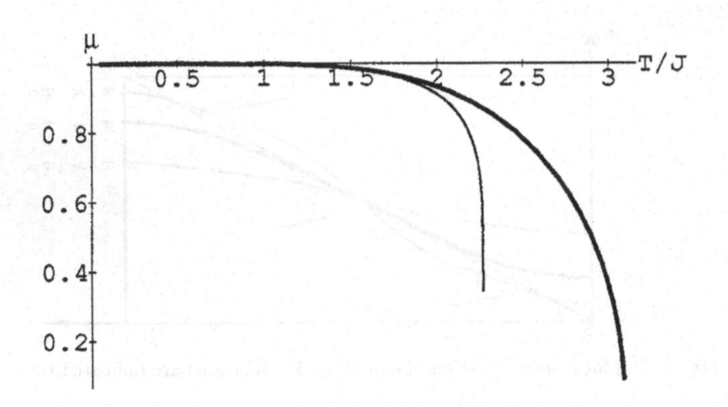

Fig. 2 Spontaneous magnetization as a function of T:
This work (bold) and Yang exact result (thin) our $T_c/J \approx 3.063$ whereas $T_c^{Yang}/J \approx 2.269$

3. Conclusion

General assumptions made here allow on the one hand to account for the high and low temperature behaviour of the 2D Ising model with zero magnetic field, and on the other hand to yield a good approximation of the critical constant K_c. Our approach gives a better approximation than that of Mean Field Theory where : $K_c = 2J_c = \frac{1}{2}$ and $T_c/J = 4.0$. This results from our decorrelation hypothesis between neighbouring spins of a given site, a finer one than in Mean Field Theory where a site is uncorrelated with each of its neighbors. The model could be extended by considering a non zero magnetic field and also the general Potts model, whose critical behaviour strongly depends on the number of spin values [7]. Studies are currently done on the subject.

References

[1] J. Besag "Spatial Interaction and the Statistical Analysis of Lattice Systems" *Journal of the Royal Statistical Society* Vol. B-36 pp. 192-236 (1974)

[2] J. Combe "Champs de Gibbs et Physique Statistique", TELECOM Paris, Memoire de fin d'etudes, 1991

[3] S. Geman and D. Geman "Stochastic relaxation, Gibbs Distribution, and the Bayesian Restoration of Images", *IEEE Transactions on Pattern Analysis and Machine Intelligence*, Vol.6, N.6, Nov.1984

[4] L. Landau and E. Lifschitz Physics Course Vol. 5 - Statistical Physics

[5] L. Onsager, *Physical Review* 65, 117 (1944)

[6] M. Sigelle and R. Ronfard "Relaxation of previously classified Images by a Markov Field Technique and its Relationship to Statistical Physics" 7th Scandinavian Image Analysis Congress, Aalborg, Danmark, 1991

[7] F.Y. Wu, "The Potts model" *Review of Modern Physics* Vol. 54, N. 1 (1982)

[8] C.N. Yang "The Spontaneous Magnetization of a Two-Dimensional Ising Model", *Physical Review*, Vol. 85, 5 pp. 808-814 (1952)

A COMPARISON OF DIFFERENT PRIOR LAWS FOR BAYESIAN IMAGE RECONSTRUCTION IN POSITRON EMISSION TOMOGRAPHY

P. Desmedt, I. Lemahieu
University of Ghent,
Laboratory of Electronics and Information Systems
St.-Pietersnieuwstraat 41
9000 Gent, Belgium

ABSTRACT. In this paper the prior law introduced by Djafari [2] is compared to the prior law proposed by Gull [5] for the image reconstruction from projections. Furthermore, an approximation is proposed to simplify the calculation of the prior law introduced by Gull, in the case of data obtained from projections.

1. Introduction

A Positron Emission Tomography (PET) scan yields information about the activity distribution of a tracer in a cross-section through the body of a patient, see [6]. This activity distribution must be recovered from a number of projection measurements. Different measurement inaccuracies encourage the interest in statistical reconstruction techniques in PET. Statistical reconstruction techniques generally maximize the posterior probability of the image ($p(\mathbf{f}|\mathbf{d}I) \sim p(\mathbf{f}|I)\,p(\mathbf{d}|\mathbf{f}I)$) given the observed data, where $\mathbf{f}$ is the image and $\mathbf{d}$ is the observed data set.

The form of the direct probability $p(\mathbf{d}|\mathbf{f}I)$ is determined by the measurement process and the statistics of the data. The noiseless measurement process is given by $\mathbf{d} = \mathbf{L}\mathbf{f}$, with $\mathbf{L}$ a known transition matrix.

2. Djafari's prior versus ICF prior

The classes of Djafari's prior laws were derived on an axiomatic base in [2]. The particular prior law used in this paper has the form:

$$p(\mathbf{f}|I_{\mathrm{Dja}}) \sim \exp(-\lambda \sum_i f_i - \mu \sum_i \ln(f_i)), \tag{1}$$

with f_i the value of the pixel i.

On the other hand Skilling and Gull used the well known monkey argument [4] to find the prior law . Using an elaborated monkey scheme for the creation of images, one finds the prior law [5]:

$$p(\mathbf{f}|I_{\mathrm{ICF}}) \sim \exp(\alpha S(\mathbf{f},\mathbf{m}) + \beta S(\tilde{\mathbf{m}},\mathbf{flat})), \tag{2}$$

with $\tilde{\mathbf{m}}$ the premodel, $\mathbf{m}$ the model and $\mathbf{flat}$ a flat model. The entropy $S(\mathbf{a},\mathbf{b})$ is given by $S(\mathbf{a},\mathbf{b}) = \sum_i(a_i - b_i - a_i \log \frac{a_i}{b_i})$ [7],[8]. We call this prior law the ICF prior (Intrinsic Correlation Function prior).

A. Mohammad-Djafari and G. Demoments (eds.), Maximum Entropy and Bayesian Methods, 393–398.

Thus we have two different prior laws with each containing two parameters: (α, β) for the ICF prior and (λ, μ) for Djafari's prior. Djafari uses the method of moments to determine these parameters [3]. During successive iterations, the first two moments of the present estimation of the image are used to estimate the λ and μ parameter values.

Gull [5] uses a marginalization procedure. In the original article [5] this marginalization procedure was developed for convolution data. In the next section a simplification is proposed to allow the application of the marginalization procedure in the case of projection data.

3. A marginalization procedure for projection data

The projection measurements of a PET scan are stored in an array, called the sinogram. Each horizontal line of the sinogram corresponds to projections along one specific view angle through the patient. The successive data points on one line of the sinogram match different translation locations of the projection stripes. This implies that a scan of a point source results in half a period of a sine in the sinogram [6].

If one follows the marginalization procedure presented in [5] to determine the parameters α and β, one needs the eigenvalues of a huge matrix $\mathbf{A}$ with elements:

$$A_{ij} = \sum_k \frac{1}{\sigma^2} L_{ki} L_{kj}, \tag{3}$$

where L_{ki} stands for the influence of the ith pixel on the kth data point. Here the data was supposed to follow Gaussian statistics with variance σ. The variances are considered equal for all data points.

When dealing with a spatially-invariant circulant convolution operation, one can perform the calculations with Fourier transforms as proposed in [5].

However, this approach can not be used for projection data. We propose some simplifications to allow the calculation of the eigenvalues. By assuming that the L_{ki}'s can only be 0 or 1, we suppose that a pixel is either seen or not seen at all by the detector pairs. If we ignore for a moment the σ^2 term, the term A_{ij} represents the sum of the data points on which pixel i AND pixel j have a simultaneous influence. But, as explained above, this is looking at the intersection of two half periods of a sine in the sinograms. Only two possible cases occur: the sines are identical or there is only one intersecting point. Thus the matrix $\mathbf{A}$ can be filled in with

$$\begin{aligned} A_{ij} &= 1, i \neq j & (4)\\ &= \theta, i = j, & (5) \end{aligned}$$

with θ the number of angles under which the patient is studied. For the marginalization procedure the eigenvalues of this matrix are needed. Fortunately the eigenvalues γ_i of this matrix $\mathbf{A}$ are easy to calculate:

$$\begin{aligned} \gamma_1 &= \theta + (n-1) & (6)\\ \gamma_i &= \theta - 1, \quad i = 2, \ldots, n, & (7) \end{aligned}$$

with n the number of pixels in the image. With these eigenvalues known, the α and β parameters can be calculated.

To conclude this section we mention some shortcomings of the proposed simplification.

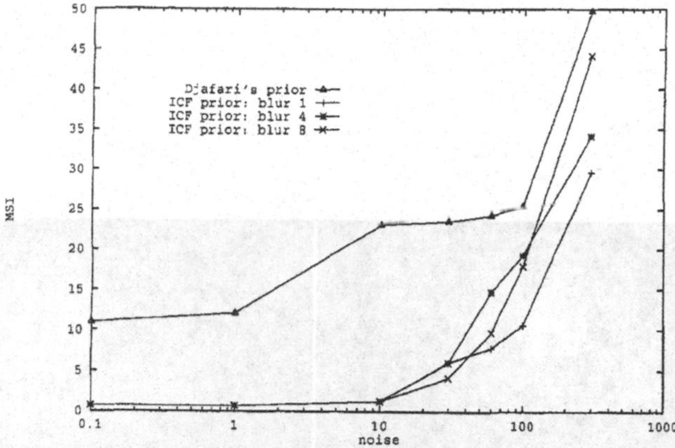

Figure 1: The mean squared difference MS between reconstructed image and reference image in function of different noise levels for different priors.

1. The intersection between two different half periods of a sine is only one point in the continuous case. When working with a discrete number of projections this number may well increase or decrease.

2. The marginalization procedure above was based on the Gauss statistics of the measurements. However, to model the PET-measurements we need Poisson statistics.

3. As mentioned before the transition matrix **L** was supposed known. However, the transition matrix **L** must generally account for other operations than the sole projection operation (convolution steps, attenuation correction [1],...), and is not completely determined.

4. Experiments

A 64x64 digitized image of a Shepp and Logan phantom is reconstructed from a sinogram with 64x64 elements. The phantom is represented in fig. 2. The pixels in the image have activities between 0 and 55 units. The projections have activities between 0-1100 units. Different amounts of Gaussian noise were added to this data. The reconstruction was done with a gradient descent method. (Similar observations can be made for the conjugate gradient method.) To achieve convergence a large number (50) of iterations was chosen.

Different blurring widths were tested. In fig. 1 the mean squared differences $MS = \sum_{i=1}^{n} \frac{(f_i - \hat{f}_i)^2}{n}$ between the reconstructed and the original image are represented, with $\hat{f}_i$ the reconstructed image, f_i the reference image and n the number of pixels in the image.

It is observed that with low noise levels the different blurs do not provide substantially different reconstructions. For higher noise levels a low blurring width (1 pixel) often produces acceptable images. Djafari's prior law produces images which are rather sensitive to noise added to the data. These images are worse than any of the images obtained with the ICF prior. The different images for 60 units of Gaussian noise added were represented in fig. 2. The image reconstructed with the Djafari's prior shows the main features of the

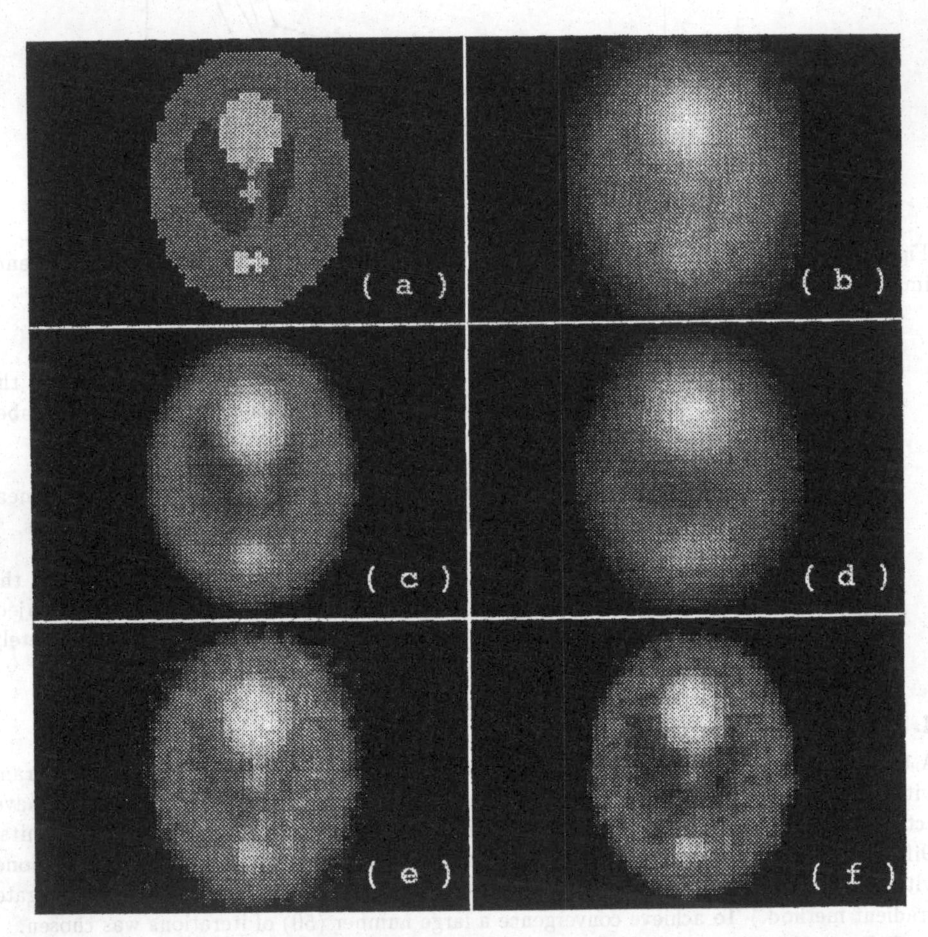

Figure 2: Images reconstructed with different priors for noise level of 60 units. The different priors are (a) Reference image (b) Djafari's prior with method of moments (c) ICF prior with blur width of 1 pixel (d) ICF prior with blur width of 4 pixel (e) ICF prior with blur width of 8 pixel (f) Djafari's prior with manually set μ and λ coefficients

phantom but is too uniform. When using the ICF prior we observe the following evolution with increasing blurring widths: a one pixel blurring width gives a rather smooth image which reproduces quite good the phantom. A blurring width of four pixels produces very smooth images only revealing the most striking features of the phantom. Finally a blurring width of 8 pixels yields images with a lot of features of the phantom, but the image is also rather noisy. This experimental behavior was predicted in [5].

5. Discussion and conclusion

By interpreting the following results one should be aware that the used program is not identical to the program of Djafari or the Cambridge program.

For all blurring widths studied here, the ICF prior with the proposed simplification for the marginalization procedure, performs better than Djafari's prior with the method of moments. The relative failure of the Djafari prior might be explained by the use of the method of moments with the wrong samples (posterior distribution). We were indeed able to reconstruct better images by manually setting the λ and μ parameters, see fig. 2(f).

Though the simplification of the marginalization procedure apparently gives satisfactory results for the applied reconstructions, some refinements are needed to deal with the shortcomings mentioned previously.

6. Acknowledgements

I would like to thank Dr. A. Mohammad-Djafari for the use of his gradient descent algorithm and for useful information provided by E-mail. Dr. Van Neck provided help for the determination of the regularization parameters in the ICF prior. One of the authors (P.D.) has a grant from "Instituut tot aanmoediging van het Wetenschappelijk Onderzoek in de Nijverheid en Landbouw" (IWONL, Brussels) and the other (I.L.) is research associate at the Belgian National Fund for Scientific Research (NFWO, Brussels).

References

[1] Desmedt P., Thielemans K., Lemahieu I., Vermeulen F., Vogelaers D., Colardyn F., *Measured attenuation correction using the Maximum Likelihood algorithm*, Medical Progress through Technology, 1991 , pp. 199-204

[2] Mohammad-Djafari A., Idier J., *Maximum entropy prior laws of images and estimation of their parameters*, in Maximum entropy and Bayesian methods, Laramie, Reidel , 1991, pp. 285-293

[3] Mohammad-Djafari A., *Maximum Likelihood estimation of the Lagrange parameters of the maximum entropy distribution*, to be published in Maximum entropy and Bayesian methods, 1992

[4] Frieden B.R., *Restoring with maximum likelihood and maximum entropy*, J. Opt. Soc. Am. 62, 1972, pp. 511-518

[5] Gull S.F., *Developments in entropy data analysis*, in Maximum entropy and Bayesian methods, Cambridge, Reidel , 1989, pp. 53 - 73

[6] Herman G.T., *Topics in applied physics Vol. 32: Image reconstruction from projections*, Ed. G.T. Herman, Springer-Verlag, 1979

[7] Skilling J., *The axioms of maximum entropy*, in Maximum entropy and Bayesian methods in science and engineering 1, Reidel , 1988, pp. 173 - 187

[8] Skilling J., *Classic maximum entropy*, in Maximum entropy and Bayesian methods, Cambridge, Reidel, 1989, pp. 45 - 52

FLEXIBLE PRIOR MODELS IN BAYESIAN IMAGE ANALYSIS

K. M. Hanson
Los Alamos National Laboratory, MS P940
Los Alamos, New Mexico 87545 USA
email: kmh@lanl.gov

ABSTRACT. A new class of prior models is proposed for Bayesian image analysis. This class of priors provides an inherent geometrical flexibility, which is achieved through a transformation of the coordinate system of the prior distribution or model into that of the object under analysis. Thus prior morphological information about the object being reconstructed may be adapted to various degrees to match the available measurements. An example of tomographic reconstruction illustrates the potential of this approach.

1. Introduction

One often encounters problems in the image analysis of objects that are ill-posed because of a lack of data. Bayesian methods help select the best of many possible solutions on the basis of the characteristics of the object that are known *a priori*. The prior knowledge concerning the object is incorporated in terms of prior probability distributions on the appropriate physical parameters. We wish to develop a method to take into account an object's approximate shape, when it is known beforehand. The model for the object being analyzed is allowed to alter its geometry to accommodate the data by warping the coordinate system of the prior model onto the coordinate system of the actual object. Thus the character of the edges of the model, as well as its morphology, are preserved. The proposed extension to the standard MAP technique overcomes its static definition of the prior providing flexibility in geometry and other characteristics. For a fuller description of the use of this flexible model, see [1, 2].

The power of this new approach to prior models is illustrated with an example of computed tomographic reconstruction in which the coordinate transformations are restricted to low-order polynomials.

2. Basic Formulation

We are given M discrete measurements that are linearly related to the amplitudes of the original image vector $\mathbf{f}$ of length N, degraded by additive noise: $\mathbf{g} = \mathbf{Hf} + \mathbf{n}$,

where $\mathbf{n}$ is the random noise vector, and $\mathbf{H}$ is the measurement matrix. From Bayes' law the negative logarithm of the posterior probability density is given by

$$-\log[p(\mathbf{f}|\mathbf{g})] = \phi(\mathbf{f}) = \Lambda(\mathbf{f}) + \Pi(\mathbf{f}) \,, \tag{1}$$

where the first term comes from the likelihood and the second term from the prior probability. For simplicity, a Gaussian distribution is chosen for the prior probability,

A. Mohammad-Djafari and G. Demoments (eds.), Maximum Entropy and Bayesian Methods, 399–406.

which is characterized by the known quantities $\bar{\mathbf{f}}$, the ensemble mean, and $\mathbf{R_f}$, the ensemble covariance matrix. The negative logarithm of the prior probability on $\mathbf{d}$, the deviation from the model $\mathbf{d} = \mathbf{f} - \bar{\mathbf{f}}$, may be written as

$$\Pi(\mathbf{d}) = \tfrac{1}{2}\mathbf{d}^{\mathrm{T}}\mathbf{R}_{\mathbf{d}}^{-1}\mathbf{d} \; , \tag{2}$$

Assuming additive Gaussian noise with a known covariance matrix $\mathbf{R_n}$, the negative log(likelihood) is just one half chi-squared

$$-\log[p(\mathbf{g}|\mathbf{d},\mathbf{a})] = \Lambda(\mathbf{d},\mathbf{a}) = \tfrac{1}{2}\chi^2 = \tfrac{1}{2}[\mathbf{g} - \mathbf{H}(\mathbf{d}+\bar{\mathbf{f}})]^{\mathrm{T}}\mathbf{R}_{\mathbf{n}}^{-1}[\mathbf{g} - \mathbf{H}(\mathbf{d}+\bar{\mathbf{f}})] \; , \tag{3}$$

which is quadratic in the residuals.

The full Bayesian solution is characterized by the posterior probability $p(\mathbf{f}|\mathbf{g})$. However, to represent the result with a single image, an appropriate choice is the image that maximizes the *a posteriori* probability, called the MAP estimate [3], which equivalently minimizes ϕ. The choice of the relative weights of the log-likelihood (3) and the log-prior (2) is critical, as it affects how well the information contained in the data is transferred to the observer of the image [4].

minimum of $\phi(\mathbf{f})$, that is, at the MAP solution $\nabla_{\mathbf{f}}\phi(\mathbf{f}) = 0$.

The problem with the preceding standard Bayesian formulation is that the model for the prior is usually considered to be geometrically fixed [5, 6]. To build flexibility into the prior, we consider $\bar{\mathbf{f}}$ to be a function of several parameters, represented by the vector $\mathbf{a}$, that is, $\bar{\mathbf{f}}(\mathbf{a})$. The parameters in vector $\mathbf{a}$ control the position, size, and shape of the prior distribution in a manner yet to be specified. Under the assumption that $\mathbf{a}$ and $\mathbf{d}$ are statistically independent, the prior on these new parameters $\Pi(\mathbf{a})$ is simply added to $\Pi(\mathbf{d})$. We take $\mathbf{d}$ and $\mathbf{a}$ to be the independent variables in the problem.

In reconstruction, we seek to estimate all pixel values in the original scene, that is $\mathbf{f}$. If there is no interest in the actual values of the parameters $\mathbf{a}$ and $\mathbf{d}$, probability theory says we must marginalize over these nuisance parameters. As the integration over so many parameters is computationally very difficult, the solution for $\mathbf{f}$ is calculated from the $\mathbf{a}$ and $\mathbf{d}$ that minimize ϕ, for which $\nabla_{\mathbf{d}}\phi = 0$ and $\nabla_{\mathbf{a}}\phi = 0$, provided the solution is not otherwise constrained. The MAP solution can be found by the method of steepest descent or some better method such as that of conjugate gradients. It is unlikely that one can guarantee that the uniqueness of the solution for ϕ. We must rely on a knowledgeable choice of initial parameters to guide the solution to a meaningful result.

3. Elastic Sheet Analogy

Suppose that the prior on the image amplitude, specified in terms of $\bar{\mathbf{f}}$ and $\mathbf{R_d}$ in the present case, is given as a function of the spatial coordinates (x', y'). Geometrical flexibility of the prior is accomplished by transforming the (x, y) coordinates of the final image $\mathbf{f}$ to the original (x', y') coordinates:

$$x' = x + u(x,y) \; ; \quad y' = y + v(x,y) \; , \tag{4}$$

where u and v are the displacements of x and y. The warp could equally well be accomplished by means of the inverse transformation, that is, from (x', y') to (x, y). The choice is mostly a matter of computational convenience and depends on the specific situation. Obviously this coordinate mapping can be arbitrary. However, it should be restricted in some way to reflect the realistic range of possibilities for the warped shape of the prior model.

The warp should meet the following general requirements. The coordinate transformation should be one-to-one in the domains of interest, which implies the transformation is invertible. Without any previously known preference for orientation or position, the transformation should be isotropic and stationary (homogeneous). The formulation for the warp should be independent of the rectilinear coordinate system in which it is stated. Neighborhood relationships should be maintained so the transformation should be continuous. Homogeneous translations and rotations should be unimportant. We will propose a mechanism to control the warp that meets all these criteria.

It is natural to draw an analogy between a 2D warp and the distortion of a sheet of elastic material that is constrained to lie in a plane. Then the constraints placed on the warp are analogous to the elastic properties of the material being distorted. In material mechanics the strain corresponds to the first derivative of the mapping. For example, $\frac{\partial u}{\partial x}$ specifies the amount of linear expansion in the x-direction producing normal tensile strain. For deformations in linear materials obeying Hooke's law, the stress induced in the material is proportional to the strain. So the strain energy density, found by integrating the stress with respect to the strain, is proportional to the square of the strain.

We would like to generalize this notion of strain energy density for application to coordinate transformations. We employ tensor notation to achieve independence of the rectilinear coordinate system and allow easy extension to 3D. The displacement vector comprised of u and v is designated u_i. The subscripts run over the coordinate indices, i.e. $1 \sim x, 2 \sim y$. The strain is expressed as a symmetric tensor $e_{ij} = \frac{1}{2}(u_{i,j} + u_{j,i})$, where the comma denotes a derivative with respect to the appropriate coordinate.

plane. Exceptions relevant to 3D will be given in parentheses. Therefore, $i = 1, 2(3)$ and similarly for j.

Thus, the strains are

$$e_{11} = \frac{\partial u}{\partial x}, \; e_{22} = \frac{\partial v}{\partial y}, \; e_{12} = e_{21} = \frac{1}{2}\left(\frac{\partial u}{\partial y} + \frac{\partial v}{\partial x}\right) \tag{5}$$

The first two quantities are the normal strain in the x- and y-directions and the last one is the usual expression for shear strain.

For small strains the general expression for the strain energy density in a linear isotropic medium is [7]

$$\begin{aligned} w &= \tfrac{1}{2}\lambda(e_{ii})^2 + \mu e_{ij} e_{ij} \\ &= \tfrac{1}{2}\lambda(e_{11} + e_{22})^2 + \mu(e_{11}^2 + e_{22}^2 + 2e_{12}^2), \end{aligned} \tag{6}$$

where λ and μ are the Lamé constants of elasticity. As usual, repeated subscripts within an expression imply summation over them. As μ controls the change in angles induced by the transformation, it the called the modulus of rigidity or shear modulus. We note that by construction the expression for w is invariant under rotation or translation of the (x, y) coordinate system.

The quantity $e_{ii} = \vartheta$ is called the area dilation, or volume dilation in 3D. It is linearly related to the Jacobian of the transformation from (x, y) to (x', y'), which is the determinant

$$J \equiv \frac{\partial(x',y')}{\partial(x,y)} = \frac{\partial x'}{\partial x}\frac{\partial y'}{\partial y} - \frac{\partial x'}{\partial y}\frac{\partial y'}{\partial x} = |\delta_{ij} + u_{i,j}| \, , \tag{7}$$

where δ_{ij} is the usual Kronecker delta. The Jacobian gives the ratio of the change in area of a differential element produced by the transformation. For small strains, $J \approx 1 + \vartheta$. For the transformation to be invertible (one-to-one), it is necessary that the Jacobian be nonzero.

The Lamé constants are a property of the material. Their relation to Young's modulus E and Poisson's ratio ν, more typically used in engineering, is given by

$$E = \frac{\mu(3\lambda + 2\mu)}{\lambda + \mu} \, , \quad \nu = \frac{\lambda}{2(\lambda + \mu)} \, . \tag{8}$$

Both λ and μ are nonnegative for real materials, which assures $w \geq 0$. Poisson's ratio specifies the relative amount of contraction in the direction perpendicular to an applied tension. In 2D the upper limit of $\nu = 1$ is attained by incompressible materials. Its upper limit in 3D is $\frac{1}{2}$.

4. Prior on the Warp

We propose to use a Gibbs' distribution [8] for the prior probability on the warp, which is proportional to $\exp(-W)$, where W is the total strain energy of the warp, given for small deformations by

$$W = \int w \, d\tau \, , \tag{9}$$

where $d\tau = dx\, dy$ is the differential area in (x, y). Then the negative log-prior on the warp is simply W. The role of the Lamé constants in Eq. (6) is then clearly identified as that of specifying the strength of the log-prior on the warp relative to the log-prior probability on the amplitude and relative to the log-likelihood. While the above expressions hold only for small strains, valid evaluation for large strains may be achieved by numerically integrating (6) and (9).

It should be emphasized that the elastic constants for the conceptual physical model of the warp are not related to those of the material from which the object being studied is actually composed. Indeed, the choices for the Lamé constants are not restricted by the usual constraints regulating physical systems [7]. Instead, their selection should reflect the range of reasonable configurations the prior distribution can assume for the class of objects being imaged. Poisson's ratio might be set to zero, if a stretch in one direction is not expected to influence what happens in the other direction. Even 'unphysical' values are legitimate. For example, $\nu = -1$, would correspond to a similarity transformation, which maintains shapes, because that would indicate that an expansion in one direction is most likely accompanied by an equal expansion in the orthogonal direction. Also note that, because material is not actually being distorted in this warping process, it is not necessary to scale the amplitude of the prior by the Jacobian of the transformation to conserve mass.

If it were deemed desirable to maintain angles in the warping, as in conformal mapping, then no shear would be allowed, even locally. This constraint could be enforced by requiring the shear (e_{12}) to be zero, which would implicitly place constraints on the parameters of the warp. Alternatively, conformality can be achieved by making μ very large compared to λ [9].

Note that, because (6) is based on a linear stress-strain relationship, there is nothing that precludes the Jacobian from going to zero. By assuming that the stress σ is proportional to a relative change in length of a differential element, that is $\sigma \propto \epsilon/(1+\epsilon)$, where ϵ is the strain, the strain energy density would be proportional to $\epsilon - \log(1+\epsilon)$, which avoids the dreaded condition $\epsilon = -1$. This expression is obviously related to the Burg entropy.

The constraint given in Eq. (9) is expressed in a general form that does not imply a representation for the warp. Wherever representation is used, the parameters involved correspond to the vector $\mathbf{a}$ in Sec. 2. Ultimate control over local distortion can be had through the use of a finite-element representation to describe the mapping [10]. Such an approach, taken by Brackbill and Saltzman [9], seems particularly well suited to the present method, because images are typically represented by pixels that give the image values on Cartesian grids. Of course, the finite elements used to represent the image need not coincide with the pixel representation. However, such a representation for the warp would preserve line elements located between pixels, and thus would be useful in maintaining a prior defined on those line elements [8].

Some simplification is achieved if we express the coordinate transformation (4) as a polynomial expansion

$$u = \sum_{mn} a_{mn} x^m y^n \; ; \quad v = \sum_{mn} b_{mn} x^m y^n \, , \tag{10}$$

where the coefficients a_{mn} and b_{mn} are represented as elements in the parameter vector $\mathbf{a}$ introduced in Sec. 2. Use of (10) results in expressions for the strains that are likewise polynomials with coefficients that are quadratic in the a_{mn} and b_{mn} parameters. Then the total strain energies W will also be quadratic in these warp parameters, consistent with a Gaussian prior probability distribution. Although the polynomial expansion for the warp is convenient, it suffers from a few fundamental difficulties. First, it does not provide much local flexibility without including high orders. Second, the mapping will inevitably cease to be invertible at some values of (x, y), when nonzero second- or higher-order terms are included.

5. The Knockwurst Example

In this example we assume that we are given five sets of parallel projections of the sausage-shaped object shown in Fig. 1. The projections contain 128 samples each and are taken at 36° increments in angle. Note that an assumption of right-left symmetry makes two of the five views redundant so these reconstructions are effectively based on only three distinct views!

The tomographic reconstructions are shown in Fig. 1. The 128×128 reconstruction obtained with the unconstrained Algebraic Reconstruction Technique (ART) [11] is predictably very poor. To exaggerate the extent to which flexibility can be incorporated in the reconstruction process, we use a circle of unit amplitude for $\bar{\mathbf{f}}$, the mean of the prior

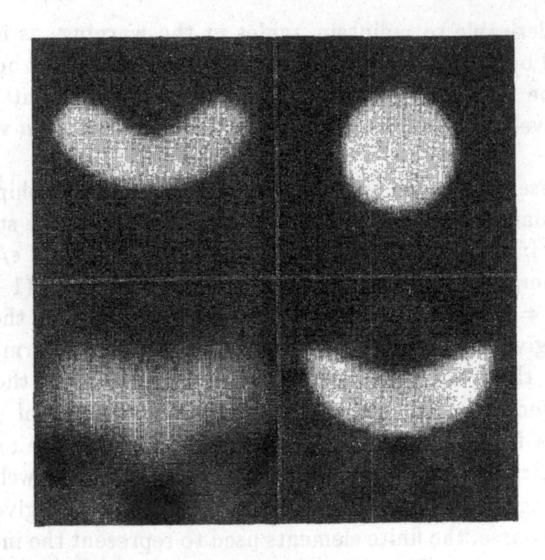

Figure 1:

The ART reconstruction (lower-left) obtained from five noiseless, parallel views taken at angular increments of 36° poorly reproduces the original sausage-shaped object (upper-left). The MAP reconstruction (lower-right) obtained from the same data is based on the circular prior distribution (upper-right) subject to a polynomial warp of third order.

probability distribution on amplitude. The important aspects that we wish to include in the reconstruction are the expected sharp boundary of the object and its constant amplitude. To permit a fair degree of flexibility, a third-order polynomial coordinate transformation is employed. The assumed right-left symmetry reduces the number of warp coefficients from 20 to 11. In the present example, the full expression given by (6) and (9) is replaced by the sum of the squared values of the a_{mn} and b_{mn}. The coefficients in this sum are chosen to be very small to allow maximum warping. The resulting 32×32 MAP reconstruction (LR) reasonably matches the original, given that the available data consist of only three distinct views.

6. Discussion

The choice of the strength of the priors on the warp parameters is critical in determining the shape of the solution. One way to determine the Lamé constants is through the posterior probability, in the same way as Gull suggests for finding the strength of the entropic prior [12]. A better method would be to base the choice on *bona fide* prior knowledge about the objects being imaged. More detailed prior information might even allow specification of the elastic constants as a function of position.

The linear-material approach taken here may be extended in many ways. The assumption of stress being proportional to relative change in dimension, mentioned in Sec. 4, leads to an energy density of entropic form. Other possibilities include endowing the fictitious material being warped with nonlinear, ductile, or even elasto-viscous behavior. The flexible structures may be 1D (or 2D in a 3D problem) as it may be most appropriate to warp lines or surfaces.

The notion of introducing geometrical flexibility into the priors used in Bayesian analysis clearly extends to all types of image analysis. Flexible models have been used for several years in computer vision [13, 14] and to match MRI brain images to generic shapes from a brain atlas [15, 16].

7. Acknowledgments

I acknowledge many helpful prior discussions with Robert F. Wagner, Kyle J. Myers, Jerry U. Backbill, James C. Gee, and David R. Wolf. This work was supported by the United States Department of Energy under contract number W-7405-ENG-36.

References

[1] K. M. Hanson. Reconstruction based on flexible prior models. *Proc. SPIE*, 1652:183–191, 1992.

[2] K. M. Hanson. Bayesian reconstruction based on flexible prior models. *to be published in J. Opt. Soc. Amer.*, 1992.

[3] A. P. Sage and J. L. Melsa. *Estimation Theory with Applications to Communications and Control.* Robert E. Krieger, Huntington, 1979.

[4] R. F. Wagner, K. J. Myers, and K. M. Hanson. Task performance on constrained reconstructions: human observers compared with suboptimal Bayesian performance. *Proc. SPIE*, 1652:352–362, 1992.

[5] B. R. Hunt. Bayesian methods in nonlinear digital image restoration. *IEEE Trans. Comp.*, C-26:219–229, 1977.

[6] K. M. Hanson. Bayesian and related methods in image reconstruction from incomplete data. In Henry Stark, editor, *Image Recovery: Theory and Application*, pages 79–125. Academic, Orlando, 1987.

[7] I. S. Sokolnikoff. *Mathematical Theory of Elasticity.* McGraw-Hill, New York, 1956.

[8] S. Geman and D. Geman. Stochastic relaxation, Gibbs distributions, and the Bayesian restoration of images. *IEEE Trans. Pattern Anal. Machine Intell.*, PAMI-6:721–741, 1984.

[9] J. U. Brackbill and J. S. Saltzman. Adaptive zoning for singular problems in two dimensions. *J. Comput. Phys.*, 46:342–368, 1982.

[10] R. D. Cook. *Concepts and Applications of Finite Element Analysis.* John Wiley & Sons, New York, 1974.

[11] R. Gordon, R. Bender, and G. Herman. Algebraic reconstruction techniques for three-dimensional electron microscopy and x-ray photography. *J. Theor. Biol.*, 29:471–481, 1970.

[12] S. F. Gull. Developments in maximum-entropy data analysis. In J. Skilling, editor, *Maximum Entropy and Bayesian Methods*, pages 53–71. Kluwer Academic, 1989.

[13] R. Szeliski. Probabilistic modeling of surfaces. *Proc. SPIE*, 1570:154–165, 1991.

[14] R. Szeliski and D. Terzopoulos. Physically-based and probabilistic models for computer vision. *Proc. SPIE*, 1570:140–152, 1991.

[15] R. Bajcsy and S. Kovačič. Multiresolution elastic matching. *Comput. Vision*, 46:1–21, 1989.

[16] J. C. Gee, M. Reivich, L. Bilaniuk, D. Hackney, R. Zimmerman, S. Kovačič, and R. Bajcsy. Evaluation of multiresolution elastic matching using MRI data. *Proc. SPIE*, 1445:226–234, 1991.

GENERALIZED MAXIMUM ENTROPY. COMPARISON WITH CLASSICAL MAXIMUM ENTROPY

A.T. Bajkova
Institute of Applied Astronomy
Russian Academy of Sciences
St.Petersburg 197042 Russia

ABSTRACT. In this paper a generalized algorithm of maximum entropy method for reconstruction of functions of any type (not only real non-negative, but real functions with alternating signs and complex ones as well) proposed in (Bajkova, 1991, 1992), is considered. The generalized maximum entropy method is based on an assumption about independence of real and imaginary parts of a complex function, representation of sequences with alternating signs as difference between positive and negative parts and appropriate modification of the optimized entropy functional in order to avoid overlapping between positive and negative parts. Generalized maximum entropy method is compared with classical maximum entropy for reconstruction of real non-negative distribution from incomplete spectrum data. Much higher quality of reconstruction is demonstrated.

1. Introduction

Maximum entropy method (MEM) in its classical form can be used for reconstruction of real non-negative functions such as intensity distributions in astronomy, optics, tomography etc. But in different areas of physics and engineering there exist problems of reconstruction of functions with alternating signs as well as complex ones. One of these problems arises, for example, in radio holography in the cases when it is required to reconstruct coherent source field distribution from incomplete and noisy hologram (spectrum) data. In this paper a generalized algorithm of MEM (generalized maximum entropy method (GMEM)) for reconstruction of functions of any type is considered.

Firstly, let us consider classical MEM in Shannon formulation in discrete form for real non-negative two-dimensional sequences

$$\min \sum_m \sum_l x_{ml} \ln(x_{ml}), \tag{1}$$

$$\sum_m \sum_l x_{ml}\, a_{ml}^{nk} = A_{nk},$$

A. Mohammad-Djafari and G. Demoments (eds.), Maximum Entropy and Bayesian Methods, 407–414.

$$\sum_m \sum_l x_{ml}\, b_{ml}^{nk} = B_{nk}, \tag{2}$$

$$x_{ml} \geq 0, \tag{3}$$

where a_{ml}^{nk}, b_{ml}^{nk} are constants, A_{nk} and B_{nk} are the data measured (in radio astronomy they are real and imaginary parts of visibility function respectively).

The problem (1)-(3) is a nonlinear optimization problem with constraints in the form of equality (2) and inequality (3). Using Lagrange method it is easy to obtain the solution for x_{ml}

$$x_{ml} = \exp\,(-\sum_n \sum_k (\alpha_{nk}\; a_{ml}^{nk} + \beta_{nk}\; b_{ml}^{nk}) - 1), \tag{4}$$

where α_{nk} and β_{nk} are Lagrange or dual factors which can be found by optimization of dual functional without supplementary conditions:

$$\max \sum_m \sum_l x_{ml} + \sum_n \sum_k (\alpha_{nk} A_{nk} + \beta_{nk} B_{nk}) \rightarrow$$

$$\max \sum_m \sum_l \exp\,(-\sum_n \sum_k (\alpha_{nk}\; a_{ml}^{nk} + \beta_{nk}\; b_{ml}^{nk}) - 1) + \sum_n \sum_k (\alpha_{nk}\; A_{nk} + \beta_{nk}\; B_{nk}). \tag{5}$$

Solution for x_{ml} in accordance with (4) is always non-negative. The purpose of this paper is to show how the classical maximum entropy method (1) - (3) can be modified in order to obtain a generalized algorithm suitable for reconstruction of sequences of any type (not only real non-negative, but real sequences with alternating signs as well as complex ones), and comparison with the classical MEM for reconstruction of real non-negative distributions.

2. Reconstruction of real function with alternating signs

In this case optimization problem can not be written as (1), because x_{ml} can take negative values. Therefore it is proposed to modify (1) in the following way

$$\min \sum_m \sum_l |x_{ml}| \ln |x_{ml}|, \tag{6}$$

where $|*|$ is the absolute value of $*$.

To avoid the absolute value in (6) and reduce the optimization problem to the conventional one, let us represent sequence x_{ml} as difference between positive and negative parts:

$$x_{ml} = y_{ml} - z_{ml}, \tag{7}$$

where both y_{ml} and z_{ml} are non-negative. In addition, let the sequences y_{ml} and z_{ml} satisfy the following conditions

$$\begin{aligned} &\text{if } x_{ml} > 0 \quad \text{then } z_{ml} \rightarrow 0 \quad \text{and} \quad x_{ml} \approx y_{ml}, \\ &\text{if } x_{ml} < 0 \quad \text{then } y_{ml} \rightarrow 0 \quad \text{and} \quad x_{ml} \approx -z_{ml}. \end{aligned} \tag{8}$$

Then optimization problem (6) can be rewritten as follows

$$\min \sum_m \sum_l (y_{ml} \ln(y_{ml}) + z_{ml} \ln(z_{ml})). \tag{9}$$

Let (9) be modified as

$$\min \sum_m \sum_l (y_{ml} \ln(a y_{ml}) + z_{ml} \ln(a z_{ml})), \tag{10}$$

where a is a parameter which can be chosen, as will be shown below, so that the conditions (8) are satisfied.

The supplementary conditions (2) can be rewritten in the following way

$$\sum_m \sum_l (y_{ml} - z_{ml})\, a_{ml}^{nk} = A_{nk}, \tag{11}$$

$$y_{ml},\ z_{ml} \geq 0. \tag{12}$$

The analysis of the solution of the optimization problem (10)-(12) shows that the value of the parameter a influences the realization of the conditions (8) and therefore the reconstruction quality.

Let us represent this problem of conditional optimization (10)-(12) as a simpler dual nonconditional optimization problem.

Constructing a Lagrange functional using (10) and (11) we obtain

$$L = \sum_m \sum_l (y_{ml} \ln(a y_{ml}) + z_{ml} \ln(a z_{ml})) + \sum_n \sum_k \alpha_{nk} (\sum_m \sum_l (y_{ml} - z_{ml})\, a_{ml}^{nk} - A_{nk}),$$

where α_{nk} are Lagrange or dual factors.

Let us find the minimum of L from the necessary extremum existence condition

$$\frac{\partial L}{\partial y_{ml}} = 0, \quad \frac{\partial L}{\partial z_{ml}} = 0:$$

$$\frac{\partial L}{\partial y_{ml}} = 1 + \ln(a) + \ln(y_{ml}) + \sum_n \sum_k \alpha_{nk}\, a_{ml}^{nk} = 0,$$

$$\frac{\partial L}{\partial z_{ml}} = 1 + \ln(a) + \ln(z_{ml}) - \sum_n \sum_k \alpha_{nk}\, a_{ml}^{nk} = 0 \quad \rightarrow$$

$$\ln(y_{ml}) = -\sum_n \sum_k \alpha_{nk} a_{ml}^{nk} - 1 - \ln(a), \quad \ln(z_{ml}) = \sum_n \sum_k \alpha_{nk} a_{ml}^{nk} - 1 - \ln(a) \quad \rightarrow$$

$$y_{ml} = \exp(-\sum_n \sum_k \alpha_{nk} a_{ml}^{nk} - 1 - \ln(a)), \tag{13}$$

$$z_{ml} = \exp(\sum_n \sum_k \alpha_{nk} a_{ml}^{nk} - 1 - \ln(a)). \tag{14}$$

As can be seen from (13), (14), conditions (12) are satisfied.

By substituting (13), (14) into Lagrange functional we obtain dual nonconditional optimization problem

$$\min \sum_m \sum_l (y_{ml}(-\sum_n \sum_k \alpha_{nk} a_{ml}^{nk} - 1 - \ln(a) + \ln(a))+$$

$$+z_{ml}(\sum_n \sum_k \alpha_{nk} a_{ml}^{nk} - 1 - \ln(a) + \ln(a)))+$$

$$+\sum_n \sum_k \alpha_{nk} \sum_m \sum_l (y_{ml} - z_{ml}) a_{ml}^{nk} - \sum_n \sum_k \alpha_{nk} A_{nk} \rightarrow$$

$$\min \sum_m \sum_l (-y_{ml} - z_{ml}) - \sum_n \sum_k \alpha_{nk} A_{nk} \rightarrow$$

$$\max \sum_m \sum_l (y_{ml} + z_{ml}) + \sum_n \sum_k \alpha_{nk} A_{nk}. \tag{15}$$

Dual factors α_{nk} are determined by solving (15).

The solutions (13), (14) show a peculiarity: the product of y_{ml} and z_{ml} depends on the parameter a in the following way

$$y_{ml}\, z_{ml} = \exp(-2 - 2\ln(a)) = K(a). \tag{16}$$

From the above expression it is clear that $K(a)$ influences the accuracy of the realization of conditions (8) and, consequently, the reconstruction quality. The increase of a results in the decrease of $K(a)$ and the improvement of the reconstruction quality.

3. Reconstruction of complex functions

Since the real and imaginary parts of the complex function are real functions with alternating signs, let us represent corresponding unknown complex sequence in the following way

$$r_{ml} + jq_{ml} = (x_{ml} - y_{ml}) + j(z_{ml} - v_{ml}), \tag{17}$$

where

$$x_{ml},\ y_{ml},\ z_{ml},\ v_{ml} \geq 0. \tag{18}$$

By analogy with (8) let the sequences x_{ml}, y_{ml}, z_{ml} and v_{ml} satisfy the following conditions

$$\begin{aligned}
&\text{if } r_{ml} > 0 \quad \text{then } y_{ml} \rightarrow 0 \quad \text{and} \quad r_{ml} \approx x_{ml},\\
&\text{if } r_{ml} < 0 \quad \text{then } x_{ml} \rightarrow 0 \quad \text{and} \quad r_{ml} \approx -y_{ml},\\
&\text{if } q_{ml} > 0 \quad \text{then } v_{ml} \rightarrow 0 \quad \text{and} \quad q_{ml} \approx z_{ml},\\
&\text{if } q_{ml} < 0 \quad \text{then } z_{ml} \rightarrow 0 \quad \text{and} \quad q_{ml} \approx -v_{ml}.
\end{aligned} \tag{19}$$

Assuming that r_{ml} and q_{ml} are independent of one another it is proposed to minimize the following functional

$$\min \sum_m \sum_l (|r_{ml}| \ln |r_{ml}| + |q_{ml}| \ln |q_{ml}|). \tag{20}$$

If conditions (19) are satisfied, the optimization problem can be rewritten by analogy with (10)-(12) in the following way

$$\min \sum_m \sum_l (x_{ml} \ln(a x_{ml}) + y_{ml} \ln(a y_{ml}) + z_{ml} \ln(a z_{ml}) + v_{ml} \ln(a v_{ml})), \tag{21}$$

$$\sum_m \sum_l ((x_{ml} - y_{ml})\, a_{ml}^{nk} - (z_{ml} - v_{ml})\, b_{ml}^{nk}) = A_{nk},$$

$$\sum_m \sum_l ((x_{ml} - y_{ml})\, b_{ml}^{nk} + (z_{ml} - v_{ml})\, a_{ml}^{nk}) = B_{nk}, \tag{22}$$

$$x_{ml},\; y_{ml},\; z_{ml},\; v_{ml} \geq 0, \tag{23}$$

where a_{ml}^{nk}, b_{ml}^{nk} are constants; A_{nk}, B_{nk} are real and imaginary parts of the measured complex samples respectively.

The solution of the optimization problem (21)-(23) is expressed as

$$x_{ml} = \exp(-\sum_n \sum_k (\alpha_{nk} a_{ml}^{nk} + \beta_{nk} b_{ml}^{nk}) - 1 - \ln(a)),$$

$$y_{ml} = \exp(\sum_n \sum_k (\alpha_{nk} a_{ml}^{nk} + \beta_{nk} b_{ml}^{nk}) - 1 - \ln(a)),$$

$$v_{ml} = \exp(-\sum_n \sum_k (\alpha_{nk} b_{ml}^{nk} - \beta_{nk} a_{ml}^{nk}) - 1 - \ln(a)), \tag{24}$$

$$z_{ml} = \exp(\sum_n \sum_k (\alpha_{nk} b_{ml}^{nk} - \beta_{nk} a_{ml}^{nk}) - 1 - \ln(a)).$$

Dual factors α_{nk} and β_{nk} can be found by optimization of the following dual functional without supplementary conditions

$$\max \sum_m \sum_l (x_{ml} + y_{ml} + z_{ml} + v_{ml}) + \sum_n \sum_k (\alpha_{nk} A_{nk} + \beta_{nk} B_{nk}). \tag{25}$$

As can be seen from (24), solutions for x_{ml}, y_{ml}, z_{ml}, v_{ml} are connected by

$$x_{ml}\, y_{ml} = z_{ml}\, v_{ml} = \exp(-2 - 2\ln(a)) = K(a)$$

similar to (16).

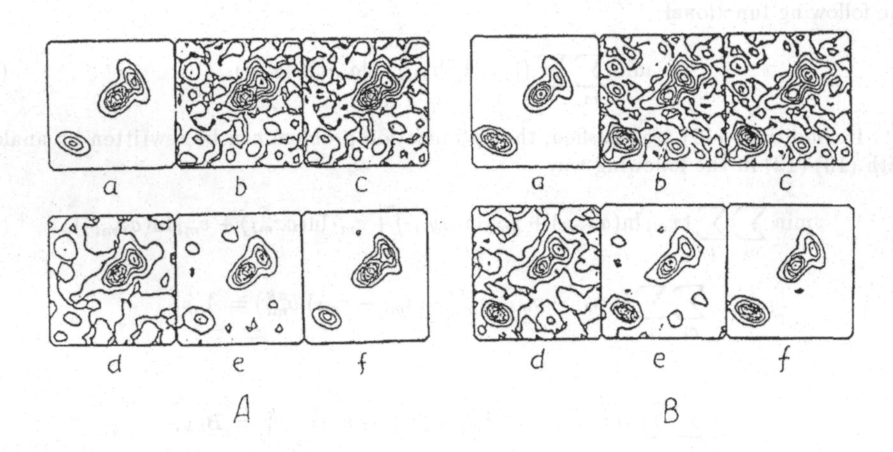

Fig. 1. Reconstruction of complex distribution with three gaussian-like components. (A) absolute value of the real part; (B) absolute value of the imaginary part; (a) object; (b) "dirty" map; (c) reconstructed map: $a = 1$; (d) reconstructed map: $a = \exp(2)$; (e) reconstructed map: $a = \exp(5)$; (f) reconstructed map: $a = \exp(7)$.

By changing a it is possible to control the reconstruction quality. The increase of a leads to the decrease of $K(a)$ and to closer realization of conditions (19), and therefore to improving the reconstruction quality.

Simulation results for complex object with gaussian-like components for different values of the parameter a are shown in Fig. 1 and 2.

4. Comparison of GMEM and classical MEM solutions

Comparison of the classical MEM and the GMEM is a question of great interest for the problem of image reconstruction from incomplete and noisy complex spectrum data. Let us compare the MEM and GMEM solutions for real non-negative distributions.

Using the GMEM for reconstruction of real non-negative distributions we seek unknown sequences in a more general complex form.

From expressions (5) and (25) it is seen that in the cases of the MEM and the GMEM we have different dual optimization problems. Therefore it is expected that corresponding solutions for $x_{m,l}$ are also different. Difference between the MEM and GMEM approaches was successfully proved by simulation technique. One of the numerous results, namely, for an object with two continuous gaussian-like components is shown in Fig. 1. As is seen, the classical MEM sharpens the smooth component of the object. In dual Fourier domain this fact relates to over-reconstruction of the spectrum. As is seen in the pictures, the GMEM allows to reconstruct the image much closer to the object than the classical MEM.

This phenomenon can be explained by the fact of the appearance in the expression for minimized functional (15) of the additional term for $z_{m,l}$ (or $v_{m,l}$) where dual variables α_{nk} and β_{nk} enter differently as compared with x_{ml} (or $y_{m,l}$). It allows to include additional

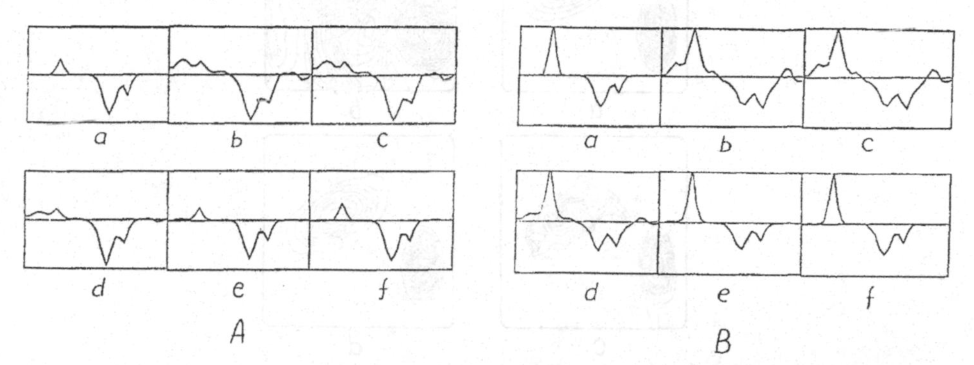

Fig. 2. Reconstruction of complex distribution with three gaussian-like components (diagonal cross-sections corresponding to Fig.1). (A) real part; (B) imaginary part; (a) object; (b) "dirty" map; (c) reconstructed map: $a = 1$; (d) reconstructed map: $a = \exp(2)$; (e) reconstructed map: $a = \exp(5)$; (f) reconstructed map: $a = \exp(7)$.

information (about imaginary part of an object) for finding α_{nk} and β_{nk} into the process of solving optimization problem, which leads to more correct determination of the dual parameters. On the contrary, in the case of the classical MEM the information about the imaginary part of an object is not taken into account explicitly.

There is analogy with Fourier transform here. Indeed, expressions for $x_{m,l}$ and $z_{m,l}$ can be rewritten as (here $a = e^{-1}$):

$$Lx_{ml} = \ln(x_{ml}) = -\sum_n \sum_k (\alpha_{nk} a_{ml}^{nk} + \beta_{nk} b_{ml}^{nk}),$$

$$Lz_{ml} = \ln(z_{ml}) = \sum_n \sum_k (\alpha_{nk} b_{ml}^{nk} - \beta_{nk} a_{ml}^{nk}).$$

In the limiting case, when $m, l, n, k = 1, ..., N$, complex two-dimensional sequence (Lx_{ml}, Lz_{ml}) is coupled with complex two-dimensional sequence $(\alpha_{nk}, \beta_{nk})$ via Fourier transform and $(\alpha_{nk}, \beta_{nk})$ is determined uniquely by (Lx_{ml}, Ly_{ml}) and vice versa.

Thus, for image reconstruction from incomplete Fourier spectrum it is more correct to seek solution in complex form (using the GMEM) even in the case of a real object, because Fourier transform is one-to-one transform of functions determined generally in complex domain.

9. Conclusions

In this paper a generalized algorithm of maximum entropy method suitable for reconstruction of complex distributions is considered. The generalized maximum entropy

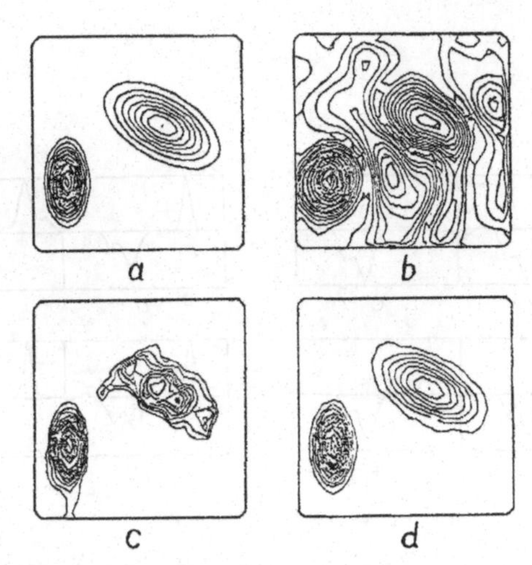

Fig. 3. Comparison of GMEM and classical MEM solutions. (a) object; (b) "dirty" image; (c) image reconstructed by MEM; (d) image reconstructed by the GMEM.

method is based on assumption about independence of real and imaginary parts of complex distribution sought for, representation of sequences with alternating signs as difference between positive and negative parts and appropriate modification of the optimized entropy functional in order to avoid overlapping between positive and negative parts. Comparison with the classical maximum entropy for reconstruction of real non-negative images from incomplete spectra data is made. It is shown that the generalized maximum entropy method ensures much more exact reconstruction of extensive details of an object and that application of the generalized maximum entropy to image reconstruction from incomplete and noisy complex spectrum data is more motivated than the classical one even in the case of real non-negative objects.

ACKNOWLEDGMENTS. I would like to thank Prof. B.R.Frieden for discussion of the method and simulation results.

REFERENCES

Bajkova, A.T.: 1991, 'The reconstruction algorithm of two-dimensional field distribution in radio holography based on maximum entropy method', in *Proceedings of the International Workshop on Holography Testing of Large Radio Telescopes*, Nauka, Leningrad.

Bajkova, A.T.: 1992, 'The generalization of maximum entropy method for reconstruction of complex functions', *Astron. and Astrophys. Tr.* 1, 313.

BINARY TASK PERFORMANCE ON IMAGES RECONSTRUCTED USING MEMSYS 3: COMPARISON OF MACHINE AND HUMAN OBSERVERS

K. J. Myers and R. F. Wagner
Center for Devices & Radiological Health/FDA
HFZ-142, Rockville MD 20857 USA

K. M. Hanson
Los Alamos National Laboratory, MS P940
Los Alamos, New Mexico 87545 USA
email: kmh@lanl.gov

ABSTRACT. We have previously described how imaging systems and image reconstruction algorithms can be evaluated on the basis of how well binary-discrimination tasks can be performed by a machine algorithm that "views" the reconstructions [1, 2]. The present work examines the performance of a family of algorithmic observers viewing tomographic images reconstructed using the Cambridge Maximum Entropy software, MEMSYS 3. We investigate the effects on the performance of these observers due to varying the parameter α, which controls the strength of the prior in the iterative reconstruction technique. Measurements on human observers performing the same task show that they perform comparably to the best machine observers in the region of highest machine scores, i.e., smallest values of α. For increasing values of α, both human and machine observer performance degrade. The falloff in human performance is more rapid than that of the machine observer, a behavior common to all such studies of the so-called psychometric function.

1. Introduction

It has been recognized for several decades that the assessment of medical images or medical imaging systems requires the specification of a task to be performed using the images. It has also been recognized that the study of task performance may be expensive and time consuming because of the need for "ground truth" against which to judge the performance of the task, and the need for a sufficient number of images and/or observers to obtain statistical significance in the results. These considerations have led to the study of task performance by machine or algorithmic observers. The question of the comparative performance of such machine observers relative to the performance of the human observer then naturally arises.

In this work images are obtained from reconstructions derived from simulations of limited-angle two-dimensional tomography. The assessment of the images proceeds according to the paradigm presented by Hanson [1]: A large number of images are generated according to a Monte Carlo technique; a binary task is specified and performed by either a machine or a human observer; and the performance is scored according to either the method of the receiver operating characteristic (ROC) curve or the method of the two-alternative-forced-choice (2AFC) [3, 4].

A. Mohammad-Djafari and G. Demoments (eds.), Maximum Entropy and Bayesian Methods, 415–421.

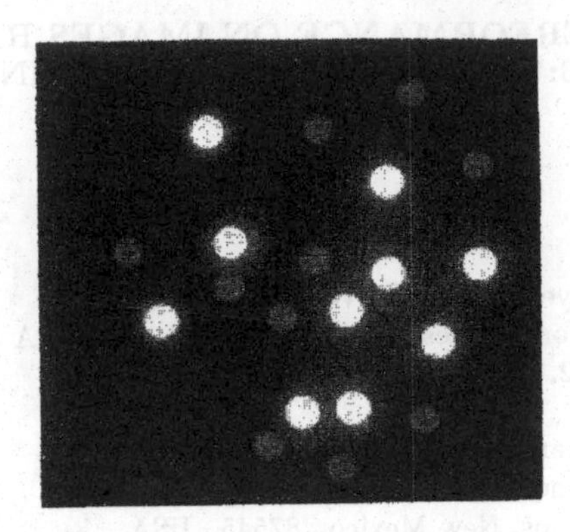

Figure 1: a) Sample scene containing 10 high-contrast disks and 10 low-contrast disks randomly placed on a zero background.

2. The Scene and the Task

The object class consists of a set of 10 scenes, each containing 20 randomly placed, non-overlapping disks on a zero background. Ten of the disks are low-contrast (amplitude = 0.1) and 10 are high-contrast (amplitude = 1.0). They are all 8 pixels in diameter in an overall field of 128 pixels in diameter. An example taken from the ensemble is shown in Figure 1. The task is the detection of the low-contrast disks. Here we have used just 8 equally spaced views, and parallel projections each containing 128 samples that include additive, zero-mean Gaussian noise with a standard deviation of 2, which is about twice the peak projection value of the low-contrast disks. The noise in the data is pre-smoothed prior to reconstruction by a triangular window with a FWHM of 3 pixels, reducing the rms noise level by a factor of 0.484.

3. The Reconstruction Algorithm

The reconstruction algorithm used here, named MEMSYS 3 [6], minimizes the expression

$$\frac{1}{2}\chi^2 - \alpha S \tag{1}$$

where χ^2 is chi-squared, the exponent in the likelihood function that expresses the probability of the data given the object scene under the assumption of Gaussian additive noise, and $-\alpha S$ is the exponent of the entropic prior probability distribution on the reconstruction [5]. Minimizing chi-squared is equivalent to finding the maximum likelihood (ML) reconstruction. Minimizing $-\alpha S$ is equivalent to maximizing the entropy S, which can be considered a measure of the degeneracy of the image; a uniformly gray image achieves the unconstrained maximum entropy. Minimizing the expression in Eq. 1 amounts to finding the "least committal image" consistent with the data.

The factor α selects one possible member of an infinite family of entropic priors; the smaller its value, the less one enforces the prior distribution, and the closer one approaches the ML solution. Several techniques for determining α have evolved over the last decade. Since many early authors picked α so that chi-squared equaled the number of measurements, this has been referred to as "historic" maximum entropy. The more recent "classic" MaxEnt determines α from the data itself. The MEMSYS 3 software also allows the user to specify an arbitrary ("ad hoc") value of the final or aimed for value of chi-squared. Reconstructions of the object scene shown in Figure 1 are given in Figure 2 for 4 values of α.

4. Algorithmic Decision Functions

The machine decision functions are various approximations to decision functions that arise in the study of Bayesian statistical decision theory:

(a) The difference in the log of the posterior probability for each hypothesis given the data, $p(\mathbf{f} \mid \mathbf{g})$.

(b) Same as in (a), but using a quadratic approximation obtained by expanding the expression for the log posterior probability in a Taylor series about the maximum (the reconstruction) [6, 2].

(c) The non-prewhitening matched filter (NPWMF) output, formed by summing all the pixels within the region of the expected signal [7, 8].

(d) The non-prewhitening matched filter, modified to include the background in an annular region centered on the location of the expected signal. The decision function, referred to as the disk contrast, is the difference between the activity in the central disk region and the estimated activity in the surrounding disk.

(e) The difference in the mean-squared-difference between the reconstruction and the expected object calculated under each of the hypotheses (disk present and absent).

To determine each machine observer's figure of merit, the decision function is applied to 100 subregions (16 pixels in diameter) in the reconstructions that contain background plus a disk. The decision function is also applied to 100 regions in the reconstructions that contain only background. The decision function outputs are histogrammed separately for the known signal and the known background locations and the receiver operating characteristic (ROC) curve is then generated [3]. The figure of merit, d_a, is derived from the area under the ROC curve via an inverse error function.

5. The Human Observer

The human observers used the same 100 realizations of the signal-plus-background images and background-alone images. Each 16-pixel diameter test region was centered in a 16×16 square, then bilinearly interpolated twice to form 64×64 pixel images for display. These images were presented to the observer in pairs in the usual two-alternative-forced-choice (2AFC) paradigm [3]. The observer's percentage correct in a 2AFC experiment corresponds to the area under the curve in the ROC paradigm when the same images are used. Thus, the detectability figure of merit for the human observers is derived via an inverse error function from the percent correct, and can be compared directly to the machine d_a.

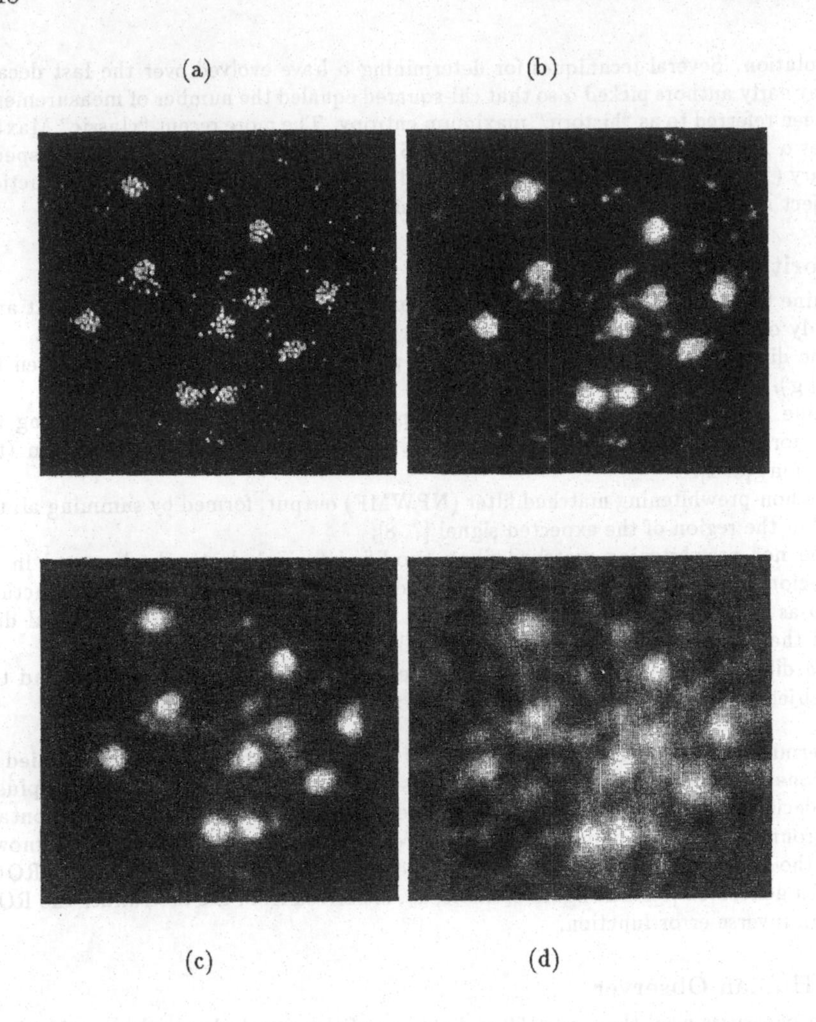

Figure 2: Reconstructions of the object scene in Figure 1 for values of α equal to a)0.002, b)0.2, c)1.8, d)20.

6. Results

In Figure 3, d_a is plotted for each of the algorithmic observers. We see that the performance of the algorithmic observers is a function of the parameter α. Generally the figure of merit is stable at small values of α and falls off at high values of α. Arrows indicate the values of α corresponding to the historic and classic MaxEnt solutions. As can be seen from the figure, the classic reconstructions have a smaller value of α (and hence χ^2) than the historic ones. For the historic run, α=01.8 and χ^2=1024; the classic run gave α=0.2 and χ^2=473. All of the decision variables except the Gaussian approximation to the posterior probability

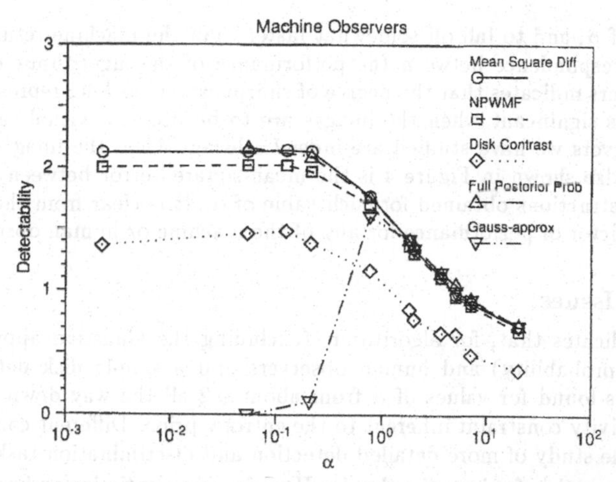

Figure 3: The detectability d_a as a function of the parameter α for each of the machine or algorithmic observers.

function perform better using the classic MaxEnt reconstructions over using the historic solution. It can be seen from Figure 3 that the decision variable based on the quadratic approximation to the log posterior probability fails catastrophically for small values of α.

The results for two human observers are presented in Figure 4.

They are seen to follow the best machine observer results to within the error bars for

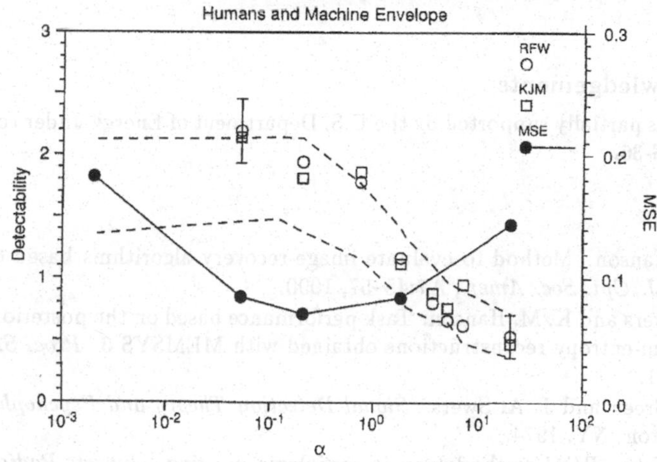

Figure 4: Detectability as a function of α for two human observers (circles and squares), bracketed by the envelope of the machine observer performance functions from Figure 3 (neglecting the Gaussian approximation to the posterior probability). The mean-squared error (MSE) between the original scene and the reconstructions is also given as a function of α.

lower values of α, and to fall off somewhat faster than the machine results as α increases. The close correspondence between the performance of the algorithmic observers and the human observers indicates that the degree of sharpness/smoothing represented by the variation over α is significant when the images are to be used for visual tasks, and that the machine observers we have studied are indeed relevant when the images are intended for human use. Also shown in Figure 4 is the mean-squared-error between the original scene and the reconstructions obtained for each value of α. It is clear from the figure that MSE is a poor predictor of performance for any of the machine or human observers.

7. Future Issues

This work indicates that, for algorithmic (excluding the Gaussian approximation to the log posterior probability) and human observers and a simple disk detection task, high detectability is found for values of α from about 0.2 all the way down to the ML limit–with the positivity constraint inherent to the entropy prior. Different conclusions might be drawn from the study of more detailed detection and discrimination tasks.

A general question for investigation is: How does one optimize an image reconstruction algorithm when that estimation step is to be followed by an image classification step? At present, most optimizers of image reconstruction routines use a figure of merit related to the MSE or rms pixel noise. Although such figures of merit can be related to certain detectability measures used here (at least for linear reconstruction schemes) [9], the relationship is neither direct nor necessarily monotonic. And, for this iterative reconstruction method, we have found that MSE does not predict human or machine detection performance. A more complete understanding of the steps that lead from estimation or reconstruction, through a machine or human observer, to a final detection or classification decision is required in order to optimize the procedure for the performance of the task for which the image was acquired.

8. Acknowledgements

This work was partially supported by the U.S. Department of Energy under contract number W-7405-ENG-36.

References

[1] K. M. Hanson. Method to evaluate image-recovery algorithms based on task performance. *J. Opt. Soc. Amer.*, A7:45–57, 1990.

[2] K. J. Myers and K. M. Hanson. Task performance based on the posterior probability of maximum-entropy reconstructions obtained with MEMSYS 3. *Proc. SPIE*, 1443:172–182, 1991.

[3] D. M. Green and J. A. Swets. *Signal Detection Theory and Psychophysics.* Krieger, Huntington, NY, 1974.

[4] C. E. Metz. ROC methodology in radiologic imaging. *Invest. Radiol.*, 21:720–733, 1986.

[5] S. F. Gull and J. Skilling. Maximum entropy method in image processing. *IEE Proc*, 131(F):646–659, 1984.

[6] S. F. Gull and J. Skilling. *Quantified Maximum Entropy - MEMSYS 3 Users' Manual.* Maximum Entropy Data Consultants Ltd., Royston, England, 1989.

[7] R. F. Wagner and D. G. Brown. Unified SNR analysis of medical imaging systems. *Phys. Med. Biol.*, 30:489–518, 1985.

[8] K. J. Myers, J. P. Rolland,H. H. Barrett, and R. F. Wagner. Aperture optimization for emission imaging: effect of a spatially varying background. *J. Opt. Soc. Amer.*, A7:1279–1293, 1990.

[9] H. H. Barrett. Objective assessment of image quality: effects of quantum noise and object variability. *J. Opt. Soc. Amer.*, A7:1266–1278, 1990.

3D RECONSTRUCTION FROM ONLY TWO PROJECTIONS?

D.E. Simpson, G.J.Daniell,
Physics Department,
University of Southampton
Southampton, SO9 5NH, U.K.

J.S. Fleming
Nuclear Medicine Department,
Southampton General Hospital,
Southampton, SO9 4XY, U.K.

ABSTRACT. In medical planar scintigraphy a two dimensional image is taken of a three dimensional gamma-ray emitting tracer distribution. In the past, the image has been considered the convolution of an ideal 2D image with an instrumental point spread function responsible for blurring. The extent of this blurring, due to the imperfect geometry of the camera collimator and Compton scatter of gamma-rays by patient tissue, varies with the depth of the emitting tracer, thus a single P.S.F. is inadequate to describe it.

We now model the tracer distribution as a series of slices parallel to the plane of the camera, each slice having an associated P.S.F. The image is now the sum of the convolutions of each slice with its P.S.F. Using maximum entropy data analysis, we reconstruct this 3D distribution from two opposed images. A sum of resulting distribution over the depth dimension weighted to simulate the effect of tissue attenuation, shows a lower number of artifacts than a simple 2D deconvolution.

1. Introduction

In medical planar scintigraphy (hereafter called gamma camera imaging) we obtain a 2D image of a 3D distribution of a gamma- ray emitting tracer distribution. A radio-nuclide labelled pharmaceutical is injected into a patient and images of the subsequent distribution give useful information about the function of the patient's organs or the presence of tumours.

The gamma camera, an ADAC Genesys, has two detector heads which can be rotated to collect several angular views or moved along the length of the patient to collect two whole body images. Work has been done on sections of these whole body images of a bone-seeking radio-pharmaceutical methylenediphosphonate labelled with Technetium 99m.

2. The Imaging Process

2.1. Principles

As gamma rays cannot be focused, a gamma camera relies on collimation to form a 2D image (figure 1). A lead collimator ensures that only those photons travelling perpendicular to the face of the camera are allowed to strike the scintillation crystal. Compton scatter of gamma-rays by patient tissues can result in rays, initially emitted at a different angle being accepted by the collimator, resulting in a blurring of the image. As photons lose energy in

A. Mohammad-Djafari and G. Demoments (eds.), Maximum Entropy and Bayesian Methods, 423–429.

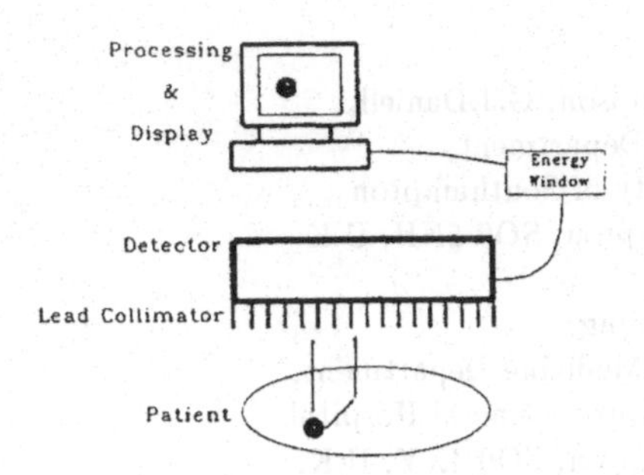

Figure 1: The Gamma Camera

the scattering process this effect can be minimised by counting only those photons whose energy is close to that known to be emitted by the radio-nuclide.

As both methods of improving resolution involve rejecting photons, a balance must be struck between sensitivity and resolution. Thus for a practical gamma camera some blurring will always be present.

2.2. Modelling

In order to correct gamma-camera images for these effects, blurring has, in the past, been represented as the result of an instrumental point spread function. The image is then considered as the convolution of this P.S.F. with a 2D distribution of radioactivity [2]. In this model, deconvolution of the image with respect to the P.S.F. should result in the recovery of a blur free 2D distribution.

A single P.S.F. cannot, however, account for the variation of these blurring processes with the distance between source and camera. The lead collimator, and energy window impose only angular constraints on accepting photons, thus both contributions to blurring increase as the source moves away from the camera face (figure2). This means a 2D model of the imaging process is an inadequate description and any data processing based on such a model will lead to artifacts in the result.

We now model the imaging process in 3D. To do this we approximate the tracer distribution by a series of 2D slices parallel to the plane of the camera. If we assume the blurring processes are constant for activity within a single slice then convolution with a single P.S.F. will accurately evaluate the contribution of each slice to the image. As patient tissues strongly attenuate gamma rays, each slice's P.S.F. must be appropriately normalised to take account of this. The image is now seen as the sum of the convolutions of each slice with its own P.S.F.

$$IMAGE = PSF_1 \otimes SLICE_1 + PSF_2 \otimes SLICE_2 + ... + PSF_n \otimes SLICE_n. \quad (1)$$

As a single image contains little or no information about the distribution of activity in the depth dimension, we use two views, from the patient's front and back, collected

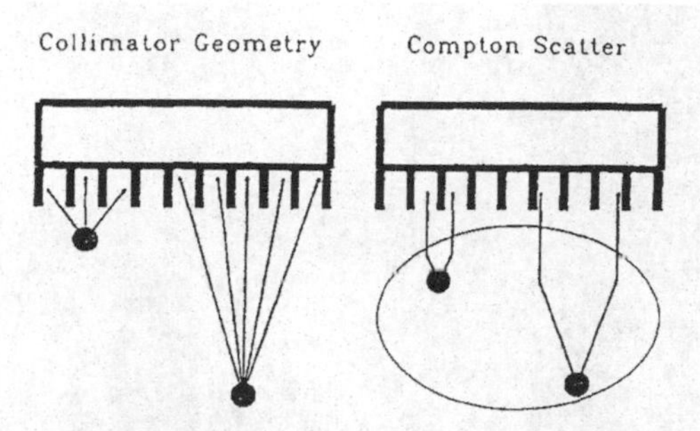

Figure 2: Dependence of Blurring Processes on Depth

simultaneously as data. Features of the distribution appear attenuated and scattered to a different degree in these opposed views and this gives some information about the depth of their source. As both images are of the same distribution, by extension of our previous argument we have:

$$IMAGE_1 = PSF_1 \otimes SLICE_1 + PSF_2 \otimes SLICE_2 + \ldots + PSF_n \otimes SLICE_n, \quad (2)$$

$$IMAGE_2 = PSF_n \otimes SLICE_1 + PSF_{n-1} \otimes SLICE_2 + \ldots + PSF_1 \otimes SLICE_n. \quad (3)$$

Here $PSF_1 \rightarrow PSF_n$ may need to be different for the two views to account for differences in the set up of the two detectors. For our camera one detector must image the patient through a bed, with its own attenuation and scatter characteristics. Patient thickness must also be known in order to assign P.S.F.s correctly.

3. Data Collection

We measured the P.S.F.s by imaging a point source through several different thicknesses of "Mix D", a material with the same attenuation and scatter characteristics as tissue. The point source was a 2mm diameter well in a piece of perspex, filled with Technetium 99m. The imaging time was extended so as to obtain P.S.F.s with good statistics.

Anterior and posterior whole body views were collected and from these a region was selected extending from the breastbone to the hips, over which the patient was of approximately uniform thickness. These views are shown in figure 3. Note the data collected was of unusually poor quality. An improved posterior view collected using a different camera, with the patient closer to the camera face, is shown in figure 7.

4. Processing

We used maximum entropy to solve the above data processing problem. Our version of maximum entropy maximises $S - \lambda\chi^2$, where

$$S = -\sum_{i=1}^{N} f_i \log\left(\frac{f_i}{b_i}\right). \quad (4)$$

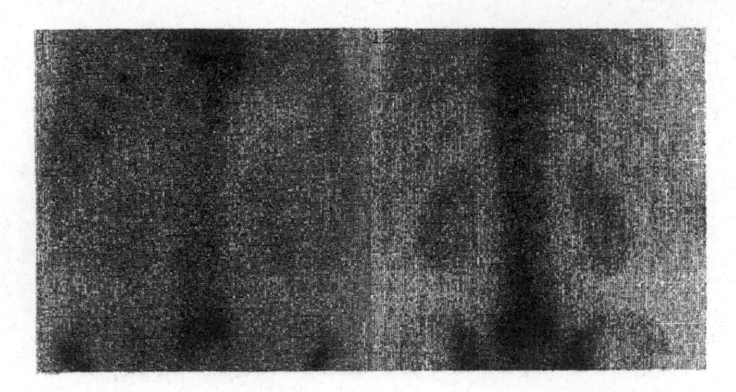

Figure 3: Original Anterior and Posterior Bone Images

Here f is the map of tracer distribution, b is the default solution and N is the total number of points in the distribution, and

$$\chi^2 = \sum_{j=1}^{P} \left(\frac{d_j - y_j}{\sigma_j} \right)^2 . \tag{5}$$

In the above, d is the image data, σ_j is the standard deviation of d_j and P is the total number of data points. The array y contains the data we would expect to get when transforming the distribution f using equations 2 and 3.

We choose the Lagrangian multiplier λ such that $\chi^2 = P$ where P is the total number of data points. Justification for the above method can be found in ref [1]. The algorithm for its implementation is based on one found in [3].

As the clinically significant features of bone images are generally of high intensity, we chose b, the default solution, to be low. This choice should minimise the number of false positive features.

5. Results

Initial results showed the maximum entropy algorithm preferentially allocated activity to surface slices at the expense of central ones. Due to the lower attenuation of gamma-rays emitted from surface levels, this strategy allows a close fit to the data whilst keeping the solution as close as possible to the low default solution. To avoid this tendency we chose the default solution to be higher for slices closer to the centre of the patient.

To simplify display of the solution in this paper we present a solution with four slices approximating the 3D distribution (figure 4). Calculations with a larger number of slices have also been performed. There is a distinct separation of features between slices. The surface anterior slice (top left) shows breastbone, ribs and hips with reduced 'shine through' from the spine and kidneys. The anterior central slice (top right) shows features originating from breastbone, spine, kidneys and anterior ribs. The central posterior slice (bottom left) shows spine, posterior ribs, hips and kidneys while the posterior surface slice (bottom right) shows features originating from the same areas though more sharply delineated.

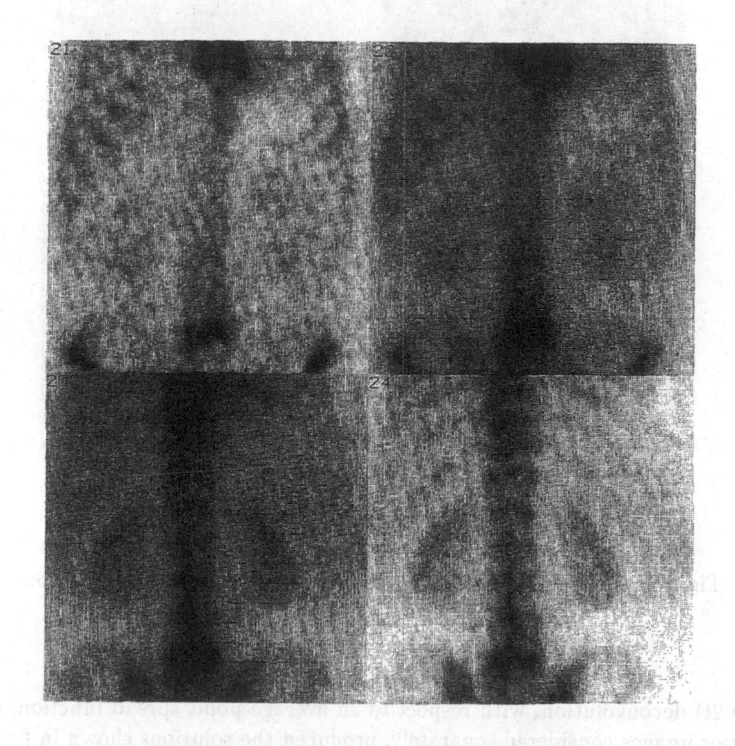

Figure 4: 3D Maximum Entropy Solution

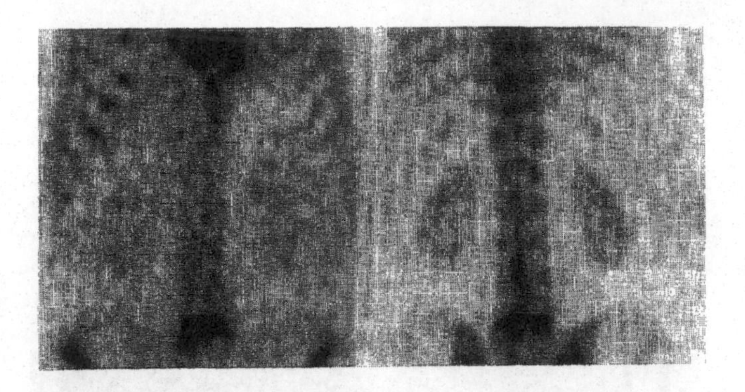

Figure 5: 2D Maximum Entropy Solutions

Figure 6: Projections of the 3D Solution Adjusted for Attenuation

Simple 2D deconvolution, with respect to an average point spread function, of anterior and posterior images considered separately, produced the solutions shown in figure 5

In order to make a comparison with the 2D deconvolution solutions (figure 5) we summed the slices of the 3D solution, weighting each level to simulate tissue attenuation. This gave us the two projections shown in figure 6

Comparing figures 5 and 6 we can see that the simple 2D deconvolution version of both views appear to follow the original data more closely than the solution produced by the 3D model. This tendency to over-fit, has produced an increase in blotchy noise over the 2D solutions. In particular, many small, high amplitude features are visible on the 2D model's solution of the posterior image along the spine and on the kidneys. These do not appear as strongly on the 3D models solution. A comparison with a better quality image of the same patient 7 shows these features to be artifacts.

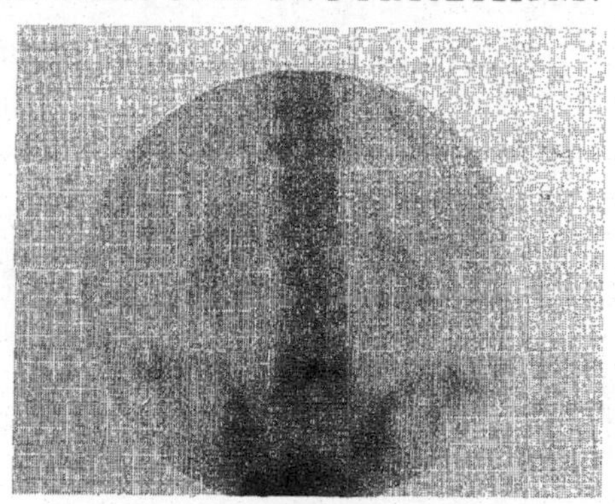

Figure 7: Posterior Image Under Improved Imaging Conditions

6. Conclusions

Maximum entropy data analysis does a creditable job in assigning activity to different depth slices, given the paucity of depth information present in the data.

Overall the 2D deconvolution solutions appear to be fitting the data too closely. This is because the limitations of the model mean that data generated from any trial solution will contain systematic errors, thus in order to satisfy the χ^2 constraint, random errors must be more closely fitted.

A more accurate model of the imaging process has resulted in an improvement in maximum entropy solutions.

References

[1] G.J. Daniell, "Of map and monkeys: an introduction to the maximum entropy method", in "Maximum Entropy in Action" ed. Buck and Macaulay 1991, Clarendon Press, Oxford.

[2] J.Skilling "Fundamentals of MaxEnt in data analysis", in "Maximum Entropy in Action" (as above)

[3] J. Skilling and R.K. Bryan "Maximum entropy image reconstruction: general algorithm", *Mon. Not. R. astr Soc.* 211 (1984) 111–124

A CLOSED FORM MAXIMUM ENTROPY APPROACH FOR RECONSTRUCTING IMAGE FROM INCOMPLETE PROJECTIONS

Yong Yuan and Tariq S. Durrani
Signal Processing Division,
Department of Electronic & Electrical Engineering,
University of Strathclyde, 204 George Street,
Glasgow G1 1XW, United Kingdom

ABSTRACT. The model of maximum entropy based on the power spectrum has been studied in the problem of image reconstruction from a few projections. A closed form relation linking the maximum entropy solution and the known prior information has been developed and geometrically explained. Some properties and limitations of the model are discussed. Simulation results are provided.

1. Introduction

The technique of image reconstruction from incomplete data by using maximum entropy methods (MEM) has been applied to many areas [1]-[3], including image reconstruction from incomplete projections in tomographic imaging [4]-[5]. The widely accepted configurational entropy model as an optimization function under constraints of given information has been used regularly, and encouraging results have been achieved through improved iterations[3]. Others such as [6]-[8] have adopted the generalized power spectrum based maximum entropy model which has good resolution performance in one dimensional spectrum estimation[9], and have studied its extendability and properties in higher dimensional cases. Justice[10] summarized those efforts and indicated some technical difficulties that still remain, such as the fundamental theorem of algebra breaks down for more than one variable, and the ability to factor polynomials is no longer guaranteed. These difficulties have eluded people from reaching a generalized solution to the problem of multi-dimensional maximum entropy analysis and giving precise relationship of prior information and maximum entropy optimized solution.

Recently, 2-d maximum entropy spectral estimation has been studied in the area of tomographic imaging by using power spectrum based maximum entropy optimization[11], a closed form relation linking the MEM spectrum and the known prior information has been developed. In this paper, further attempt has been made to provide a closed form solution to the problem of image reconstruction from incomplete projections based on the same optimization model. Some properties of MEM model are discussed. simulation results are provided.

2. Mathematical Approach

Approaching maximum entropy solution to the problem of image reconstruction from incomplete projections in computerised tomography involves determining an image function

A. Mohammad-Djafari and G. Demoments (eds.), Maximum Entropy and Bayesian Methods, 431–436.

$f(x,y)$ whose power spectrum density $S(u,v)$ maximizes the entropy rate:

$$H = \int_{-\infty}^{\infty}\int_{-\infty}^{\infty} \ln S(u,v)\,du\,dv \tag{1}$$

subject to constraint that the projections of image $f(x,y)$ match the known projections $p_{\theta_i}(D)$ at given angles $\{\theta_i\}$:

$$\int_{L_1}\int f(x,y)\,\delta(x\cos\theta_i + y\sin\theta_i - D)\,dx\,dy = p_{\theta_i}(D) \qquad i = 0,1,2,\ldots,I-1 \tag{2}$$

where I is the number of projections available, and the delta line $\delta(x\cos\theta_i + y\sin\theta_i - D)$ shifts out the desired line in $f(x,y)$ to provide an effective line integration. p_{θ_i} is the known projection data at θ_i, According to the central section theorem[12] – the Fourier transform of a projection at angle forms a line in 2-d Fourier plane at the same angle, a limited number of incomplete known projections only form a limited number of lines in Fourier plane (each crossing the origin), and allows the chosen methods to interpolate between lines. The aim of MEM is to seek a particular Fourier transform distribution between those given lines so that entropy rate of (1) reaches its maximum. Let:

$$J = H - Q \tag{3}$$

where

$$Q = \sum_{i=0}^{I-1} \int_{-\infty}^{\infty} \lambda_i(D)\,w_i(x,y)\,dD \times \left[\int_{-\infty}^{\infty}\int_{-\infty}^{\infty} f(x,y)\,\delta(x\cos\theta_i + y\sin\theta_i - D)\,dx\,dy - p_{\theta_i}(D)\right] \tag{4}$$

in which $\{\lambda_i(D)\}$ are Lagrangian parameters, and $w_i(x,y)$ is the effective window of the ith projection:

$$w_i(x,y) = w_{\theta_i}(x\cos\theta_i + y\sin\theta_i) \times \delta\left(-x\sin\theta_i + y\cos\theta_i\right) \tag{5}$$

Let $F(u,v)$ denotes the Fourier transform of image function $f(x,y)$, and put power spectrum density $S(u,v) = F(u,v)F^*(u,v)$ into (1), then the equation:

$$\frac{\partial J}{\partial F} = 0 \tag{6}$$

will give the solution equation of (1) subject to (2) as

$$\frac{1}{F^*} + \frac{1}{F} \times \frac{\partial F}{\partial F^*} = \Lambda(u,v) \tag{7}$$

It can be shown that the MEM solution form, in a polar coordinate system, is

$$\hat{F}(\xi,\theta) = 2.0/\Lambda^*(\xi,\theta) \tag{8}$$

or:

$$\hat{F}(\xi,\theta) = 2.0/\sum_{i=0}^{I-1} \Lambda_i^*[\xi\,\cos(\theta-\theta_i)] \tag{9}$$

where,

$$\Lambda_i(\xi) = \int_{-\infty}^{\infty} \lambda_i(D)\,w_{\theta_i}(D)e^{-j2\pi\xi D}\,dD$$

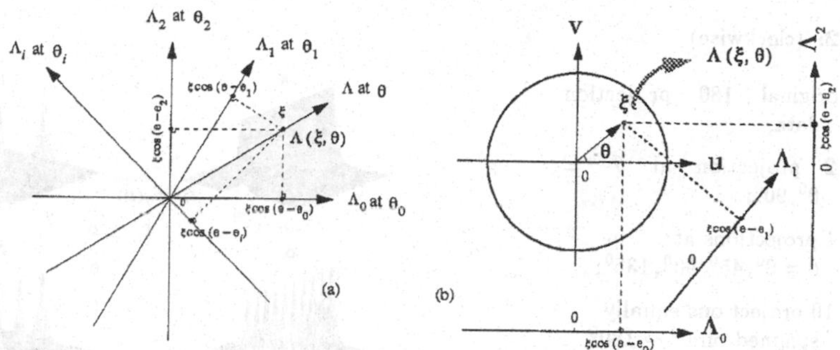

Fig 1: (a) $\hat{F}(\xi,\theta)$ is the conjugate inverse of $\Lambda(\xi,\theta)$ which is the sum of contribution of each individual $\Lambda_i(\xi)$ at $\xi\cos(\theta-\theta_i)$; (b) $\hat{F}(\xi,\theta)$ is the conjugate inverse of $\Lambda(\xi,\theta)$ which is the sum of backprojection of each individual $\Lambda_i(\xi)$.

Equation (9) suggests that it is the conjugate of $\Lambda(\xi,\theta)$, which is the summation of each individual $\Lambda_i(\xi)$ at $\xi\cos(\theta-\theta_i)$, that forms the maximum entropy solution in Fourier plane to the problem of (1) subject to (2). Fig1(a) shows the relation of $\Lambda(\xi,\theta)$ with each contribution $\Lambda_i(\xi)$.

An alternative geometrical explanation of relation (9) is that (see Fig 1b): it is the backprojection of each $\Lambda_i(\xi)$ at angle θ_i onto Fourier frequency domain that provides contribution to the maximum entropy solution (inversely) which can be seen as the conjugate Fourier transform of the MEM image.

In order to determine $\{\Lambda_i(\xi), i=0,1,\ldots,I-1\}$, solution (9) should match the known Fourier transform lines which, according to the central section theorem, is:

$$F(\xi,\theta_i)=\int_{-\infty}^{\infty}p_{\theta_i}(D)\,w_{\theta_i}(D)\,e^{-j2\pi\xi D}\,dD \qquad i=0,1,2,\ldots,I-1 \tag{10}$$

So the equation determining $\Lambda_i(\xi)$, from (9), is:

$$F^{*}(\xi,\theta_k)\sum_{i=0}^{I-1}\Lambda_i[\xi\cos(\theta_k-\theta_i)]=2 \qquad k=0,1,2,\ldots,I-1 \tag{11}$$

or in space domain,

$$\sum_{i=0}^{I-1}\int_{-\infty}^{\infty}p_{\theta_k}[D+D'\cos(\theta_k-\theta_i)]\,\lambda_i(D')\,dD'=2\,\delta(D) \tag{12}$$

MEM solution could be obtained either in Fourier domain by bringing $\{\Lambda_i(\xi)\}$ into (9), or in space domain where a full range of projection data could be approached from $\{\lambda_i(D)\}$, and the MEM image can be reconstructed without difficulty.

3. Simulation Results

Fig 2(a) shows an original full (180) projection data equally spanned with 1^0 interval angle, while 2, 4 and 10 equally spanned projections shown in Fig.2(b,c,d), respectively, are known for image reconstruction.

Fig 2: (clockwise)

(a) original 180 projection data;

(b) 2 projection at $\theta = 0^0, 90^0$;

(c) 4 projections at $\theta = 0^0, 45^0, 90^0, 135^0$;

(d) 10 projections equally spanned on $0^0 \sim 180^0$

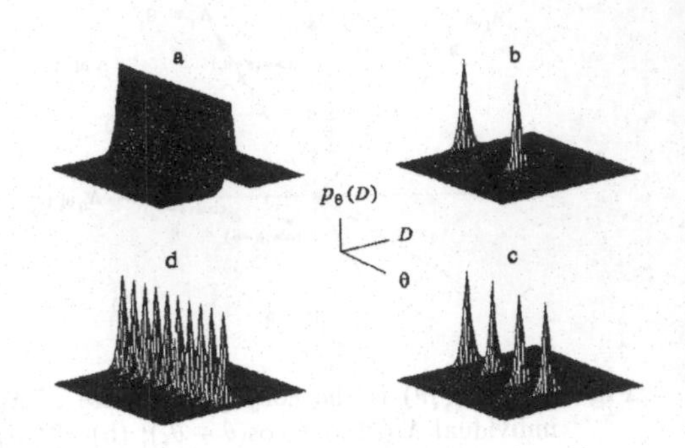

Fig 3,4,5 show the image reconstruction from 2, 4 and 10 projections, respectively. Reconstruction results obtained by ART and MART algorithms are also provided for comparison, both algorithms give results after 10 iterations with relaxation consideration. Clockwise, in each figure, (a) is the original image, and (b), (c) show reconstruction by ART and MART, (d) gives reconstruction by proposed MEM algorithms.

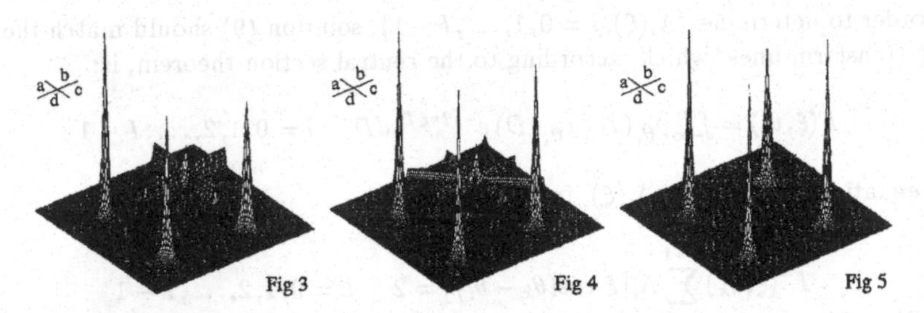

Fig 3,4,5: image reconstruction from 2, 4 and 10 projections, respectively; and in each figure, a) original image, b) using ART, c) using MART, d) using MEM

Several phenomena could be pointed out immediately from Fig 3-5. First, both MEM and MART give far better reconstruction than ART when only a few projections available. Second, in all three cases, MEM and MART reach very similar results closing to the original image. Finally, the advantage of both MEM and MART algorithms over ART algorithms is getting lost when more information – the projection data – is available.

It is quite interesting to see how the proposed MEM algorithm in space domain to predict (interpolate) missing projection data laterally. Fig 6 shows full projecting data achieved from given 2, 4 and 10 projections in Fig 2, respectively. Fig 6(a) is the original projection data. MEM always gives better prediction for the missing projecting data which are close to the known projection data, and the further away from the known information, the worse prediction approached. But it is improved when number of projection data increased.

Fig 6: MEM projection reconstruction from:

(b) 2 projections

(c) 4 projections

(d) 10 projections

and (a) is original full projections

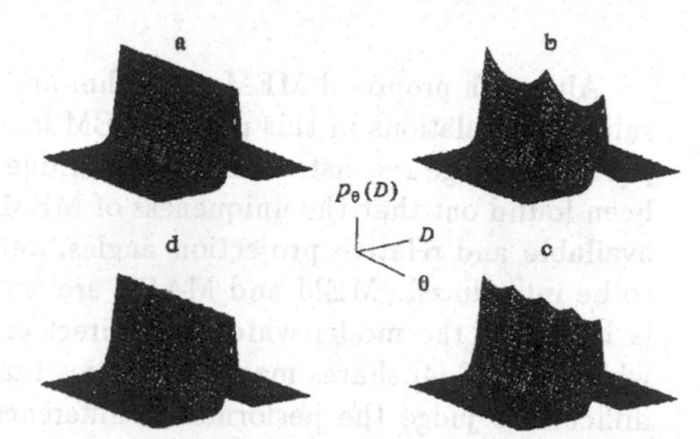

Fig7(a) shows the errors of MEM reconstructing full projection data from 2, 4 and 10 known projection data, respectively on ($0^0 \sim 180^0$). Error calculation is based on:

$$error(\theta) = \left\{ \sum_m [ReconstructP_\theta(m) - OriginalP_\theta(m)]^2 \;/\; \sum_m OriginalP_\theta^2(m) \right\}^{1/2} \tag{13}$$

Error amplitude declines dramatically when number of known projections increase. Small errors happen around the known projection angles, and no errors at given known projection angles.

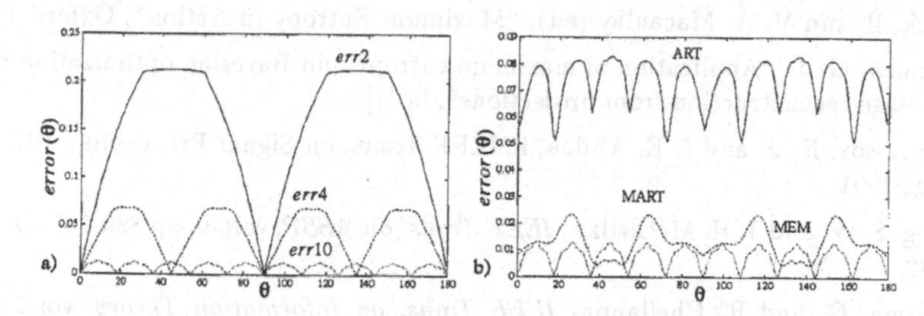

Fig 7: (a) Errors of MEM projection reconstructions from 2, 4 and 10 known projections at different angles. (b) Errors of projection reconstructions from 10 known projections by using ART, MART and MEM

Fig 7(b) shows errors of reconstructing projection from 10 known projection data by using ART, MART and proposed MEM algorithms, respectively. Results from ART and MART are obtained after 10 iterations. It can be seen that errors from both MART and MEM are very small and are comparable, compared with errors from ART.

4. Conclusions and Discussions

In this paper, a new approach of reconstructing image from incomplete projection data by using MEM based on maximizing (1) subject to (2) is presented. A closed form solution is developed, and the relationship between MEM solution and existed known information presented by a group of parameters is explained geometrically. From the particular problem discussed in this paper, we can easily see how the known data propagate their information to the surrounding area and form a distribution as a whole to reach MEM solution. Simulation results suggest that using MEM based on (1) subject to (2) is possible and accessible.

Although proposed MEM algorithm and MART algorithm approached very similar results in simulations in this paper, MEM is more computationally intensive in dealing with practical image reconstruction problem due to its algorithm's complexity. Besides, it has been found out that the uniqueness of MEM solution is influenced by both projection data available and relative projection angles, and the approximation technique sometimes has to be introduced. MEM and MART are actually based on totally different models. MEM is based on the model which is a direct extension from 1d maximum entropy model[9], while MART [4] shares many properties that the configurational entropy model has. It is difficult to judge the performance difference between these two algorithms based on the simple simulations provided in this paper, however they have some general properties in common, although MART algorithm does not take the neighbourhood information (the autocorrelation) into consideration.

References

[1] Smith, C. R. and W. T. Grandy, Jr. (ed.), "Maximum-Entropy and Bayesian Methods in Inverse Problems", Reidel Publishing Company, Holland, 1985.

[2] Skilling, J. (ed), "Maximum Entropy and Bayesian Methods", Kluwer Academic Publisher, the Netherlands, 1989.

[3] Buck, B. and V. A. Macaulay (ed), "Maximum Entropy in Action", Oxford 1991.

[4] Herman, G. T. "Application of maximum entropy and Bayesian optimization methods to image reconstruction from projections", in [1].

[5] Dusaussoy, N. J. and I. E. Abdou, it IEEE Trans. on Signal Processing, vol.39, No.5, May, 1991.

[6] Lang, S. W. and J. H. McClellan, *IEEE Trans. on ASSP*, vol.30, pp 880-887, December, 1982.

[7] Sharma, G. and R. Chellappa, *IEEE Trans. on Information Theory*, vol IT-31, pp 90-99, January, 1985.

[8] Lev-Ari, H., S. R. Parker and T. Kailath, *IEEE Trans on Information Theory*, vol. 35, no. 3 May 1989.

[9] Burg, J. P. "Maximum Entropy Spectral Analysis", PhD Thesis, Stanford University, 1979.

[10] Justice, J. H., "Multivariate Extensions of Maximum Entropy Methods", pp 339-349 in [1], 1985

[11] Yuan, Y. and T. S. Durrani, "2-d Maximum Entropy Spectrum Estimation In Computerized Tomography – An

Approach of Closed Form Solution", *Proc. of EUSIPCO'92*,

Brussels, Aug. 24-27, 1992.

[12] Macovski, A., "Medical Imaging Systems", Prentice-Hall Inc., Englewood Cliffs, New Jersey, 1983.

KEY WORDS INDEX

Index

AUTHORS INDEX

Index

Fundamental Theories of Physics

Series Editor: Alwyn van der Merwe, *University of Denver, USA*

1. M. Sachs: *General Relativity and Matter*. A Spinor Field Theory from Fermis to Light-Years. With a Foreword by C. Kilmister. 1982 ISBN 90-277-1381-2
2. G.H. Duffey: *A Development of Quantum Mechanics*. Based on Symmetry Considerations. 1985 ISBN 90-277-1587-4
3. S. Diner, D. Fargue, G. Lochak and F. Selleri (eds.): *The Wave-Particle Dualism*. A Tribute to Louis de Broglie on his 90th Birthday. 1984 ISBN 90-277-1664-1
4. E. Prugovečki: *Stochastic Quantum Mechanics and Quantum Spacetime*. A Consistent Unification of Relativity and Quantum Theory based on Stochastic Spaces. 1984; 2nd printing 1986 ISBN 90-277-1617-X
5. D. Hestenes and G. Sobczyk: *Clifford Algebra to Geometric Calculus*. A Unified Language for Mathematics and Physics. 1984 ISBN 90-277-1673-0; Pb (1987) 90-277-2561-6
6. P. Exner: *Open Quantum Systems and Feynman Integrals*. 1985 ISBN 90-277-1678-1
7. L. Mayants: *The Enigma of Probability and Physics*. 1984 ISBN 90-277-1674-9
8. E. Tocaci: *Relativistic Mechanics, Time and Inertia*. Translated from Romanian. Edited and with a Foreword by C.W. Kilmister. 1985 ISBN 90-277-1769-9
9. B. Bertotti, F. de Felice and A. Pascolini (eds.): *General Relativity and Gravitation*. Proceedings of the 10th International Conference (Padova, Italy, 1983). 1984 ISBN 90-277-1819-9
10. G. Tarozzi and A. van der Merwe (eds.): *Open Questions in Quantum Physics*. 1985 ISBN 90-277-1853-9
11. J.V. Narlikar and T. Padmanabhan: *Gravity, Gauge Theories and Quantum Cosmology*. 1986 ISBN 90-277-1948-9
12. G.S. Asanov: *Finsler Geometry, Relativity and Gauge Theories*. 1985 ISBN 90-277-1960-8
13. K. Namsrai: *Nonlocal Quantum Field Theory and Stochastic Quantum Mechanics*. 1986 ISBN 90-277-2001-0
14. C. Ray Smith and W.T. Grandy, Jr. (eds.): *Maximum-Entropy and Bayesian Methods in Inverse Problems*. Proceedings of the 1st and 2nd International Workshop (Laramie, Wyoming, USA). 1985 ISBN 90-277-2074-6
15. D. Hestenes: *New Foundations for Classical Mechanics*. 1986 ISBN 90-277-2090-8; Pb (1987) 90-277-2526-8
16. S.J. Prokhovnik: *Light in Einstein's Universe*. The Role of Energy in Cosmology and Relativity. 1985 ISBN 90-277-2093-2
17. Y.S. Kim and M.E. Noz: *Theory and Applications of the Poincaré Group*. 1986 ISBN 90-277-2141-6
18. M. Sachs: *Quantum Mechanics from General Relativity*. An Approximation for a Theory of Inertia. 1986 ISBN 90-277-2247-1
19. W.T. Grandy, Jr.: *Foundations of Statistical Mechanics*. Vol. I: *Equilibrium Theory*. 1987 ISBN 90-277-2489-X
20. H.-H von Borzeszkowski and H.-J. Treder: *The Meaning of Quantum Gravity*. 1988 ISBN 90-277-2518-7
21. C. Ray Smith and G.J. Erickson (eds.): *Maximum-Entropy and Bayesian Spectral Analysis and Estimation Problems*. Proceedings of the 3rd International Workshop (Laramie, Wyoming, USA, 1983). 1987 ISBN 90-277-2579-9

Fundamental Theories of Physics

22. A.O. Barut and A. van der Merwe (eds.): *Selected Scientific Papers of Alfred Landé. [1888-1975]*. 1988 ISBN 90-277-2594-2
23. W.T. Grandy, Jr.: *Foundations of Statistical Mechanics.* Vol. II: *Nonequilibrium Phenomena.* 1988 ISBN 90-277-2649-3
24. E.I. Bitsakis and C.A. Nicolaides (eds.): *The Concept of Probability.* Proceedings of the Delphi Conference (Delphi, Greece, 1987). 1989 ISBN 90-277-2679-5
25. A. van der Merwe, F. Selleri and G. Tarozzi (eds.): *Microphysical Reality and Quantum Formalism, Vol. 1.* Proceedings of the International Conference (Urbino, Italy, 1985). 1988 ISBN 90-277-2683-3
26. A. van der Merwe, F. Selleri and G. Tarozzi (eds.): *Microphysical Reality and Quantum Formalism, Vol. 2.* Proceedings of the International Conference (Urbino, Italy, 1985). 1988 ISBN 90-277-2684-1
27. I.D. Novikov and V.P. Frolov: *Physics of Black Holes.* 1989 ISBN 90-277-2685-X
28. G. Tarozzi and A. van der Merwe (eds.): *The Nature of Quantum Paradoxes.* Italian Studies in the Foundations and Philosophy of Modern Physics. 1988 ISBN 90-277-2703-1
29. B.R. Iyer, N. Mukunda and C.V. Vishveshwara (eds.): *Gravitation, Gauge Theories and the Early Universe.* 1989 ISBN 90-277-2710-4
30. H. Mark and L. Wood (eds.): *Energy in Physics, War and Peace.* A Festschrift celebrating Edward Teller's 80th Birthday. 1988 ISBN 90-277-2775-9
31. G.J. Erickson and C.R. Smith (eds.): *Maximum-Entropy and Bayesian Methods in Science and Engineering.* Vol. I: *Foundations.* 1988 ISBN 90-277-2793-7
32. G.J. Erickson and C.R. Smith (eds.): *Maximum-Entropy and Bayesian Methods in Science and Engineering.* Vol. II: *Applications.* 1988 ISBN 90-277-2794-5
33. M.E. Noz and Y.S. Kim (eds.): *Special Relativity and Quantum Theory.* A Collection of Papers on the Poincaré Group. 1988 ISBN 90-277-2799-6
34. I.Yu. Kobzarev and Yu.I. Manin: *Elementary Particles. Mathematics, Physics and Philosophy.* 1989 ISBN 0-7923-0098-X
35. F. Selleri: *Quantum Paradoxes and Physical Reality.* 1990 ISBN 0-7923-0253-2
36. J. Skilling (ed.): *Maximum-Entropy and Bayesian Methods.* Proceedings of the 8th International Workshop (Cambridge, UK, 1988). 1989 ISBN 0-7923-0224-9
37. M. Kafatos (ed.): *Bell's Theorem, Quantum Theory and Conceptions of the Universe.* 1989 ISBN 0-7923-0496-9
38. Yu.A. Izyumov and V.N. Syromyatnikov: *Phase Transitions and Crystal Symmetry.* 1990 ISBN 0-7923-0542-6
39. P.F. Fougère (ed.): *Maximum-Entropy and Bayesian Methods.* Proceedings of the 9th International Workshop (Dartmouth, Massachusetts, USA, 1989). 1990 ISBN 0-7923-0928-6
40. L. de Broglie: *Heisenberg's Uncertainties and the Probabilistic Interpretation of Wave Mechanics.* With Critical Notes of the Author. 1990 ISBN 0-7923-0929-4
41. W.T. Grandy, Jr.: *Relativistic Quantum Mechanics of Leptons and Fields.* 1991 ISBN 0-7923-1049-7
42. Yu.L. Klimontovich: *Turbulent Motion and the Structure of Chaos.* A New Approach to the Statistical Theory of Open Systems. 1991 ISBN 0-7923-1114-0

Fundamental Theories of Physics

43. W.T. Grandy, Jr. and L.H. Schick (eds.): *Maximum-Entropy and Bayesian Methods.* Proceedings of the 10th International Workshop (Laramie, Wyoming, USA, 1990). 1991 ISBN 0-7923-1140-X
44. P.Pták and S. Pulmannová: *Orthomodular Structures as Quantum Logics.* Intrinsic Properties, State Space and Probabilistic Topics. 1991 ISBN 0-7923-1207-4
45. D. Hestenes and A. Weingartshofer (eds.): *The Electron.* New Theory and Experiment. 1991 ISBN 0-7923-1356-9
46. P.P.J.M. Schram: *Kinetic Theory of Gases and Plasmas.* 1991 ISBN 0-7923-1392-5
47. A. Micali, R. Boudet and J. Helmstetter (eds.): *Clifford Algebras and their Applications in Mathematical Physics.* 1992 ISBN 0-7923-1623-1
48. E. Prugovečki: *Quantum Geometry.* A Framework for Quantum General Relativity. 1992 ISBN 0-7923-1640-1
49. M.H. Mac Gregor: *The Enigmatic Electron.* 1992 ISBN 0-7923-1982-6
50. C.R. Smith, G.J. Erickson and P.O. Neudorfer (eds.): *Maximum Entropy and Bayesian Methods.* Proceedings of the 11th International Workshop (Seattle, 1991). 1993 ISBN 0-7923-2031-X
51. D.J. Hoekzema: *The Quantum Labyrinth.* 1993 ISBN 0-7923-2066-2
52. Z. Oziewicz, B. Jancewicz and A. Borowiec (eds.): *Spinors, Twistors, Clifford Algebras and Quantum Deformations.* Proceedings of the Second Max Born Symposium (Wrocław, Poland, 1992). 1993 ISBN 0-7923-2251-7
53. A. Mohammad-Djafari and G. Demoment (eds.): *Maximum Entropy and Bayesian Methods.* Proceedings of the 12th International Workshop (Paris, France, 1992). ISBN 0-7923-2280-0
54. M. Riesz: *Clifford Numbers and Spinors* with Riesz' Private Lectures to E. Folke Bolinder and a Historical Review by Pertti Lounesto. E.F. Bolinder and P. Lounesto (eds.). 1993 ISBN 0-7923-2299-1

KLUWER ACADEMIC PUBLISHERS – DORDRECHT / BOSTON / LONDON